Indexed by Scopus

The series Springer Proceedings in Physics, founded in 1984, is devoted to timely reports of state-of-the-art developments in physics and related sciences. Typically based on material presented at conferences, workshops and similar scientific meetings, volumes published in this series will constitute a comprehensive up to date source of reference on a field or subfield of relevance in contemporary physics. Proposals must include the following:

- Name, place and date of the scientific meeting
- A link to the committees (local organization, international advisors etc.)
- Scientific description of the meeting
- List of invited/plenary speakers
- An estimate of the planned proceedings book parameters (number of pages/articles, requested number of bulk copies, submission deadline).

Please contact:

For Americas and Europe: Dr. Zachary Evenson; zachary.evenson@springer.com
For Asia, Australia and New Zealand: Dr. Loyola DSilva; loyola.dsilva@springer.com

Springer Proceedings in Physics

Volume 296

Indexed by Scopus

The series Springer Proceedings in Physics, founded in 1984, is devoted to timely reports of state-of-the-art developments in physics and related sciences. Typically based on material presented at conferences, workshops and similar scientific meetings, volumes published in this series will constitute a comprehensive up to date source of reference on a field or subfield of relevance in contemporary physics. Proposals must include the following:

- Name, place and date of the scientific meeting
- A link to the committees (local organization, international advisors etc.)
- Scientific description of the meeting
- List of invited/plenary speakers
- An estimate of the planned proceedings book parameters (number of pages/articles, requested number of bulk copies, submission deadline).

Please contact:

For Americas and Europe: Dr. Zachary Evenson; zachary.evenson@springer.com
For Asia, Australia and New Zealand: Dr. Loyola DSilva; loyola.dsilva@springer.com

Olena Fesenko · Leonid Yatsenko
Editors

Nanostructured Surfaces, Nanocomposites and Nanomaterials, and Their Applications

Selected Proceedings of the 10th International Conference on Nanotechnologies and Nanomaterials (NANO2022), 25–27 August 2022, Ukraine

Springer

Editors
Olena Fesenko
Institute of Physics
National Academy of Sciences of Ukraine
Kyiv, Ukraine

Leonid Yatsenko
Institute of Physics
National Academy of Sciences of Ukraine
Kyiv, Ukraine

ISSN 0930-8989 ISSN 1867-4941 (electronic)
Springer Proceedings in Physics
ISBN 978-3-031-42703-9 ISBN 978-3-031-42704-6 (eBook)
https://doi.org/10.1007/978-3-031-42704-6

This Springer imprint is published by the registered company Springer Nature Switzerland AG
The registered company address is: Gewerbestrasse 11, 6330 Cham, Switzerland

Paper in this product is recyclable.

Preface

This book highlights the most recent advances in nanoscience from leading researchers in Ukraine, Europe, and beyond. It features contributions from participants of the 10th International Research and Practice Conference "Nanotechnology and Nanomaterials" (NANO-2022), held in Lviv, Ukraine, on August 25 to 27, 2022. This event was organized jointly by the Institute of Physics of the National Academy of Sciences of Ukraine, Lviv Polytechnic National University (Ukraine), University of Tartu (Estonia), University of Turin (Italy), and Pierre and Marie Curie University (France). Internationally recognized experts from a wide range of universities and research institutions share their knowledge and key findings on material properties, behavior, synthesis, and their applications. The book reflects cutting-edge research from Ukraine, Europe, and beyond and will be interesting for leading scientists and advanced undergraduate and graduate students in material and nanoscience.

Today, nanotechnology is one of the most actively developing and promising fields of science. Numerous nanotechnology investigations are already producing practical results that can be applied in various areas of human life from science and radical innovations. Nanotechnology has the potential to revolutionize various industries by enabling the development of new materials and composites with unique properties and applications.

The aim of these books is to highlight the latest investigations from different areas of nanoscience and to stimulate new interest in this field. This volume of the two-volume work covers such important topics as structural, morphological, magnetic and optical properties of nanomaterials, provides new technologies for nanostructure synthesis and achievements in the field of nanomaterials and nanocomposites, new effects in nanoscale physics and its applications.

This book is divided into two sections: Part I—Nanostructured Surfaces, Part II— Nanocomposites and Nanomaterials. Sections covering Nanochemistry and Nanobiotechnology, Nanoelectronics and nanooptics can be found in Volume II: **Nanoelectronics, Nanooptics, Nanochemistry and Nanobiotechnology, and Their Applications**.

The papers published in these five sections fall under the broad categories of nanomaterial preparation and characterization, nanobiotechnology, nanodevices and quantum structures, and spectroscopy and nanooptics. The book will help readers to familiarize with current research and practical applications in nanoscience and thus to promote further implementation of nanotechnologies into innovations according to public needs. We hope that both volumes will be equally useful and interesting for young scientists or PhD students and mature scientists alike.

Kyiv, Ukraine
15 December 2023

Olena Fesenko
Leonid Yatsenko

Contents

Contributors

O. I. Aksimentyeva Ivan Franko National University of Lviv, Lviv, Ukraine

M. Anoshenko Faculty of Chemistry, Macromolecular Chemistry Department, Taras Shevchenko National University of Kyiv, Kyiv, Ukraine

N. Atamas Taras Shevchenko National University of Kyiv, Kyiv, Ukraine; Institute of Physical Chemistry Polish Academy of Sciences, Warsaw, Poland; Vienna University, Wien, Austria

Denys Baklan Department of Chemical Technology of Composite Materials, Chemical Technology Faculty, Igor Sikorsky Kyiv Polytechnic Institute, Kyiv, Ukraine

T. Bazylyuk Faculty of Chemistry, Macromolecular Chemistry Department, Taras Shevchenko National University of Kyiv, Kyiv, Ukraine

Anna Bilousova Department of Chemical Technology of Composite Materials, Chemical Technology Faculty, Igor Sikorsky Kyiv Polytechnic Institute, Kyiv, Ukraine

Vitalii L. Borblik V. Lashkaryov Institute of Semiconductor Physics, Kyiv, Ukraine

I. V. Brazhnyk Gimasi SA, Mendrisio, Switzerland

V. Budzinska Institute of Macromolecular Chemistry of National Academy of Sciences of Ukraine, Kyiv, Ukraine

Borys Chaikovs'kyi Stepan Gzhytskyi National University of Veterinary Medicine and Biotechnologies of Lviv, Lviv, Ukraine

Xiaohong Chen Physics Department, East China Normal University, Shanghai, China

V. Chornovol Department of Chemistry, Kyiv National University of Construction and Architecture, Kyiv, Ukraine

Tetiana Dontsova Kyiv, Ukraine

Z. A. Duriagina Department of Materials Science and Engineering, Lviv Polytechnic National University, Lviv, Ukraine;
Department of Materials Engineering, The John Paul II Catholic University of Lublin, Lublin, Poland

O. M. Fesenko Institute of Physics of National Academy Sciences of Ukraine, Kyiv, Ukraine

A. I. Gab University «Ukraine», Kyiv, Ukraine

I. Grechanyuk Department of Chemistry, Kyiv National University of Construction and Architecture, Kyiv, Ukraine

M. Grechanyuk Department of Abrasion- and Corrosion-Resistant Powder Construction, Materials Institute for Problems of Material Sciences, NAS of Ukraine, Kyiv, Ukraine

V. Grechanyuk Department of Chemistry, Kyiv National University of Construction and Architecture, Kyiv, Ukraine

Volodymyr Gurey Lviv Politechnic National University, Lviv, Ukraine

V. V. Honcharov Institute for Sorption and Problems of Endoecology, National Academy of Sciences of Ukraine, Kyiv, Ukraine

A. V. Honcharova Junior Academy of Sciences of Ukraine, Rubizhne, Ukraine

Yu. Yu. Horbenko Ivan Franko National University of Lviv, Lviv, Ukraine

N. Iukhymenko Faculty of Chemistry, Macromolecular Chemistry Department, Taras Shevchenko National University of Kyiv, Kyiv, Ukraine

Roman Iurkevych Hetman Petro Sahaidachnyi National Army Academy, Lviv, Ukraine

Iryna Ivasenko Karpenko Physico-Mechanical Institute of the NASU, Lviv, Ukraine;
Lviv Polytechnic National University, Lviv, Ukraine

Yurii Kanyuk Karpenko Physico-Mechanical Institute of the NASU, Lviv, Ukraine

H. Klym Lviv Polytechnic National University, Lviv, Ukraine;
Ivan Franko National University of Lviv, Lviv, Ukraine

Mariia Kontseva Kyiv, Ukraine

A. M. Korol National University for Food Technologies, Kyiv, Ukraine;
Laboratory on Quantum Theory in Linkoping, International Society for Independent ReSearch (ISIR), Linkoping, Sweden

A. M. Korostil Institute of Magnetism of NAN of Ukraine and MES of Ukraine, Kyiv, Ukraine

Ihor Koval'chuck Karpenko Physico-Mechanical Institute of the NASU, Lviv, Ukraine

Yulia Kovalchuk Department of Chemistry, Kyiv National University of Construction and Architecture, Kyiv, Ukraine

T. M. Kovbasiuk Department of Materials Science and Engineering, Lviv Polytechnic National University, Lviv, Ukraine

A. Kozyrev Department of Chemistry, Kyiv National University of Construction and Architecture, Kyiv, Ukraine

M. M. Krupa Institute of Magnetism of NAN of Ukraine and MES of Ukraine, Kyiv, Ukraine

Gryhoriy Krymets Kyiv, Ukraine

V. V. Kulyk Department of Materials Science and Engineering, Lviv Polytechnic National University, Lviv, Ukraine

Yuriy Kulyk Ivan Franko National University of Lviv, Lviv, Ukraine

N. Kutsevol Taras Shevchenko National University of Kyiv, Kyiv, Ukraine

Volodymyr Kyryliv Karpenko Physico-Mechanical Institute of the NAS of Ukraine, Lviv, Ukraine

O. M. Lavrynenko I. Frantsevich Institute for Problems in Material Science, NAS of Ukraine, Kyiv, Ukraine

V. M. Ledovskykh National Aviation University, Kyiv, Ukraine

Tetiana A. Len Departments of Physics, Taras Shevchenko National University of Kyiv, Kyiv, Ukraine

S. V. Levchenko National Aviation University, Kyiv, Ukraine

E. A. Lysenkov Petro Mohyla Black Sea National University, Mykolaiv, Ukraine

P. Ya. Lyutyy Department of Materials Science and Engineering, Lviv Polytechnic National University, Lviv, Ukraine

A. Makhinia International European University, Kyiv, Ukraine

Olha Maksymiv Karpenko Physico-Mechanical Institute of the NAS of Ukraine, Lviv, Ukraine

Vladimir Malyshev Institute of High Technologies, Kyiv National Taras Shevchenko University, Kyiv, Ukraine

A. Matsenko Department of Chemistry, Kyiv National University of Construction and Architecture, Kyiv, Ukraine

Lyudmila Yu. Matzui Departments of Physics, Taras Shevchenko National University of Kyiv, Kyiv, Ukraine

N. V. Medvid National University for Food Technologies, Kyiv, Ukraine

Ilgar Mirzoiev B. Verkin Institute for Low Temperature Physics and Engineering of NAS of Ukraine, Kharkiv, Ukraine

Oleksiy Myronyuk Department of Chemical Technology of Composite Materials, Chemical Technology Faculty, Igor Sikorsky Kyiv Polytechnic Institute, Kyiv, Ukraine

O. Nadtoka Taras Shevchenko National University of Kyiv, Kyiv, Ukraine

Yevgen Oberemok Department of Radiophysics, Kyiv National Taras Shevchenko University, Kyiv, Ukraine

O. I. Olifan I. Frantsevich Institute for Problems in Material Science, NAS of Ukraine, Kyiv, Ukraine

Iryna V. Ovsiienko Departments of Physics, Taras Shevchenko National University of Kyiv, Kyiv, Ukraine

O. Yu. Pavlenko I. Frantsevich Institute for Problems in Material Science, NAS of Ukraine, Kyiv, Ukraine

Volodymyr Posuvailo Karpenko Physico-Mechanical Institute of the NASU, Lviv, Ukraine

Oleksandr Prokopenko Institute of High Technologies, Kyiv National Taras Shevchenko University, Kyiv, Ukraine

Yuriy I. Prylutskyy Departments of Biophysics, Taras Shevchenko National University of Kyiv, Kyiv, Ukraine

Sergey Savenkov Department of Radiophysics, Kyiv National Taras Shevchenko University, Kyiv, Ukraine

D. B. Shakhnin University «Ukraine», Kyiv, Ukraine

O. Shevchenko National University for Food Technologies, Kyiv, Ukraine

Denys O. Shpylka Departments of Physics, Taras Shevchenko National University of Kyiv, Kyiv, Ukraine

V. M. Sidak Department of Biomedical Physics and Informatics, Dnipro State Medical University, Dnipro, Ukraine

Andrii S. Sizhuk Institute of High Technologies, Kyiv National Taras Shevchenko University, Kyiv, Ukraine

T. V. Stetsyuk Frantsevich Institute for Problems of Materials Science of National Academy Sciences of Ukraine, Kyiv, Ukraine

O. V. Strutskyi Institute of Macromolecular Chemistry, NAS of Ukraine, Kyiv, Ukraine

Zhuo Sun Physics Department, East China Normal University, Shanghai, China

A. M. Trostianchyn Department of Materials Science and Engineering, Lviv Polytechnic National University, Lviv, Ukraine

M. P. Trubitsyn Institute for Energy Efficient Technologies and Materials Sciences, Oles Honchar Dnipro National University, Dnipro, Ukraine

Tatiana L. Tsaregradskaya Departments of Physics, Taras Shevchenko National University of Kyiv, Kyiv, Ukraine

O. Vashchenko Institute for Scintillation Materials of NAS of Ukraine, Kharkiv, Ukraine

B. D. Vasyliv Department of Hydrogen Technologies and Alternative Energy Materials, Karpenko Physico-Mechanical Institute of the NAS of Ukraine, Lviv, Ukraine

V. I. Vavrukh Department of Materials Science and Engineering, Lviv Polytechnic National University, Lviv, Ukraine

V. V. Vira Department of Strength of Materials and Structural Mechanics, Lviv Polytechnic National University, Lviv, Ukraine

V. A. Vynar Department of Corrosion and Corrosion Protection, Karpenko Physico-Mechanical Institute of the NAS of Ukraine, Lviv, Ukraine

Yu. P. Vyshnevska National Technical University of Ukraine "Igor Sikorsky Kyiv Polytechnic Institute", Kyiv, Ukraine

Ivan Yaroshovych Stepan Gzhytskyi National University of Veterinary Medicine and Biotechnologies of Lviv, Lviv, Ukraine

Kovalchuk Yu Department of Chemistry, Kyiv National University of Construction and Architecture, Kyiv, Ukraine

V. O. Zazhigalov Institute for Sorption and Problems of Endoecology, National Academy of Sciences of Ukraine, Kyiv, Ukraine

Zhenjie Zhao Physics Department, East China Normal University, Shanghai, China

Olha Zvirko Karpenko Physico-Mechanical Institute of the NAS of Ukraine, Lviv, Ukraine

Nanostructured Surfaces

Changes in Structure and Micelle Formation in Aqueous Solutions of Alcohols

N. Atamas and A. Makhinia

Abstract In this work, the change in the structure of aqueous solutions of monohydric alcohols (ethyl and propyl alcohols) in a wide range of concentrations at $T = 300$ K was analyzed by the Monte Carlo method. It has been shown that in an aqueous solution of ethyl and propyl alcohols at infinitesimal concentrations (below $X_{et} < 0.05$ ppm in the case of ethyl alcohol and $X_{pr} < 0.04$ ppm in the case of propyl alcohol), the alcohol molecule does not change the percolation properties of the network of HBs between water molecules; cardinal rearrangement of the structure of the studied systems is observed in the $X_{alcohol} = (0.2–0.25)$ ppm concentration range, which results in the formation of micelles from four alcohol molecules at concentrations of alcohol in water over $X_{alcohol} > 0.3$ ppm. In this case, the structure of the alcohol molecule determines the number of water molecules surrounding the micelle. It has also been shown that at concentrations $X_{alcohol} = (0.6–0.8)$ ppm rupture of micelles formed by alcohol molecules occurs. At these concentrations, there can also be free alcohol molecules and half-chains of four alcohol molecules.

1 Introduction

Recently, more and more attention has been drawn to the study of the mechanisms that regulate hydrophobic and hydrophilic interactions in water, since they determine the main causes of the behavior of aqueous solutions of biological macromolecules. The study of processes in bio-solutions is in close connection with studies

N. Atamas (✉)
Taras Shevchenko National University of Kyiv, Kyiv, Ukraine
e-mail: atamasphys@mail.univ.kiev.ua

Institute of Physical Chemistry Polish Academy of Sciences, Kasprzaka 44/52, 01-224 Warsaw, Poland

Vienna University, Campus-Vienna-Biocenter 1, 1030 Wien, Austria

A. Makhinia
International European University, Kyiv, Ukraine

O. Fesenko and L. Yatsenko (eds.), *Nanostructured Surfaces, Nanocomposites and Nanomaterials, and Their Applications*, Springer Proceedings in Physics 296,
https://doi.org/10.1007/978-3-031-42704-6_1

of their structure, macroscopic behavior, and thermodynamic properties. The presence of hydrophilic (hydroxyl –OH groups) and hydrophobic groups (methyl and ethyl groups) in monohydric alcohols makes it possible to consider them as simple model systems for obtaining information about the behavior of biomolecules in water. It is known that the hydrophobic properties of alcohol molecules are closely related to the size of the alkyl chain of the amphiphilic molecule and vary depending on the number of $-CH_2-$ fragments in it [1, 2]. The hydrophobic properties of alcohol molecules in an aqueous solution are one of the most important factors that determine the specifics of the physical and chemical properties of water-alcohol solutions at different concentrations and the formation of complex, micelle-like structures from alcohol molecules in them [3, 4]. At the same time, the influence of the hydrophobic properties of alcohol molecules can be divided into the following two components: first, the hydrophobic effect as a result of the interaction of attraction between the alkyl chains of molecules due to van der Waals or London forces [5], which is effective in the interaction of organic molecules that have one or more $-CH_2-$ fragments. The second manifestation of the hydrophobic effect is hydrophobic solvation, which occurs due to the interaction of the alkyl chain of the alcohol molecule with water molecules. The study of the specifics of the concentration behavior of aqueous solutions of ethyl and propyl alcohols makes it possible to study at the molecular level the influence of these two factors, as well as the influence of the length of the alkyl chain on the properties of aqueous alcohol solutions using molecular modeling.

On the other hand, although water is the object of many studies [6–8] its statistical and dynamic properties are currently not fully understood. Therefore, water-solvent systems, for example alcohols, which are quite complex systems with hydrogen bonds, are the objects of many experimental and theoretical studies [9–14]. In molecular aqueous solutions, the structure of the liquid changes depending on the concentration and properties of the dissolved molecule, which leads to a special behavior of the physicochemical properties of the solutions depending on the concentration of the foreign substance [15–17]. Experimental studies [18–20] confirmed the existence of a special behavior of the physical and chemical properties of aqueous alcohol solutions depending on the concentration of the introduced substance. The neutron experiment made it possible to obtain information about the existence of a highly heterogeneous structure of water-alcohol solutions at concentrations at which there is anomalous behavior of the thermodynamic properties of the liquid system [21–23]. But unfortunately, the neutron scattering experiment cannot provide data at the atomic level, which is necessary to understand the processes and mechanisms that cause these anomalies.

The use of molecular modeling of aqueous solutions of alcohols makes it possible, on the one hand, to develop a methodology for studying the behavior of biomolecules in water. On the other hand, the analysis of the energy and structural characteristics of alcohol solutions at different concentrations makes it possible to determine the causes of the anomalous behavior of the physicochemical properties of alcohols at different concentration. In addition to the above, when carrying out studies of the concentration-structural features of aqueous solutions of monohydric alcohols by the Monte Carlo method, we achieved the following goals: to study the effect of the hydrophilic properties of alcohols on the change in the structural characteristics of aqueous solutions at various concentrations of alcohol and build model representations for the structure of aqueous solutions of alcohols at various concentrations.

2 Methods

Aqueous solutions of alcohols were modeled by the Monte Carlo method in the NVT ensemble at densities with the corresponding experimental solution densities at given concentrations of alcohols at $T = 300$ K. The system consists of 256 molecules in the cubic box with the periodic boundary conditions. Initial configuration is formed in cubic box of 18.6 Å length. Intermolecular interactions are calculated according to (1). Maximum trial displacements are ± 0.15 Å, maximum rotations are ± 15°, and 2×10^6 of Monte Carlo steps for the equilibration of the system, 10^6 Monte Carlo steps for obtaining structural properties, and 5×10^5 of steps for obtaining energy properties were used. Short-range intra-atomic interactions in liquid can be specified by Lennard–Jones potential and long-range interactions, which are described by Columbic potential:

$$U = U_{LJ} + U_{COUL} = \sum_{ij} 4\varepsilon_{ij}\left[\left(\frac{\sigma_{ij}}{r_{ij}}\right)^{12} - \left(\frac{\sigma_{ij}}{r_{ij}}\right)^{6}\right] + \sum_{ij} \frac{q_i q_j}{r_{ij}} \quad (1)$$

where U_{LJ}—Lennard–Jones potential, U_{COUL}—Columbic potential, r—distance between the interacting sites, and q_{ij}—charge on a site.

OPLS potentials were used to describe the atom–atom interaction between alcohol molecules in an aqueous solution of alcohol [24, 25]. Within this representation, the alcohol molecule is described within the framework of the model of three power centers. In this case, the Coulomb interaction is taken into account as follows: it is assumed that the positively charged atom q_H is located on the hydrogen atom of the hydroxyl group and on a separate "effective" atom with which the methyl and ethyl groups of the alcohol molecule are identified, while the q_O atom has a negative charge. Electrical neutrality is achieved by placing a certain charge on a model "effective" atom.

The SPC/E model is used for water molecules. The potential parameters (1) ε_{ij} and σ_{ij} for the interaction between the atoms of the solvent molecules and the atoms of the solute molecules are calculated using the Lorentz-Bartlot combination rule [26]:

$$\sigma_{ij} = \frac{\sigma_i + \sigma_j}{2} \quad \varepsilon_{ij} = \sqrt{\varepsilon_i \varepsilon_j} \quad (2)$$

Table 1 presents the geometric parameters for the solute and solvent molecules. Table 2 shows the values of parameters ε and σ and charges q of the OPLS potential on the force centers of molecules.

The total energy of intermolecular interaction $\langle E_{total}\rangle$ of solution can be represented by the sum of three components: interaction energy between the molecules of solvent, interaction energy between the molecules of the dissolved substance, and interaction energy between the molecules of solvent and dissolved substance. In the case of the aqueous solution of propanol, the total average energy of intermolecular interaction in the solution is:

$$\langle E_{total}\rangle = \langle E_{LJ}\rangle\langle E_{COUL}\rangle = \langle E_{alcohol\text{-}alcoholl}\rangle + \langle E_{water\text{-}alcohol}\rangle + \langle E_{water\text{-}water}\rangle, \quad (3)$$

where $\langle E_{alcohol\text{-}alcohol}\rangle$—the total energy of interaction between the molecules of alcohol, $\langle E_{water\text{-}water}\rangle$—the total energy of interaction between the molecules of water, and $\langle E_{water\text{-}alcohol}\rangle$—the total energy of interaction between the molecules of water and alcohol.

Table 1 Geometric parameters for ethanol and propanol

Molecule	Interaction	Length, Å	Angle	Angle, deg
$H–O–CH_2–CH_3$	OH	0.9450	∠ (HO–CH_2)	108.5
	O–C_2H_5	2.9420	∠ (HO–C_2H_5)	107.8
	OH	0.9450	∠ (HO–CH_2)	108.5
$H–O–C_2H_4–CH_3$	O–C_3H_7	3.3005	∠ (HO–C_3H_7)	107.8

Table 2 Charges q and values of σ and ε for ethanol and propanol atoms

Molecule	Atom	q, e	σ, Å	ε, kcal/mol
	H	0.435	0	0
$H–O–CH_2–CH_3$	O	−0.7	3.071	0.71162
	$CH_2–CH_3$	0.265	3.840	0.6542
	H	0.4350	0	0
$H–O–CH_2–CH_2–CH_3$	O	−0.7	3.071	0.71162
	C_3H_7	0.2650	3.905	0.86664

3 Results and Discussion

3.1 Energetics Properties of an Aqueous Solution of Alcohols at Different Concentrations

The energy dependences for water-alcohol systems presented in Fig. 1 correspond to the general trends characterizing the behavior of binary organic solutions [26]. In particular: the contribution of each component of the solution to the total energy of the system increases with an increase in the concentration of this component, and the decrease in the values of the average energy of interaction between water molecules $\langle E_{water\text{-}water}\rangle$ with an increase in the concentration of alcohol in water is compensated by changes, which leads to an increase in the values of the total energy of interaction in the system water-alcohol. This behavior of the energy characteristics of the alcohol-water molecular system can be characterized using Le Chatelier's principle: a decrease in the average water-alcohol interaction energy leads to an increase in aggregation between water molecules. Despite the increased interaction between water molecules in an aqueous solution, at high concentrations of alcohol in water, the hydrophobic properties of alcohol molecules play a dominant role in forming the properties of the solution [22].

First of all, the analysis of changes in the structure of water clusters in a water-alcohol solution was carried out depending on the average values of the interaction energy of water molecules $\langle E_{water\text{-}water}\rangle$. In this case, it was taken into account that the change in the energy values between different structures of water molecules correlates with the rearrangement of clusters of water molecules due to the formation or destruction of HBs in the middle of the cluster itself and the interaction of water molecules with alcohol molecules [27]. Based on such an analysis, one can make assumptions not only about the interaction between water molecules in solution, but also about the interaction between water molecules and alcohol molecules. The obtained values of the energy of the intermolecular interaction of water at a certain

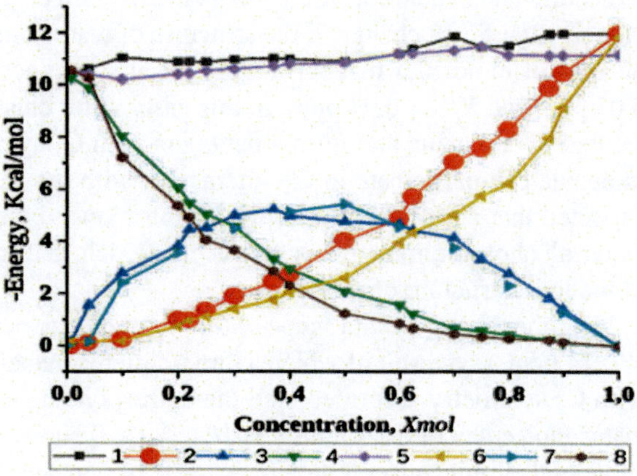

Fig. 1 Concentration dependence $\langle E_{sum}\rangle$ (1-ethanol, 5-propanol), $\langle E_{alcohol\text{-}alcohol}\rangle$ (2-ethanol, 7-propanol), $\langle E_{wat\text{-}wat}\rangle$ (4-ethanol, 8-propanol), and $\langle E_{water\text{-}alcohol}\rangle$ (3-ethanol, 7-propanol) in water-alcohol solution at $T = 300$ K

concentration were compared with the values of the energy of water clusters of different structures, according to the ideas proposed in the works of Xantheats S.S [27]. We will analyze the change in the structure of water clusters in alcohol solutions of ethyl and propyl alcohol depending on the average values of the interaction energy between water molecules $\langle E_{water\text{-}water} \rangle$. Let us assume that the change in energy values between different structures made of water molecules correlates with the rearrangement of clusters made of water molecules due to the formation or breaking of HBs in the middle of the cluster itself and the interaction of water molecules with alcohol molecules [27]. Such an analysis allows us to make assumptions not only about the interaction between water molecules in the solution but also about the interaction between water molecules and alcohol molecules.

In the case of the water-ethyl alcohol system, the value of the intermolecular interaction energy in water is $\langle E_{water\text{-}water} \rangle = (10.34 \pm 0.3)$ kcal/mol. According to the data given in the work of Xantheats [27], the value of the energy of intermolecular interaction in water is equal to $\langle E_{water\text{-}water} \rangle = 10.5$ kcal/mol, while water molecules form planar clusters of six molecules, which resemble a "book" in shape. With an increase in the content of ethanol in water, the energy of interaction between water molecules decreases $\langle E_{water\text{-}water} \rangle$, and at a concentration of $X_{et} = 0.05$ ppm, the total energy of intermolecular interaction between water molecules in an aqueous solution of ethyl alcohol has the value $\langle E_{water\text{-}water} \rangle = -9.2$ kcal/mol. In this case, there is a rearrangement of clusters of six water molecules that form a "book" to a voluminous water cluster that has the shape of a "prism".

With an increase in the alcohol content in water, the energy of interaction between water molecules $\langle E_{water\text{-}water} \rangle$ decreases, and at concentrations of $X_{et} = 0.05$ ppm and $X_{pr} = 0.04$ ppm, $\langle E_{water\text{-}water} \rangle$ has a value of ~ -9.2 kcal/mol. In this case, there is a rearrangement of clusters of six water molecules that form a "book" to a voluminous water cluster that has the shape of a "prism". A change in the shape of clusters of water molecules in the area of concentrations lower than $X_{et} = 0.05$ ppm and $X_{pr} = 0.04$ ppm does not lead to a change in the number of water molecules in the clusters. That is, in this range of concentrations, alcohol molecules do not take part in breaking HBs between water molecules, but only distort the network of HBs between water molecules and can be considered as a kind of "defect" in the local structure of water. A more significant change in the structure of water clusters, namely a change in the number of molecules that form them, occurs at concentrations greater than $X_{et} = 0.05$ ppm and $X_{pr} = 0.04$ ppm. In this case, water clusters consist of five molecules located in one plane and form a pentagon with C_1 symmetry. Thus, the sixth water molecule can participate in the interaction with the alcohol molecule. It should be expected that starting from these concentrations, the rearrangement of HBs in the water-alcohol solution will begin to occur, which, in turn, will lead to a drastic change in the local structure of the solution.

A further change in the shape of the water cluster occurs at a concentration of $X_{et} = 0.14$ ppm, as a result of which a three-dimensional cluster of five water molecules with Cs symmetry is formed, and the corresponding energy of interaction between water molecules in the solution is $\langle E_{water\text{-}water} \rangle = -7.0$ kcal/mol. The established value of the limit concentration of $X_{et} = 0.14$ ppm and $X_{pr} = 0.10$ ppm, at which the local structure of aqueous solutions of alcohols changes and correlates with the data obtained by the authors of the work [28] based on studies of ultrasound propagation and neutron experiments [29]. Note that the water-alcohol solution is a typical representative of non-ideal molecular solutions, the physical and chemical properties of which depend on the strength of interaction between the molecules of the various components of the solution. When such forces are the same (for $X_{et} = 0.42$ ppm and $X_{pr} = 0.40$ ppm in Fig. 1), the solution is an ideal solution and its behavior can be described using Raoult's law. Deviations from this law [28] are observed in the special behavior of the partial molar volume and thermodynamic functions, such as entropy and free energy of the mixture. Thus, it should be expected that in the vicinity of concentrations of $X_{alcohol} = (0.10–0.25)$ ppm, a radical restructuring of the local structure of the solution must occur, the consequence of which is a significant change in the characteristics of water clusters. This assumption is confirmed by the results of the study of diffusion in an aqueous solution of ethyl alcohol, according to which in the vicinity of concentrations of $X_{et} = 0.2$ ppm, the minimum value of the self-diffusion coefficient is observed [30]. According to these studies, the concentration of $X_{et} = 0.22$ ppm in the ethanol–water solution is characterized by a decrease in the diffusion movement of molecules, which is caused by clustering processes in the system.

3.2 Local Structure of an Aqueous Solution of Alcohols at Different Concentrations

The obtained experimental values give only the averaged characteristics of the movement and interaction of molecules. To verify and confirm the model representations proposed above regarding the existence of various local structures in an aqueous solution of propyl alcohol, RDFs were calculated and analyzed for the probabilities of the arrangement of atoms in a wide range of concentrations. Based on the RDFs analysis, an analysis of the effect of alcohol concentration on the change in the probabilities of interactions between water molecules, water and alcohol, and between alcohol molecules was carried out.

First of all, we will consider interactions between atoms of water molecules in an aqueous solution of alcohol at different concentrations at $T = 300$ K. Special attention was paid to the analysis of interatomic interactions that determine the network of hydrogen bonds in the system [31], namely interactions between oxygen atoms and hydrogen atoms of water molecules $O^{water} \ldots H^{water}$ (Fig. 2).

The analysis shows that RDFs values increase with an increase in the alcohol content in the solution, which indicates an increase in the probability of the formation of hydrogen bonds between water molecules. The position of the first maximum of the $G^{water-water}(r)$ RDFs, the value of which is equated with the length of the hydrogen bond, is fixed at a distance of ~ 2.2 Å. The radius of the first hydration sphere does not depend on the characteristics of the alcohol on its concentration in the solution and

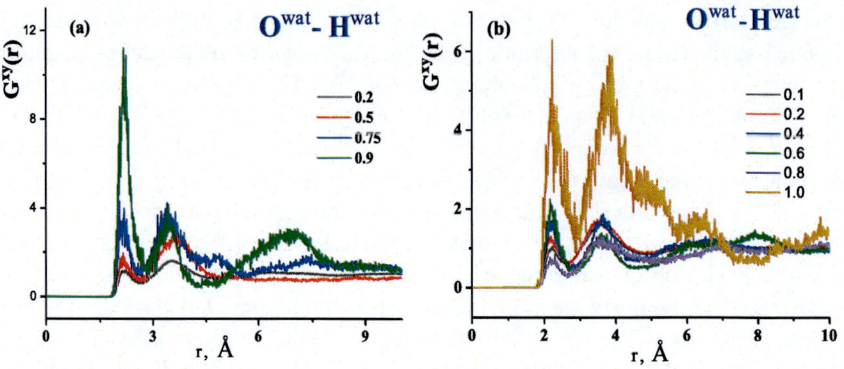

Fig. 2 RDFs for the probability of the location of atoms of O^{wat}–H^{wat} water molecules in an aqueous solution of alcohol (a-ethanol, b-propanol) at different concentrations at $T = 300$ K

is ~ 2.6 Å. With an increase in the alcohol content of the solution, the RDFs values for the maxima increase, but the functions themselves cease to be clear and smooth, which indicates an increase in the role of fluctuation phenomena with a decrease in the water content in the system.

Special attention is drawn to the values of the concentration $X_{et} = 0.85$ ppm and $X_{pr} = 0.7$ ppm, in the vicinity of which there is a sharp increase in RDFs $G^{Owat-Hwat}(r)$ values for the first maximum. The obtained changes indicate the rearrangement of clusters of water molecules surrounded by propanol molecules and an increase in the role of long-range interactions. Note that with an increase in the content of propanol in the RDFs liquid, they become more structured, namely the sizes of the second and third coordination spheres increase, which indicates an increase in the role of long-range interactions. The obtained RDFs $G^{Owat-Hwat}(r)$ shows that there is no change in the position of the first and second RDFs maxima. The obtained result confirms the fact that the lengths of hydrogen bonds (HB) between water molecules in the water-alcohol system are independent of the alcohol-water content, are ~ 2.4 Å, and correlate with the data, which presents the results of dielectric studies and found that that the range of concentrations $X_{pr} = (0.7–0.8)$ ppm is limiting for all monoatomic alcohols. It was found that in this range of concentrations, the dependence of the shape parameter, which characterizes the asymmetric representation of the relaxation time, has a minimum. They also found that at concentrations ~ 0.8 ppm in an aqueous solution of alcohol, a change in the local structure of the solution occurs.

Let us consider the mechanism of changing the structure of an aqueous solution of alcohol: a sharp increase in the RDFs value of $G^{Halcohol-Oalcohol}(r)$ is recorded at $X_{et} = 0.22$ ppm and $X_{pr} = 0.20$ ppm, and their analysis shows that due to $H^{alcohol}\ldots O^{alcohol}$ interaction, no hydrogen bond is formed. RDFs $G^{Calcohol-Oalcohol}(r)$ records an increase in values and expansion of the first maximum at $X_{et} = 0.22$ ppm and $X_{pr} = 0.20$ ppm. That is, the dominant role in the interaction between alcohol molecules belongs to $C^{alcohol}\ldots O^{alcohol}$ interactions, due to which HB bonds are formed between alcohol molecules. The position of the first RDF maximum $G^{Oet-Oet}(r)$ determines the distance between adjacent ethanol molecules in the solution. A decrease in the distance between the nearest ethanol molecules with an increase in the local density of the solution is shown by RDFs $G^{Oalcohol-Oalcohol}(r)$, for which there is a slight shift in the position of the maximum of the first RDFs peak. In turn, a decrease in the distance between alcohol molecules in the solution by $X_{et} = 0.22$ ppm and $X_{pr} = 0.20$ ppm leads to an increase in the role of repulsive interactions, as a result of which there is a decrease in the potential barrier for two alcohol molecules separated by a water molecule, and an increase in the role of fluctuations associated with transitions between contact and separated pairs of molecules. On the other hand, according to the ideas of Franks, the change in the network of HBs is a consequence of the formation of clusters of hydrophobic molecules.

That is, the obtained results allow us to state that at concentrations lower than $X_{et} = 0.22$ ppm and $X_{pr} = 0.20$ ppm, the interaction between water molecules and individual alcohol molecules is dominant; for $X_{et} = 0.22$ ppm and $X_{pr} = 0.20$ ppm, a significant role is played by the hydrophobic properties of alcohol molecules, due to the interaction between which clusters of alcohol molecules are formed. In the case when the limit value of the probability of the formation of HBs is reached, the network of HBs between water molecules loses its infinite connectivity, and at this stage, the solution consists of two types of clusters, one of which is dominated by water molecules and the other alcohol molecules, as well as free water and alcohol molecules.

Thus, it was obtained that for the $X_{et} = 0.22$ ppm and $X_{pr} = 0.20$ ppm due to the hydrophobic interaction between alcohol molecules, clusters of alcohol molecules are formed, which are surrounded by water molecules. Also, in solution there continue to cluster consisting of not only water molecules.

In the area of concentration $X_{et} = (0.38–0.45)$ ppm and $X_{pr} = (0.3–0.45)$ ppm, a further micelle from alcohol molecules and a change in the structure of the solution occur due to the rearrangement of water clusters. Clusters of alcohol molecules at $X_{alcohol} > 0.3$ ppm consist of four alcohol molecules surrounded by 12 water molecules. Clusters of alcohol molecules are similar in structure to micelles, in the middle of which there are hydroxyl groups and outside there are alkyl groups of alcohol molecules. Eight water molecules are around the alkyl chain of ethanol molecules, and four are around the hydroxyl group of the alcohol molecule (Fig. 3).

We recall that for the concentration of $X_{et} = 0.41$ ppm and $X_{pr} = 0.45$ ppm, the values of the energy of interaction between water molecules and the energy of interaction between alcohol molecules coincide, and the water-alcohol system is an ideal solution in which clusters of alcohol molecules resemble micelles in their structure. The obtained data correlate with the data [32] obtained using the method of dielectric relaxation spectroscopy, according to which the existence of a rearrangement of the local structure of the water-propyl alcohol system at $X_{pr} = 0.14$ ppm was revealed. The numbers of water and alcohol molecules that form systems in solutions at different concentrations were calculated from the analysis of the number of nearest neighbors obtained based on RDFs analysis (Table 3).

It is known [33, 34] that in the vicinity of $X_{pr} = (0.1–0.25)$ ppm, there is a maximum of fluctuations in alcohol concentrations in the solution and anomalous

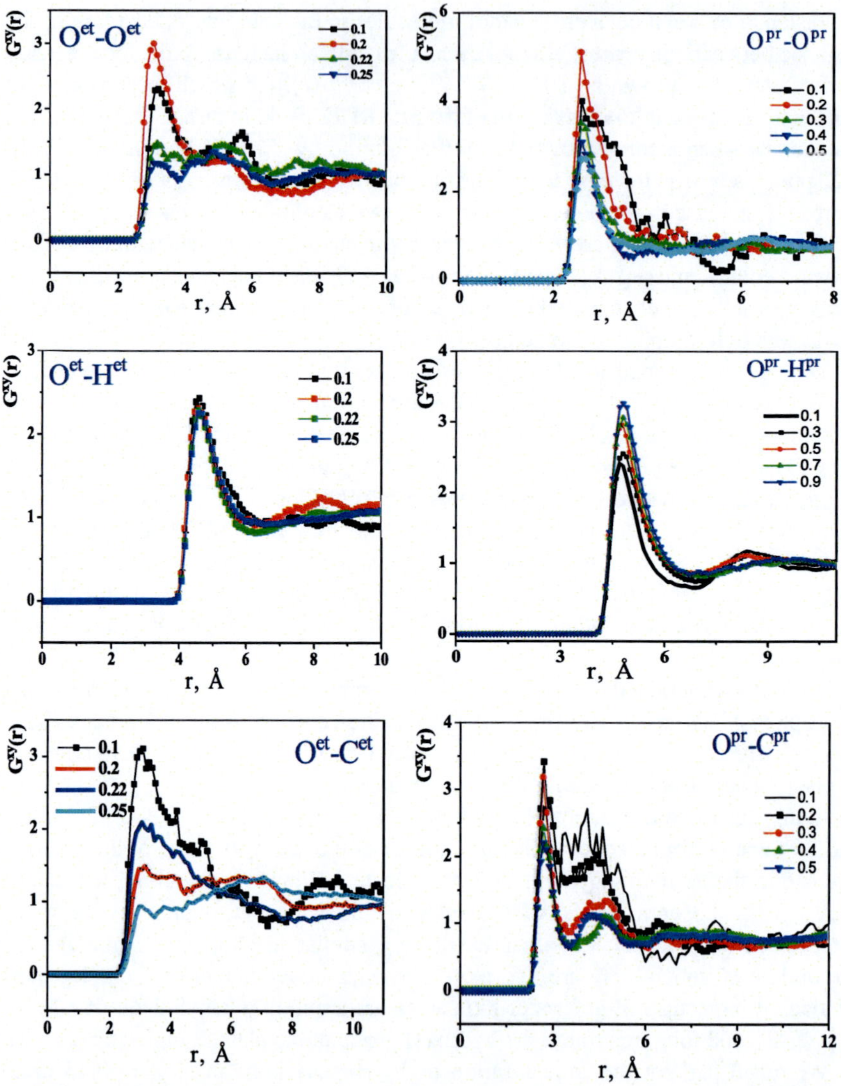

Fig. 3 RDFs for the probability of the arrangement of atoms of alcohol molecules (left panel—ethyl alcohol, right panel—propyl alcohol) in an aqueous solution at different concentrations at $T = 300$ K

behavior of the heat capacity of the aqueous propanol solution. This behavior of alcohol molecules may be associated with an increase in the role of hydrophobic interactions between propanol molecules and the formation of clusters from propanol molecules. For concentration values at which the values of the average energies of water-water and water-alcohol interactions coincide, the shape of the water cluster changes without changing the number of water molecules that form it. In the vicinity

Table 3 Ranges of concentrations of water-alcohol solutions at which the structure of the solutions changes

Aqueous solution of ethyl alcohol	Aqueous solution of propyl alcohol
1. $X_{et} < 0.055$ ppm: The introduction of ethanol molecules does not lead to a rupture of the network of HB between water molecules. Around the ethanol molecule introduced into the solution are five water molecules forming a pentagon. Interaction between ethanol molecules at long distances does not lead to the formation of complexes from ethanol molecules, and clusters from water molecules consist of six molecules	1. $X_{pr} < 0.04$ ppm: The introduction of propanol molecules does not break the network of hydrogen bonds between water molecules. Around the propanol molecule introduced into the solution are (7–8) water molecules, and the interaction between propanol molecules at long distances does not lead to the formation of complexes with ethanol molecules. Clusters of water molecules consist of six molecules
2. $X_{et} = (0.055–0.14)$ ppm: The shape of the water clusters is rearranged, and the number of molecules in the cluster decreases to five. Interaction between ethanol molecules at long distances does not lead to the formation of complexes of ethanol molecules surrounded by water molecules	2. $X_{pr} = (0.04–0.10)$ ppm: The shape of the water clusters is rearranged, and the number of molecules in the cluster decreases to five. Together with this, systems consisting of at least one molecule of water and one molecule of propanol are formed. There are also systems consisting of one propanol molecule surrounding (7–8) water molecules
3. $X_{et} = (0.14–0.18)$ ppm: There is a further restructuring of the shape of water clusters and a reduction in the number of molecules that form them to four (Fig. a). Due to this, free systems consisting of two water molecules and at least one ethanol molecule are formed. Systems consisting of one water and ethanol molecules and systems of water molecules surrounding one ethanol molecule also continue to exist	3. $X_{pr} = (0.10–0.14)$ ppm: The shape of the water clusters is rearranged and the number of molecules in the cluster decreases to five. Together with this, systems consisting of at least one molecule of water and one molecule of propanol are formed. There are also systems consisting of one propanol molecule surrounding (7–8) water molecules
4. $X_{et} = (0.18–0.22)$ ppm: The following stages can be distinguished: first, at the concentration of $X_{et} = 0.22$ ppm due to the hydrophobic properties of ethanol molecules, their interaction with each other, the interaction between water and ethanol molecules, structureless clusters of seven ethanol molecules surrounded by 18 water molecules are formed (Fig. a) (a) 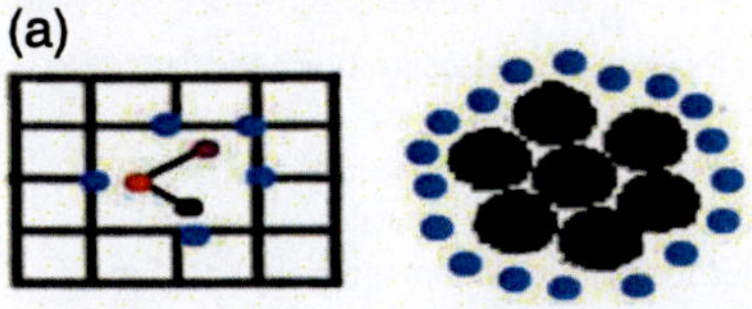 There also continue to exist systems consisting of water molecules surrounding one ethanol molecule analogous to systems found in infinitely dilute solutions	4. $X_{pr} = (0.14–0.25)$ ppm: The shape of water clusters is further rearranged, and the number of molecules that form them is reduced to four (Fig. a); systems consisting of two water molecules and at least one propanol molecule are formed. Some systems are formed from one water molecule and one propanol molecule, and systems from (7–8) water molecules surround one propanol molecule. When the maximum concentration of $X_{pr} = (0.18–0.25)$ ppm is reached due to the hydrophobic properties of propanol molecules and their interaction with each other and the interaction between water and propanol molecules, structureless clusters of (5–6) propanol molecules surrounded by (18–20) water molecules are formed (like Fig. a)

(continued)

Table 3 (continued)

Aqueous solution of ethyl alcohol	Aqueous solution of propyl alcohol
5. $X_{et} = (0.22–0.38)$ ppm: The formation of the structure of clusters from ethanol molecules takes place in the vicinity of concentrations of $X_{et} > 0.3$ ppm and consists of four ethanol molecules surrounded by 12 water molecules. Clusters of ethanol molecules are similar in structure to micelles, in the middle of which there are OH groups and outside there are alkyl groups of ethanol. Eight water molecules are around the alkyl chain of ethanol molecules, and four are around the OH group of the ethanol	5. $X_{pr} = (0.25–0.3)$ ppm: For $X_{pr} = 0.3$ ppm, the values of the average energies of water-water and water-propanol interactions coincide. At this concentration, the shape of the water cluster changes without changing the number of water molecules that make it up
6. $X_{et} = (0.38–0.45)$ ppm: There is a gradual transition from the scheme of the local structure of the solution presented in Fig. (a) to the scheme of the local structure of the solution, which is presented in Fig. (b), which ends at $X_{et} =$ 0.45 ppm. Within this range of concentrations, the formation of micelles from 4 ethanol molecules surrounded by water molecules continues, and there are clusters consisting of three water molecules	6. $X_{pr} = (0.3–0.45)$ ppm: The local structure of the solution is realized, which was formed in the vicinity of the concentrations $X_{pr} = (0.25–0.4)$ ppm, and there are clusters consisting of 3 water molecules. The change in the number of molecules making up the cluster is observed at $X_{pr} = 0.45$ ppm, which can be considered as an ideal solution consisting of relatively large clusters of propanol molecules and resemble inverted micelles in shape and clusters of water molecules. Propanol molecules in micelle-like formations are located externally by alkyl chains that interact with clusters of water molecules as with one molecule
7. $X_{et} = (0.45–0.6)$ ppm: Within this range of concentrations, the number of micelles from four ethanol molecules surrounded by water molecules increases, and there are clusters consisting of three water molecules	7. $X_{pr} = (0.45–0.8)$ ppm: There is a gradual transition from the scheme of the local structure of the solution presented in Fig. (a) to the scheme of the local structure of the solution, which is presented in Fig. (b), which ends at $X_{pr} =$ 0.45 ppm (b) 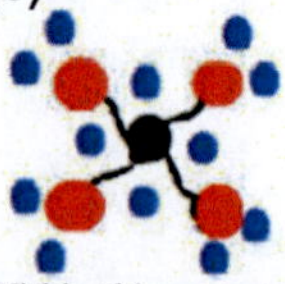 Within this range of concentrations, the formation of micelles from 4 ethanol molecules surrounded by water molecules continues, and there are clusters consisting of three water molecules. Micelles formed by alcohol molecules are broken, there may also be free alcohol molecules, and half-chains are formed from four alcohol molecules At $X_{pr} > 0.6$ ppm, the van der Waals interaction begins to play a dominant role; that is, around this concentration, there is a transition to the layered structure of the liquid, which is a characteristic of alcohols

(continued)

Table 3 (continued)

Aqueous solution of ethyl alcohol	Aqueous solution of propyl alcohol
8. $X_{et} = (0.6–0.8)$ ppm: There is a transition from the bulk structure of the solution inherent in water to the layered structure of the solution inherent in alcohols	8. $X_{pr} = (0.8–1.0)$ ppm: At $X_{pr} > 0.8$ ppm single water molecules can be considered as "defects" in the structure of propyl alcohol, and formations from alcohol molecules will resemble chains of 4 alcohol molecules. At $X_{pr} > 0.8$ ppm, micelles formed by alcohol molecules are broken; free alcohol molecules and half-chains formed from four propyl alcohol molecules may exist
9. $X_{et} = (0.8–1.0)$ ppm: Single water molecules can be considered as "defects" in the structure of ethyl alcohol, and formations from alcohol molecules will resemble chains of four alcohol molecules. At $X_{et} > 0.8$ ppm, micelles formed by alcohol molecules are broken; free alcohol molecules and half-chains formed from 4 ethanol molecules may exist	

of this concentration, the values of the intermolecular interaction energies of water-water and alcohol-alcohol coincide, and the aqueous solution of alcohol at this concentration is azeotropic. Thus, an aqueous solution of propanol of this concentration can be considered an ideal solution consisting of relatively large clusters of alcohol molecules, which resemble inverted micelles in shape and clusters of water molecules. Propanol molecules in micelle-like formations are located externally by alkyl chains that interact with clusters of water molecules as with one molecule. The existence of micelles in water-alcohol solutions is also indicated by the data of small-angle X-ray scattering [35], according to which in the vicinity of $X_{pr} = 0.2$ ppm of an aqueous solution of propyl alcohol, the value of the derivative of the partial pressure of the saturated vapor has relatively small values, and the behavior of the concentration dependence of the solution on the content of propanol is typical for the processes of micelle formation.

Note that the obtained result confirms the existence of two types of water in alcohol-water aqueous solutions. Namely, the first type of water system is located on the fractal surface of water clusters or an isolated alcohol molecule. The second type is not located on the fractal surface of the cluster formed from alcohol molecules and does not interact with alcohol molecules. Thus, water molecules in the region of dilute solutions can be classified into two types. It should be noted that two of the water molecules located on the surface of clusters of alcohol molecules are located near the hydroxyl group of the alcohol molecule. On the fractal surface formed by clusters of alcohol molecules, there are (12–13) water molecules, and their diffuse movement resembles the traditional idea of the movement of water, which forms a clathrate structure.

4 Conclusions

Based on Monte Carlo studies, the following has been established:

- In an aqueous solution of ethyl and propyl alcohols at infinitesimal concentrations (below $X_{et} < 0.05$ ppm in the case of ethyl alcohol and $X_{pr} < 0.04$ ppm in the case of propyl alcohol), the alcohol molecule does not change the percolation properties of the network of HBs between water molecules.
- The value of the dipole moment of an alcohol molecule affects the structural characteristics of clusters of alcohol molecules in water at concentrations of $X_{alcohol} = (0.2–0.25)$ ppm.
- At concentrations of alcohol in water over $X_{alcohol} > 0.3$ ppm, the structure of clusters is rearranged from alcohol molecules into structures similar to reverse micelles. The number of alcohol molecules forming reverse micelles does not depend on the structure and dipole moment of alcohol molecules and is equal to 4. In this case, the structure of the alcohol molecule determines the number of water molecules surrounding the micelle.
- At concentrations $X_{alcohol} = (0.6–0.8)$ ppm, rupture of micelles formed by alcohol molecules occurs. At these concentrations, there can also be free alcohol molecules and half-chains of four alcohol molecules.

References

1. S. Choi, P. Saravanan, Understanding alcohol aggregates and the water hydrogen bond network towards miscibility in alcohol solutions: graph theoretical analysis. Phys. Chem. Chem. Phys. **22**(30), 17181 (2020)
2. I. Bakó et al., Water–methanol mixtures: topology of hydrogen bonded network. Phys. Chem. Chem. Phys. **10**(32), 5004 (2008)
3. S. Javadian et al., Determination of the physico-chemical parameters and aggregation number of surfactant in micelles in binary alcohol–water mixtures. J. Mol. Liq. **137**(1–3), 74 (2008)
4. R.A. Abdel-Rahem et al., Rheological behavior of wormlike micelles (WLMs) in alcohol/water mixed solvent: in fluence of alcohol chain length. Colloid Polym. Sci. **299**(8), 1337 (2021)
5. M.A. Strauss, A.H. Wegner, Molecular systems for the quantification of London dispersion interactions. Eur. J. Organic Chem. **2**, 295 (2019)
6. D. Gavryushenko, K.S. Yablochkova, G. Taranyik, Temperature and temporal heterogeneities of water dynamics in the physiological temperature range. J. Mol. Liq. **340**, 117201 (2021)
7. Y. Tominaga, Dynamical structure of water by Raman spectroscopy. Fluid Phase Equilib. **133**, 323 (1992)
8. V. Bardik, Water dynamics and stability of major blood proteins at pre-denaturation stage. Atti della Accademia Peloritana dei Pericolanti **97**, A16 (2019)
9. L.A. Bulavin, Calculations of the structural parameters of aqueous solutions of propanol by the Monte Carlo method. Rus. J. Phys. Chem. A **83**(5), 763 (2009)
10. G. Aziz, F. Artzner, P. Malfreyt, Physical properties and hydrogen-bonding network of water–ethanol mixtures from molecular dynamics simulations. J. Phys. Chem. B **120**, 793 (2016)
11. K. Sharp, Water structure changes induced by hydrophobic and polar solutes revealed by simulations and IR spectroscopy. J. Chem. Phys. **114**, 1791 (2001)

12. L. Bulavin, A Monte Carlo study of the structure of highly dilute ethanol solutions and its variation with temperature. Rus. J. Phys. Chem. **79**, 1260 (2005)
13. A. Staib, Theoretical study of hydrogen bond dynamics of methanol in solution. J. Chem. Phys. **108**(11), 4554 (1998)
14. C. Ye, R. Ye, J. Liu, Effects of different concentrations of ethanol and isopropanol on physicochemical properties of zein-based films. Ind. Crops Products **53**, 140 (2014)
15. S. Nigan, Multicomponent calibration and analysis in liquid. Anal. Chem. **72**, 1956 (2000)
16. B. Ladanyi, Wave vector-dependent dielectric relaxation of methanol-water mixtures. J. Phys. Chem. **100**, 1368 (1996)
17. D. Fioretto, Dielectric relaxation in water-tert-butanol mixtures. The water rich region. J. Chem. Phys. **99**(15), 8115 (1993)
18. D.S. Venables, Spectroscopy and dynamics of mixtures of water with acetone and methanol. J. Chem. Phys. **113**(24), 11222 (2000)
19. V.I. Kuchuk, I.Yu. Shirokova, E.V. Golikova, Physicochemical properties of water-alcohol mixtures of a homological series of lower aliphatic alcohols. Glass Phys. Chem. **38**(5), 4605 (2012)
20. J.R. Elliott, *Introductory Chemical Engineering Thermodynamics* (Prentica Hall PTR, USA, 1996), p. 670
21. M.M. Lazarenko, Topological solitons in chain molecular crystals with stoichiometric obstacles and hydrogen bonds. J. Phys. Chem. Solids **144**, 109514 (2000)
22. J.B. Ott, *Chemical Thermodynamics, Principles and Applications* (Academic Press, USA, 2000), p. 664
23. A.A. Atamas, L.A. Bulavin, Structure effects in concentrated aqueous solutions of ethanol. Ukr. J. Phys. **49**(6), 552 (2004)
24. W.L. Jorgensen, J.D. Madura, C.J. Swenson, Optimized intermolecular potential functions for liquid alcohols. J. Am. Chem. Soc. **106**, 6638 (1984)
25. M. Allen, *Computer Simulation of Liquids* (Clarenton Press, Oxford, 2001), p. 432
26. J.R. Elliott, *Introductory Chemical Engineering Thermodynamics* (Prentica Hall PTR, USA, 1999), p. 670
27. S.S. Xantheats, Cooperativity and hydrogen bonding network in water clusters. Chem. Phys. **258**, 225 (2000)
28. T. Sato, Cooperative and molecular dynamics of alcohol/water mixtures: a view of dielectric spectroscopy. J. Mol. Liqs. **117**, 23 (2005)
29. S. Lenton, Temperature-dependent segregation in alcohol–water binary mixtures is driven by water clustering. J. Phys. Chem. B **122**(32), 78844 (2018)
30. D.W. Sharp, *Dictionary of Chemistry* (Penguin Books, London, 2018), p. 440
31. N. Atamas, Clustering in water-propanol solutions, in *2021 IEEE 11th International Conference Nanomaterials: Applications & Properties (NAP)*. IEEE, vol. 1 (2006)
32. B.G. Oliveira, Vasconcellos ML Hydrogen bonds in alcohols: water complexes: a theoretical study about new intramolecular interactions via CHELPG and AIM calculations. J Mol. Struct.: THEOCHEM **774**(1), 83 (2006)
33. S. Mashimo, Structure and primary alcohol studies by microwave dielectric analyses. J. Chem. Phys. **95**, 6257 (1999)
34. B. Khalfaouli, Thermodynamics properties of water-normal alcohols and vapor equilibria for binary system of methanol or 2-propanol with water. Fluid Phase Equilib. **127**, 181 (1997)
35. H. Hayashi, Small-angle X-ray scattering study of fluctuations in 1-propanol-water and 2-propanol-water systems. J. Phys. Chem. **94**, 8334 (1990)

Thermal Efficiency of Nanoscale Aluminum Layer Modified by Ionic Implantation

V. V. Honcharov, V. O. Zazhigalov, and A. V. Honcharova

1 Introduction

Global challenges of today require scientists and engineers to constantly look for new ways to improve heating devices, radiators, convectors, electric furnaces, etc. in order to increase the heat transfer of these devices. In some cases, it is necessary to intensify the heat transfer process, and in others, on the contrary, to slow it down [1–4].

Material scientists are practically interested in thermal conductivity and radiation, since the material of the heat exchanger device is most important for these parameters. Therefore, modern heating radiators are most often made of metals. They have the best combination of characteristics, the main of which is the heat transfer coefficient.

One of the best metals in terms of thermal conductivity is silver, but it is used where its significant price is justified. Instead, aluminum is already widely used. Although the heat transfer of aluminum is almost two times lower than copper, it is significantly high compared to other metals. Aluminum is light, heats up quickly, and transfers heat efficiently [5]. But it is also quite reactive and can interact with other metals in the system. As a result, corrosion begins, and metals are quickly destroyed.

In addition to the material, the heat transfer coefficient is significantly affected by the geometric parameters of the surface, such as the value of the surface area of heat and mass transfer. Therefore, modification of the heating surface to artificially change the roughness is one of the ways to intensify heat transfer [6].

V. V. Honcharov (✉) · V. O. Zazhigalov
Institute for Sorption and Problems of Endoecology, National Academy of Sciences of Ukraine, Kyiv, Ukraine
e-mail: milostiprosim@i.ua

A. V. Honcharova
Junior Academy of Sciences of Ukraine, Rubizhne, Ukraine

O. Fesenko and L. Yatsenko (eds.), *Nanostructured Surfaces, Nanocomposites and Nanomaterials, and Their Applications*, Springer Proceedings in Physics 296,
https://doi.org/10.1007/978-3-031-42704-6_2

In this sense, the technologies of surface modification through irradiation of solids by ion and plasma flows, which cause various changes in the relief, are promising [7–12]. Depending on the parameters of the irradiation flux and conditions on the surface, these changes are manifested both in the development and smoothing of the relief. Surface modification can be a direct result of ion exposure, and it is not surprising that the problem of aluminum modification by ion implantation has received considerable attention [13–17].

In particular, it was found that argon bombardment [14] leads to an improvement in the wettability of aluminum with water and redistribution of structural defects and, as a result, makes it more reactive. Instead, nitrogen implantation [13, 15] leads to the formation of nitrides, which can significantly increase the hardness of the material. When implanting titanium into aluminum [16], there is a decrease in the number and size of cavities, which reduces the risk of corrosion.

The most attractive from a scientific point of view is the effect of implantation on thermal properties [17, 18]. In particular, the authors of [17] claim a decrease in thermal stresses in aluminum when implanted with nitrogen and titanium, and the results of studies [18] indicate that the surface film created by implanting nitrogen on aluminum reduces the heat transfer coefficient by almost 40%.

Taking into account that implantation leads to the destruction of crystal structures [9, 11, 12], the information from the source [5] on increasing the energy saving efficiency by 40% when using amorphous materials is encouraging.

Thus, it can be concluded that the use of ion implantation technology allows to change the surface texture, affecting the average roughness and, as a consequence, changes the specific surface area of heat transfer. In addition, the saturation of the surface with nitrogen and titanium ions changes the electrical and thermal conductivity of the surface of the treated element, which allows us to recommend ion implantation for modifying the materials of heat exchange equipment, in particular aluminum.

For a more accurate understanding of the processes occurring in the surface layer of the material, in addition to physical and chemical methods of analysis, mathematical and computer modeling methods can be used. Quite effective and plausible models of the processes occurring during ion bombardment are described in [19–24].

Thus, it can be argued that it is expedient to use titanium ions implantation for modification of aluminum and to use computer simulation for research. Moreover, the study of properties should be carried out for two cases of using implanted aluminum as a heat conductive material and as a screen for radiation heat transfer.

2 The Experimental Part

Aluminum foil according to GOST 32582-2013 was used for the study (Fig. 1).

To treat the surface with titanium ions, an ionic implantation unit, which is described in [9], was used.

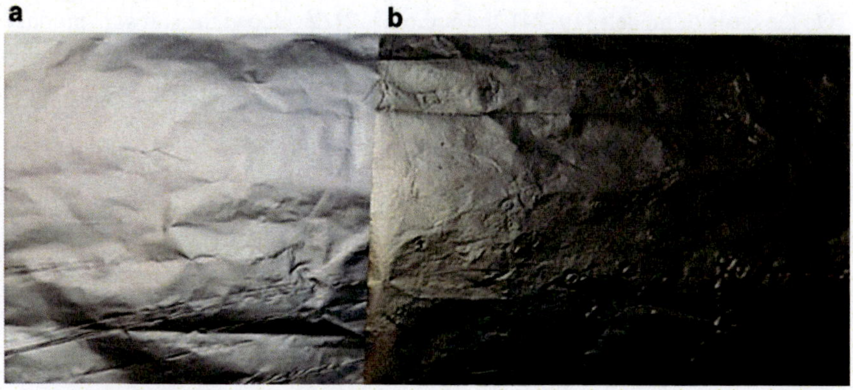

Fig. 1 Samples: **a** the initial Al, **b** Ti/Al

Its operation mode was as follows. A vacuum with a pressure of about 10^{-5} Pa was created in the chamber. The electric voltage between the ion source and the sample was about 20 kV. The processing time was 60 min (selected so that the doping dose was 5×10^{17} ions/cm^2).

As a result, two types of samples were obtained (Fig. 1): the original foil (initial Al) and the one treated with titanium ions (Ti/Al).

Thermophysical properties were investigated by measuring temperature parameters. Measurements were carried out for two cases. Firstly, the samples were fixed on the heating radiator in the same place, and thus, the temperatures were obtained due to thermal conductivity. Secondly, the samples were placed next to the radiator without contact with it and thus measured the absorption capacity of the surface to thermal radiation.

For both schemes, an infrared thermometer "HoldPeak HP-1300" with a measuring range from 50 to + 1300 °C was used to measure temperatures.

The ability to shield electromagnetic waves, i.e., reflection during radiation heat transfer, was evaluated using the Color Picker program, which is an application for determining some optical parameters from a smartphone camera with many additional functions. In particular, with the help of the Color Picker program, it is possible to determine the color temperature of a body, that is, the equivalent temperature of an absolutely black body, according to which the emissivity is determined through the Stefan-Boltzmann law.

Using the microscope "USB Microscope Digital microscope", which has a magnification of 500x, the surface microstructure of the samples was studied. This high-quality microscope allows to analyze images and videos, as well as to measure surface inhomogeneities.

On the basis of models [19–24], the authors [12] developed the software product "RIO", which is a convenient and effective tool for calculating the implantation parameters (ion travel, ion deflection, ion energy distribution, sputtering coefficient, film thickness, etc.) depending on the target material and process parameters. It is useful to visualize the concentration profiles of the implanted ions distribution in the surface layer of the material.

In this work, the study of implantation parameters was carried out according to the following procedure: select the Yudin implantation model, determine the physicochemical parameters for titanium and aluminum, assign the processing mode, and obtain data and graphs.

3 Results and Discussion

The results of measuring the surface temperatures (Fig. 2) showed a difference in the absolute values of temperatures between the two samples along the entire measuring circuit of about 3 °C, which is more than 10%. That is, the difference is significant, and it indicates that the thermal conductivity is greater in the treated sample. The temperature difference between the two samples along the entire measurement contour was about 1.5 °C or 4% in favor of the treated sample, which indicates its greater absorption capacity of thermal radiation.

The results of comparison of the ability to reflect electromagnetic waves, including infrared, are shown in Fig. 3.

Based on the diagram in Fig. 3, a, raw aluminum has a higher emissivity or radiative heat transfer capability.

In addition, treatment with titanium ions leads to deterioration of optical properties. In particular, the brightness of light in the optical range (Fig. 3b) is significantly higher for untreated aluminum foil (53.8% vs. 37.4%).

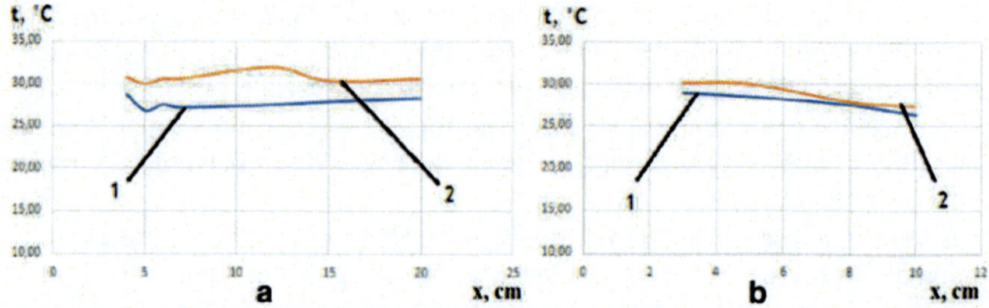

Fig. 2 Diagrams of surface temperatures of samples under conditions of heat transfer by thermal conductivity (**a**) and the samples under the conditions for the absorption capacity (**b**) for the initial Al (1) and Ti/Al (2)

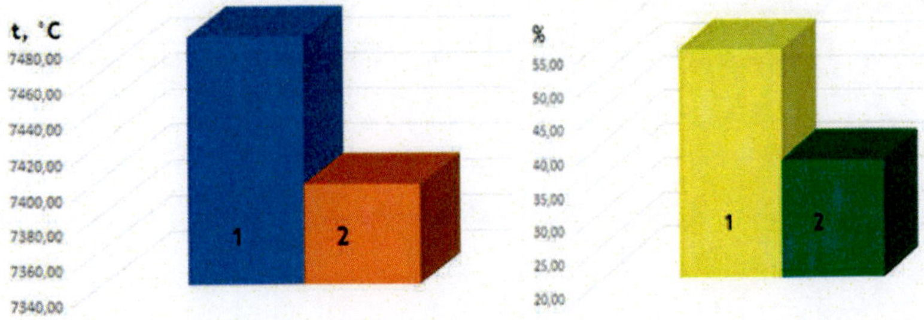

Fig. 3 Diagrams of color temperatures (**a**) and of reflected light brightness (**b**) for the initial Al (1) and Ti/Al (2)

Thus, it can be argued that after implantation of titanium ions, aluminum increases its thermal conductivity and absorption property, worsening the ability to reflect electromagnetic radiation.

Microscopic analysis (Fig. 4) showed that when titanium is implanted in aluminum, relief elements are smoothed.

In particular, the technological strips obtained during foil rolling by the manufacturer disappear. However, it is difficult to say about deeper and smaller changes in microrelief, as the size of the irregularities is very small.

This is confirmed by the results of modeling the processes occurring during implantation. If we consider the implantation of titanium ions in aluminum foil, it can be seen from the distribution profile (Fig. 5) that the modification of the material occurs mainly in the layer with a thickness of 30 nm.

The maximum concentration is near the surface, and the sputtering coefficient is 2.97 ion/atom. Accordingly, we have smoothing of the upper layer (especially the tops of bumps) due to sputtering of the material (according to the model, the thickness of the sputtered layer is 24.5 nm).

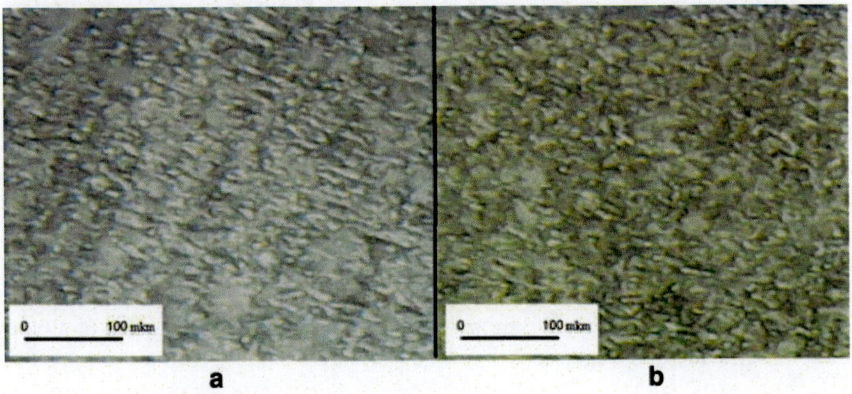

Fig. 4 Microphotographs of the initial Al (**a**) and Ti/Al (**b**)

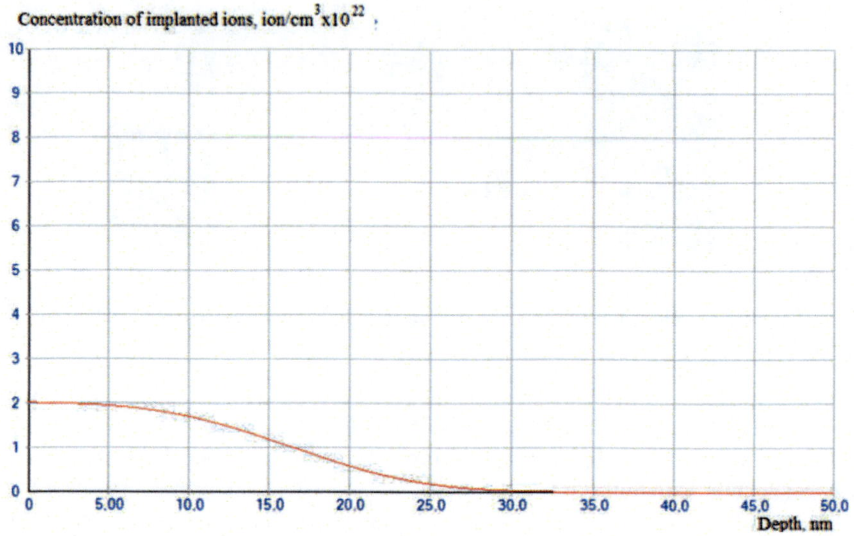

Fig. 5 Distribution of Ti ions by depth

However, it should be taken into account that the emission of titanium ions in the ion source occurs due to the active nitrogen plasma, the ions of which also enter the flow and, accordingly, bombard the surface.

Therefore, the implantation of nitrogen ions into the aluminum surface was additionally simulated. The results of modeling the distribution of nitrogen ions (Fig. 6) show that in this case the surface modification occurs at a much greater depth. The thickness of the layer of nitrogen ions penetration is slightly more than 100 nm.

The results of modeling the distribution of nitrogen ions (Fig. 6) show that in this case the surface modification occurs at a much greater depth. The thickness of the layer of nitrogen ions penetration is slightly more than 100 nm. The maximum concentration of the introduced ions is observed at a depth of 44 nm. In contrast to titanium ions, nitrogen bombardment is accompanied by deposition of an additional layer on the surface, especially in the depressions (sputtering coefficient is 0.38 ion/atom, and deposited layer thickness is 43.8 nm).

Thus, the changes that occur in the microgeometry of the surface layer are nanoscale in nature. Therefore, it can be assumed that the thermal properties of aluminum implanted with titanium ions are affected by two main factors.

Firstly, the surface is saturated with titanium ions, nitrogen, and their compounds to a depth of about 100 nm and, accordingly, the physical, chemical, and, consequently, thermal properties of aluminum foil change. Secondly, the microrelief is changed by spraying the tops of bumps with titanium ions and deposition of atoms in the hollows due to nitrogen ions.

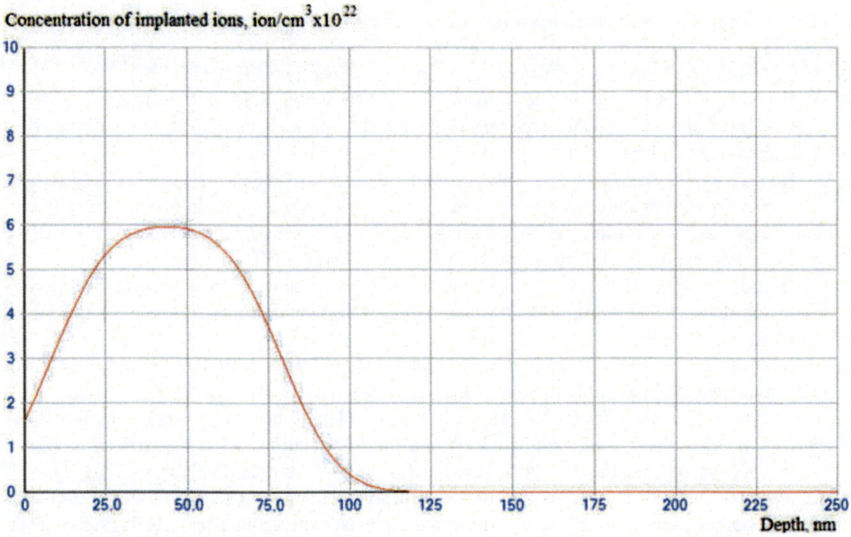

Fig. 6 Distribution of N ions by depth

4 Conclusions

It was established that the ionic implantation of titanium increases the thermal conductivity of aluminum, which makes it possible to recommend it for the creation of contact heat exchange devices.

Based on the simulation results, it is shown that nanoscale modification of aluminum occurs during ionic implantation.

The study of the surface microstructure showed that its treatment with titanium ions leads to a change in the microrelief. This testifies to the benefit of using ionic implantation to modify elements operating under conditions of not only thermal but also mechanical loads, in particular burners, catalysts, filters, etc.

Therefore, ionic implantation has significant prospects in ecological catalysis, energetics, chemical, and pharmaceutical industries.

References

1. F.F. Czvetkov, B.A. Grigor'ev, Teplomassoobmen: Izdatel'stvo MEI, Moskva (2005)
2. I.M. Kuzmenko, Experimental research of heat transfer in cooling tower. Am. J. Energy Res. **3**(2), 47–52 (2014)
3. G. Changsheng, Y. Shaopan, Numerical methods for temperature field about random heat source model of ground source heat pump. Am. J. Ind. Eng. **3**(1), 20–27 (2013)
4. G.A. Frolov, V.P. Solncev, S.M. Soloni et al., Vysokotemperaturnye ispytaniya metallicheskoj teplovoj zashhity dlya raketno-kosmicheskoj texniki. Aviacionno-kosmicheskaya texnika i texnologiya **10**(57), 62–64 (2008)

5. T.B. Bobrova, Osnovy materialoznavstva. Navchalnyi posibnyk. Resursnyi tsentr HURT, Kyiv (2019)
6. M.A. Mixeev, I.M. Mixeeva, Osnovy teploperedachi. Energiya, Moskva (1977)
7. S. Kunihiro, *Ion Implantation and Activation*, vol. 2 (Bentham Science Publishers, 2013)
8. L.B. Begrambekov, Modifikaciya poverhnosti tverdyh tel pri ionnom i plazmennom vozdejstvii. Moskovskij inzhenerno-fizicheskij institut, Moskva (2001)
9. V. Honcharov, V. Zazhigalov, Z. Sawlowicz, R. Socha, J. Gurgol, Structural, catalytic, and thermal properties of stainless steel with nanoscale metal surface layer, in *Nanophysics, Nanomaterials, Interface Studies, and Applications*, ed. by O. Fesenko, L. Yatsenko. NANO 2016. Springer Proceedings in Physics, vol. 195 (Springer, Cham, 2017)
10. T.J. Kang, J.-G. Kim, H.-Y. Lee et al., Modification of optical and mechanical surface properties of sputter-deposited aluminum thin films through ion implantation. Int. J. Precis. Eng. Manuf. **15**(5), 889–894 (2014)
11. D.M. Pout, G. Foti, D.K. Dzhekobson, Modifitsirovanie i Legirovanie Lazernyimi, Ionnyimi i Elektronnyimi Puchkami. Mashynostroenie, Moskva (1987)
12. A.A. Cherny, S.V. Maschenko, V.V. Honcharov et al., Nanodimension layers on stainless steel surface synthesized by ionic implantation and their simulation, in *Nanoplasmonics, Nano-Optics, Nanocomposites and Surface Studies.* (Springer, Switzerland, 2015), pp. 203–213
13. D. Heyden, D. Müller, G.K. Wolf et al., Modification of stainless steel and aluminium with pulsed energetic ion beams in the millisecond regime. Nucl. Instrum. Methods Phys. Res. Sect. B **175–177**, 403–409 (2001)
14. C. Niu, J. Han, S. Hu, Surface modification and structure evolution of aluminum under argon ion bombardment. Appl. Surf. Sci. **536**, 147819 (2021)
15. M. Okumiya, Y. Tsunekawa, H. Sugiyama et al., Surface modification of aluminum using ion nitriding and barrel nitriding. Surf. Coat. Technol. **200**(1–4), 35–39 (2005)
16. M.X.Y. Yao, C.S. Kumai, T.M. Devine et al., The pitting corrosion behavior of aluminum ion implanted with titanium. Nucl. Instrum. Methods Phys. Res. Sect. B **80–81**(1), 267–270 (1993)
17. J.X. Liao, L.F. Xia, M.R. Sun et al., Structural characteristics of 2024 aluminum alloy plasma-based ion implanted with nitrogen then titanium. Appl. Surf. Sci. **240**(1–4), 71–76 (2005)
18. K. Kim, J.H. Jeong, Condensation mode transition and condensation heat transfer performance variations of nitrogen ion-implanted aluminum surfaces. Int. J. Heat Mass Transf. **125**, 983–993 (2018)
19. V.A. Nikonenko, Matematicheskoe modelirovanie texnologicheskih processov: Modelirovanie v srede MathCAD. Praktikum. Moskovskij institut stali i splavov, Moskva (2001)
20. A.V. Bobyl, S.F. Karmanenko, Fiziko-himicheskie osnovy tehnologii poluprovodnikov. Puchkovye i plazmennye processy v planarnoj tehnologii: Ucheb. Posobie. Izd-vo Politehn. un-ta, SPb (2005)
21. D. Falcone, Sputtering theory. UFN **162**(1), 71–117 (1992)
22. R. Berish, *Raspylenie tverdyh tel ionnoj bombardirovkoj* (Mir, Moskva, 1984)
23. F. Keywell, Measurements and collision-radiation damage theory of high-vacuum sputtering. Phys. Rev. **97**(6), 1611–1619 (1955)
24. P. Antonetti, D. Antoniadisa, R. Dattona et al., MOP-SBIS. Modelirovanie elementov i tehnologicheskih processov. Radio i svyaz‘, Moskva (1988)

Structural Transformations in Oxide Ceramic Coatings Formed on Aluminum Alloys in Silicate Electrolyte

Volodymyr Posuvailo, Ihor Koval'chuck, Iryna Ivasenko, Yurii Kanyuk, and Roman Iurkevych

Abstract X-ray phase analysis of oxide ceramic coatings formed on the alloys D16T and AK9M2 in the process of plasma electrolyte oxidation in the electrolyte (KOH (3 g/l), Na_2SiO_3 (2 g/l) is fulfilled. It is established that at the beginning of the coating synthesis (after 5 min) there is a transformation of the crystal lattice of aluminum with the formation of the compound $Al_{3.21}Si_{0.47}$ in the surface layers of both alloys. After 1 h of synthesis, the corresponding oxide ceramic coatings on alloys D16T and AK9M2 contain the following phases: α-Al_2O_3 (corundum), γ-Al_2O_3, and $Al_2O_3{\cdot}SiO_2$ (sillimanite). When increasing the synthesis time to 2 h there is a partial conversion of sillimanite ($Al_2O_3{\cdot}SiO_2$) to mullite ($3Al_2O_3{\cdot}2SiO_2$) in the coatings on the alloy AK9M2. The study of the surface microstructure showed a uniform growth of oxide ceramic coatings and an increase of Si areas due to the formation of SiO_2 and the formation of a low-temperature substitution compound $Al_{3.21}Si_{0.47}$. The image segmentation method was used to analyze the pore sizes and the distribution and size of bit channels. It is established that the minimum pore size is approximately 0.9 μm.

V. Posuvailo (✉) · I. Koval'chuck · I. Ivasenko · Y. Kanyuk
Karpenko Physico-Mechanical Institute of the NASU, Lviv, Ukraine
e-mail: vposuvailo@gmail.com

I. Koval'chuck
e-mail: ihor-kovalchuk@i.ua

I. Ivasenko
Lviv Polytechnic National University, Lviv, Ukraine

R. Iurkevych
Hetman Petro Sahaidachnyi National Army Academy, Lviv, Ukraine

O. Fesenko and L. Yatsenko (eds.), *Nanostructured Surfaces, Nanocomposites and Nanomaterials, and Their Applications*, Springer Proceedings in Physics 296,
https://doi.org/10.1007/978-3-031-42704-6_3

1 Introduction

An urgent task of modern science and technology is the development of new technologies for the formation of highly effective, reliable coatings for the protection and strengthening of metal products. There are many ways of increasing the wear and corrosion resistance of light alloys [1–10]: anodizing, chromium plating, phosphating, oxidation, nitriding, etc. One of these methods is plasma electrolyte oxidation. Therefore, a new type of surface treatment and strengthening of metal materials, which is a variant of traditional anodizing pro-plasma electrolytic oxidation (PEO), is actively developing. It makes it possible to obtain multifunctional oxide ceramic coatings with a unique set of properties on valve metals (Al, Mg, Ti, Zr, Ta) [11–13]. However, in addition to high corrosion resistance, oxide ceramic coatings should also have good wear resistance. There are four main stages of the formation of oxide ceramic coatings [14] on valve metals: the formation of the primary oxide film at the pre-spark stage by an electrochemical mechanism; breakdown of the primary oxide film and the appearance of a plasma clot in the discharge channel; plasma chemical reactions of formation of intermediate and final products; and condensation and polymorphic transformations of oxide phases [14]. In recent years, this method has been actively used to form wear- and corrosion-resistant coatings on silumins, which are widely used in industry and contain 4…0.22 mass % of silicon [15–17]. Silumins have a low cost, and they are widely used in various industries, namely: in mechanical engineering (pistons, body parts, engine cylinders), aircraft construction (cylinder blocks, pistons for cooling, aviation assemblies), aerospace engineering (parts with a low temperature coefficient of linear expansion and a high level of mechanical properties), during the manufacture of gas turbine equipment (generators, heat exchangers), etc. Their corrosion resistance is higher than that of deformable aluminum alloys, but at the same time their wear resistance is not high. In the last decade, methods of applying PEO coatings on deformable aluminum alloys and silumins have been actively developed.

The purpose of this work was to establish the mechanism of interaction of plasma discharge channels with aluminum and silicon and the effect of silicon on the phase composition, crystal structure, and porosity of oxide ceramic coatings synthesized on aluminum alloys of the Al–Si and Al–Cu systems.

2 Methods and Materials

Oxide ceramic coatings were synthesized on aluminum alloys AK9M2 and (Al—87.6–93.6%, Si—6–8%, Cu—1.5%, Mg—0.2–0.5%, Mn—0.2–0.5%) and D16 (94.7% Al; 3.8…4.9 Cu; 1.2…1.8 Mg; 0.3…0.9% Mn). The size of the samples is 20 * 15 * 3 mm. Before synthesis, the samples were polished and washed in distilled water and ethyl alcohol. The coating was formed by alternately applying anodic and cathodic pulses to the sample. The coatings were synthesized at cathodic to anodic current densities ratio $j_c/j_a = 15/15$ A/dm^2 for time 5 min and 2 h. An aqueous solution of KOH (3 g/l), Na_2SiO_3 (2 g/l) served as the electrolyte. A DRON-3.0 diffractometer with CuK_α radiation was used for X-ray phase analysis of coatings. The content of each of the phases was determined by diffractograms using the FullProf program package, using the multi-disciplinary method of Rietveld [18]. Porosity was investigated by analyzing microphotographs of PEO coatings obtained on a Zeiss EVO-40XVP scanning electron microscope with a magnification of ×500 times according to the method [19].

3 Results and Discussions

X-ray structural analysis was used to determine the silicon content in the original AK9M2 alloy, as well as the composition of the corresponding oxide ceramic coatings after 5 min and 2 h of PEO treatment. The corresponding X-ray diffractograms of the original AK9M2 alloy and the oxide ceramic coating on it after 5 min of processing are shown in Fig. 1.

According to X-ray images, it was established that the intensity of Al, Si reflexes decreases after PEO treatment of the AK9M2 alloy. In addition, traces of impurity phases such as SiO_2 and $Al_{2.67}O_4$ were found in the coatings. It was established for the first time that already at the beginning of the synthesis of the oxide ceramic coating, aluminum and silicon form $Al_{4-x}Si_x$ substitution compounds. In particular, the substitution compound $Al_{3.21}Si_{0.47}$ has space group $Fm3m$ and lattice parameters $a = 0.5641$ nm [20]. Its presence is confirmed by the decrease in intensity of Al and Si reflexes and their increase for $Al_{4-x}Si_x$.

Figure 2 shows X-ray diffraction patterns of alloy D16 after 5 min of PEO treatment.

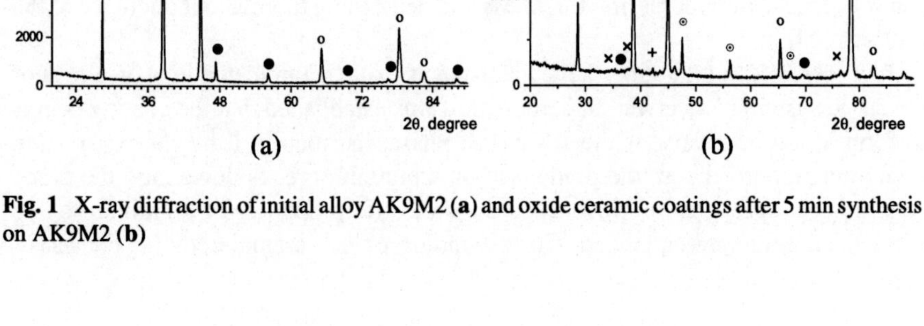

Fig. 1 X-ray diffraction of initial alloy AK9M2 (**a**) and oxide ceramic coatings after 5 min synthesis on AK9M2 (**b**)

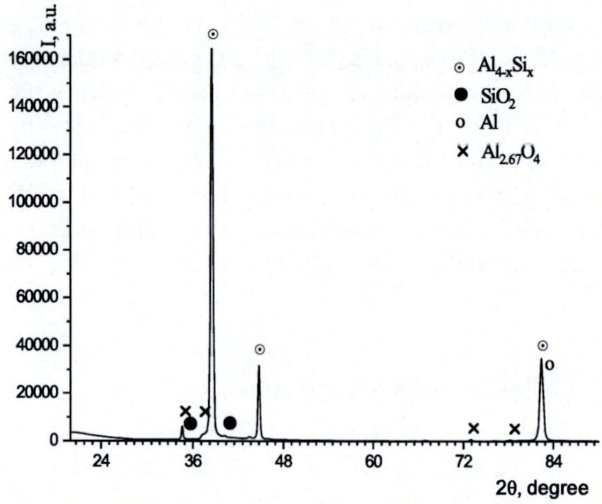

Fig. 2 X-ray diffraction of oxide ceramic coatings on D16 alloy after 5 min synthesis

On the surface of D16 alloy after 5 min of PEO treatment, in addition to Al, substitution compounds $Al_{4-x}Si_x$ as well as traces of impurity phases $Al_{2.67}O_4$ and SiO_2 were detected.

This indicates that silicon is reduced from the electrolyte (Na_2SiO_3) with subsequent synthesis of the $Al_{4-x}Si_x$ compound in the plasma discharge channels on the surface of the alloy.

The works [14, 21, 22] showed the growth mechanisms of the oxide ceramic coating on aluminum alloys in the process of plasma electrolytic oxidation. As you know, valve metals (Al, Ti, Mg, Zr, Ta) are usually covered with a thin passive oxide film. Due to the application of high voltage, this film breaks down and discharge plasma channels appear. Atomization and ionization of the electrolyte and base metal from which the oxide ceramic coating grows take place in them.

Previously, an analysis of the distribution of silicon in oxide ceramic coatings was carried out [23]. In particular, it was established that the transition layer between the aluminum base and oxide ceramics contains up to 0.5 at. % silicon and is partially amorphized. Al_2SiO_5 and $Al_6Si_2O_{13}$ phases are formed in the near-electrolyte region.

The main phases of oxide ceramic coatings are γ-Al_2O_3 (*Fm*3*m*) and α-Al_2O_3 (*R*-3*C*). The formation of α-Al_2O_3 (corundum) occurs when the temperature in the vicinity of the discharge channel rises above 1000 K to the temperature of the phase transition of γ-Al_2O_3 to α-Al_2O_3. The increase in corundum content in oxide ceramic coatings grows with increasing thickness and decreasing thermal conductivity of the coatings [24].

In works [25, 26] the growth mechanisms of oxide ceramic coatings on alloys of the Al–Si system were given. In particular, it was established that the coating begins to form at the boundary of the Al and Si phase distribution from the passivation of aluminum and silicon, the oxide film on aluminum breaks down, and the oxide ceramic coating begins to grow when the dielectric properties of aluminum oxide and silicon oxide are equalized. Corresponding oxide ceramic coatings consist of

γ-Al_2O_3 and α -Al_2O_3 and a small amount of SiO_2 and Al_2SiO_5. It should be noted that these coatings are synthesized at a much lower current density (4–6 A/dm^2) and in the presence of a phosphate electrolyte. Therefore, the increase in current density up to 15 A/dm^2 can also affect the coating formation mechanism. In Fig. 3 the following photographs of the AK9M2 alloy surface after 5 min of synthesis of the oxide ceramic coating were obtained using optical (a) and electron microscopes (b).

The alloy AK9M2 contains silicon in the form of lamellar inclusions. In Fig. 3a a photomicrograph of the AK9M2 surface is shown, where the inclusions have a characteristic dendritic structure. Based on the X-ray phase analysis of the oxide ceramic coating (Fig. 1b), this can be explained by the formation of SiO_2 oxide on silicon. In Fig. 3b the given micrograph of the same surface was made with the help of an electron microscope at a magnification of ×500. It has elongated (elongated) pores (black areas) of large size, which were formed at the Al–Si interface.

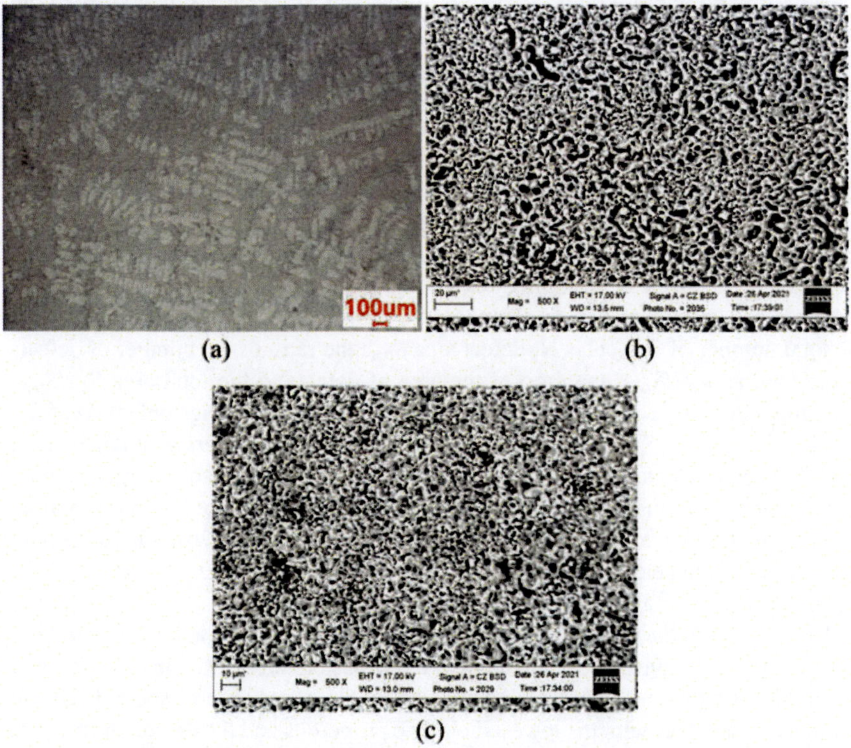

Fig. 3 Surfaces of oxide ceramic coatings on AK9M2 (**a**, **b**) and D16T (**c**) alloy after 5 min PEO synthesis

Table 1 Segmentation results of surface defects

	D16 X500_BSD	AK9M2 X500_BSD
N	9881	13,062
S_{def}, m^2	1.98×10^{-8}	3.10×10^{-8}
N/S_{surf}, 1/m^2	4.72×10^{10}	6.24×10^{10}
S_{def}/S_{surf}	0.095	0.15
S_{min}, m^2	5.16×10^{-14}	2.07×10^{-13}
S_{max}, m^2	4.14×10^{-11}	9.16×10^{-11}
S_{av}, m^2	2.01×10^{-12}	2.27×10^{-12}

The surface of the D16 alloy (Fig. 3c) under the same conditions of synthesis of the oxide ceramic coating has a significantly different structure. Larger pores, which are formed within the first 5 min of synthesis, are caused by the presence of $CuMgAl_2$ and $CuAl_2$ intermetallics in the D16 alloy. They are significantly smaller in size and round in shape, unlike the lamellar silicon inclusions in the AK9M2.

Image segmentation was performed by thresholding of image intensity function. The operation of thresholding consists in comparing the value of the image intensity function of each pixel of the image with a given threshold value. The selection of the appropriate value makes it possible to select areas of a certain type on the image. In the binarization process, the initial grayscale image, which has 256 Gy levels, is transformed into black and white, where the background pixels have a value of 1, and the pore pixels have a value of 0 (see Fig. 3). In the case of a noisy background, a filtering procedure can be applied before segmentation [27–29].

The defect parameters were determined by the segmentation procedure, where the total number of defects is N; defect area S_{def}; the ratio of the number of defects to the total area N/S_{pov}; the ratio of the area of defects to the total area S_{def}/S_{pov}; minimum defect size S_{min}; maximum defect size S_{max}; and average defect size S_{avg}. Square of surface is $S_{pov} = 2.09 \times 10^{-7}$ m^2. Defect parameters are given in Table 1.

Oxide ceramic coatings on the AK9M2 alloy have a significantly larger number of pores and more than twice their maximum size compared to the corresponding coatings on the D16 alloy. The obtained results correlate with [26], where the pore sizes in the oxide ceramic coating on Z10L9 silumin are 3.61 ± 1.31 μm and are much larger than on aluminum (1.06 ± 0.47 μm).

The analysis of the emission spectra of the electrolytic plasma at the initial time of synthesis indicates the presence in them of the emission lines of Cu on the D16 alloy and Si on the silumins. This can be explained by the occurrence of spark discharge channels in the areas with the lowest dielectric properties and the defective structure that occurs in the places where the inclusions come to the surface.

Figure 4 shows the diffractograms of oxide ceramic coatings on AK9M2 and D16 alloys after 2 h of coating synthesis. The quantitative content of phases calculated by the complete profile method of Rietveld using the FullProf program is given in Table 2.

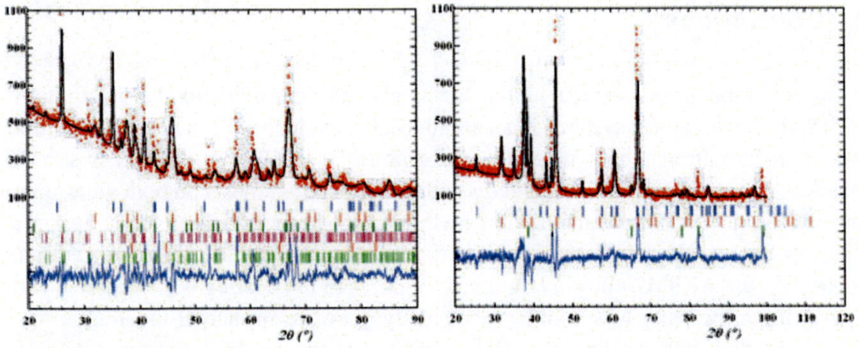

Fig. 4 X-ray diffraction patterns of ceramic oxide coatings obtained on AK9M2 and D16 alloys after 2 h

Table 2 Phase composition of oxide ceramic coatings obtained on AK9M2 and D16 alloys

AK9M2		D16
Phase	Content, mass%	Content, mass%
α-Al_2O_3	11.28	17.48
γ-Al_2O_3	44.45	73.29
SiO_2	3.08	–
$Al_2O_3 \cdot SiO_2$	40.12	–
Al	1.07	9.23

The main phases in coatings on AK9M2 alloy are γ-Al_2O_3 and sillimanite—$Al_2O_3 \cdot SiO_2$, and on D16 alloy γ-Al_2O_3 and α-Al_2O_3. It should be noted that the content of corundum and sillimanite in oxide ceramic coatings increases with increasing time of coating synthesis.

Based on the phase analysis of oxide ceramic coatings and their surface, the effect of silicon on structural transformations in aluminum alloys of the Al–Si and Al–Cu systems was established. In particular, in both cases, at the beginning of the synthesis of the oxide ceramic coating, a transition compound $Al_{4-x}Si_x$ and SiO_2 is formed, which contributes to the formation of a uniform coating and the reduction of the breakdown voltage.

The content of these phases is significantly higher in the alloy with a high silicon content than on the alloy of the Al–Cu system. As the synthesis time increases, its influence changes significantly. In silumins, silicon is included in the composition of sillimanite and silicon oxide, and in alloys of the Al–Cu system of silicon, only traces of silicon are observed in the near-electrolyte region of the coating.

4 Conclusions

X-ray structural analysis revealed that during plasma electrolytic oxidation of alloys of the Al–Si and Al–Cu systems in an alkaline silicate electrolyte, a number of structural transformations occur in the oxide ceramic layers. In particular, it is shown that $Al_{4-x}Si_x$ compounds form already within 5 min of synthesis on both aluminum and silicon alloys. The segmentation method was used to analyze oxide ceramic coatings on two doping systems. After 5 min of synthesis of the oxide ceramic coating on the AK9M2 alloy, a larger size of pores in the coating was recorded, and its maximum porosity was 14.8%. Most likely, they are formed at the aluminum–silicon interface and have a characteristic oblong shape. In the coating on the D16 alloy under the same synthesis conditions, the porosity is 9.5%. During 2 h of synthesis, the influence of silicon on alloys increases significantly. Only traces of silicon are observed in the oxide ceramic coating on the D16 alloy, while $Al_2O_3{\cdot}SiO_2$ (sillimanite) and $3Al_2O_3{\cdot}2SiO_2$ (mullite) are formed on AK9M2. Therefore, in the plasma discharge channels on the surface of both alloys, silicon is reduced from the electrolyte (Na_2SiO_3) with subsequent synthesis of $Al_{4-x}Si_x$ compounds.

References

1. M.M. Student et al., Influence of the composition of electrolyte for hard anodizing of aluminium on the characteristics of oxide layer. Mater. Sci. **57**, 240–247 (2021)
2. V.I. Pokhmurskii, I.M. Zin, V.A. Vynar, L.M. Bilyy, Contradictory effect of chromate inhibitor on corrosive wear of aluminium alloy. Corros. Sci. **53**, 904–908 (2011)
3. V.I. Pokhmurskii, I.M. Zin, V.A. Vynar, O.P. Khlopyk, L.M. Bilyy, Corrosive wear of aluminium alloy in presence of phosphate. Corros. Eng. Sci. Technol. **47**, 182–187 (2012)
4. V. Hutsaylyuk, M. Student, K. Zadorozhna, P. Maruschak, H. Pokhmurska, Improvement of wear resistance of aluminum alloy by HVOF method. J. Market. Res. **9**, 16367–16377 (2020)
5. V. Kyryliv, B. Chaikovs'kyi, O. Maksymiv, B. Mykytchak, Fatigue and corrosion fatigue of roll steels with surface nanostructure. J. Nano Res. **51**, 92–97 (2018)
6. I.M. Zin, O.P. Khlopyk, M.Ya. Holovchuk, Protective action of inorganic inhibitors on mechanically activated aluminum surfaces. Mater. Sci. **49**, 298–303 (2013)
7. I. Pohrelyuk, O. Yaskiv, O. Tkachuk, D.B. Lee, Formation of oxynitride layers on titanium alloys by gas diffusion treatment. Met. Mater. Int. **15**, 949–953 (2009)
8. K. Szymkiewiccz et al., Effect of nitriding conditions of Ti6Al7Nb on microstructure of TiN surface layer. J. Alloy. Compd. **845**, 156320 (2020)
9. I.M. Pohrelyuk, O.V. Tkachuck, R.V. Proskurnyak, Corrosion resistance of the Ti–6Al–4V titanium alloy with nitride coatings in 0.9% NaCl. JOM **63**, 35–40 (2011)
10. I. Pohrelyuk, J. Morgel, O. Tkachuk, K. Szymkiewiccz, Effect of temperature on gas oxynitiding of Ti–6Al–4V alloys. Surf. Coat. Technol. **360**, 103–109 (2019)
11. Y.G. Gutsalenko, E.K. Sevidova, I.I. Stepanova, Evaluation of technological capability to form dielectric coatings on AK6 alloy, using method of microarc oxidation. Surf. Eng. Appl. Electrochem. **55**, 602–606 (2019)
12. M.M. Student, V.M. Posuvailo, H.H. Veselivs'ka, Y.Y. Sirak, R.A. Yatsyuk, Corrosion resistance of plasma-electrolytic layers on alloys coatings of Al–Cu–Mg system for various modes of heat treatment. Mater. Sci. **53**, 789–795 (2018)

13. H.M. Nykyforchyn, V.S. Agarwala, M.D. Klapkiv, V.M. Posuvailo, Simultaneous reduction of wear and corrosion of titanium, magnesium and zirconium alloys by surface plasma electrolytic oxidation treatment. Adv. Mater. Res. **38**, 27–35 (2008)
14. M.D. Klapkiv, Simulation of synthesis of oxide-ceramic coatings in discharge channels of a metal-electrolyte system. Mater. Sci. **35**, 279–283 (1999)
15. H.V. Кaraкurkchi, M.D. Sakhnenko, M.V. Ved, S.I. Zyubanova, I.I. Stepanova, Corrosion and physicomechanical properties of the coatings on Ak12m2mgn alloy formed by plasma-electrolytic oxidation. Mater. Sci. **55**, 693–702 (2020)
16. G. Li, F. Ma, Z. Li, Y. Xu, F. Gao, L. Guo, J. Zhu, G. Li, Y. Xia, Influence of applied frequency on thermal physical properties of coatings prepared on Al and AlSi alloys by plasma electrolytic oxidation. Coatings **11**, 1439 (2021)
17. A.V. Polunin, P.V. Ivashin, I.A. Rastegaev, E.D. Borgardt, M.M. Krishtal, Wear resistance of the oxide layers formed on AK9pch silumin by microarc oxidation in an electrolyte modified by silicon dioxide nanoparticles, in *Russian Metallurgy (Metally)* (2016), pp. 385–388
18. J. Rodrı́guez-Carvajal, *Program FullProf.2k. Version 2.20* (Laboratoire Léon Brillouin (CEA–CNRS), France, 2002).
19. I. Ivasenko, V. Posuvailo, M. Klapkiv, V. Vynar, S. Ostap'yuk, Express method for determining the presence of defects of the surface of oxide-ceramic coatings. Mater. Sci. **45**, 460–464 (2009)
20. M. Senoo, H. Mii, I. Fujishiro, T. Fujikawa, Precise measurements of lattice compression of Al, Si and Al-Si alloys by high pressure X-ray diffractometry. Jpn. J. Appl. Phys. **15**, 871–880 (1976)
21. R.O. Hussein, X. Nie, D.O. Northwood, An investigation of ceramic coating growth mechanisms in plasma electrolytic oxidation (PEO) processing. Electrochim. Acta **112**, 111–119 (2013)
22. T.W. Clyne, S.C. Troughton, A review of recent work on discharge characteristics during plasma electrolytic oxidation of various metals. Int. Mater. Rev. **64**, 127–162 (2018)
23. F. Monfort et al., Development of anodic coatings on aluminium under sparking conditions in silicate electrolyte. Corros. Sci. **49**, 672–693 (2007)
24. P. Leone, A. Nominé, D. Veys-Renaux, G. Henrion, T. Belmonte, J. Martin, Influence of electrolyte ageing on the plasma electrolytic oxidation of aluminium. Surf. Coat. Technol. **269**, 36–46 (2015)
25. L. Wang, X. Nie, Silicon effects on formation of EPO oxide coatings on aluminum alloys. Thin Solid Films **494**, 211–218 (2006)
26. H. Yu, Q. Dong, Y. Chen, C. Chen, Influence of silicon on growth mechanism of micro-arc oxidation coating on cast Al–Si alloy. R. Soc. Open Sci. **5**, 172428 (2018)
27. R.M. Palenichka, P. Zinterhof, I.B. Ivasenko, Adaptive image filtering and segmentation using robust estimation of intensity, in *Lecture Notes in Computer Science*, vol. 1876 (2000)
28. R.M. Palenichka, P. Zinterhof, Y.B. Rytsar, I.B. Ivasenko, Structure-adaptive image filtering using order statistics. J. Electron. Imaging **7**, 339–349 (1998)
29. R. Vorobel, I. Ivasenko, O. Berehulyak, Automatized computer system for evaluation of rust using modified single-scale retinex, in *Proceedings of 1st Ukraine Conference on Electrical and Computer Engineering*, UKRCON 2017, Kyiv, Ukraine (2017), pp. 1002–1006

Development of Unified Technology for Nanostructural Metal Oxides Synthesis

Mariia Kontseva, Tetiana Dontsova, and Gryhoriy Krymets

Abstract This work considers the synthesis of nanostructural metal oxides by the unified technology. Intensive development of the world production of metal oxides nanoparticles makes the development of the latest chemical technologies for the synthesis of metal oxide nanomaterials an urgent issue. It is proposed to create a unified scheme for the synthesis of nanostructured metal oxides (Fe_3O_4, TiO_2, ZrO_2), which is economically efficient due to the possibility of obtaining three metal oxides simultaneously. The method of homogeneous deposition was used for the synthesis of magnetite, titanium (IV) and zirconium (IV) oxides. It allowed creating one scheme for production due to the proximity of concentration and temperature conditions of obtaining these metal oxides. The required weights of reagents for 100 kg of magnetite, 100 kg of titanium (IV) oxide and 100 kg of zirconium (IV) oxide, which are produced in one cycle, were calculated. The main energy consumption for obtaining all metal oxides was determined, and it is 28.15$ per cycle for each metal oxide. The developed unified scheme provides a significant reduction in the cost of purchasing the necessary equipment and significantly reduces the required area for its placement in comparison with individual schemes for their synthesis. That is why the proposed unified technology for the synthesis of nanostructured metal oxides can be a worthy alternative for the production of metal oxide nanomaterials existing in the world.

M. Kontseva (✉) · T. Dontsova · G. Krymets
Kyiv, Ukraine
e-mail: marishka95@ukr.net

T. Dontsova
e-mail: dontsova@ua.fm

G. Krymets
e-mail: krimets@xtf.kpi.ua

O. Fesenko and L. Yatsenko (eds.), *Nanostructured Surfaces, Nanocomposites and Nanomaterials, and Their Applications*, Springer Proceedings in Physics 296,
https://doi.org/10.1007/978-3-031-42704-6_4

1 Introduction

Research of the American consulting company "Grand View Research" shows a steady increase in global production of metal oxide nanoparticles (NPMeO) since 2010 due to increasing demand from various industries. Industries that need NPMeO include the automotive industry, energy, electronic and optical electronics, aerospace and more. Also, modern innovations promote the use of NPMeO in cosmetic, pharmaceutical, medical, biological and environmental fields. It is expected that this will further increase the demand for nanodispersed metal oxide powders on the market [1].

The world's major manufacturers of metal oxide nanopowders are Reinste Nanoventures, American Elements, NanoScale Corporation, Nanoparticles & Microspheres Eprui Co., Sigma Aldrich, Altair Nanomaterials, US Research Nanomaterials Inc. and Access Business Group [2]. On an industrial scale, NPMeOs such as copper oxide, aluminum oxide, iron oxide, magnesium oxide, cerium oxide, silicon oxide, manganese oxide, titanium oxide, zinc oxide, zirconium oxide, etc., are produced. In practice, these products are mainly used in construction materials or for biomedical purposes. Therefore, the production of metal oxide nanomaterials is in the stage of intensive development. Accordingly, this makes it extremely relevant and promising in terms of developing economic and environmental schemes for the production of metal oxide nanomaterials.

2 Analysis of Literature Data and Problem Statement

Besides to use in composite materials and biomaterials, NPMeO also have a high potential for use in applied ecology, as adsorbents and photocatalysts [3]. Crucial in the development of nanomaterials, especially for their specific applications, is the method of synthesis, which plays a decisive role in the physicochemical properties of the material. The results of research in [4] indicate that the method of synthesis affects the shape and size of the resulting particles, nanostructurality, degree of crystallinity, morphology, structural-sorption characteristics and so on.

It is shown that for the production of metal oxide nanomaterials in the literature consider such methods as chemical deposition (heterogeneous and homogeneous), sol–gel technology, hydrothermal synthesis, gas-phase synthesis and others [5]. But the wide application in the literature of "soft chemistry" methods (chemical deposition, sol–gel technology, hydrothermal synthesis) is associated with the relative simplicity of their implementation, a wide range of control parameters. In addition, these methods can be easily combined and use various additional processing methods during synthesis, which allows obtaining products with more diverse properties [6].

Critical analysis of "soft chemistry" methods in [7] gives grounds to assert the prospects of a homogeneous method of analysis for the production of adsorption materials based on nanostructured metal oxides. The results of research in [8] (on the synthesis of iron (III) oxide) indicate higher adsorption properties in sorbents of iron oxide, which are obtained by homogeneous precipitation. The authors in [9] showed that the use of the method of chemical deposition allows obtaining iron-containing magnetically controlled sorbents. They not only have a high adsorption capacity to various pollutants but also add additional functional properties.

Data analysis of the photocatalytic metal oxide materials production, in particular on the titanium (IV) oxide basis, indicates the perspectivity of the hydrothermal synthesis method, which is realized at elevated temperatures [10]. However, other research [11] indicates the perspectivity of sol–gel technology due to its simplicity and lower temperatures for use in the synthesis of TiO_2 photocatalysts. At the same time, other researchers claim that ultrafine titanium (IV) oxide powders synthesized by homogeneous precipitation can vary in phase composition due to changes in synthesis parameters [12]. As a result, the analysis of these methods leads to the conclusion that all methods allow obtaining the desired characteristics of photocatalysts based on titanium (IV) oxide. An option for obtaining the desired characteristics in powders may be the use of hydrothermal synthesis, which allows obtaining powders with different porous structures in a wide range [13]. This approach was used in [14], and the obtained data indicate a smaller crystallite size for ZrO_2, which increases its catalytic activity.

On the other hand, in [15], zirconium (IV) oxide powders were obtained by homogeneous precipitation using urea. The obtained powders had a homogeneous microstructure and a narrow pore size distribution, which is highly desirable for sorption and catalytic materials. Another work [16] provides data on the production of ZrO_2 powders in the form of spherical particles consisting of metastable tetragonal zirconium (IV) oxide. Therefore, the use of a homogeneous method allows obtaining ultrathin, monodisperse particles with a narrow size distribution [17].

Despite advances in laboratory research, the development of industrial technology for the synthesis of nanostructured metal oxides by homogeneous deposition remains an unresolved issue. The reason for this is the presence of existing technologies for metal oxide nanomaterials production, which are based on other more energy-intensive schemes for obtaining metal oxides, such as high-temperature combustion, PVD and CVP (physical and chemical deposition from the gas phase, respectively) [18]. A way to reduce energy consumption for the production of nanostructured metal oxides can be a creation of a technology based on homogeneous deposition method. In addition, the proximity of concentration and temperature conditions allows creating a unified scheme for obtaining several metal oxides [19]. This, in turn, provides a significant reduction in the cost of purchasing the necessary equipment and significantly reduces the required area for placement of this equipment. Thus, all of the above gives grounds to assert that it is expedient to conduct research on quantitative calculations and justification of the basic equipment choice to develop a unified technology for obtaining nanostructured metal oxides.

3 The Purpose and Objectives of the Study

The aim of the research is to develop an economically efficient unified technology for the synthesis of nanostructured metal oxides, which allows producing magnetite, titanium (IV) oxide and zirconium (IV) oxide.

To achieve this purpose it was necessary to solve the following tasks:

- calculate the amount of the initial reagents for the production of magnetite (Fe_3O_4), titanium (IV) oxide and zirconium (IV) oxide with a capacity of 100 kg per cycle;
- estimate the main energy consumption for the corresponding synthesis implementation;
- develop the unified technological scheme for the sequential production of magnetite, titanium (IV) oxide and zirconium (IV) oxide, and choose the main apparatus for the synthesis of metal oxides;
- design a 3D model of the developed scheme of Fe_3O_4, TiO_2 and ZrO_2 production in the Blender environment, and calculate capital costs for the unified technology.

4 Research Materials and Methods

Reagents, the price and characteristics of which are shown in Table 1, were used to calculate the material balance of the unified technology for the production of metal oxides.

The productivity of the unified scheme for obtaining three different nanodispersed metal oxides (Fe_3O_4, TiO_2, ZrO_2), which is developed, is 100 kg per cycle for each metal oxide. The mode of production for all metal oxides is periodic. The material balance calculation was performed according to the standard method [22].

Table 1 Initial reagents and their characteristics [20]

Reagent/chemical formula	Qualification	Solubility, g per 1 L (25 °C)	Harmfulness	Price (average), $ per 1 ton
$FeSO_{4\,cr}$ / $FeSO_4 \cdot 7H_2O$	Technical	295	Warning	50
$FeCl_{3\,cr}$/ $FeCl_3 \cdot 6H_2O$		912	Danger	200
$TiOSO_{4\,cr}$ / $TiOSO_4 \cdot H_2O$		740	Corrosive	500
$ZrOCl_{2\,cr}$ / $ZrOCl_2 \cdot 7H_2O$		520	Harmful and corrosive	2500
$(NH_2)_2CO$		1080	Harmful	200
H_2O	Drinking	–	–	500 [21]

Energy consumption was estimated according to the method described in [23]. A calculation model for RIA-P batch reactors was used for operating on the adiabatic regime [24]. Based on the fact that the synthesis of the corresponding oxides is carried out in highly dilute media, some simplifications were introduced. The amount of heat released during chemical reactions was neglected due to the insignificant contribution to the thermal balance of processes. The amount of heat required for 1 h of holding the reaction medium at 95 °C was also neglected due to the process under adiabatic conditions. Therefore, to estimate the energy consumption for the implementation of syntheses in an isolated batch reactor was calculated only the amount of heat required to heat the solution from 20 to 95 °C [25]:

$$Q = V_r \cdot C_p \cdot (T_{\text{fin}-\text{init}}), \qquad (1)$$

where

V_r—reactor volume, m^3;
C_p—average volumetric heat capacity in the temperature range from T_{init} to T_{fin}, J/(kg K);
T_{init}—the temperature from which the solution begins to heat, K;
T_{fin}—the temperature to which the solution must be heated, K.

The 2D drawing and 3D scheme of the spatial arrangement of the apparatuses of the developed unified technology are executed in modern software complexes for computer chemistry: computer-aided design and drafting system KOMPAS-3D (Russia) and 3D creation suite Blender (Netherlands), respectively. KOMPAS-3D was chosen because its functionality allows creating projects which correspond to the standards for CAD/CAM designs. Blender was chosen because it is free (open source), accessible and provides a fairly wide range of functionality [26].

5 The Results of Calculations, Development and Design of a Unified Scheme for the Synthesis of Nanodispersed Metal Oxides

5.1 Calculation of the Initial Reagents Number for Magnetite, Titanium (IV) Oxide and Zirconium (IV) Oxide

The production of metal oxides in a unified scheme is carried out according to the following chemical equations. Production of Fe_3O_4 is carried out by reactions:

$$(NH_2)_2CO + 3H_2O = 2NH_4OH + CO_2, \qquad (2)$$

Table 2 Mass of reagents (kg) required for the synthesis of 100 kg of nanodispersed metal oxides

Me_xO_y	$FeSO_{4\ cr}$	$FeCl_{3\ cr}$	$TiOSO_{4\ cr}$	$ZrOCl_{2\ cr}$	$(NH_2)_2CO$	H_2O
Fe_3O_4	119.8	233.2	–	–	229.9	1707.4
TiO_2	–	–	222.5	–	166.7	1872.5
ZrO_2	–	–	–	247.2	108.4	1814.6

$$FeSO_4 + 2FeCl_3 + 8NH_4OH = Fe_3O_4 + (NH_4)_2SO_4 + 4H_2O + 6NH_4Cl. \quad (3)$$

The synthesis of TiO_2 and ZrO_2 is carried out by reactions (2), (4) and (2), (5), respectively:

$$TiOSO_4 + 2NH_4OH = TiO_2 + (NH_4)_2SO_4, \quad (4)$$

$$ZrOCl_2 + 2NH_4OH = ZrO(OH)_2 + 2NH_4Cl. \quad (5)$$

The presented chemical transformations are realized under the following conditions: synthesis temperature 95 °C, concentration in terms of metal oxides in the reactor 50 g/dm^3.

Calculations of the initial reagents for one cycle of nanodispersed metal oxides Fe_3O_4, ZrO_2, TiO_2 synthesis were performed according to the equations of reactions (1)–(4) and the standard method of calculating the material balance. The results of the calculations are summarized in Table 2.

The mass of carbamide, which is necessary for the implementation of all processes, is calculated based on the fact that the practical degree of hydrolysis of $(NH_2)_2CO$ is 45% [27]. In addition, to maintain pH 9 during Fe_3O_4 synthesis, an additional 0.75 kg $(NH_2)_2CO$ must be added.

5.2 *The Main Energy Consumption for the Implementation of the Synthesis of Metal Oxides*

The temperature modes of all three chemical processes are identical, so the heat consumption and, consequently, the electricity consumption will be close in value. Based on the fact that the initial mixture in the reactor is mainly water (~ 95%), the heat required to increase the temperature of the solution in the reactor from 20 to 95 °C is calculated based on the average logarithmic heat capacity of water. According to Formula 1, the amount of heat required to heat the solution in the reactor from 20 °C to 95 °C is:

$$Q = 2.5 \cdot 4.2 \cdot (368–293) = 787.5\,\text{MJ}.$$

To calculate the required amount of electricity, the obtained amount of heat was converted from J to W h. Since 1 W h = 3600 J, then 787.5 MJ = 218.75 kW h. As of 2022 and at consumption of up to 20 thousand kW h, the price per 1 kW is \$0.1287 (UAH 5.20) [28]. Therefore, the following formula calculates the cost of electricity required to heat the solution in the reactor from 20 to 95 °C:

$$P = 218.75 \cdot 0.1287 = 28.15\$.$$

Therefore, \$28.15 is required to carry out the synthesis in an ideal mixing reactor under adiabatic conditions.

5.3 Development of a Unified Basic Technological Scheme for the Production of Magnetite, Titanium (IV) Oxide and Zirconium (IV) Oxide

The creation of a unified basic technological scheme is based on the principle of modularity of equipment, which ensures the production of a wide range of products without significant technological and production costs. The main modules of the scheme are preparation of initial reagents, which is carried out in the system of circulating reactors; implementation of the hydrolysis process; separation of the obtained suspensions by magnetic separation or filtration; and further heat treatment (if necessary). The unified scheme developed according to this approach is presented in Fig. 1.

According to the presented principal technological scheme (Fig. 1), the process of obtaining metal oxides begins with the preparation of reagents in tanks 1 and 4, in which intensive mixing is provided by circulating pumps 2 and 3. After the preparation of working solutions, they are dosed into reactor 5, where synthesis takes place. The process at high temperatures in reactor 5 provides immersion heat exchange elements. The prepared reaction mass from apparatus 5 is transported by compressed air (indicated in the diagram by the sign P). In the case of nanomagnetite synthesis, the suspension from apparatus 5 is moved to the magnetic sedimentation tank 6 and in the case of nanostructured oxides of titanium and zirconium to the bag filter 7. In the magnetic sedimentation tank 6, the magnetite is flushed, after which it is transported to the warehouse. The bag filter 7 is the filtration of suspensions of titanium or zirconium oxides through the sleeve with a pore size of 1 μm. The required pressure in apparatus 7 is created by compressed air.

The washed wet oxide precipitates are then mechanically unloaded onto a heated tray 8 for drying. And in the case of obtaining zirconium (IV) oxide after drying, the final heat treatment is carried out to crystallize it in the muffle furnace 9. The main apparatus of the unified scheme is reactor 5, which is an autoclave with a volume of

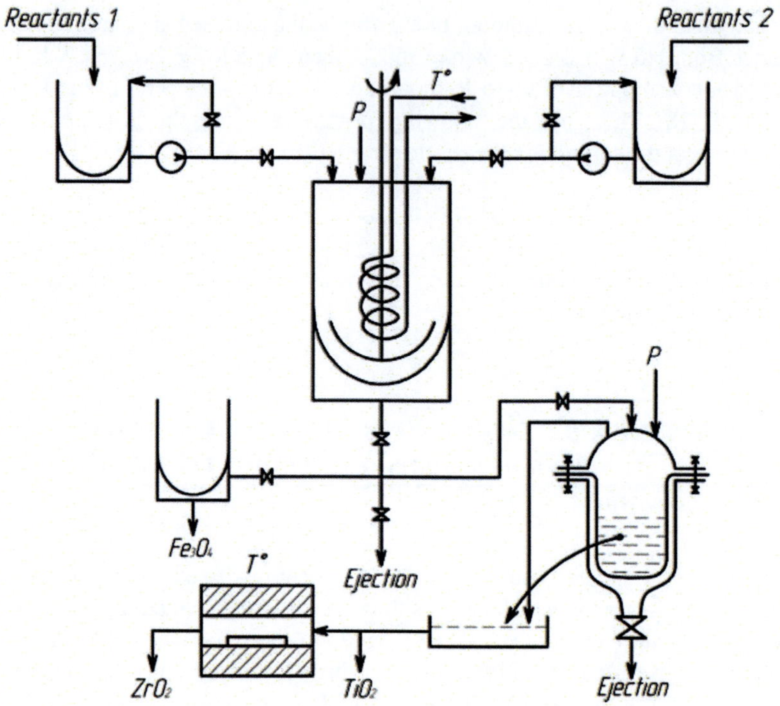

Fig. 1 Schematic technological scheme for obtaining nanostructured metal oxides: 1, 4—tanks for the preparation of initial working solutions; 2, 3—circulating pumps; 5—autoclave reactor with temperature control and stirring; 6—magnetic sedimentation tank; 7—hose filtration system with a filtration rating of 1 μm; 8—thermostatic pallet; and 9—muffle furnace

3.2 m^3. This volume is sufficient to load a mass of up to 2–3 tons (according to the material balance), taking into account the allowable degree of filling of the reactor [29].

5.4 3D Model of the Developed Scheme of Production of Fe_3O_4, TiO_2 and ZrO_2

Figure 2a, b shows a 3D model of the proposed unified scheme for the production of oxides of Fe_3O_4, TiO_2, ZrO_2, which was created in the Blender environment

According to Fig. 2, this 3D model includes the main components of the process, lifting equipment and control system. In addition, it can be seen that the worksite of metal oxide production provides areas for storage of raw materials and areas for fast movement of finished products.

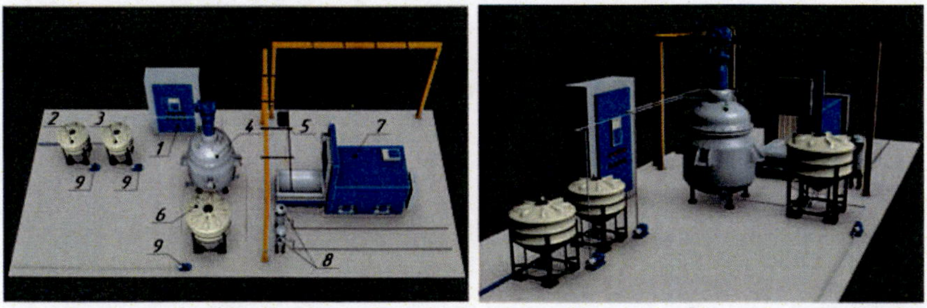

Fig. 2 3D model of the proposed unified scheme of Fe_3O_4, TiO_2, ZrO_2 production: **a** top view; **b** general view; 1—automatic process control panel; 2, 3—tanks for the preparation of initial working solutions; 4—autoclave reactor with temperature control and stirring; 5—thermostatic pallet; 6—magnetic sedimentation tank; 7—muffle furnace; 8—hose filtration system with a filtration rating of 1 μm; and 9—circulating pumps

The proposed unified scheme can be fully automated for the purpose of automatic control and management of technological processes that take place during the production of metal oxide. To facilitate working conditions, means of industrial mechanization (monorail cranes) are provided. As you can see (Fig. 2b) the scheme is quite compact, so it can be easily implemented both in existing production and separately.

Table 3 shows the cost of equipment required for the operation of a unified technology for obtaining three metal oxides. The calculated capital expenditures for the unified scheme indicate the absence of significant expenditures in this part.

For comparison, the analysis of the cost of individual productions of nanodispersed oxides Fe_3O_4, TiO_2, ZrO_2 is presented in Table 4.

Table 3 Cost of equipment for a unified scheme [30]

Position in Fig. 2	Equipment	Cost, $	Quantity	Total cost, $
1	Automatic process control panel	1000	1	1000
2, 3	Tanks for the preparation of initial working solutions	200	2	400
4	Autoclave reactor with temperature control and stirring	10,000	1	10,000
5	Thermostatic pallet	1000	1	1000
6	Magnetic sedimentation tank	500	1	500
7	Muffle furnace	2500	1	2500
8	Hose filtration system with a filtration rating of 1 μm	900	1	900
9	Circulating pumps	400	3	1200
Cost of the whole scheme				17,500

Table 4 Cost of technology for three separate schemes

Scheme	Required equipment items (from the unified scheme)	Cost of the required equipment, $
Scheme Fe_3O_4	1, 2, 3, 4, 6, 9, 9, 9	13,100
Scheme TiO_2	1, 2, 3, 4, 5, 8, 9, 9	14,100
Scheme ZrO_2	1, 2, 3, 4, 5, 7, 8, 9, 9	16,600
Total cost of three separate schemes		43,800

Tables 3 and 4 show that the cost of equipment for a particular scheme is commensurate with the cost of equipment in the development of unified technology. This indicates the economic feasibility of the proposed scheme and the rationality of this approach in creating new industries.

6 Discussion of the Obtained Unified Technology for the Production of Magnetite, Titanium (IV) Oxide and Zirconium (IV) Oxide

The presented results of material and energy calculations indicate insignificant costs for working capital of the proposed unified technology for obtaining nanostructured metal oxides. A comparison of the capital costs of the unified technology scheme and the total cost of the three separate schemes shows significant savings in the case of unified technology. It is important to note that the proposed scheme of production of metal oxides due to its combinedness and unification allows saving significantly on equipment that is typical, suitable for the creation of each of these productions. In addition, this technology is quite easy to regulate and automate, which makes it even more cost-effective.

The approach feature of creating a unified scheme is a significant cost saving for equipment, which is, according to Tables 3 and 4, 43,800 17,500$ = 26,300$. Additional development of 3D model of the developed technology allows visualizing the spatial location of all equipment that is necessary for the implementation in practice of this production. In addition, the 3D model of the unified scheme (Fig. 2) shows that there is a significant reduction in the area for equipment placement in the implementation of unified technology in comparison with the implementation of all productions for the synthesis of metal oxides. The full automation possibility of the Fe_3O_4, TiO_2 and ZrO_2 synthesis unified scheme testifies to its rationality and economic expediency. In general, this approach is profitable in the creation of other productions of metal oxides, which is why it is promising, which is also confirmed in the literature [31].

Another feature of creating the unified scheme for the synthesis of nanostructured metal oxides is the use of an autoclave reactor consisting of stainless steel (steel 20 C-B58, Russia) [32], which is typical for this reactor type. An important parameter of the reactor is the wall thickness, which is calculated based on the threshold of strength and operating temperature, which is 90 °C and corresponds to the temperature of synthesis of metal oxides. Calculations of the wall thickness, taking into account the service life of 5 years and the corrosion rate, indicate that it must be at least 0.7 mm [33]. In the selected typical autoclave reactor, it is equal to 3.2 mm, which is satisfactory [34]. Also, this reactor is equipped with a jacket, which must supply 787.5 MJ of thermal energy in the form of saturated water vapor with a temperature of 100–102 °C. For this purpose, it is advisable to use a steam generator with a capacity of 5 kW and a capacity of up to 40 kg of steam per hour [35].

The disadvantage of using this unified scheme is the need to wash some equipment used for the production of several metal oxides to avoid contamination of the product in the next cycle with substances from previous cycles. However, the value of the cost of washing these devices is significantly less than the savings in general from the implementation of a unified scheme for the production of three different nanodispersed metal oxides (Fe_3O_4, TiO_2, ZrO_2).

In the future, when implementing this technology, it is necessary to pay attention to its environmental friendliness. As a result of using a large amount of water at the beginning of each metal oxide production process, this technology produces a large amount of wastewater. Wastewater produced during technological processes contains chlorides and ammonium sulfates. The most rational approach may be to treat the generated wastewater and return it to the cycle.

7 Conclusions

1. The calculated initial masses for obtaining 100 kg of nanostructured Fe_3O_4, TiO_2, ZrO_2, even taking into account the degree of carbamide hydrolysis and achieving the required pH, indicate the economic feasibility of the proposed unified technology.
2. The main energy consumptions calculation for the unified technology implementation indicates a small cost of synthesis (namely the need for $28.15 to implement the process of synthesis of the corresponding metal oxide).
3. The unified principal technological scheme of obtaining nanostructured metal oxides by the method of homogeneous deposition is developed, and the equipment for metal oxides synthesis is substantiated. It was shown that the created unified scheme of obtaining different metal oxides is realized due to the proximity of concentration and temperature conditions for obtaining these oxides.
4. The 3D model of the unified scheme of Fe_3O_4, TiO_2 and ZrO_2 production is designed in Blender environment, and capital costs are calculated. The implementation of the proposed unified scheme in comparison with three separate schemes of these oxides production allows to significantly reduce capital costs

for the purchase of necessary equipment (namely $26,300). It is shown that the 3D model significantly reduces the area for the necessary equipment. The presented 3D model is a ready-to-build spatial placement plan of the developed nanodispersed metal oxides Fe_3O_4, TiO_2 and ZrO_2 production. Thus, the proposed scheme for obtaining nanostructured metal oxides due to its unification is rational and economically feasible.

Acknowledgements We thank the National Research Foundation of Ukraine for the program "Support for the research of leading and young scientists", within which the work was done (registration number of the project 2020.02/0024).

References

1. N. Zafar, A. Madni, A. Khalid, T. Khan, R. Kousar, S.S. Naz, F. Wahid, *Pharmaceutical and Biomedical Applications of Green Synthesized Metal and Metal Oxide Nanoparticles*. National Library of Medicine (2020). https://doi.org/10.2174/1381612826666201126144805
2. Metal Oxide Nanoparticles Market Size, Status Report 2020 by Manufacturers, Type and Application, Forecast to 2026. https://www.wfmj.com/story/42359962/metal-oxide-nanoparticles-market-size-status-report-2020-by-manufacturers-type-and-application-forecast-to-2026
3. S. Singh, K.C. Barick, D. Bahadur, *Functional Oxide Nanomaterials and Nanocomposites for the Removal of Heavy Metals and Dyes*. SAGE Journals (2013). https://doi.org/10.5772/57237
4. S. Nagirnyak, V. Lutz, T. Dontsova, I. Astrelin, The effect of the synthesis conditions on morphology of tin (IV) oxide obtained by vapor transport method, in *Nanophysics, Nanophotonics, Surface Studies, and Applications* (2016), pp. 331–341. https://doi.org/10.1007/978-3-319-30737-4_28
5. E.A. Trusova, K.V. Vokhmintcev, I.V. Zagainov, Wet-chemistry processing of powdery raw material for high-tech ceramics. Natl. Center Biotechnol. Inf. U. S. National Library Med. **7**(1), 51 (2012). https://doi.org/10.1186/1556-276X-7-58. https://www.ncbi.nlm.nih.gov/pmc/articles/PMC3275523/
6. T. Dontsova, I. Ivanenko, I. Astrelin, Synthesis and characterization of titanium (IV) oxide from various precursors, in *Nanoplasmonics, Nano-Optics, Nanocomposites, and Surface Studies*, vol. 167 ed. by O. Fesenko, L. Yatsenko. Springer Proceedings in Physics (2015), pp. 275–293. https://doi.org/10.1007/978-3-319-18543-9_19
7. R.E. Adam, E. Mustafa, S. Elhag, O. Nur, M. Willander, Photocatalytic properties for different metal-oxide nanomaterials. Society of Photo-Optical Instrumentation Engineers (SPIE) (2019). https://doi.org/10.1117/12.2517436
8. M. Litynska, T. Dontsova, O. Yanushevska, V. Tarabaka, Development of iron-containing sorption materials for water purification from arsenic compounds. Eastern-Eur. J. Enterprise Technol. **2**(110), 35–42 (2021). https://doi.org/10.15587/1729-4061.2021.230216
9. N.O. Mykhailenko, O.V. Makarchuk, T.A. Dontsova, S.V. Horobets, I.M. Astrelin, Purification of aqueous media by magnetically operated saponite sorbents. Eastern-Eur. J. Enterprise Technol. **4**(10)(76), 13–20 (2015). https://doi.org/10.15587/1729-4061.2015.46573
10. O. Yanushevska, T. Dontsova, S. Nahirniak, V. Alisova, TiO2–ZnO nanocomposites for photodegradation of dyes in water bodies, in *Nanomaterials and Nanocomposites, Nanostructure Surfaces, and Their Applications*, vol. 246, ed. by O. Fesenko, L. Yatsenko. Springer Proceedings in Physics (2020), pp. 719–731. https://doi.org/10.1007/978-3-030-51905-6_49
11. A.S. Kutuzova, T.A. Dontsova, Synthesis, characterization and properties of titanium dioxide obtained by hydrolytic method, in *2017 IEEE 7th International Conference Nanomaterials:*

Application & Properties (NAP), 01NNPT02-1-01NNPT02-5 (2017). https://doi.org/10.1109/NAP.2017.8190182
12. S.J. Kim, S.D. Park, Y.H. Jeong, S. Park, Homogeneous precipitation of TiO_2 ultrafine powders from aqueous $TiOCl_2$ solution. J. Am. Ceram. Soc. **4**(82), 927–932 (2004). https://doi.org/10.1111/j.1151-2916.1999.tb01855.x
13. A. Kutuzova, T. Dontsova, W. Kwapinski, TiO_2–SnO_2 nanocomposites: effect of acid-base and structural-adsorption properties on photocatalytic performance. J. Inorg. Organomet. Polym Mater. **30**, 3060–3072 (2020). https://doi.org/10.1007/s10904-020-01467-z
14. J. Baneshi, M. Haghighi, H. Ajamein, M. Abdollahifar, Homogeneous precipitation and urea-nitrate combustion preparation of nanostructured $CuO/CeO_2/ZrO_2/Al2O_3$ oxides used in hydrogen production from methanol for fuel cells, in *Particulate Science and Technology* (2020). https://doi.org/10.1080/02726351.2018.1455778
15. F. Tong, Y. Zhao, M.H. Wang, Synthesis of monodisperse ZrO_2 microspheres via urea homogeneous precipitation and its effect on the sintering properties. Micro Nano Lett. **14**(8), 823–931 (2019). https://doi.org/10.1049/mnl.2018.5328
16. M. Déchamps, B. Djuriçić, S. Pickering, Structure of zirconia prepared by homogeneous precipitation. J. Inorg. Organomet. Polym Mater. **78**(11), 2867–3168 (1995). https://doi.org/10.1111/j.1151-2916.1995.tb09058.x
17. D.M. Bezerra, E.M. Assaf, Influence of the preparation method on the structural properties of mixed metal oxides. Sci. Technol. Mater. **30**(3), 131–188 (2018). https://www.elsevier.es/en-revista-science-tecnology-materials-395-articulo-influence-preparation-method-on-structural-S2603636318300344
18. N. Patelli, A. Migliori, V. Morandi, L. Pasquini, One-step synthesis of metal/oxide nanocomposites by gas phase condensation. Nanomaterials **9**(2), 219 (2019). https://doi.org/10.3390/nano9020219 https://www.ncbi.nlm.nih.gov/pmc/articles/PMC6409555/
19. T.C. Quinby, *Method of Producing Homogeneous Mixed Metal Oxides and Metal-Metal Oxide Mixtures*. CA PATENT DOCUMENT 1085657/A/. Available from Micromedia Ltd., 165 Hotel de Ville, Hull, Quebec, Canada J8X 3X2; U.S. pat. appl. 787128 611, 12 (19) (1980). https://inis.iaea.org/search/searchsinglerecord.aspx?recordsFor=SingleRecord&RN=12627851
20. Weifang Zhongqing Fine Chemical Co., Ltd. https://zqchemical.en.made-in-china.com/
21. Тарифи, абонплата. ПрАТ АК Київводоканал. https://vodokanal.kiev.ua/rozraxunki-%D1%96-tarifi
22. Яворський, В.Т. Загальна хімічна технологія: підручник /В.Т. Яворський, Т.В. Перекупко, З.О. Знак, Л.В. Савчук. – Львів: Видавництво Національного університету «Львівська політехніка» (2005). 552 с. ISBN: 978-966-553-822-6
23. A.E. Morris, G. Geiger, H.A. Fine, *Handbook on Material and Energy Balance Calculations in Material*, 3rd edn (Wiley, 2011). https://doi.org/10.1002/9781118237786.fmatter
24. A. Kayode Coker, *Non-Isothermal Reactors. Modeling of Chemical Kinetics and Reactor Design* (2001), pp. 424–551. https://doi.org/10.1016/B978-088415481-5/50008-5
25. D.E. Boldrini, Monolithic stirrer reactor for vegetable oil hydrogenation: a technical and economic assessment. Chem. Eng. Process. Process Intensification **132**, 229–240 (2018). https://doi.org/10.1016/j.cep.2018.09.001
26. Blender Studio. https://www.blender.org/download/
27. A. Lundstrom, T. Snelling, P. Morsing, P. Gabrielsson, E. Senar, L. Olsson, Urea decomposition and HNCO hydrolysis studied over titanium dioxide, Fe-Beta and γ-Alumina. Appl. Catal. B **106**, 273–279 (2011). https://doi.org/10.1016/j.apcatb.2011.05.010
28. Ціни, тарифи. Київська обласна енергопостачальна компанія. https://koec.com.ua/page?root=23
29. M. Müller, M. Kutscherauer, S. Böcklein, G.D. Wehinger, T. Turek, G. Mestl, Modeling the selective oxidation of n-butane to maleic anhydride: From active site to industrial reactor. Catalysis Today 376376 (2021) (in press). https://doi.org/10.1016/j.cattod.2021.04.009
30. Industrial and Chemical Processing Equipment. https://www.leeind.com/industries/industrial-chemical-processing-equipment

31. W. Gao, Z. Sun, H. Cao, H. Ding, Y. Zeng, P. Ning, G. Xu, Y. Zhang, Economic evaluation of typical metal production process: a case study of vanadium oxide production in China. J. Clean. Prod. **256**, 120217 (2020). https://doi.org/10.1016/j.jclepro.2020.120217
32. W.L. Luyben, *Chemical Reactor Design and Control* (Wiley, 2006). https://onlinelibrary.wiley.com/doi/book/10.1002/9780470134917
33. C. Sun, R. Hui, W. Qu, S. Yick, Progress in corrosion resistant materials for supercritical water reactors. Corros. Sci. **51**(11), 2508–2523 (2009). https://doi.org/10.1016/j.corsci.2009.07.007
34. G. Yang, V. Pointeau, E. Tevissen, A. Chagnes, A review on clogging of recirculating steam generators in pressurized-water reactors. Progr. Nucl. Energy **97**, 182–196 (2017). https://doi.org/10.1016/j.pnucene.2017.01.010
35. M.S.S. Danish, A. Bhattacharya, D. Stepanova, A. Mikhaylov, M.L. Grilli, M. Khosravy, T. Senjyu, A systematic review of metal oxide applications for energy and environmental sustainability. Mach. Learn. Methods Sustain. Dev. Metal Oxides Multilayer Metal-Oxides **10**(12), 1604 (2020). https://doi.org/10.3390/met10121604

Transport Properties of Surface-Modified Single-Walled Carbon Nanotubes

Denys O. Shpylka, Iryna V. Ovsiienko, Tetiana A. Len, Lyudmila Yu. Matzui, Yuriy I. Prylutskyy, Ilgar Mirzoiev, and Tatiana L. Tsaregradskaya

Abstract The paper presents the results of studies of transport properties of single-walled carbon nanotubes modified with cobalt-containing complexes. It is revealed that for source single-walled carbon nanotubes the main mechanism of conductivity is the hopping conductivity with the variable hopping length for 3D system. Magnetoresistance of as-sprepared carbon nanotubes is caused by two mechanisms, such as mechanism of compression of the wave function of localized charge carriers and spin-orbital mechanism. The thermopower of as-prepared SWCNTs is linearly dependent on temperature. It is shown that modification of single-walled carbon nanotubes by cobalt-containing complexes results into a change in the character of the conductivity for bulk specimens of single-walled carbon nanotubes. For bulk specimens of modified single-walled carbon nanotubes conductivity is described in the terms of power temperature law that is typical for individual single-walled carbon nanotubes. The negative magnetoresistance observed for modified SWCNTs at low temperatures is related with the manifestation of the effect of charge carriers' weak localization for a one-dimensional system. Thermopower of modified carbon nanotubes is the sum of the diffuse and phonon drag components of thermopower.

1 Introduction

Investigations of the single-walled carbon nanotubes (SWCNTs) have attracted much interest due to their unique properties and great application potentials . One of

D. O. Shpylka (✉) · I. V. Ovsiienko · T. A. Len · L. Yu. Matzui · T. L. Tsaregradskaya
Departments of Physics, Taras Shevchenko National University of Kyiv, Volodymyrska 64/13, Kyiv 01601, Ukraine
e-mail: denys8600@ukr.net

Y. I. Prylutskyy
Departments of Biophysics, Taras Shevchenko National University of Kyiv, Volodymyrska 64/13, Kyiv 01601, Ukraine

I. Mirzoiev
B. Verkin Institute for Low Temperature Physics and Engineering of NAS of Ukraine, Nauky Ave. 47, Kharkiv 61103, Ukraine

O. Fesenko and L. Yatsenko (eds.), *Nanostructured Surfaces, Nanocomposites and Nanomaterials, and Their Applications*, Springer Proceedings in Physics 296,
https://doi.org/10.1007/978-3-031-42704-6_5

the many aspects of investigations of SWCNTs is the study of the possibilities of changing their properties through surface functionalization and modification [1–5]. On the one hand, functionalization is a necessary step providing the formation of homogeneous dispersion of SWCNTs in solvents and, on the other hand, causes a significant number of defects in SWCNTs structure.

Another important direction of SWCNTs chemistry is chemical functionalization of SWCNTs with metal-containing molecular complexes. Such functionalization process not only opens the area of metal–organic chemistry to nanotubes, but also suggests potential applications in catalysis and molecular electronics.

A significant number of works is devoted to the issues of covalent and non-covalent functionalization of CNTs by metal-containing complexes [6–8]. In these papers methods for obtaining such complexes on the surface of CNTs and their properties are considered. Thus, the effect of the surface functionalization with complexes containing magnetic metals on the magnetic properties of CNTs has been theoretically considered in [9]. In [10] influence of surface functionalization with transition-metal phthalocyanines on reduction ability has been studied. The main focus in the articles is paid to the chemical properties of the obtained modified CNTs, in particular chemical stability of functionalized SWCNTs [11], their catalytic activity and the possibility of using them as catalysts in chemical synthesis [11, 12], increasing the solubility of SWCNTs in various solvents and the possibility of forming colloidal solutions and dispersions of CNTs [13]. However, the questions of changing the physical properties of modified on the surface with metal-containing complexes SWCNTs, especially the properties associated with charge transfer, are discussed in these works much less. This especially applies to experimental studies of the properties of functionalized CNTs.

The main goal of presented work is to establish the possibility of surface modification of SWCNTs by cobalt-containing complexes, to identify the structural and morphological state of cobalt on the surface of SWCNTs, as well as to study the influence of surface modification on the transport properties of bulk specimens of modified SWCNTs.

2 Experimental

2.1 As-Prepared Carbon Nanotubes

As source for modification single carbon nanotubes (as-prepared SWCTs) have been chosen. As-prepared SWCNTs with mean value of tubes diameter ~ 1.4 nm have been produced by catalytic decomposition of acetylene with use yttrium and nickel as catalysts.

Figure 1 presents the fragment of X-ray diffraction pattern for as-prepared SWCTs.

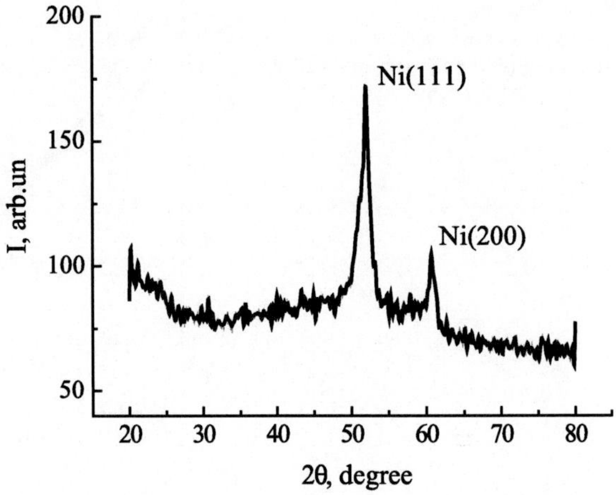

Fig. 1 Fragment of X-ray diffraction pattern for as-prepared SWCTs

As can be seen from the figure the X-ray diffraction pattern contains only intensive peaks that correspond to nickel. The graphite lines are absent on the diffractogram. Thus, there are no ordered multi-walled carbon structures in the source nanocarbon material.

The fragment of transmission electron microscopy image of as-prepared SWCNTs is presented in Fig. 2.

As it is seen from the figure long strands from 3 nm up to 8 nm in diameter, which apparently consist of thinner CNTs, as well as catalyst metal particles and disordered carbon particles up to 5 nm in size, are visualized in the image.

The presence of nanosized nickel particles is also confirmed by the results of thermo-magnetometric studies, which are presented in Fig. 3.

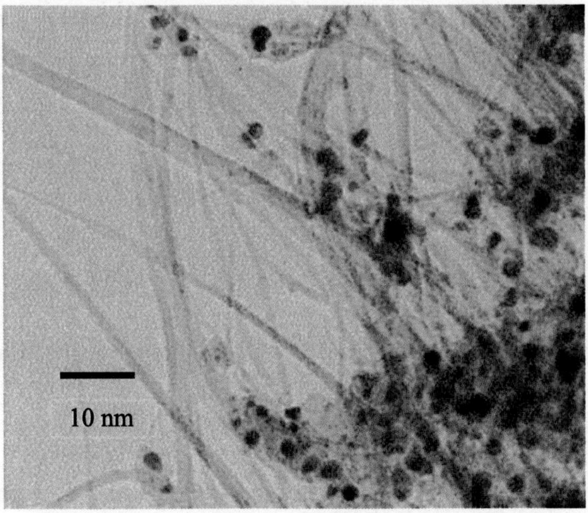

Fig. 2 Fragment of TEM image of as-prepared SWCNTs

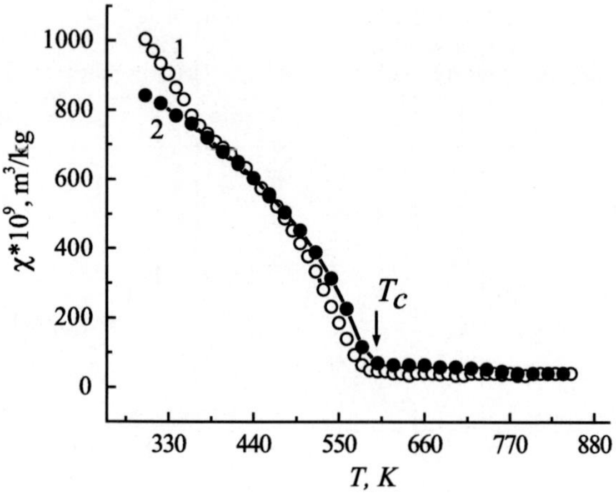

Fig. 3 Temperature dependence of magnetic susceptibility $\chi(T)$ for as-prepared SWCNTs

As it follows from the figure, character of temperature dependence $\chi(T)$ for as-prepared SWCNTs is typical as for material containing magnetic admixture nickel. However, the Curie temperature according to the experimental results is slightly shifted toward low temperatures $T_C = 580$ K in comparison with the Curie temperature for bulk nickel ($T_{Cb} = 630$ K).

Such shifting of Curie point can be explained by nanodispersive character of nickel particles distribution.

Estimating the size L of nickel particles according to [14]

$$L = 3d\left[\frac{T_C}{2(T_{Cb} - T_C)} - 1\right], \tag{1}$$

where d is the atomic or molecular diameter of magnetic phase, gives a value of $L \sim 2.2$ nm, which correlates well with the size of the catalyst metal particles according to electron microscopy data.

Thus, the source for modification as-prepared SWCNTs is nanocarbon material containing only SWCNTs in the form of strands and separate tubes and some portion of the metal catalyst in the form of nanosized particles.

2.2 *Modification of Source SWCNTs*

The chemical modification of as-prepared SWCNTs has been carried out in four stages.

At the first stage, the tubes were opened from the ends (the «caps» on the ends of tubes were torn off) and cut into smaller pieces, i.e., the size and structure of the tubes changed. For this, the as-prepared SWCNTs were treated with a 35% aqueous solution of hydrogen peroxide for 18 h.

The task of the second stage of processing was to remove metal catalyst residues from the as-prepared SWCNTs. For this, the as-prepared SWCNTs were boiled in a hydrochloric acid solution for 10 h, after which the water-soluble nickel chloride was washed out.

At the third stage, the SWCNTs surface was functionalized to create conditions for attachment of cobalt cations to the tubes surface, since cobalt cations cannot directly attach to the SWCNTs surface. Non-covalent functionalization was carried out with three different surfactants: 1,3diaminopropane (DAP, specimen #1), monoethanolamine (MEA, specimen #2) and 1,3diaminopropane in combination with benzophenone (specimen #3).

And at the last stage treated SWCNTs were heated at temperature 60 °C in cobalt (II) chloride solution, while the cobalt cations were attached through functional groups to the surface of the SWCNTs.

Thus, three specimens of modified SWCNTs which differ in the substance used to functionalize the surface were obtained.

2.3 The Structure and Phase Composition of Modified SWCNTs

The X-ray diffraction, tunnel electron microscopic and thermo-magnetometric investigations of modified SWCNTs have been carried out for definition of structure and phase composition.

Figure 4 presents the fragments of X-ray diffraction patterns for specimens of modified with different methods SWCNTs.

As it is follows from the figure none of the fragments of diffractograms for different specimens contain bands that correspond to reflections from graphite planes. Also, reflections corresponded to cobalt were not detected at the X-ray diffraction patterns.

Fig. 4 Fragments of the X-ray diffraction patterns for specimens of modified SWCNTs. The specimen number is indicated in the figure

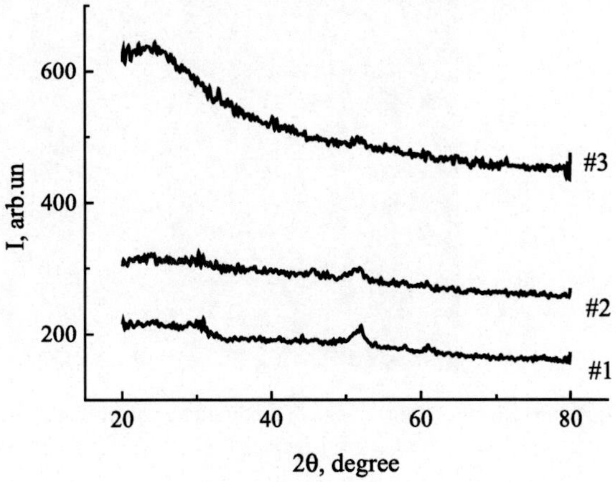

But all diffractograms contain weak bands, which can be detected as reflections from nickel catalyst.

The most intensive nickel line is observed for specimen #1; for specimen #3 the intensity of the nickel band is the weakest.

Thus, all specimens of modified SWCNTs do not contain any multilayer carbon structures and cobalt particles large than 0.3 nm. However, the cleaning methods that were used in the above schemes do not lead to the total deleting of the metal catalysts particles. Therefore, specimens of modified SWCNTs may contain small amounts of metal catalyst residues.

The fragment of TEM image of modified SWCNTs (specimen #1) is shown in Fig. 5.

As can be seen from the figure the chemical treatment of as-prepared SWCNTs results into separation of CNTs bunches into individual tubes and their shortening. Also, a significant decrease in the number of catalyst metal particles and disordered carbon particles is also observed.

The sizes of the remaining particles became significantly smaller compared to the source material.

Figure 6 shows the temperature dependences of magnetic susceptibility for modified with cobalt-containing complexes SWCNTs (specimen #1) for the alternating heating/cooling cycles.

As can be seen from the figure, the dependence $\chi(T)$ is very complicated. This complexity of dependence $\chi(T)$ reflects these complex processes that occur in the modified material during heating.

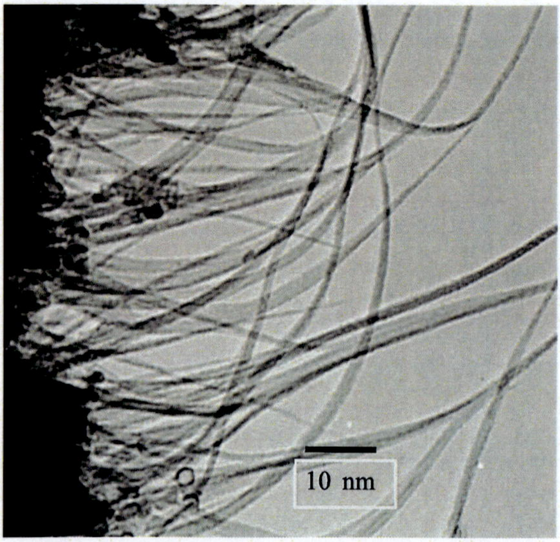

Fig. 5 Fragment of TEM image of modified by cobalt-containing complexes SWCNTs (specimen #1)

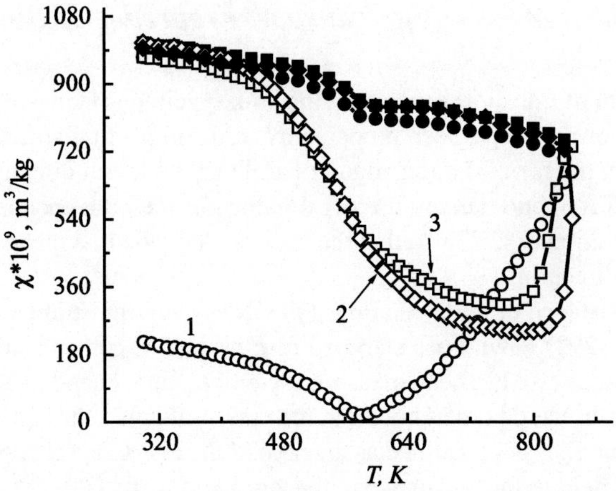

Fig. 6 Temperature dependences of magnetic susceptibility $\chi(T)$ for modified by cobalt-containing complexes SWCNTs (specimen #1) for the sequenced heating–cooling cycles, the curve number corresponds to the cycle number, the open labels are heating, and the closed ones are cooling

In the first heating cycle at a temperature of ~ 580 K, there is a minimum in the dependence, which is obviously associated with the transition from the ferromagnetic to the paramagnetic state of the remaining particles of the catalyst nickel, which, according to the X-ray diffraction data, is contained in small amounts in the modified SWCNTs.

Note that the transition temperature for modified SWCNTs coincides with the transition temperature for the as-prepared SWCNTs; however, the susceptibility value for the as-prepared SWCNTs below the Curie temperature is 6 times greater than for modified SWCNTs in the same temperature range. Upon further heating of the modified SWCNTs, the magnetic susceptibility begins to increase, which is uncharacteristic for a substance in a paramagnetic state. Such dependence $\chi(T)$ can be explained as follows. Cobalt, which is contained in modified SWCNTs as a cation in the composition of complex metal ion complexes and does not affect the magnetic state of the modified SWCNTs at temperatures below 580 K, begins to be reduced to pure cobalt when the temperature increases above 580 K. Small cobalt particles form large conglomerates as a result of thermally stimulated diffusion. This leads to a significant increase in the value of susceptibility since cobalt is a ferromagnet at temperatures up to 1388 K. When the specimen of modified SWCNTs is cooled, the value of the susceptibility practically does not change, and at a temperature of 580 K there is a weak jump in dependence $\chi(T)$. In more detail, the structural and phase composition of modified by cobalt-containing complexes SWCNTs is described in [15].

2.4 *Methods of Measuring Transport Properties of Bulk*

For measurement of transport properties, the bulk specimens from source and modified SWCNTs powders have been prepared by cold compacting without binder. The specimens are in the form of a rectangular parallelepiped with dimensions of 2 mm × 3.5 mm × 15 mm and density up to 1.6 g/cm^3. In the bulk specimens there is a prevailing orientation of CNTs so that the axes of CNTs lie in a plane perpendicular to the pressure direction.

Electrical resistance of bulk specimens of CNTs has been studied in the temperature range (4.2–293) K with use standard four-probe DC compensation technique. Electrical resistance also has been measured at temperature 77 and 293 K in magnetic field up to 2.2 T directed perpendicular to the current flowing through the specimen. Magnetoresistance has been defined as the ratio of the resistance change $\rho(B) - \rho(0)$ in the magnetic field to the resistance in the zero field $\rho(0)$: $\frac{\rho(B)-\rho(0)}{\rho(0)} = \frac{\Delta\rho}{\rho}$.

The measurements of thermopower have been made in the temperature range from 4.2 K up to 300 K. The specimens were cooled to a temperature of 4.2 K at a constant rate of 0.30 K/min. A temperature gradient (0.5–1.0 K) was created in the specimen and was measured with the help of a differential thermocouple Cu–CuFe (the sensitivity in the temperature range (4.2–30) K is (8–16) mcV/K). The measuring circuit was preliminarily calibrated with the superconducting ceramic YBaCuO which has a superconducting transition temperature $T_c = 80$ K and zero thermopower below T_c.

The error of the measurements did not exceed 0.5% for resistance and 1% for thermopower.

3 Results and Discussion

3.1 *Resistivity of Modified SWCNTs*

Figure 7 presents the temperature dependence of resistivity for bulk specimens of source and functionalized SWCNTs.

As it follows from the figure, for all specimens the character of resistivity temperature dependence is similar. At low temperature resistivity decreases sharply and then it is weakly dependent on temperature. However, the resistivity values themselves, as well as the ratio of resistivity at 4.2 K to resistivity at room temperature for specimens of modified SWCNTs, change significantly after modification. So, for the bulk specimen of modified SWCNTs #1, the values of resistivity at low temperatures become even slightly larger compared to the as-prepared SWCNTs. For bulk specimen of modified SWCNTs #3 resistivity at low temperature is somewhat lower, but only slightly. Therefore, for both these specimens, the ratio $\rho_{4.2}/\rho_r$ is close to this ratio in the as-prepared SWCNTs. Recall that both specimens (#1 and #3) were modified using 1,3diaminopropane as a functionalizing agent with minor differences

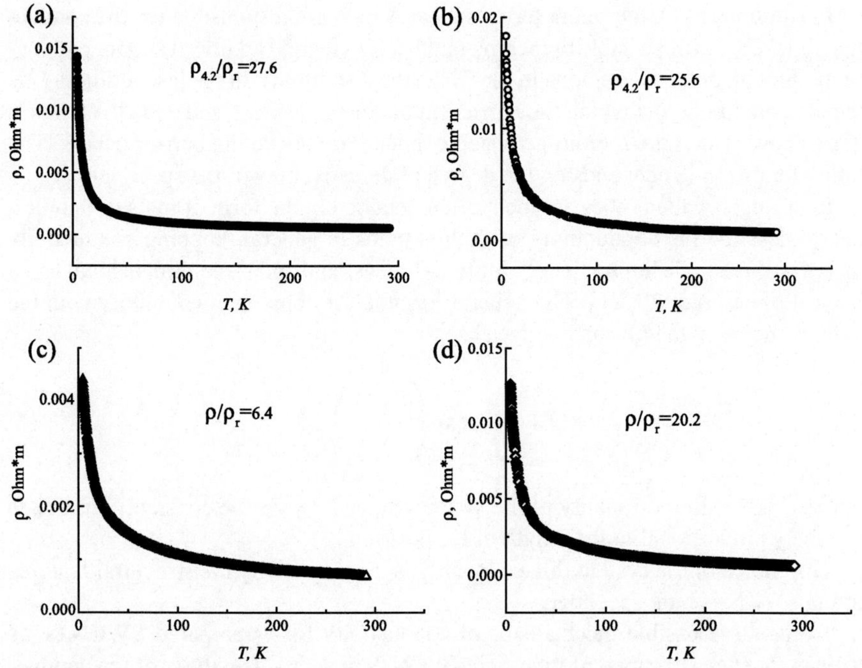

Fig. 7 Temperature dependences $\rho(T)$ for bulk specimens of as-prepared **a** and modified SWCNTs: **b** specimen #1, **c** specimen #2 and **d** specimen #3. The ratio of resistivity at 4.2 K and room temperature $\rho_{4.2}/\rho_r$ for each specimen is indicated in the figures

in the modification process. On the other hand, significant changes in the value of low-temperature resistivity (almost three times) occur for specimen #2 of SWCNTs modified with the use of MEA as a functionalizing substance. That leads to a significant decrease of resistivity ratio $\rho_{4.2}/\rho_{293}$ from 26 for source SWCNTs to 6.3 for modified SWCNTs.

Let us consider possible conduction mechanisms for SWCNTs. The conductivity of a single-walled CNT is usually considered within the terms of a model of a one-dimensional Luttinger liquid of strongly interacting electrons. Within the framework of this model, there is a power-law temperature dependence of conductivity [16]:

$$\sigma(T) = aT^{\alpha}, \alpha = \frac{g + \frac{1}{g} - 2}{8}, \quad (2)$$

where a and α are the constants and α is related with Luttinger parameter g that characterizes the degree of charge carrier's interaction in the system. For defective SWCNTs, equation for conductivity takes on a more complex form [17–19]:

$$\sigma(T) = aT^{\alpha} + bT^{-1} + ce^{-\frac{T_l}{T+T_t}}. \quad (3)$$

In addition to (2), two more terms appear, which are responsible for the conductivity in the areas of possible branching of the CNT (second addition) and for conductivity through defective regions in the CNT (third addition). In the last addition T_1 is the temperature below which the barrier tunneling takes place and T_t is the temperature above which the thermo-activated conductivity above the barrier occurs. The ratio of terms in (3) depends on the degree of defectiveness of the specimen.

Real single-walled tubes are most often obtained in the form of mats or bundles, and to describe the conductivity of such systems in general, hopping conductivity model with variable hoping length for two-dimensional and three-dimensional cases is most often used [20, 21]. The general formula for hopping conductivity with the variable hopping length can be given as:

$$\sigma(T) = \sigma_0 \exp\left(-\left(\frac{T_0}{T}\right)^{\frac{1}{d}}\right), \tag{4}$$

where d is the dimensionality of the system, σ_0 and T_0 are the constants, and T_0 is inversely proportional to the length of localization.

The choice of the system dimensionality is determined by the structural features of the SWCNTs bulk specimens.

To identify possible mechanisms of conductivity for as-prepared SWCNTs, as well as for each specimen of modified SWCNTs, a detailed analysis of the conductivity temperature dependence has been carried out. The results of analysis are shown in Fig. 8. For each SWCNTs specimen, the figure shows the temperature dependence of conductivity in coordinates $\ln \sigma = f(T^{-1/4})$, which corresponds to the hopping mechanism of conductivity with a variable hoping length (3), and in coordinates $\ln \sigma = f(\ln T)$, which respectively corresponds to conductivity in the terms of model of a one-dimensional Luttinger liquid of strongly interacting electrons.

As it follows from presented dependences for as-prepared SWCNTs the main conductive mechanism is hopping with variable hopping lengths for the three-dimensional case. Such a conductivity mechanism is usually the main one precisely for the as-prepared SWCNTs, which are obtained in the form of mats or felt and which contain, in addition to CNTs, a sufficient number of disordered carbon particles [22–24]. Also, this conductivity mechanism is typical for disordered graphite, in particular so-called amorphous carbon. The characteristic temperature T_0 determined from the dependence $\ln \sigma = f(T^{-1/4})$ equals to 2902 K. This value of characteristic temperature T_0 is much larger than corresponding value T_0 for mats of SWCNTs in [21], but close to value T_0 for amorphous carbon in [25].

As can be seen from the figures above, modification of SWCNTs by cobalt-containing complexes did not lead to a dramatic change in the conductivity of the

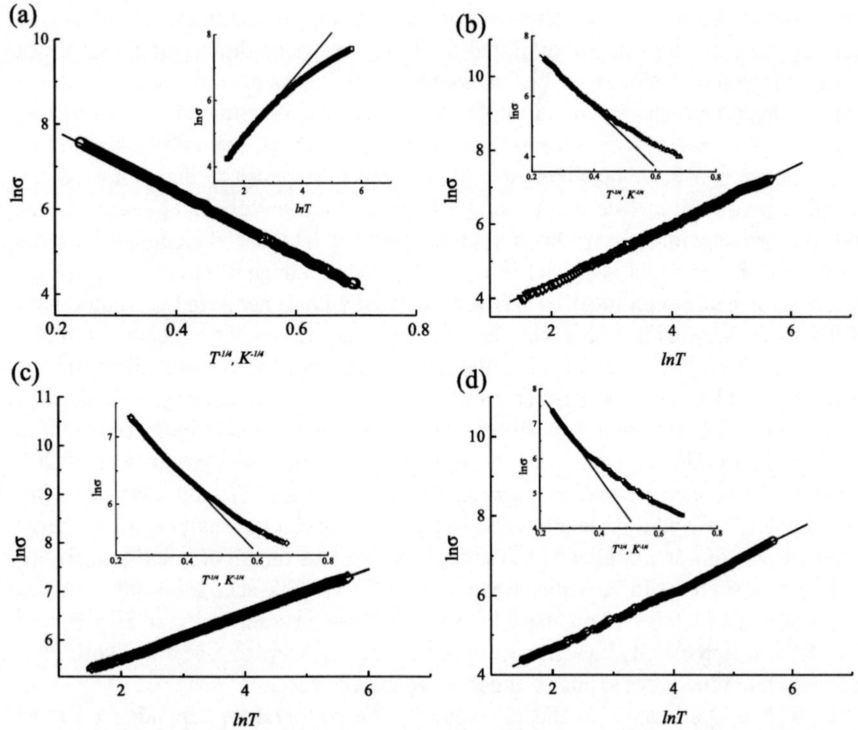

Fig. 8 Temperature dependences of conductivity for bulk specimens of as-prepared **a** and modified, **b** specimen #1, **c** specimen #2 and **d** specimen #3 SWCNTs in coordinates $\ln\sigma = f(T^{-1/4})$ (**b**, **c**, and **d**—inset) and in coordinates $\ln\sigma = f(\ln T)$ (**a**—inset)

CNTs bulk specimens. However, such modification significantly affected the character of the temperature dependence of conductivity for these specimens. For all specimens of modified SWCNTs conductivity temperature dependence is well described in the terms of power temperature law that is typical for individual SWCNTs (1).

In Table 1, calculated from dependence $\ln\sigma = f(\ln T)$, parameters of conductivity for bulk specimens of modified SWCNTs are presented.

As can be seen from Table 1, for specimens of modified SWCNTs #1 and #3, for which the conductivity does not change significantly after modification, the values of the coefficient α are close, while for specimen of modified SWCNTs #2, for which the conductivity increases after modification, the value of α is almost two

Table 1 Calculated parameters α, g and a for modified SWCNTs

Specimen	α	g	a, S/m
#1	0.79	0.12	16.6
#2	0.46	0.18	110.0
#3	0.73	0.13	23.6

times smaller. However, the values of parameter g for all specimens are close and indicate the formation in the modified SWCNTs of a one-dimensional system of strongly interacting electrons. Let us dwell on this moment in more detail. As is known for pressed specimens, the total resistance R is determined as sum of two additions: $R = R_{CNT} + r_C$, where R_{CNT} is the resistance of separate CNTs and r_C is the contact resistance between tubes. The contact resistance r_C depends on many factors among them surface condition, size of contact spot and contact pressure. Since bulk-pressed specimens have been obtained without a binder, the contact between individual tubes is direct, without a layer of polymer. A change in the conductivity of a bulk specimen after modification can be associated with a change in the conductivity of the tubes R_{CNT} and with a change in the contact resistance r_C between them, as well as these two processes simultaneously. Indeed, the functionalization and modification of the CNTs' surface can both reduce the conductivity of individual CNTs due to the destruction of the graphite layer π-system and increase it due to charge redistribution as a result of the attachment of functional groups to the CNT surface. The contact resistance between the separate tubes usually increases after functionalization and modification due to the appearance of a small negative charge on the surface of functionalized CNTs caused by the attachment of functional groups to the tubes surface. In addition, treatment of CNTs with acid solutions provides to remove disordered carbon particles from the primary networks of as-prepared SWCNTs, which usually have a developed surface and serve as a peculiar «bridges» between individual CNTs, increasing their contact surface.

Thus, it is the change in the character of the temperature dependence of the conductivity of the SWCNTs bulk specimens after chemical treatment indicates that the SWCNTs surface has been modified.

3.2 *Magnetoresistance of Modified SWCNTs*

Figure 9 presents the dependence of magnetoresistance $\frac{\Delta\rho}{\rho}$ on magnetic field for as-prepared SWCNTs at room temperature and $T = 77$ K.

As it is seen from the figure, magnetoresistance is symmetrical with respect to the direction of the magnetic field. At room temperature, the magnetoresistance is positive, and its values do not exceed 0.035% at the maximum magnetic field of measurement. When the temperature decreases, the magnetoresistance changes its sign, while the absolute value of the magnetoresistance increases to 0.15% in the maximum measurement field.

For materials in which the hopping conduction mechanism is implemented, the effect of compression of the wave function of the localized state in the magnetic field takes place in the magnetoresistance [26–29]. According to this mechanism, within the limits of weak magnetic fields ($\xi < l_B$), the magnetoresistance is described by the expression:

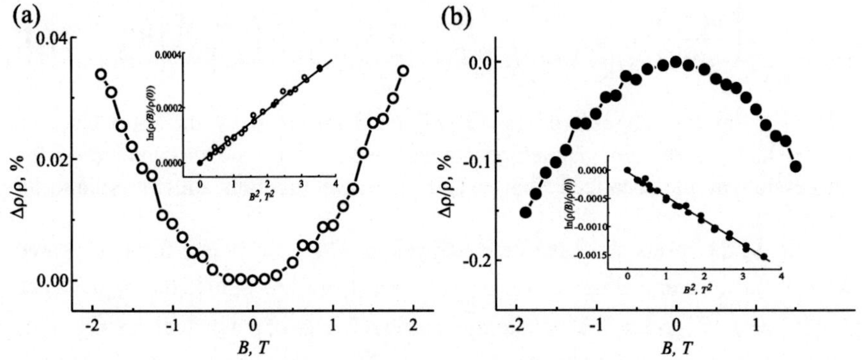

Fig. 9 Dependence $\frac{\Delta\rho}{\rho}(B)$ for bulk specimen of as-prepared SWCNTs at room temperature (**a**) and T = 77 K (**b**). Corresponding dependences in coordinates $\ln\left(\frac{\rho(B)}{\rho(0)}\right) = f\left(B^2\right)$ are shown in the inset

$$ln\left(\frac{\rho(B)}{\rho(0)}\right) = t_d\left(\frac{\xi}{l_B}\right)^4\left(\frac{T_0}{T}\right)^{3\alpha}, \tag{5}$$

where ξ is the wave function localization radius, l_B is the magnetic length, $l_B = \sqrt{\frac{h}{2\pi eB}}$, t_d and α are the numerical coefficients, and T_0 is the characteristic temperature.

So, implementation of the mechanism of compression of the localized state wave function in the magnetic field in disordered systems involves a positive magnetoresistance that depends quadratically on the magnetic field. However, as can be seen from Fig. 9, as the temperature decreases, the magnetoresistance of as-prepared SWCNTs becomes negative. Therefore, to describe the magnetoresistance of as-prepared SWCNTs, it is not enough to consider only the model of the wave function compression in a magnetic field [29]. As shown in works [30, 31], the negative component of the magnetoresistance can be related to the implementation in carbon materials of the spin-polarized magnetoresistance mechanism [31].

The spin-polarized mechanism of magnetoresistance leads to the following magnetoresistance expression in the region of a weak magnetic field [30]:

$$ln\left(\frac{\rho(B)}{\rho(0)}\right) = \alpha A\left(\frac{T_0}{T}\right)^{\alpha}\left(\frac{\mu B}{k_b T}\right)^2, \tag{6}$$

where μ is the effective magnetic moment of an electron and A is the some parameter that is related to the relationship between concentrations and localization radii of localized states.

Thus, it follows from (5) and (6) that both mechanisms of magnetoresistance are quadratic with respect to magnetic induction within weak magnetic fields

$$\ln\left(\frac{\rho(B)}{\rho(0)}\right) = \left[t_d\left(\frac{2e^2}{h}\right)^2\left(\frac{T_0}{T}\right)^{3\alpha} + \alpha A\left(\frac{T_0}{T}\right)^{\alpha}\left(\frac{\mu}{k_b T}\right)^2\right]B^2. \quad (7)$$

The sign of the magnetoresistance will be determined by the ratio between the contributions of two magnetoresistance mechanisms, the mechanism of the compression of the localized charge carriers wave function and the spin–orbit mechanism.

In Fig. 9, the insets show the dependences $\ln\left(\frac{\rho(B)}{\rho(0)}\right) = f(B^2)$ for as-prepared SWCNTs at two temperatures. As can be seen from the figure, the dependence $\ln\left(\frac{\rho(B)}{\rho(0)}\right) = f(B^2)$ is linear at both temperatures. Change of magnetoresistance signs at different temperatures indicates the change of primary mechanisms of magnetoresistance in different temperature ranges. At room temperature, the compression mechanism of the wave function of localized charge carriers is prevailing. When the temperature decreases, the spin-orbital magnetoresistance makes a more significant contribution to the magnetoresistance, which causes changes in the sign of the magnetoresistance.

Figure 10 shows the dependences $\frac{\Delta\rho}{\rho}(B)$ for bulk specimens of modified by cobalt-containing complexes SWCNTs (specimens #2 and #3) at two temperatures: $T = 293$ K and $T = 77$ K.

As it is follows from the figures, the character of the temperature and field dependences of magnetoresistance for both specimens of modified SWCNTs changes significantly compared to the as-prepared SWCNTs. For both specimens of modified SWCNTs, the magnetoresistance is negative at room temperature and at a temperature of 77 K. As the temperature decreases, the absolute value of the magnetoresistance for both specimen increases. However, the absolute values of magnetoresistance for specimen #2 are approximately 10 times smaller compared to the absolute values of magnetoresistance for specimen #3. In addition, as can be seen from the figure, for both specimens of modified SWCNTs at $T = 77$ K, with increasing magnetic

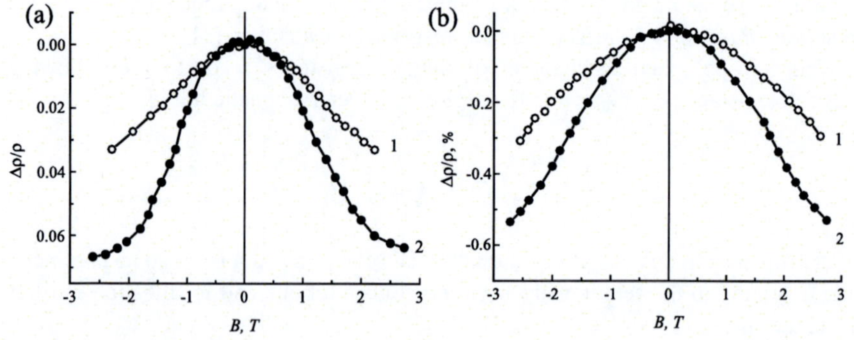

Fig. 10 Dependences $\frac{\Delta\rho}{\rho}(B)$ for bulk specimens of modified SWCNTs (#2 (**a**) and #3 (**b**)) at $T = 293$ K (curve 1) and $T = 77$ K (curve 2)

field the deviations from the quadratic dependence of the magnetoresistance on the magnetic field are observed.

Let us consider in more detail the mechanisms of magnetoresistance for modified SWCNTs. As was shown above, the temperature dependence of conductivity for modified SWCNTs is described by a power law, which is characteristic for one-dimensional conductors. Moreover, the values of the power exponent are close to the corresponding values of the power exponent within the terms of the conduction model of a one-dimensional Luttinger liquid of strongly interacting electrons. As it is known, for CNTs of a perfect structure, the conductivity of which is considered within the terms of a two-dimensional model, a negative magnetoresistance associated with the manifestation of the charge carriers' weak localization effect is observed. Let us evaluate whether the negative magnetoresistance for modified SWCNTs can be a manifestation of the effect of charge carriers' weak localization, but for a one-dimensional system.

As is well known weak localization occurs as a result of the interference of direct and elastically scattered electron waves on inhomogeneities of the system. For one-dimensional system, conductivity in magnetic field taking into account the additive due to weak localization is given by equation [32, 33]

$$\Lambda(B) = \Lambda_0 - \frac{2e^2}{hL}\left(\frac{1}{L_\varphi^2} + \frac{e^2B^2w^2}{3\hbar}\right)^{-\frac{1}{2}}, \tag{8}$$

where $\Lambda(B)$ is the conductivity in magnetic field B, Λ_0 is classical Drude conductance without the localization addition, L_φ is the coherence length of the wave function, and L and w, respectively, the length and diameter of the nanotube. Note that (8) is valid only for case $L_\varphi > w$. In the absence of a magnetic field, the conductivity can be written as

$$\Lambda(0) = \Lambda_0 - \frac{2e^2L_\varphi}{hL}. \tag{9}$$

Let us find the difference between conductivities in a magnetic field and in the absence of a magnetic field

$$\Lambda(\mathrm{B}) - \Lambda(0) = \frac{2e^2L_\varphi}{hL}\left[1 - \frac{\sqrt{3}\hbar}{\sqrt{3\hbar^2 + L_\varphi^2e^2B^2w^2}}\right]. \tag{10}$$

Accordingly, the difference between specific conductivities in a magnetic field $\sigma(B)$ and in the absence of a magnetic field $\sigma(0)$ has the form

$$\sigma(B) - \sigma(0) = \Delta\sigma(B) = \frac{8e^2L_\varphi}{\pi w^2h}\left[1 - \frac{\sqrt{3}\hbar}{\sqrt{3\hbar^2 + L_\varphi^2e^2B^2w^2}}\right] \tag{11}$$

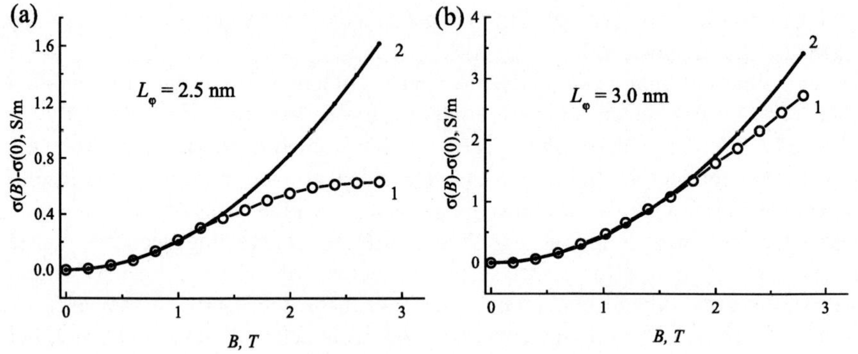

Fig. 11 Experimental (1) and calculated by (10) (2) dependences of specific conductivity on magnetic field for specimens of modified SWCNTs #2 (**a**) and #3 (**b**). The values of the fitting parameter L_φ are indicated in the figure

Equation (11) contains only one unknown parameter L_φ which was fitted.

The obtained experimental dependences $\Delta\sigma(B)$ as well as calculated by (11) using the fitting parameter for both specimens of modified SWCNTs are shown in Fig. 11.

As can be seen from the figure, the best match between the calculated dependence $\Delta\sigma(B)$ and the experimental one for SWCNTs with a fixed diameter $w = 1.4$ nm is observed at the values of the fitting parameter L_φ, 2.5 nm for specimens #2 and 3 nm for specimen #3, respectively. Note that condition $L_\varphi > w$ is fulfilled. However, the coincidence is observed not in the entire interval of the magnetic field in which the measurements have been carried out, but only when $B < 1.3$ T for specimen #2 and for $B < 1.7$ T for specimen #3.

Thus, it can be assumed that the positive conductivity in the magnetic field for specimens of modified SWCNTs below the specified magnetic fields is associated with the manifestation of the effect of charge carriers' weak localization for a one-dimensional system.

3.3 *Thermopower of Modified SWCNTs*

Measurements of thermopower temperature dependence $S(T)$ have been carried out for as-prepared SWCNTs in temperature interval from 77 K up to 293 K and for modified SWCNTs (specimen #2) in temperature interval from 4.2 K up to 293 K. The results of measurements are presented in Fig. 12.

As can be seen from the figure, for as-prepared SWCNTs the linear temperature dependence of thermopower is observed. For modified SWCNTs there is linear dependence of thermopower on temperature at low and high temperatures. However, in the temperature interval from ~ 50 K up to 250 K significant deviation from linear dependence occurs.

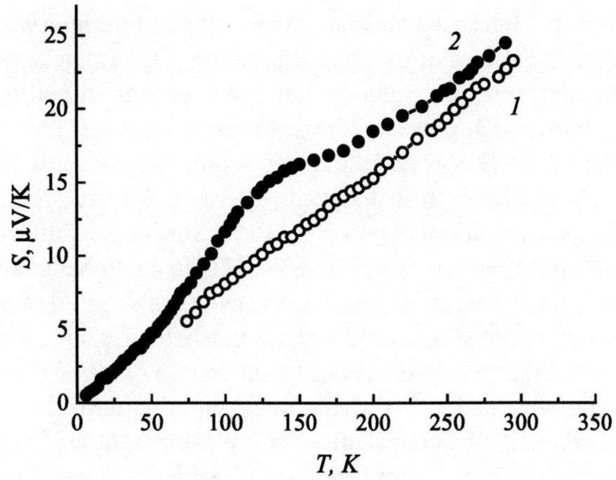

Fig. 12 Temperature dependences of thermopower $S(T)$ for as-prepared SWCNTs (1) and for SWCNTs modified with cobalt-containing complexes (2)

As is known, thermopower is very sensitive to the features of the band structure of the material, its electronic and phonon spectrum, as well as to the scattering mechanisms of charge carriers and phonons.

For graphite materials with a perfect structure, the dependence of thermopower on temperature has a very complex form. Such dependence of thermopower on temperature is caused by two factors. First, thermopower is determined by the contribution of two components: diffuse thermopower and thermopower associated with phonon drag effect. The phonon drag effect leads to the appearance of additional thermopower, caused by additional charge carriers dragged from the hot specimen end to the cold end by phonon flux via momentum transfer. Secondly, diffuse thermopower for graphite with a perfect structure, which has two types of main charge carriers with equal concentration, is characterized by a complex, nonlinear temperature dependence. However, the situation changes significantly for defective graphite materials. For such materials, the temperature of maximum the phonon drag effect shifts to low temperatures, and the magnitude of the effect itself becomes very small. As for the diffuse thermopower, it acquires a linear dependence on temperature under conditions of predominant hole conductivity. This temperature dependence of thermopower is typical for metals. So, metallic diffusion thermopower can be written as

$$S_d = \frac{\pi^2 k_b^2 T}{3eE_F}(1 - p), \quad (12)$$

where E_F is the Fermi energy and p is the coefficient associated with the prevailing charge carriers scattering mechanism. Similar temperature dependences of thermopower were observed both for SWCNTs' mats and for individual SWCNTs in a number of papers.

Modification of as-prepared SWCNTs with cobalt-containing complexes leads to a change in the temperature dependence of the thermopower, namely to a deviation from the linear dependence in the temperature interval of (50–250) K. At the same time, the values of thermopower at low and high temperatures almost coincide with the corresponding values of thermopower for unmodified SWCNTs.

Figure 13 presents the difference between the experimental values of the thermopower S_{exp} and the linear approximation $S_{\mathrm{d}} = cT$ (12).

As it follows from the figure, the curve $S_{\mathrm{exp}} - S_{\mathrm{d}}$ has the form of a broad band with a maximum at temperature ~ 130 K. Similar deviations from the linear dependence $S(T)$ have been observed for SWCNTs in a number of studies. However, in the literature, there is no single view on the nature of such a deviation from linear dependence. Several effects that could lead to such behavior have been discussed in the literature. In [34] thermopower is considered as sum of a linear metallic term and an exponentially weighted $T^{1/2}$ variable range hopping term that reflects the "freezing-out" of semiconducting contributions at low temperatures. For doped SWCNTs, the nonlinearity in thermopower is associated with the appearance of localized states near the Fermi level due to the introduction of boron or nitrogen atoms into the CNTs wall structure. Another reason that can lead to the deviation of thermopower from a linear dependence is the Kondo effect. The thermopower investigations on SWCNTs mats in [35] suggest that the observed giant thermopower comes from the contributions of Kondo state induced by magnetic transition-metal impurities. For the investigated SWCNTs, the cobalt ion is contained in the composition of complexes, and cobalt does not present as individual particles. Obviously, for these SWCNTs, the Kondo effect cannot cause deviation of the thermopower from the linear dependence.

In the [36] the authors assume that the origin of the deviation of the thermopower temperature dependence from the linear one is the manifestation of the effect of phonon drag of charge carriers. The carried-out calculations by Scarola and Mahan in [37] show that the phonon drag thermopower in SWCNTs increases rapidly with T at low temperatures and approaches approximately a constant for $T > 100$ K. So, the thermoelectric power of the SWCNTs can be considered as sum of a linear diffusion

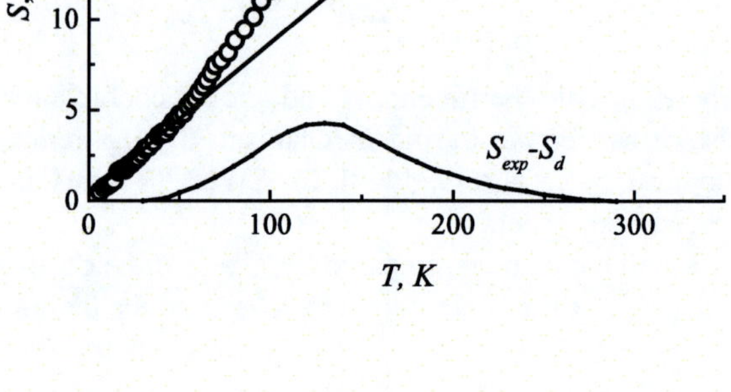

Fig. 13 Difference between the experimental values of the thermopower S_{exp} and the linear approximation $S_{\mathrm{d}} = cT$

thermopower S_d and phonon drag thermopower, which is approximately constant for temperature interval from 110 K up to 300 K.

Really, the shape of the dependence $(S_{exp} - S_d) = f(T)$ is similar to the temperature dependence of the phonon drag thermopower, which is observed in acceptor intercalated compounds based on fine crystalline graphite. The temperature at which the maximum in the dependence $(S_{exp} - S_d) = f(T)$ occurs also coincides with the temperature of the phonon drag thermopower maximum.

Let us estimate the value of the Fermi energy E_F from the linear section of the dependence $S(T)$ (Fig. 13) using (12). The calculated value of E_F is 0.29 eV. The corresponding value of the Fermi wave vector is $k_F = 4.5 \times 10^8$ m^{-1}.

Let us determine the temperature at which the maximum effect of phonon capture of charge carriers occurs, using the formula

$$T_{max} = \frac{2\hbar v_s k_F}{k_b}, \tag{13}$$

where v_s is the sound speed and $v_s = 2.1 \times 10^4$ m/s as for monocrystalline graphite. According to calculation $T_{max} = 140$ K, which correlates well with the value of the temperature $T_{max\ exp} = 130$ K, at which a maximum in the dependence $(S_{exp} - S_d) = f(T)$ is observed.

Thus, the carried-out studies of thermopower revealed that for the as-prepared SWCNTs, thermopower has a linear dependence on temperature and is determined only by the diffuse component of thermopower, while for modified SWCNTs, two mechanisms contribute to thermopower, namely diffuse thermopower and thermopower caused by phonon drag of charge carriers.

3.4 Discussion

The results of research into the transport properties of SWCNTs surface-modified with cobalt-containing complexes allow to assume that these modified SWCNTs can be considered as one-dimensional conductors, and such one-dimensionality of transport properties is preserved in bulk pressed specimens of modified SWCNTs.

The as-prepared SWCNTs are considered as 3D system. This is explained by the fact that, on the one hand, SWCNTs are contained in the source carbon nanomaterial in the form of mats, bundles of very tangled tubes. On the other hand, the source carbon nanomaterial contains a sufficiently large amount of disordered carbon phase, in other words, amorphous nanocarbon particles with a developed surface. The presence in specimen a significant amount of amorphous nanocarbon particles causes a significant decrease of contact resistance between individual nanotubes. The amorphous nanocarbon particles with a sufficiently developed surface promote the adhesion of individual CNTs to each other. It is the presence of large number of disordered carbon phase that allows to obtain bulk specimens of SWCNTs without the use of a binder. Thus, bulk specimen of as-prepared SWCNTs can be considered as 3D system.

Let us analyze the changes in the structural and phase composition of the source carbon nanomaterial and the surface morphology of the as-prepared SWCNTs in the process of their modification with cobalt-containing complexes. On the first stage the treatment of the source SWCNTs with hydrogen peroxide results in formation of a significant number of broken bonds. At the next stage as-prepared SWCNTs are treated with water solution of hydrochloric acid that leads to the removal from the SWCNT's strands of metal catalyst and amorphous nanocarbon. Then CNTs surface is functionalized by methods of non-covalent functionalization, and cobalt-containing complexes can join to carbon atoms at the CNTs surface through functional groups. So, treatment of as-prepared SWCNTs leads to the removal of amorphous nanocarbon particles and contributes to the disentanglement of CNTs. It was these two factors that ensured the "three-dimensionality" of the bulk specimens of the as-prepared SWCNTs. In addition, as is known, a small localized negative charge appears on the SWCNT surface as a result of functionalization. This also creates conditions that prevent the movement of charge carriers between the individual tubes.

Thus, all the given facts indicate that after the proposed treatment of as-prepared SWCNTs, the contact resistance between the individual CNTs in the bulk specimen has increased significantly. Therefore, the charge carriers in the bulk specimen can move only along each tube, and charge transfer between individual CNTs is unlikely. Thus, in the bulk specimens of SWCNTs, the conductivity mechanism is formed, which is characteristic for the 1D systems. This situation in the bulk specimens of SWCNTs can be compared with the processes that occur at the formation of graphite intercalation compounds (GICs). Graphite intercalation compounds are layered structures in which layers of graphite (their number S determines the stage of the GICs) and monomolecular or monoatomic layers of other substances called intercalates alternated. The condition for the formation of GICs is the transfer of charge from the intercalates layers to the graphite layers, which results in an increase in the concentration of free charge carriers in the graphite layers. At the same time intercalates layers form an electrostatic barrier from the charges localized on them. Due to this the charge carriers in the GICs can move only along the graphite layers, forming a two-dimensional electronic system. And GICs are considered as two-dimensional conducting systems which two-dimensionality is caused not by their geometrical sizes and is connected with features of their electronic structure.

Thus, it can be assumed that the "one-dimensionality" of the transport properties of modified SWCNTs bulk specimens can be explained by the features of the zone structure of SWCNTs modified with cobalt-containing complexes. The concentration of a small negative charge on the CNT surface and, obviously, the enrichment of the CNT with an additional positive charge lead to the formation of one-dimensional system in the bulk specimen, which is not related to the geometric dimensions and shape of the bulk conductor, but is caused by the peculiarities of the zone structure of individual modified SWCNT.

4 Conclusion

Thus, the carried-out investigations of structural and phase composition of modified SWCNTs have revealed that all proposed modification schemes allow to obtain SWCNTs modified on the surface by cobalt-containing complexes.

The studies of the resistivity temperature and magnetic field dependences have shown that for bulk specimens of as-prepared SWCNTs the main mechanism of conductivity is the 3D hopping mechanism with the variable hopping length. This conduction mechanism is typical for disordered graphite materials, as well as for mats and binders of SWCNTs. Magnetoresistance of as-prepared SWCNTs is caused by two mechanisms. The first of them is the mechanism of compression of the wave function of localized charge carriers that leads to positive magnetoresistance. The second mechanism is spin-orbital, which causes the change in the sign of the magnetoresistance. The thermopower of as-prepared SWCNTs has only a linear dependence on temperature.

Modification of SWCNTs by surface with cobalt-containing complexes results into a change in the character of the conductivity of SWCNTs specimens. For modified SWCNTs the bulk specimens' conductivity is described in the terms of power temperature law that is typical for individual SWCNTs. Moreover, the negative magnetoresistance observed for modified SWCNTs at low temperatures is related with the manifestation of the effect of charge carriers' weak localization for a one-dimensional system. Thermopower of modified with cobalt-containing complexes SWCNTs contains two additions. One of them is the diffuse thermopower, and other is the phonon drag thermopower.

Thus, when modifying the SWCNTs' surface with cobalt-containing complexes, a transition from a three-dimensional system to a one-dimensional system occurs. Such transition from a three-dimensional system for bulk specimen of as-prepared SWCNTs to a one-dimensional system for bulk specimen of modified SWCNTs is due to a change in both the structural and phase composition of the source carbon nanomaterial and a change in the state of the SWCNTs' surface as a result of modification. Surface modification of SWCNTs leads to the creation on the surface of CNTs small localized negative charge which acts as an electrostatic screen. The presence of such a charge makes it impossible to transfer charge between individual tubes and promotes the formation of a 1D conductive system of SWCNTs.

References

1. R. Andrews, M.C. Weisenberger, Carbon nanotube polymer composites. Solid State Mater. Sci. **8**, 31–37 (2004). https://doi.org/10.1016/j.cossms.2003.10.006
2. J. Yun, J.S. Im, Y.-S. Lee, H.-I. Kim, Effect of oxyfluorination on electromagnetic interference shielding behavior of MWCNT/PVA/PAAc composite microcapsules. Eur. Polym. J. **46**, 900–909 (2010). https://doi.org/10.1016/j.eurpolymj.2010.02.005

3. J. Zhang, H. Zou, Q. Qing, Y. Yang, Q. Li, Z. Liu, X. Guo, Z. Du, Effect of chemical oxidation on the structure of single-walled carbon nanotubes. J. Phys. Chem. B **107**, 3712–3718 (2003). https://doi.org/10.1021/jp027500u
4. M.N. Tchoul, W.T. Ford, G. Lolli, D.E. Resasco, S. Arepalli, Effect of mild nitric acid oxidation on dispersability, size, and structure of single-walled carbon nanotubes. Chem. Mater. **19**, 5765–5772 (2007). https://doi.org/10.1021/cm071758l
5. A.B. Sulong, N. Muhamad, J. Sahari, R. Ramli, B.M. Deros, J. Park, Electrical conductivity behaviour of chemical functionalized MWCNTs epoxy nanocomposites. Eur. J. Sci. Res. **29**, 13–21 (2009)
6. S.A.V. Jannuzzi, B. Martins, L.E.S.C. Huamaní, A.L.B. Formiga, Supramolecular approach to decorate multi-walled carbon nanotubes with negatively charged iron(II) complexes. J. Braz. Chem. Soc. **28**, 2–10 (2017). https://doi.org/10.5935/0103-5053.20160137
7. S.M. Alshehri, T. Ahamad, A. Aldalbahi, N. Alhokbany, Pyridylimine cobalt(II) and nickel(II) complex functionalized multiwalled carbon nanotubes and their catalytic activities for ethylene oligomerization. Adv. Polym. **35**, 21528.1–21528.10 (2016). https://doi.org/10.1002/adv.21528
8. V. Yu Evtushok, I.D. Ivanchikova, O. Yu Podyacheva, O.A. Stonkus, A.N. Suboch, Yu.A. Chesalov, O.V. Zalomaeva, O.A. Kholdeeva, Carbon nanotubes modified by Venturello complex as highly efficient catalysts for alkene and thioethers oxidation with hydrogen peroxide. Front. Chem. **7**, 858 (2019). https://doi.org/10.3389/fchem.2019.00858
9. S. Gómez-Coca, E. Ruiz, Magnetic behaviour of transition metal complexes with functionalized chiral and C_{60}-filled nanotubes as bridging ligands: a theoretical study. Magnetochemistry **1**, 62–71 (2015). https://doi.org/10.3390/magnetochemistry1010062
10. W. Orellana, Metal-phthalocyanine functionalized carbon nanotubes as catalyst for the oxygen reduction reaction: a theoretical study. Chem. Phys. Lett. **541**, 81–84 (2012). https://doi.org/10.1016/j.cplett.2012.05.048
11. S. Banerjee, S.S. Wong, Functionalization of carbon nanotubes with a metal-containing molecular complex. Nano Lett. **2**, 49–53 (2002). https://doi.org/10.1021/nl010070j
12. S. Arumugam, L.D. Chakkarapani, Metal nanoparticles functionalized carbon nanotubes for efficient catalytic application. Mater. Res. Express **6**(10), 50e3 (2019). https://doi.org/10.1088/2053-1591/ab42ff
13. A. Oki, L. Adams, Z. Luo, E. Osayamen, P. Biney, V. Khabashesku, Functionalization of single-walled carbon nanotubes with N-[3-(trimethoxysilyl)propyl]ethylenediamine and its cobalt complex. J. Phys. Chem. Solids **69**, 1194–1198 (2008). https://doi.org/10.1016/j.jpcs.2007.10.129
14. I.V. Ovsiienko, L. Yu Matzui, M.I. Zakharenko, N.G. Babich, T.A. Len, Yu.I. Prylutskyy, D. Hui, Yu.M. Strzhemechny, P.C. Eklund, Magnetometric studies of catalyst refuses in nanocarbon materials. Nanoscale Res. Lett. **3**, 60–64 (2008). https://doi.org/10.1007/s11671-007-9115-z
15. I.V. Ovsiienko, T.A. Len, L.Yu. Matzui, O.A. Golub, Yu.I. Prylutskyy, T.L. Tsaregradskaya, G.V. Saenko, The structure and transport properties of single-walled carbon nanotubes modified by cobalt-containing complexes. J. Nano Electron. Phys. **12**, 06023–1–06023–7 (2020). https://doi.org/10.21272/jnep.12(6).06023
16. M. Bockrath, D.H. Cobden, J. Lu, A.G. Rinzler, R.E. Smalley, L. Balents, P.L. McEuen, Luttinger-liquid behaviour in carbon nanotubes. Nature **397**, 598–601 (1999)
17. D.J. Bae, K.S. Kim, Y.S. Park, E.K. Suh, K.H. An, J.-M. Moon, S.C. Lim, S.H. Park, Y.H. Jeong, Y.H. Lee, Transport phenomena in an anisotropically aligned single-wall carbon nanotube film. Phys. Rev. B **64**, 233401–233411 (2001). https://doi.org/10.1103/PhysRevB.64.233401
18. M. Shiraishi, M. Ata, Conduction mechanisms in single-walled carbon nanotubes. Synth. Met. **128**, 235–239 (2002). https://doi.org/10.1016/S0379-6779(02)00013-9
19. A.B. Kaiser, G.U. Flanagan, D.M. Stewart, D. Beaglehole, Heterogeneous model for conduction in conducting polymers and carbon nanotubes. Synth. Met. **117**, 67–73 (2001). https://doi.org/10.1016/S0379-6779(00)00540-3
20. B. Liu, B. Sundqvist, O. Andersson, T. WaÊgberg, E.B. Nyeanchia, X.-M. Zhu, G. Zou, Electric resistance of single-walled carbon nanotubes under hydrostatic pressure. Solid State Commun. **118**, 31–36 (2001). https://doi.org/10.1016/S0038-1098(01)00034-5

21. I.V. Ovsiyenko, T.A. Len, L.Yu. Matzui, Yu.I. Prylutskyy, U. Ritter, P. Scharff, F. Le Normand, P. Eklund, Resistance of a nanocarbon material containing nanotubes. Molecular Cryst. Liquid Cryst. **468**, 289/[641]–297[649] (2007). https://doi.org/10.1080/15421400701231582
22. Y. Yosida, I. Oguro, Variable range hopping conduction in bulk samples composed of single-walled carbon nanotubes. J. Appl. Phys. **86**, 999 (1999). https://doi.org/10.1063/1.370838
23. R. Gaa, J.-P. Salvetat, L. Forro, Pressure dependence of the resistivity of single-wall carbon nanotube ropes. Phys. Rev. B **61**, 7320 (2000). https://doi.org/10.1103/PhysRevB.61.7320
24. M.S. Fuhrer, W. Holmes, P.L. Richards, P. Delaney, S.G. Louie, A. Zettl, Nonlinear transport and localization in single-walled carbon nanotubes. Synth. Met. **103**, 2529–2532 (1999). https://doi.org/10.1016/S0379-6779(98)00305-1
25. L. Matzui, L. Vovchenko, I. Ovsienko, Thermopower of pregraphitic carbons. Mol. Cryst. Liquid Cryst. **340**, 361–366 (2000). https://doi.org/10.1080/10587250008025493
26. G.T. Kim, E.S. Choi, D.C. Kim, D.S. Suh, Y.W. Park, K. Liu, G. Duesberg, S. Roth, Magnetoresistance of an entangled single-wall carbon-nanotube network. Phys. Rev. B **58**, 16064–16069 (1998). https://doi.org/10.1103/PhysRevB.58.16064
27. V.A. Samuilov, J. Galibert, V.K. Ksenevich, V.J. Goldman, M. Rafailovich, J. Sokolov, I.A. Bashmakov, V.A. Dorosinets, Magnetotransport in mesoscopic carbon networks. Physica B **294–295**, 319–323 (2001). https://doi.org/10.1016/S0921-4526(00)00668-2
28. S.V. Demyshev, A.A. Pronin, N.E. Sluchanko, N.A. Samarin, A.G. Lyapin, V.V. Brazhkin, T.D. Varfolomeeva, S.V. Popova, 1D-3D crossover in the hopping conductivity of carbines. Lett. JETP **72**, 547–552 (2000) (in Russian)
29. M.E. Raikh, J. Czingon, F. Koch, W. Schoepe, K. Ploog, Mechanisms of magnetoresistance in variable-range-hopping transport for two-dimensional electron systems. Phys. Rev. B **45**, 6015–6022 (1992). https://doi.org/10.1103/PhysRevB.45.6015
30. S.V. Demyshev, A.D. Bozhko, V.V. Glushkov, E.A. Kataeva, A.G. Lyapin, E.D. Obraztsova, T.V. Ishchenko, N.A. Samarin, Scaling of magnetoresistance of carbon nanomaterials in the region of Mott type hopping conductivity. FTT **50**, 1332–1337 (2008). (in Russian)
31. S.V. Demyshev, A.A. Pronin, Magnetoresistance of carbon nanomaterials. FTT **48**, 1285–1294 (2006). (in Russian)
32. B.L. Altshuler, A.G. Aronov, D.E. Khmelnitzky, Effects of electron-electron collisions with small energy transfers on quantum localization. J. Phys. C Solid State Phys. **15**, 7367 (1982). https://doi.org/10.1088/0022-3719/15/36/018
33. C.W.J. Beenakker, H. van Houten, Boundary scattering and weak localization of electrons in a magnetic field. Phys. Rev. B **38**, 3232 (1988). https://doi.org/10.1103/PhysRevB.38.3232
34. Y.-M. Choi, D.-S. Lee, R. Czerw, P.-W. Chiu, N. Grobert, M. Terrones, M. Reyes-Reyes, H. Terrones, J.-C. Charlier, P.M. Ajayan, S. Roth, D.L. Carroll, Y.-W. Park, Nonlinear behavior in the thermopower of doped carbon nanotubes due to strong, localized states. Nano Lett. **3**, 839–842 (2003). https://doi.org/10.1021/nl034161n
35. F. Xu, J.-L. Zhu, Conductance and thermoelectric power in carbon nanotubes with magnetic impurities. Mesoscale Nanoscale Phys. (2007). arXiv:cond-mat/0703373 [cond-mat.mes-hall]. https://doi.org/10.48550/arXiv.cond-mat/0703373
36. C. Yu, L. Shi, Z. Yao, D. Li, A. Majumdar, Thermal conductance and thermopower of an individual single-wall carbon nanotube. Nano Lett. **5**, 1842–1846 (2005). https://doi.org/10.1021/nl051044e
37. V.W. Scarola, G.D. Mahan, Phonon drag effect in single-walled carbon nanotubes. Phys. Rev. B. **66**, 205405 (2002). https://doi.org/10.1103/PhysRevB.66.205405

Dispersion Kinetics of Thin Double Niobium-Palladium Films Deposited onto Oxide Ceramic Materials and Annealed in Vacuum

T. V. Stetsyuk, O. M. Fesenko, D. B. Shakhnin, and A. I. Gab

Abstract The dispersion kinetics of niobium-palladium thin double films deposited onto leucosapphire, alumina, and zirconia ceramics and annealed in vacuum at temperatures up to 1200°C at different exposure intervals at each temperature (from 5 to 20 min) is studied. The double films consisted of two layers. The first one consists of metallized layer–niobium nanofilms with 150 nm thickness deposited onto oxide surface. The second–palladium layer with 1.5 μm thickness is deposited onto first one and supposed to serve as a solder when the metallized oxide samples are joined together. According to our study results, kinetic curves of the investigated thin double metallic films decomposition have been built, from which the basic process parameters (temperature and exposition time at this temperature) for brazing or pressure welding of ceramic materials can be determined. Using the data obtained, process regimes for joining by brazing and pressure welding were selected. According to these regimes, joints' prototypes with seams up to 2 μm thick were made, which strength up to 150 MPa.

1 Introduction

Joining ceramic materials with each other and with metals is carried out by two main methods:

(1) Brazing with molten metallic solders,

T. V. Stetsyuk (✉)
Frantsevich Institute for Problems of Materials Science of National Academy Sciences of Ukraine, 3, Krzhyzhanovskogo st., Kyiv 03142, Ukraine
e-mail: tvst@ukr.net

O. M. Fesenko
Institute of Physics of National Academy Sciences of Ukraine, 46, Nauky ave., Kyiv 03680, Ukraine

D. B. Shakhnin · A. I. Gab
University «Ukraine», 23 Lvivska st., Kyiv 03115, Ukraine

O. Fesenko and L. Yatsenko (eds.), *Nanostructured Surfaces, Nanocomposites and Nanomaterials, and Their Applications*, Springer Proceedings in Physics 296,
https://doi.org/10.1007/978-3-031-42704-6_6

(2) Solid-phase pressure welding.

Since ceramic materials, particularly oxide ones, are usually poorly wetted by molten metals, metal coatings are often used applied in various ways (electron beam sputtering, magnetron sputtering, chemical deposition, etc.). Adhesive-active metals such as titanium, chromium, and niobium are the most commonly used to make such coatings. Then, thus metallized ceramic parts are brazed in vacuum or an inert medium (argon, helium, etc.) by molten metallic solders on the basis of palladium, tin, silver, copper, nickel, etc. In this case, the thickness of the solder seam is from 50 to 100 μm up to several millimeters. Sometimes, multilayer metal coatings are used, but the thickness of the layers and brazed seams remains high [1–9].

At the same time, there is information that the reduction in the thickness of the brazed seam leads to a significant increase in the strength of the brazed joint [10–12]. Obtaining brazed or welded joints of metallized ceramics is possible through reducing the thickness of both the metallization coating on the ceramics and the brazing layer by itself. This can be achieved if the soldering metal or alloy is also applied in the form of a rather thin film, the thickness of which does not exceed several μm. In addition, a thin soldering seam allows you to obtain precision welded or brazed ceramic and ceramic metal units which can be used in microelectronics, radio engineering, microwave engineering, etc.

This objective can be achieved by application onto the ceramic surfaces of double metal films, one of which is 100–200 nm thick and consists of an adhesion-active metal such as Ti and Cr, and the other is slightly thicker (2–3 μm) and serves as a solder, e.g., Pd, Cu, Ag, etc., which will ensure joining of metallized ceramic materials during brazing or welding with fine (2–4 μm thick) solder seam.

The structure of such two- or multiphase coatings, determination of the optimal ratio of the thickness of each layer, processes of interaction at the phases interface, and behavior of these double films during annealing of them, is also an important area of research.

The study of dispersion kinetics two-layer niobium-nickel coatings (films) on oxide materials during annealing in vacuum and the creation of brazed and welded oxide ceramics joints based on them with super-thin brazed seam, the thickness of which does not exceed 5 μm, is the main task of the present work.

2 Materials and Experimental Methods

In this paper an electron beam method for sputtering of metal (Nb, Pd) thin films was used.

The thickness of the deposited films was measured by two methods:

(1) The metal of the given weight (calculated for films of the required thickness) is completely evaporated, then the thickness of the film can be easily calculated according to the law of Lambert [13, 14].

(2) With the help of a special quartz sensor located in the vacuum sputtering chamber near the sample on which the film is deposited.

Solid non-metallic substrates were made of sapphire, alumina, and zirconia ceramics as small thin plates 4 × 3 × 2 mm in size. One of the flat surfaces of each specimen was well polished to a roughness $R_z = 0.03 \div 0.05$ μm. After polishing, all specimens were thoroughly defatted and burned in air at 1100 °C for one hour.

As a metal deposited directly onto non-metallic surfaces, niobium was used. Metallized samples were then coated by nickel films serving as a solder when joining the samples.

The quality of all applied thin films was controlled using a XJL-17 metallographic microscope.

The specimens with deposited onto them metal films were annealed in a vacuum chamber for various periods of time (from 5 up to 20 min) and at different temperatures (from1000 °C up to 1100 °C) in the vacuum not less than 2×10^{-3} Pa.

Annealed specimens were investigated using SEM and ASM microscopy with microphotographs storing. Using these microphotographs, the areas of metal islets on the surface of non-metallic samples were determined by the planimetric method [10]. The experimental data obtained were processed in the form of graphs showing the dependence of the surface area of the samples covered with metal thin films on the annealing parameters (temperature, time).

3 Results and Discussion

Double niobium-palladium films consisted of the niobium metallization layer 150 nm thick, deposited by electron beam sputtering onto the polished surfaces of substrates made of alumina, leucosapphire, and zirconium dioxide ceramics, on which the palladium layer 1.5 μm thick was also deposited by electron beam sputtering.

All initial films were continuous and remained the same during annealing at temperatures no higher than 1000 °C. During short-time annealing (5 min) at 1000 °C, the films also remained almost unchanged, with the exception of the film at leucosapphire surface, in which minor defects appeared which almost did not violate the film integrity. With the increase of the exposure time at this temperature up to 10 min, the number of defects in the films at all oxides surfaces increased, the films began to swell and even to crack. After a twenty-minute exposure at 1000 °C, the number of defects at all oxides surfaces increased even more, and the film at the leucosapphire surface cracked noticeably (Fig. 1).

When the annealing temperature was increased up to 1100 °C, the film underwent noticeable changes already after five-minute exposure (Fig. 2). This effect is very clearly visible at images of niobium-palladium double film at zirconium dioxide ceramics surface made at high magnification degree with an atomic force microscope (Figs. 2, c; and 3). With increase of the exposure time up to 20 min, the number of defects (swelling, ruptures, cracks) in double niobium-palladium films at

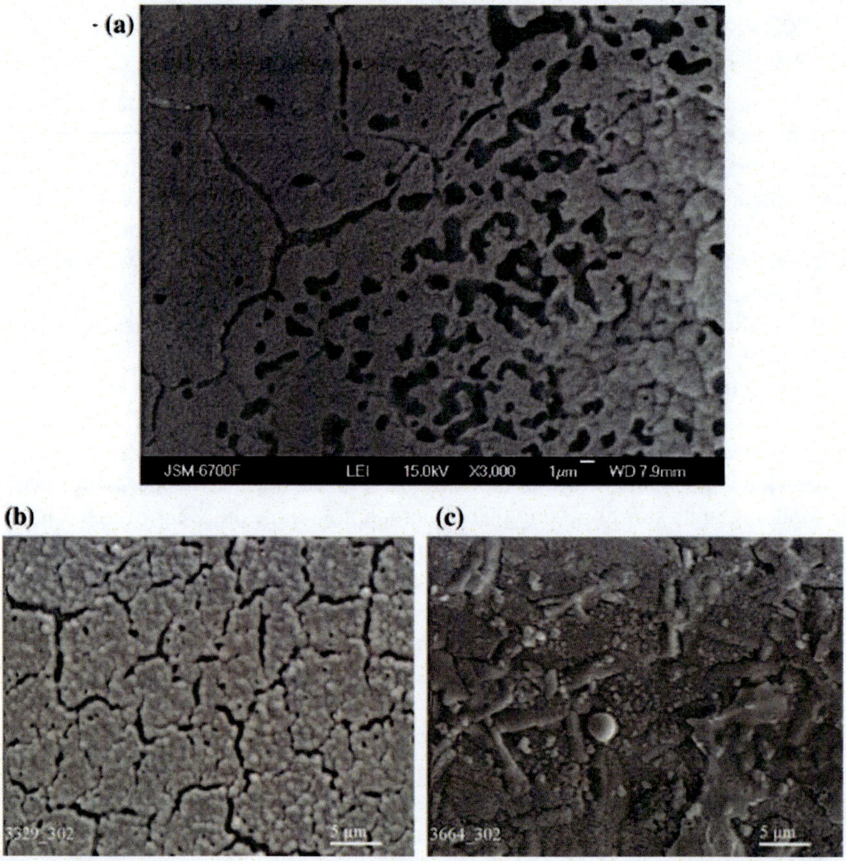

Fig. 1 SEM image of double niobium-palladium film deposited onto oxide materials and further annealed at 1000 °C during 20 min in vacuum, ×3000: **a** alumina ceramics; **b** leucosapphire; **c** zirconia ceramics

all oxides surfaces increased (Fig. 4) which is very clearly visible at the image of double niobium-palladium film at the alumina ceramics surface (Fig. 4a; 5) made at high magnification with the atomic force microscope, where you can see significant number of swellings and tears (Fig. 5). The maximum swelling height reaches 400 nm.

Finally, annealing at 1200 °C accelerated sharply the film dispersion, which started to disintegrate visibly after ten-minute exposure; twenty-minute annealing led not only to a significant films disintegration, but also to the interaction of palladium with niobium (Fig. 6). Nevertheless, the films remains have still covered 70–80% of the substrate surface area (Fig. 7), which made it possible to use double niobium-palladium films for joining oxide ceramics at temperatures up to 1200 °C.

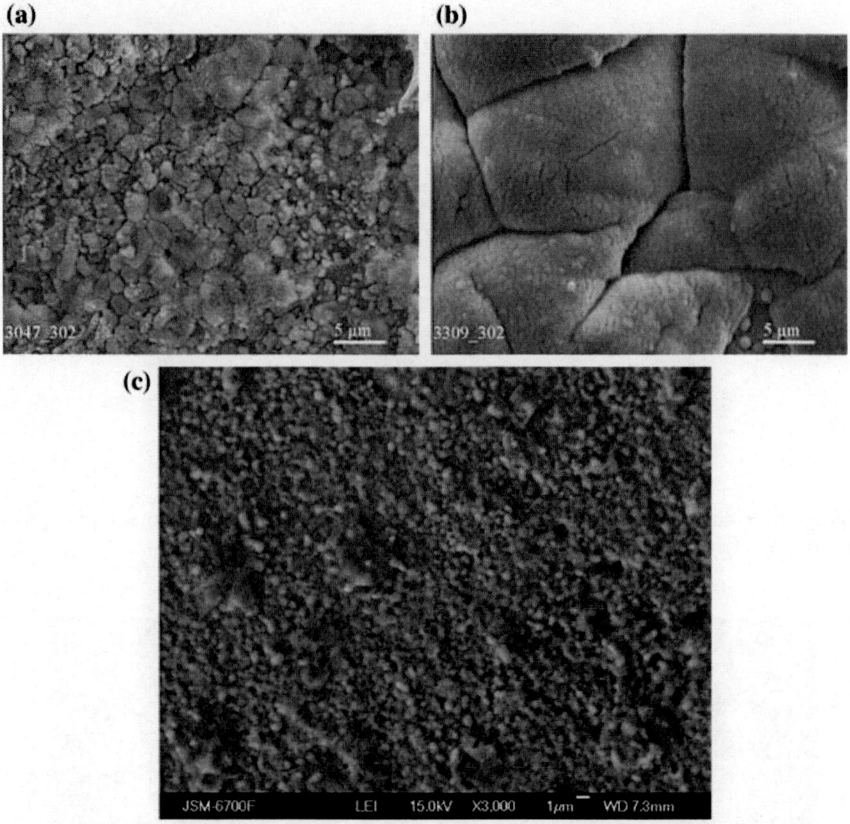

Fig. 2 SEM image of double niobium-palladium film deposited onto oxide materials and further annealed at 1100 °C during 5 min in vacuum, ×3000: **a** alumina ceramics; **b** leucosapphire; **c** zirconia ceramics

In Fig. 7, the dispersion kinetic curves are shown of the palladium films at the oxides surfaces metallized with niobium nanofilms and annealed in vacuum at temperatures 1000–1200 °C during different exposure times.

With the double niobium-palladium film use, the alumina ceramics joints were obtained by brazing at the temperature 1200 °C under load up to 2 MPa to remove solder residues from the brazing gap (Fig. 8). The brazed joint thickness was about 1.5 μm, and its shear strength reached 150 MPa.

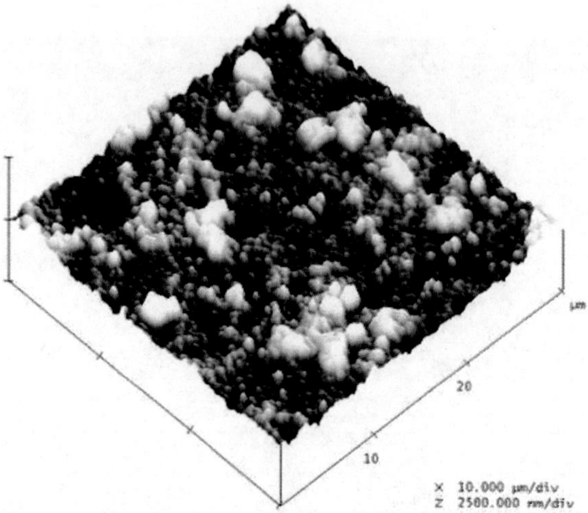

Fig. 3 Three-dimensional AFM image thin double niobium-palladium film deposited onto zirconia ceramics and further annealed at 1100 °C during 5 min in vacuum

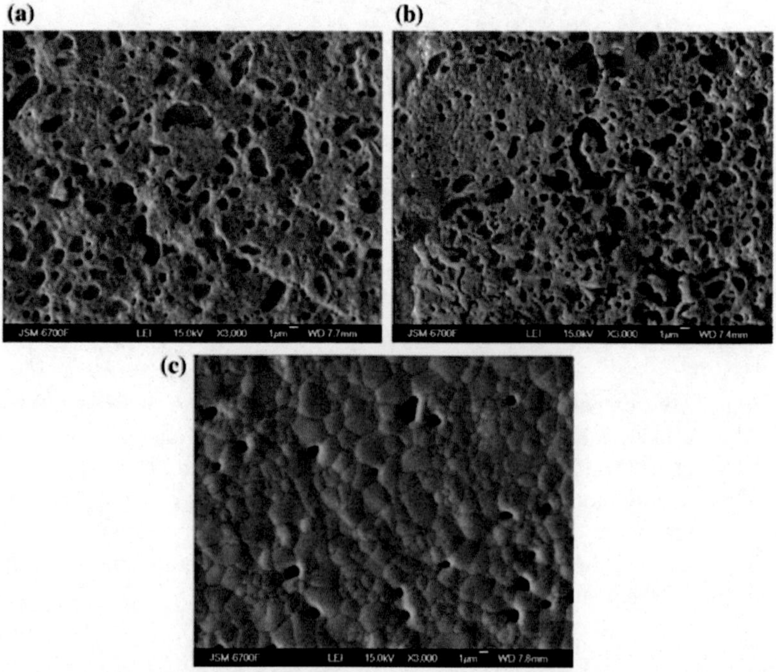

Fig. 4 SEM image of double niobium-palladium film deposited onto oxide materials and further annealed at 1100 °C during 20 min in vacuum, ×3000: **a** alumina ceramics; **b** leucosapphire; **c** zirconia ceramics

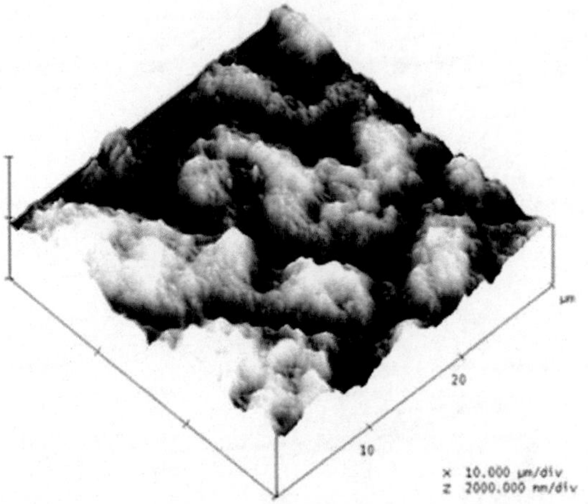

Fig. 5 Three-dimensional AFM image thin double niobium-palladium film deposited onto alumina ceramics and further annealed at 1100 °C during 20 min in vacuum

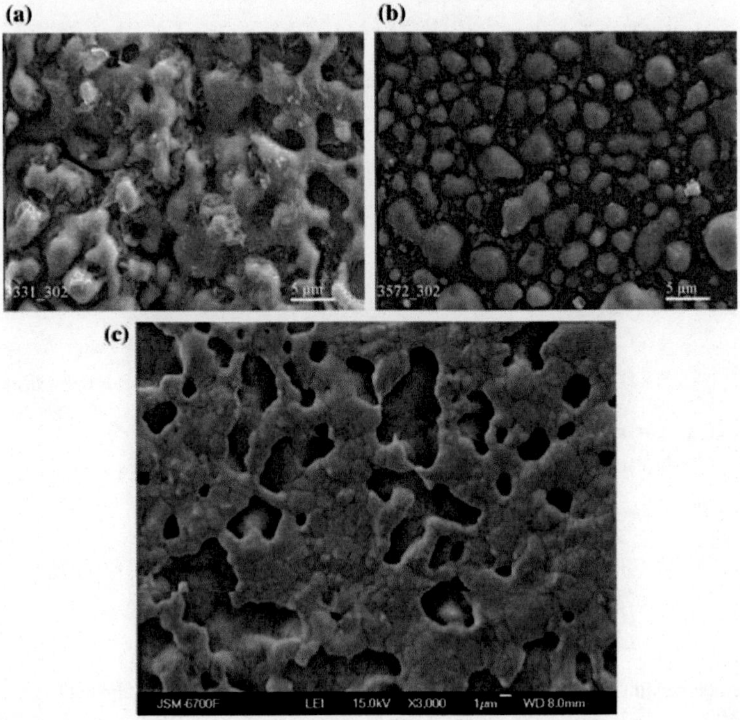

Fig. 6 SEM image of double niobium-palladium film deposited onto oxide materials and further annealed at 1200 °C during 20 min in vacuum, ×3000: **a** alumina ceramics; **b** leucosapphire; **c** zirconia ceramics

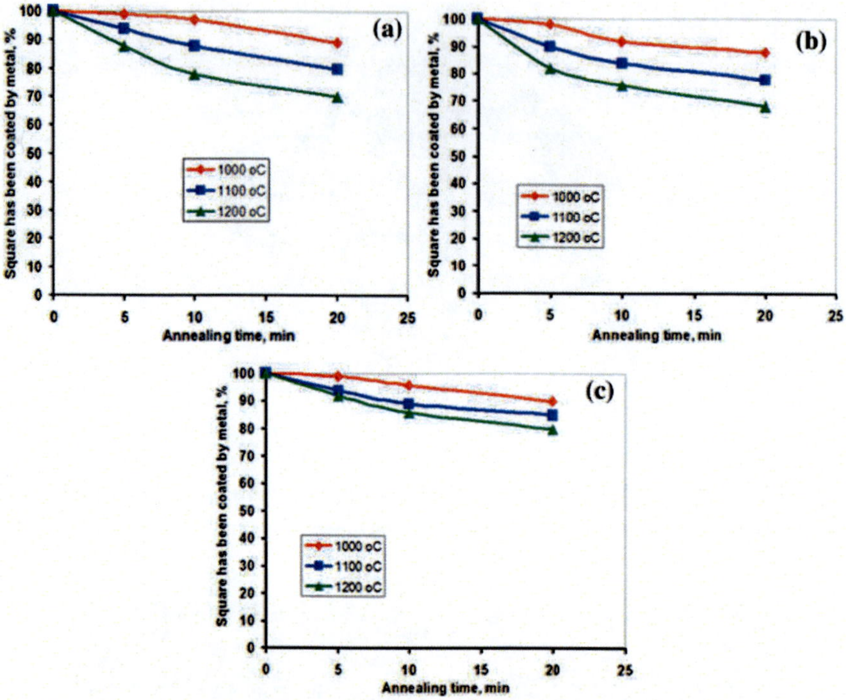

Fig. 7 Dependence of oxide materials area covered by double niobium-palladium film on annealing time at various temperatures (1000–1200 °C): **a** alumina ceramics; **b** leucosapphire; **c** zirconia ceramics

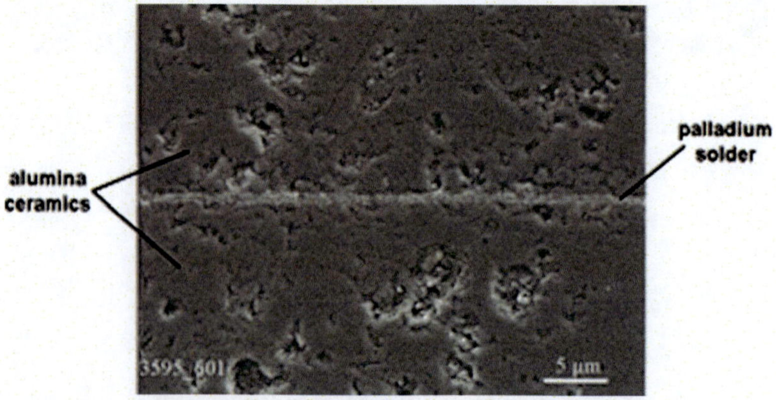

Fig. 8 Soldered joint of alumina ceramics through double niobium-palladium film with an ultrathin seam, ×600

4 Conclusions

The decay kinetics of double niobium-palladium films deposited onto leucosapphire, alumina, and zirconia ceramics during vacuum annealing was investigated. Annealing was performed in vacuum not worse than 2×10^{-3} Pa at temperatures up to 1200 °C with different exposure times (from 5 up to 20 min) at each temperature. The behavior of films at all three oxides surfaces during annealing was similar. It was established that the first minor changes in the films appear already after a ten-minute annealing at 1000 °C. A further increase in the annealing temperature intensifies the process of changing the films morphology, especially when the annealing time is increased. In particular, after a twenty-minute exposure at 1000 °C, signs of dispersion are already visible (the number of films defects at all oxides surfaces increases significantly). At 1100 °C, the films disperse significantly even after a five-minute exposure. Annealing at 1200 °C sharply accelerates the films dispersion, which starts to disintegrate visibly after a ten-minute exposure; twenty-minute annealing leads to significant film disintegration, as well as to the interaction of palladium with niobium.

According to the research results, the kinetic curves of the films dispersion were constructed, using which it is possible to choose the optimal parameters (temperature and time) for the processes of joining metallized ceramics. Based on these curves, the joining parameters of metallized alumina ceramics to each other by brazing were selected, according to which the joints themselves were made with an ultrathin seam (about 1.5 μm thick), which shear strength reached 150 MPa.

References

1. V.R. Yevdokimov, S.L. Kashtanov, L.N. Lado, S.N. Shubin, Diffuzionnaya svarka okhvatyvayushchego soyedineniya alyumooksidnoy keramiki s korrozionnostoykoy stal'yu. J Svarochnoye proizvodstvo **8**, 2 (1995). (in Russian)
2. Request 60-239373, C04B 37/02, B23K 20/00 (28 Nov 1985)
3. Request 380163, C04B 37/02, B23K 3/00 (4 Apr 1991)
4. N. Yu, High strength ceramics brazed joints: scientific and technological bases. J. Ind. Ceram. **19**(3), 162 (1999)
5. A.A. Andreyev, G.I. Kostyuk, N.A. Minayev, Effektivnyye nanostrukturnyye mnogosloynyye pokrytiya dlya rezhushchikh instrumentov i detaley meditsinskogo naznacheniya, rabotayushchikh v dinamicheskom rezhime. J Aviatsionno-kosmicheskaya tekhnika i tekhnologiya **2**(89), 28 (2012). (in Russian)
6. S.V. Zaytsev, Yu.V. Gerasimenko, M.V. Lobanov, A.M. Khoviv, Issledovaniye morfologii poverkhnosti oksidiro-vannykh plenok sistem Ti-Nb. J Kondensirovannyye sredy i mezhfaznyye granitsy **16**(2), 153 (2014). (in Russian)
7. T.A. Lobanova, A.O. Volkhonskiy, I.V. Blinkov, J Innovatsionnyye tekhnologii polucheniya iznoso-toykikh i triboadaptiruyemykh pokrytiy v industrii nanosistem **1**(10), 76 (2013). (in Russian)
8. D.V. Velikodnyy, C.I. Protsenko, I.Ye. Protsenko, Tenzoeffekt v dvukhsloynykh plenkakh Cu/Cr i Fe/Cr . J Fizicheskaya inzheneriya poverkhnosti **6**(1–2), 37 (2008). (in Russian)
9. M. Naveed, A. Obrosov, S. Weib, Investigation of the wear resistance properties of Cr/CrN multilayer coatings against sand erosion. Hindawi Publ. Corp. Conf. Papers in Sci. Volume. Article ID 873543 (2015). https://doi.org/10.1155/2015/873543

10. Yu.V. Naydich, I.I. Gab, B.D. Kostyuk, T.V. Stetsyuk, D.I. Kurkova, S.V. Dukarov, Investigation of the ceramic materials connection processes (soldering) using metal nanofilms. J. Rep. Natl. Acad. Sci. Ukr. **35**, 97 (2007). (in Ukranian)
11. Yu.V. Naydich, I.I. Gab, B.D. Kostyuk, T.V. Stetsyuk, D.I. Kurkova, S.V. Dukarov, Issledovaniye struktury, morfologii i smachiva-niya metallami metallicheskikh nanoplenok, nanesennykh na po-verkhnost' oksida alyuminiya s tsel'yu polucheniya svarnykh i paya-nykh soyedineniy materialov na yego osnove. J Tekhnika mashinostroyeniya **1**, 28 (2006). (in Russian)
12. V.M. Yakovich, V.M. Komarovskaya, Mnogosloynyye pokrytiya. Inzhenerno-pedagogicheskoye obrazovaniye v XXI veke: materialy XII Respublikanskoy nauchno-prakticheskoy konferentsii molodykh uchenykh i studentov BNTU, Minsk (2016), pp 215–218. (in Russian)
13. S. Metfessel, Tonkie plenki, sh szgotovlenie I izmerenie. Gosenergoizdat, Moscow-Liningrad (1963). (in Russian)
14. G. Khaas, P.E. Tun, *Thin Film Physics* (Mir, Moskow, 1968). (in Russian)

The Ellipsometry of Chromium–Glass, Nickel–Glass, and Nickel–Chromium–Glass Nanosystems Below and Higher the Néel Temperature

Yevgen Oberemok, Andrii S. Sizhuk, Xiaohong Chen, Zhenjie Zhao, Zhuo Sun, Sergey Savenkov, Vladimir Malyshev, and Oleksandr Prokopenko

Abstract In this work, an investigation of the optical properties for the nanoribbons of nickel on chromium nanofilm and the set of chromium nanofilms is proposed in terms of Mueller ellipsometry. It was observed that if a nickel nanoribbon with a thickness of 250 nm is produced at 250 °C on a chromium film with a thickness of 250 nm, then the experimentally observed state (internal structure) of the nickel nanoribbons is different for the nanosystems, such as nickel nanoribbon-chromium nanofilm and nickel nanoribbon-glass substrate. Despite the opacity of the nickel nanoribbons with the given thicknesses for the used laser light source, the optical properties of the outer surface are thus determined by the interface between the nanoribbon and the type of material of the next layer. For the investigated multilayer nanosystem of the nickel nanoribbons on chromium nanofilm, the global extremum (in the sense of the quadratic dependence indicated above) of the value of the ellipsometric angle is 74.5° for the incident light beam that coincides with the system of chromium nanofilm on glass substrate. At the higher temperature than the Néel temperature for a macroscopic bulk of chromium, the observed abrupt change in the ellipsometric parameters of the nickel nanoribbons on the chromium film is the pretendant to be discussed as the confirmation of the phase transition in the ordering of the internal structure of system with the ferromagnetic–antiferromagnetic interface.

Y. Oberemok · S. Savenkov
Department of Radiophysics, Kyiv National Taras Shevchenko University, Acad. Glushkova Avenue 4-g, Kyiv 03022, Ukraine

A. S. Sizhuk (✉) · V. Malyshev · O. Prokopenko
Institute of High Technologies, Kyiv National Taras Shevchenko University, Acad. Glushkova Avenue 4-g, Kyiv 03022, Ukraine
e-mail: andrii.sizhuk@gmail.com

X. Chen · Z. Zhao · Z. Sun
Physics Department, East China Normal University, Shanghai 200062, China

O. Fesenko and L. Yatsenko (eds.), *Nanostructured Surfaces, Nanocomposites and Nanomaterials, and Their Applications*, Springer Proceedings in Physics 296,
https://doi.org/10.1007/978-3-031-42704-6_7

1 Introduction

Multicomponent nanofilms, combining ferromagnetic and antiferromagnetic materials, are promising for controlling the specified magnetic and optical properties. For the comparative analysis and construction of the theoretical model, it is important to study the precision antiferromagnetic and ferromagnetic films. In this work, an investigation of the optical properties for the nanoribbons of nickel on chromium nanofilm and the set of chromium nanofilms is proposed. Despite the relatively strong discrete absorption activity of chromium in the optical range, we tested the ellipsometry technique.

Thus, one of the relevant directions of research is the determination of the magnetic properties of nanosamples in a high-frequency electromagnetic field (microwave, infrared, and optical ranges), including the possibility of exciting spin waves by the method of optical pumping.

It is known that a pulsed laser beam in the optical range can excite the propagation of a surface disturbance on a metal surface (for example, plasmon-polaritons, magnons). At the same time, as recent studies show, the spatial configuration (roughness) of the nanoscale can play a determinative role in the efficiency of the interaction between the optical field and the surface perturbation(s) (see, for example, [1–4]). The latter is actual for the systems, in which the generation of spin waves is one of the main effects or mechanisms of scattering and dissipation of electromagnetic energy (see, for example, [5–8]).

In this work, the reflective properties of a set of nickel nanoribbons, deposited on a glass surface and a chromium nanofilm, are described in terms of polarimetry. Experimentally determined ellipsometric parameters make it possible to determine the complex index of refraction of these thin films, in particular Cr–glass substrate, Ni–glass substrate, and Ni–Cr–glass substrate. If quantum transitions in films are not taken into account during light reflection/absorption, then we can say that the complex refractive index depends on the high-frequency conductivity of the multilayer system. At the same time, the considered systems are structures with interface planes (transitions) of the ferromagnet-glass and ferromagnet–antiferromagnet-glass types. It is demonstrated that a 250-nm-thick nickel nanoribbon, sputtered at a temperature of 250 °C on a 250 nm thick chromium film, has different states (internal structures depending on the interface) for such nanosystems as nickel nanoribbon-chromium nanofilm and nickel nanoribbon-glass substrate.

Since our investigation is conducted near the normal conditions, the temperature dependence of the properties of the antiferromagnetic–ferromagnetic interface is also important, since a chromium sample of macroscopic dimensions has its Néel temperature of about 35 °C. At the same time, there is a problem with the application of the concept of temperature for a nanoscale flat system: a relatively rapid change in optical characteristics, when passing through the Néel temperature for chromium, was not observed in the both cases of chromium–glass and nickel–chromium interfaces. Because the concept of temperature can be applied only to macroscopic objects (thermodynamic systems), the explanation of the observed phenomena can be of a fundamental problem. Strictly speaking, by definition, the measurement procedure of temperature establishes a thermal balance with another macroscopic body. So, in our opinion, there is an important fundamental problem of correctly formulating the state of the nanofilm. Introducing the concept of an equilibrium state, defined by such thermodynamic parameters, as pressure, temperature, and entropy, is not possible for a relatively small, low-dimensional microscopic system, where the predominant interaction is the attraction to the substrate. It is also easy to argue that it is impossible to arbitrarily divide such a low-dimensional (in the space of configurations for the geometry of nanofilms) microscopic system into subsystems without changing the conditions of local dynamic equilibrium with the third macroscopic system (substrate or a thermometer).

Because the optical properties of such nanosystems can be controlled by the temperature of a substrate, there is a potential for applications in programming hardware memory.

1.1 General References

In general, ellipsometry is considered to be an experimental technique that characterizes the composition, roughness, thickness, degree of crystallization, concentration of impurities, electrical conductivity (high-frequency), and other properties of materials (see, for example, [9–13]). Mainly, the data presented here are applicable to the study of the complex refractive index (refractive index) for the obtained chromium films (see previously obtained data and their application in [14, 15]).

The elements of the Mueller matrix M in this work are obtained for four initial Stokes vectors $S_{\rm in}^{(0)}$, $S_{\rm in}^{(1)}$, $S_{\rm in}^{(2)}$, $S_{\rm in}^{(3)}$, as a solution of the following problem:

$$S_{\rm out}^{(k)} = M\, S_{\rm in}^{(k)}, \tag{1}$$

where $S_{\rm in}^{(k)} = \left(S_{in0}^{(k)}, S_{in1}^{(k)}, S_{in2}^{(k)}, S_{in3}^{(k)}\right)^T$ are the initial Stokes vectors; $M = \left(M_{ij}\right)$, with $(i, j) = (0\ldots3, 0\ldots3)$, is the Mueller matrix. Note, in the nonstandard notation of the text below, the indexes $i,\ j$ for the elements of the Mueller matrix run over digits $(1\ldots4, 1\ldots4)$.

The initial Stoke vectors $S_{\rm in}^{(k)}$ are prepared in the following way. After passing the linear polarizer (denoted by LP in Fig. 1, see below in the text), the initial Stokes vector becomes defined by the following set of numbers (in the case of normalized calculus):

$$S^{II} = (1, -1, 0, 0)^T\,. \tag{2}$$

Then, passing through the phase plate (denoted PP) in the polarization generator GP (see the scheme in Fig. 1) the state of the light beam becomes defined as

$$S_{\text{in}} = \begin{pmatrix} 1 \\ \cos^2(2\alpha) + \cos(\Delta)\sin^2(2\alpha) \\ -\cos(2\alpha)\sin(2\alpha)(\cos(\Delta) - 1) \\ \sin(2\alpha)\sin(\Delta) \end{pmatrix}. \tag{3}$$

The sequence of angles for the relative orientation α of the phase plate PP at the fixed phase shift Δ in the polarization generator PG (see Fig. 1) are set to be:

$$\{-52°, -14°, 14°, 51°\}. \tag{4}$$

The corresponding initial Stokes vectors therefore are

$$\begin{aligned} S_{\text{in}}^{(0)} &= (1, 0.768, -0.422, -0.482)^T; \\ S_{\text{in}}^{(1)} &= (1, 0.768, 0.422, 0.482)^T; \\ S_{\text{in}}^{(2)} &= (1, 0.056, 0.230, -0.972)^T; \\ S_{\text{in}}^{(3)} &= (1, 0.056, -0.230, 0.972)^T. \end{aligned} \tag{5}$$

As we noted above, in addition to such surface properties as roughness and crystalline structure, the polarimetry can help to determine the high-frequency electrical resistance (complex impedance) if the quantum nature of absorption (or refraction) is not taken into account (see, for example, [9–17]). In particular, nanoribbons, deposited on chrome surfaces, have many optical properties, which are determined by relatively very optically active chromium. This is also confirmed by experimental data that will be discussed below. To cause elliptic polarization to a linear polarized beam is one of the known important features of metal surfaces. And, the processes of reflection can be nonlinear in intensity, that restricts the use of the essentially linear theory for the Fresnel equations. Nevertheless, the method of the Mueller matrix allows to analyze the contributions to the linear and circular polarizations as the constituents to the state of the elliptically polarized scattered light beam. The presented below ellipsometric angles Ψ and Δ determine the complex refractive index of chromium and nickel thin films (see previously obtained data and their application in [16–18]). The complex refractive index determines the complex dielectric constant. In turn, the imaginary part of the complex dielectric constant is determined by the conductivity of the light-reflecting material. A detailed study of the frequency dependence of the dielectric permittivity makes it possible to formulate a model of the internal structure of the nanosystem (see, for example, [19]). Here, we propose to focus attention on the properties of the system at a given frequency of the optical beam to determine the effect of the antiferromagnet–ferromagnet interface on the optical (impedance) characteristics of the nanosurface.

1.2 The Method of Investigation

The Mueller matrix elements for a set of chromium nanofilms at room temperature are determined by the ellipsometry of the incident and reflected beam (see Fig. 1) for the given four initial polarization states, S_{in}. For a comparison, chromium nanofilms with an average thickness of 132, 151, and 183 nm, obtained by thermal sputtering on a glass substrate, are scanned. The normalized elements of the Mueller matrix as the function of the angle of incidence are shown in the corresponding graphs. The used light source is a He–Ne laser with the wavelength of 633 nm and the power of approximately 15 mW.

The accuracy of the experimental measurements is estimated to be within 2.6%.

There is the schematic illustration of a Muller polarimeter in Fig. 1. The Muller polarimeter consists of a light source S, a polarization generator PG, a phase plate PP, a linear polarizer P, and a photodetector PD.

The sketches of the discovered here samples with the corresponding scanning points are shown in Fig. 2.

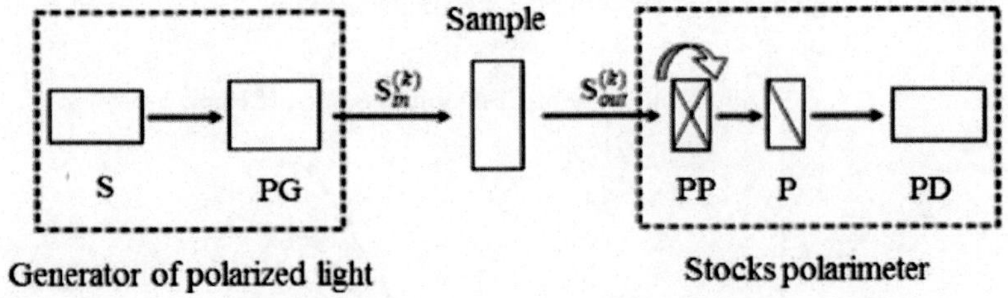

Fig. 1 Schematic illustration of a Muller polarimeter with a light source S, a polarization generator PG, a phase plate PP, a linear polarizer P, and a photodetector PD

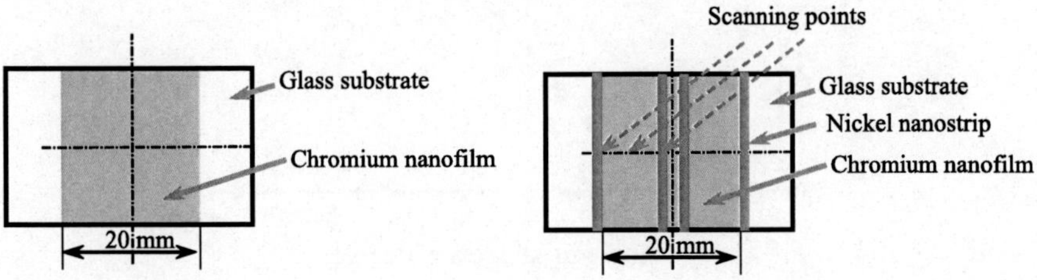

Fig. 2 Locations of the He–Ne laser beam scattering at the topographical sample with the deposited nickel nanostrips

2 The Results of the Experimental Observations and Calculations

The normalized elements M_{12} and M_{43} of the Mueller matrix as the function of the angle of incidence on the chromium nanofilm (for an estimated film thickness of 151 nm) are presented in Fig. 3.

The normalized M_{21} and M_{31} elements of the Mueller matrix as the function of the angle of incidence at the chromium nanofilm (corresponding to an estimated thickness of 132 nm) are shown in Fig. 4.

The corresponding M_{21} and M_{31} function curves for the 151 and 183 nm samples are indistinguishable at this resolution and repeat the dependence contour for the 132 nm thickness.

The normalized elements M_{24} and M_{34} are depicted in Fig. 5.

At the given resolution, the dependences on the angle of incidence are identical for the given set of the samples of different thicknesses.

The normalized elements M_{33} and M_{44} of the Mueller matrix for the thicknesses of 132, 151, and 183 nm are shown in Fig. 6.

The curves for the different thicknesses are indiscernible.

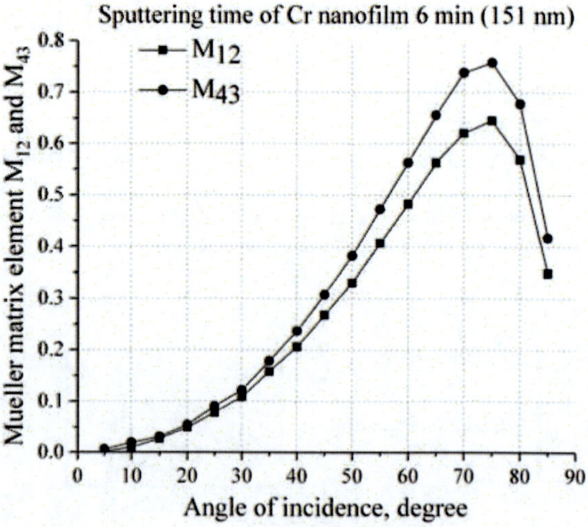

Fig. 3 Normalized elements M12 and M43 of the Mueller matrix as the function of the angle of incidence at the chromium nanofilm. The curves for the samples with the thickness of 132 and 183 nm are visually identical at this resolution and follow the 151 nm curve presented. The data were checked for the same samples through one year, see [20]

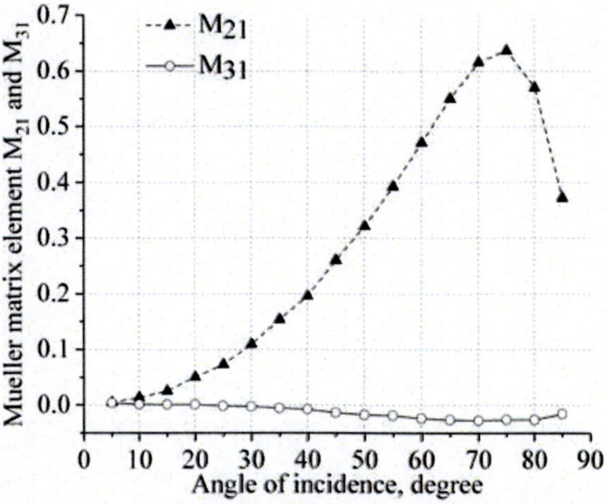

Fig. 4 Normalized elements of the Mueller matrix M_{21} and M_{31} as the functions of the angle of incidence on the chromium nanofilm (corresponding to an estimated thickness of 132 nm). The data were checked for the same samples through one year, see [20]

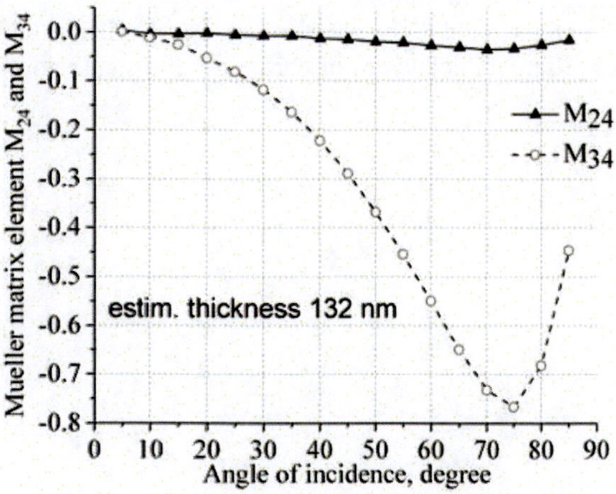

Fig. 5 Normalized elements M_24 and M_34. The data were checked for the same samples through one year, see [20]

2.1 *The Investigation of the Sample with the Deposited on the Chromium Film Nanoribbons of Nickel at Temperatures Below the Néel point for a Bulk Chromium*

The results of the polarimetry analysis for the metal thin films in the mirror geometry of the experimental setup Fig. 1 are shown in Figs. 7 and 8.

Here, the samples of nickel–chromium nanofilms on a glass substrate with a length of 20 mm, a width of 2 mm, and a thickness of approximately 250 nm were studied at room temperature (below 25 °C).

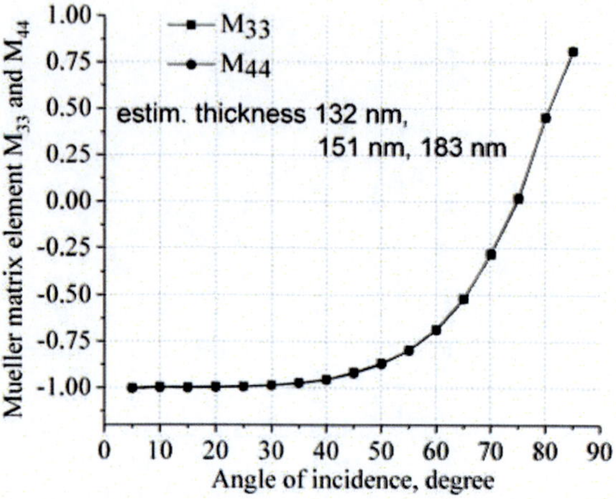

Fig. 6 Normalized elements M_{33} and M_{44}. The data are taken from [20]

As it can be seen from the graphs in Fig. 7, the angle of incidence $74.5° \pm 0.25°$ corresponds to the location of the global extremum of the dependencies for the three types of the nanofilms.

In Figs. 7 and 8, the polarization characteristics (dichroic angles Ψ and phase Δ) for different thicknesses of chromium films coincide within the specified accuracy

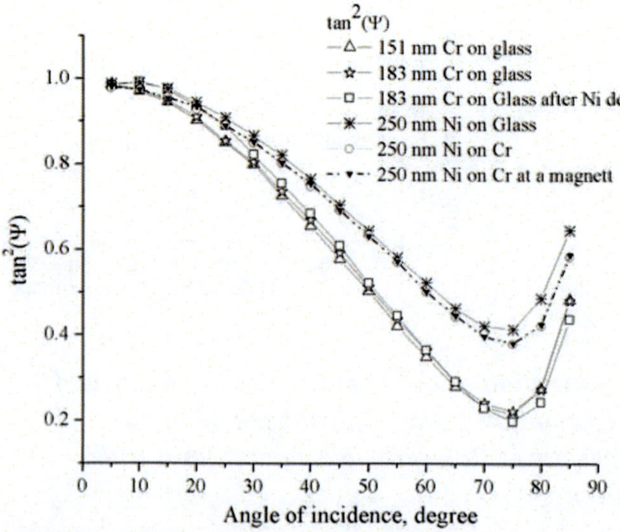

Fig. 7 Squared tangent of the ellipsometric angle Ψ as the function of incidence angle. The data were obtained from the Mueller matrices, experimentally measured for chromium nanofilms on glass, nickel on glass, and nickel on chromium nanofilm on glass. The data are from the work of authors [21]

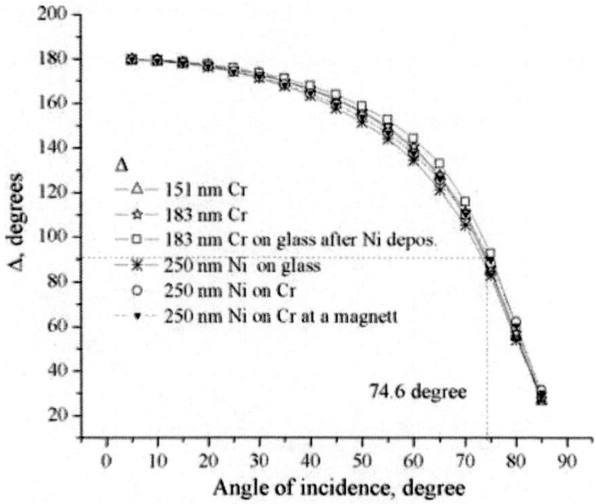

Fig. 8 Ellipsometric angle Δ as the function of incidence angle. The data were obtained from the Mueller matrices experimentally measured for chromium nanofilms on glass, nickel on glass, and nickel on chromium nanofilm on glass. The data are from the work of authors [21]

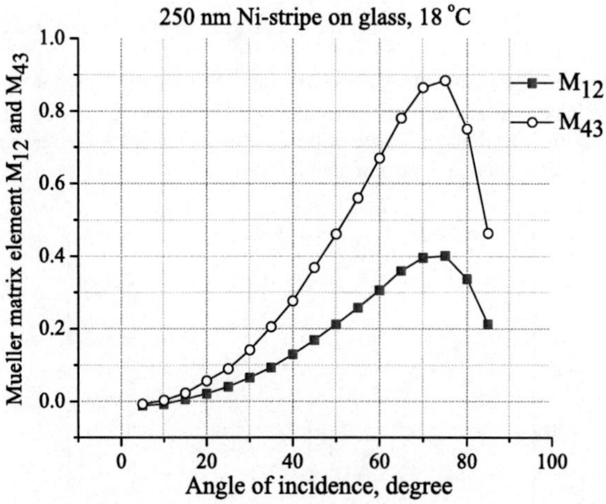

Fig. 9 Normalized (to the M11 element) element M_{12} and M_{43} of the Mueller matrix as the function of the angle of incidence at the nickel nanoribbon. The the nickel nanoribbon was thermally sputtered onto a glass plate (glass temperature during sputtering was approximately 250 °C). The thickness of the nanoribbon is approximately 250 nm

(2.6%). When reflected, the dichroic properties of the directly sputtered onto glass nickel nanoribbons differ significantly from those sputtered onto the chromium film on the glass. The difference in the phase parameters, Δ, of the studied samples is also noticeable within the angle of incidence of 30°–75°. But this difference is less pronounced compared to dichroic characteristics.

To compare the behavior of the Mueller matrix elements, characterizing the nickel, chromium, and glass substrate, the corresponding incidence angle functions of M_{12} (and M_{21}, which has a similar dependence), M_{14}, M_{43} are presented below. The matrix elements are relatively sensitive to the type of a substrate surface.

Figure 9 shows the normalized (to M_{11}) elements M_{12} and M_{43} as the function of incidence angle at a nickel nanoribbon. The nickel nanoribbon was thermally sputtered onto a glass plate (glass temperature during sputtering was approximately 250 °C). The thickness of the nanoribbon is approximately 250 nm.

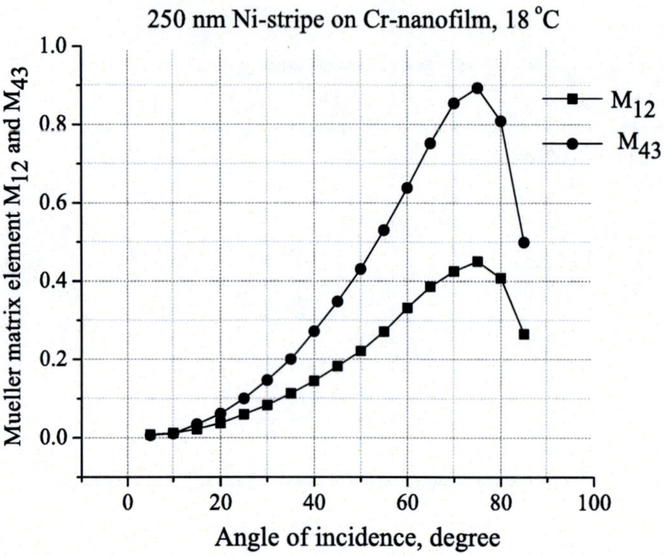

Fig. 10 Normalized element M_{12} and M_{43} of the Mueller matrix as the function of the angle of incidence at the nickel nanoribbon. The nickel layer was thermally sputtered onto chromium nanofilm (with its thickness within 250 nm) on glass at the temperature of 250 °C. The thickness of the nanoribbon is approximately 250 nm

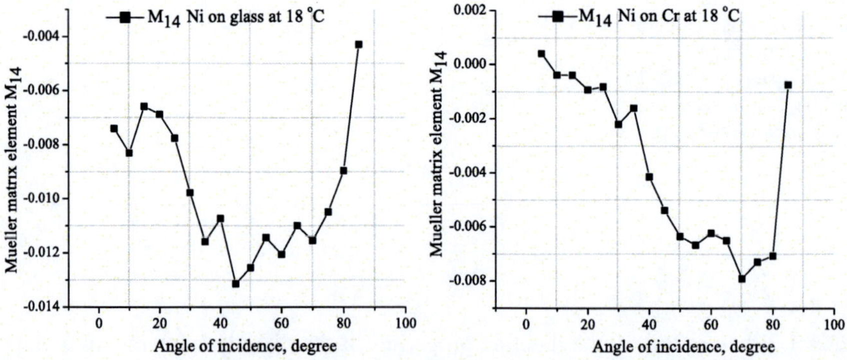

Fig. 11 Normalized element M_{14} (or/and M_{41}) of the Mueller matrix as the function of the angle of incidence at the nickel nanoribbon. The nickel layer was thermally sputtered onto glass or chromium nanofilm (with its thickness within 250 nm) on glass at the temperature of 250 °C. The thickness of the nanoribbon is approximately 250 nm

The graphs in 10 depict the dependence of the normalized element M_{12} and M_{43} of the Mueller matrix from the angle of incidence at the nickel nanoribbon. Here, the nickel layer was thermally deposited onto glass or chromium nanofilm (with a thickness within 250 nm) on glass at the temperature 250 °C. The thickness of the nickel strips is approximately 250 nm. Other Mueller matrix elements for the nanosystem of chrome film-glass substrate are presented by the authors in [20].

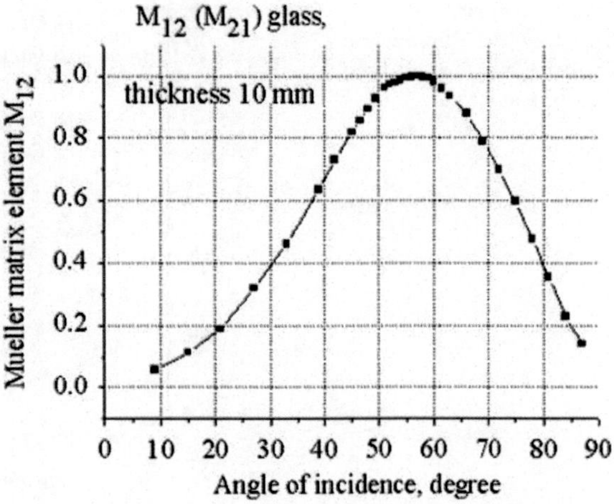

Fig. 12 Normalized element M_{12} (or/and M_{21}) of the Mueller matrix as the function of the incidence angle of a light beam at a glass plate with the thickness of 10 mm. The data are taken from the recent authors' work [21]

In Fig. 11, the normalized (to M_{11}) element M_{14} as the function of incidence angle at a nickel nanoribbon is shown.

The normalized element M_{12} (or/and M_{21}) of the Mueller matrix as the function of the incidence angle of a light beam at a glass plate with the thickness of 10 mm is shown in Fig. 12.

We experimentally investigated that the behavior (form of dependence) of these nanosurfaces of chromium and nickel chemical elements is qualitatively similar. Global extremes are in the range of 73°−75°. Global extrema for the studied materials, mainly for Cr and Ni, are relatively close to the samples with an annealed Au surface (see [21]). This allows us to think about the rather high smoothness of the specified surfaces.

At the same time, the change in the polarization state of the light beam after reflection is more pronounced at the nanosurfaces of ferro(antiferro)magnetic materials. Under certain assumptions, neglecting, for example, the scattering from surface irregularities, the angle of incidence at 74.5° corresponds to Brewster's angle.

The discussed here data are obtained from the experimentation on a sample in air. Thus, the presence of water molecules on the surface of typical chromium or nickel samples causes certain changes in the background reflectance. Some analysis of changes in the refractive index in the presence of a nanolayer of water, depending on temperature regimes, type of substrate, etc., can be found, for example, in [22–25].

2.2 *The Investigation of the Sample with the Deposited on the Chromium Film Nanoribbons of Nickel at Temperatures Higher the Néel Point for a Bulk Chromium*

Here, we describe the observation of a phase transition of the internal structural ordering for the ferromagnetic nickel nanostrip on an antiferromagnetic surface of chromium at a temperature above the Néel temperature for "macroscopic" chromium.

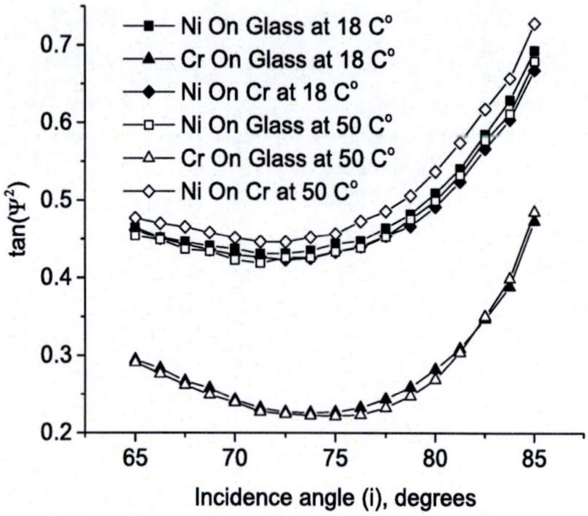

Fig. 13 Angular dependences of ellipsometric parameter Ψ at temperatures of 18 °C and 50 °C

Nickel nanostrips with a thickness of 250 nm and a width of about 2 mm on glass (the side strips in Fig. 2) and on a chrome film on glass (the central two strips in Fig. 2) were produced by the method of thermal sputtering using the VUP-5M system.

At temperatures around 20–40 °C, significant changes in the values of ellipsometric parameters were not observed for this topographic sample, even with the passage of a relatively significant period of time, which is about one year. But, at a temperature of about 50 °C, the behavior of the nickel nanoribbon on chromium changes significantly compared to room temperatures (see Figs. 13 and 14). At the same time, changes in the values of ellipsometric parameters for chromium on glass and nickel on glass are detectable but relatively small.

Moreover, unlike chromium–glass and nickel–glass systems, the nickel–chromium–glass system increases the value of the dichroic angle Ψ with increasing temperature at the extremum. Hence, at the significantly higher temperature than the Néel temperature for a macroscopic bulk of chromium, the observed abrupt change in the ellipsometric parameters is the pretendant to be discussed as the confirmation of the phase transition in the ordering of the internal structure of the ferromagnetic nickel nanostrip on the antiferromagnetic surface of chromium.

Figure 15 shows the normalized (to M_{11}) element M_{12} (or/and M_{21}) and M_{43} as the function of incidence angle at a chromium nanofilm thermally sputtered onto a glass plate. The thickness of the nanofilm is approximately 250 nm.

Figure 16 shows the normalized (to M_{11}) element M_{12} (or/and M_{21}) and M_{43} as the function of incidence angle at a nickel nanoribbon thermally sputtered onto a glass plate and chromium film on glass (glass temperature during sputtering was approximately 250 °C). The thickness of the nanoribbon is approximately 250 nm. The temperature of the glass substrate during the scanning was 50 °C.

Figure 17 shows the normalized (to M_{11}) element M_{14} as the function of incidence angle at a nickel nanoribbon thermally sputtered onto a glass plate and chromium film on glass (glass temperature during sputtering was approximately 250 °C). The thickness of the nanoribbon is approximately 250 nm. The temperature of the glass substrate during the scanning was 50 °C.

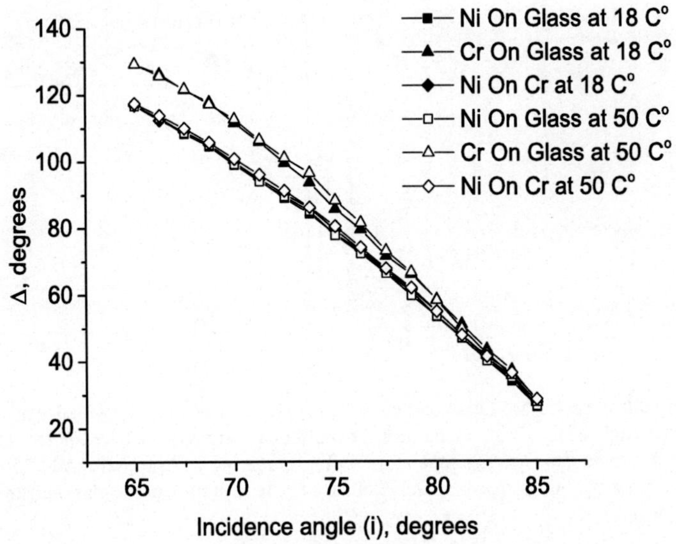

Fig. 14 Angular dependences of ellipsometric parameter Δ at temperatures of 18 °C and 50 °C

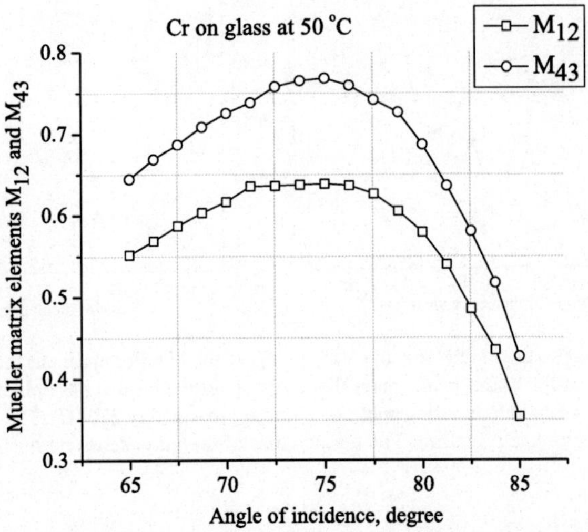

Fig. 15 Normalized (to the M11 element) element M_{12} (or/and M_{21}) of the Mueller matrix as the function of the angle of incidence at a chromium nanofilm thermally sputtered onto a glass plate. The thickness of the nanofilm is approximately 250 nm. The temperature of the glass substrate during the scanning was 50 °C

It is interesting to note that, in comparison with the other elements of the matrix, such term as M_{41}, for the multilayer systems of Ni on Cr and Ni on glass at 50 °C, shows significant difference in values (see the graphs in Fig. 18). The change in sign of the quantity can be referred to the existence of certain critical point. In case of the

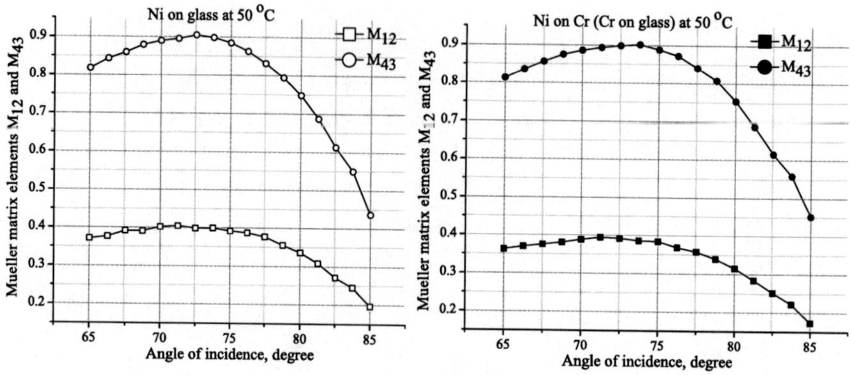

Fig. 16 Normalized (to the M11 element) element M_{12} (or/and M_{21}) of the Mueller matrix as the function of the angle of incidence at the nickel nanoribbon thermally sputtered onto a glass plate and chromium film on glass (glass temperature during sputtering was approximately 250 °C). The thickness of the nanoribbon is approximately 250 nm. The temperature of the glass substrate during the scanning was 50 °C

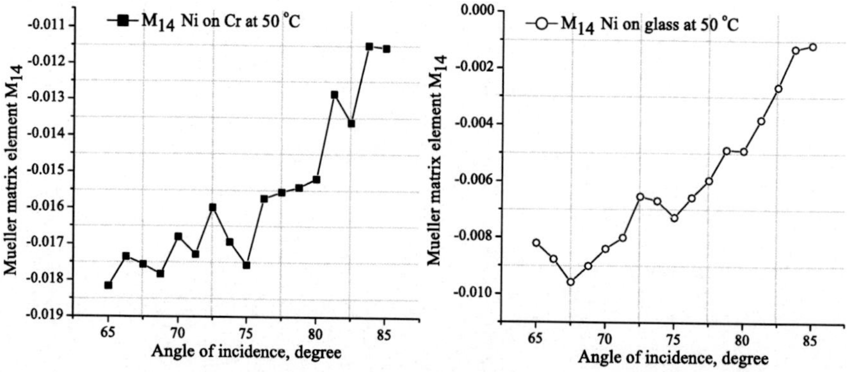

Fig. 17 Normalized (to the M11 element) element M_{14} of the Mueller matrix as the function of the angle of incidence at the nickel nanoribbon thermally sputtered onto a glass plate and chromium film on glass (glass temperature during sputtering was approximately 250 °C). The thickness of the nanoribbon is approximately 250 nm. The temperature of the glass substrate during the scanning was 50 °C

ferromagnetic–antiferromagnetic interface, the latter is supposed to be related to a phase transition in the intrinsic order of the films.

By the multiplication rules for matrix (1), in accordance with the definition, the element M_{41} contributes to the $S^{(k)}_{out3}$ element of the final Stokes vector through the additive term $M_{41} S^{(k)}_{in0}$. The term $S^{(k)}_{out3}$ describes, by the definition of the Stokes vector, the order of the circular polarization of the scattered beam, while the element $S^{(k)}_{in0}$ of the initial Stokes vector is related to the initial intensity of the incoming light beam. Thus, depending on the temperature of the substrate, the contribution to the circular polarization of the scattered beam is

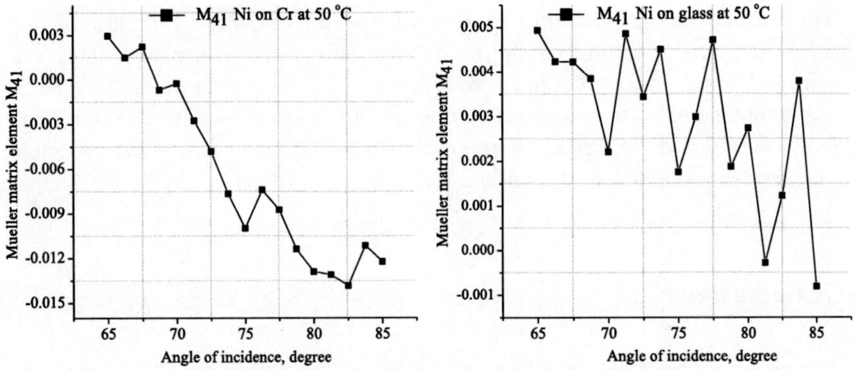

Fig. 18 Normalized (to the M11 element) element M_{41} of the Mueller matrix as the function of the angle of incidence at the nickel nanoribbon thermally sputtered onto a glass plate and chromium film on glass (glass temperature during sputtering was approximately 250 °C). The thickness of the nanoribbon is approximately 250 nm. The temperature of the glass substrate during the scanning was 50 °C

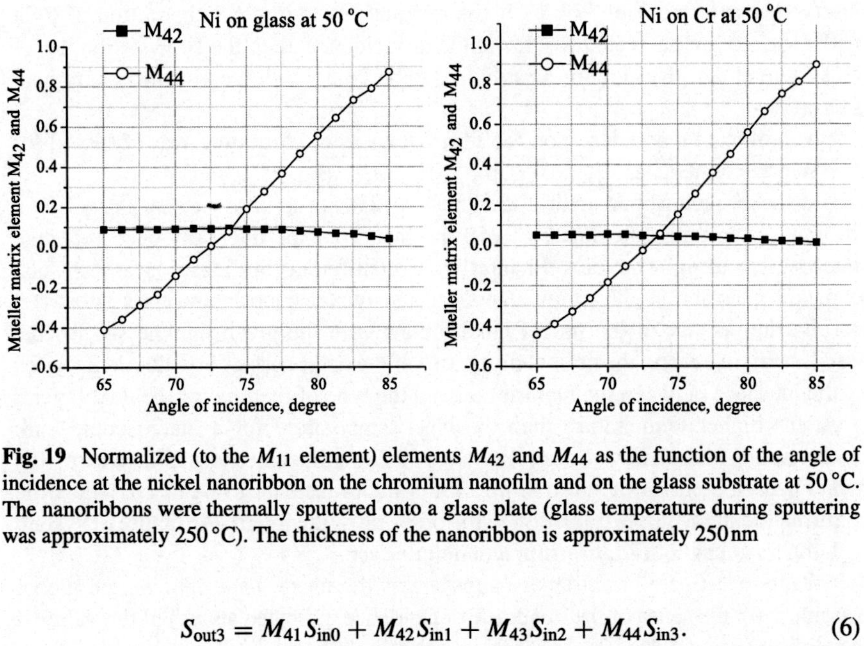

Fig. 19 Normalized (to the M_{11} element) elements M_{42} and M_{44} as the function of the angle of incidence at the nickel nanoribbon on the chromium nanofilm and on the glass substrate at 50 °C. The nanoribbons were thermally sputtered onto a glass plate (glass temperature during sputtering was approximately 250 °C). The thickness of the nanoribbon is approximately 250 nm

$$S_{out3} = M_{41} S_{in0} + M_{42} S_{in1} + M_{43} S_{in2} + M_{44} S_{in3}. \tag{6}$$

The above term S_{out3} can therefore be found using the corresponding Mueller matrix elements M_{41}, M_{42}, M_{43}, M_{44} and the initial polarization state vector (Stokes vector) S_{in}. By analogy, it can be found the rest of the final Stokes vector S_{out}. The rest of the needed for this elements M_{42} and M_{44} for the nickel nanoribbon on the chromium nanofilm on glass substrate at 50 °C are provided in Fig. 19.

The influence of the crystalline substructures of the nanolayers and the substrate on the formation of structured nanofilms of nickel and chromium also takes place (see [8]), as was observed for silicon substrates in [13, 14]. The latter strongly depends on the temperature of the substrate during the deposition process. The oxidation process (see [6] and the refractive index tables in [15]) also changes the reflective and absorptive properties of the surface.

3 Conclusions

In this research, the light reflection properties of the collection of nanoribbons, sputtered onto a glass plate, are described in terms of polarimetry. Experimentally obtained ellipsometric parameters allow to determine the complex index of refraction for Ni–Cr–glass, Cr–glass, and Ni–glass nanostructures.

The global maximum (minimum) of elements M_{12}, M_{43}, M_{21}, M_{31}, M_{24}, M_{34} for the samples of chromium with the thicknesses of 132, 151, and 183 nm was observed at approximately 74.5°. If the residual scattering and absorption of light by the surface are neglected for this angle of incidence, then the Brewster angle for the system under investigation is assumed to be 74.5 ° with the given measurement accuracy.

No significant dependence on the film thickness of chromium was observed for such matrix elements as M_{12}, M_{43}, M_{21}, M_{31}, M_{24}, M_{34}, M_{33}, and M_{44}.

If a nickel nanoribbon with a thickness of 250 nm is produced at 250 °C on a chromium film with a thickness of 250 nm, then the experimentally observed state (internal structure) of the nickel nanoribbons is different for the nanosystems, such as nickel nanoribbon-chromium nanofilm and nickel nanoribbon-glass substrate. Despite the opacity of the nickel nanoribbons with the given thicknesses for the used laser light source, the optical properties of the outer surface are thus determined by the interface between the nanoribbon and the type of material of the next layer.

At the higher temperature than the Néel temperature for a macroscopic bulk of chromium, the observed abrupt change in the ellipsometric parameters of the nickel nanoribbons on the chromium film is the pretendant to be discussed as the confirmation of the phase transition in the ordering of the internal structure of system with the ferromagnetic–antiferromagnetic interface.

For the investigated multilayer nanosystems the nickel nanoribbons, the global extremum (in the sense of the quadratic dependence indicated above) of the value of the ellipsometric angle is 74.5° for the incident light beam.

Acknowledgements This work was supported in part by the grant No. 22BF07-03 from the Ministry of Education and Science of Ukraine.

References

1. E. L. Gurevich, S. V. Gurevich, Laser induced periodic surface structures induced by surface plasmons coupled via roughness. Appl. Surf. Sci. **302**, 118–123 (2014). https://doi.org/10.1016/j.apsusc.2013.10.141
2. M. Ardron, N. Weston, D. Hand, A practical technique for the generation of highly uniform LIPSS. Appl. Surf. Sci. **313**, 123–131 (2014). https://doi.org/10.1016/j.apsusc.2014.05.154
3. M. Huang, F. Zhao, N. Xu. Y. Cheng, Origin of laser-induced near-subwavelength ripples: interference between surface plasmons and incident laser. ACS Nano **3**(12), 4062–4070 (2009)
4. K. Zhou, X. Jia, T. Jia, K. Cheng, K. Cao, S. Zhang, D. Feng, Z. Sun, The influences of surface plasmons and thermal effects on femtosecond laser-induced subwavelength periodic ripples on Au film by pump-probe imaging. J. Appl. Phys. **121**, 104301 (2017). https://doi.org/10.1063/1.4978375
5. A. Kamimaki, S. Iihama, Y. Sasaki, Y. Ando, Micro-focused pulse laser-induced propagating spin waves in permalloy films with different thicknesses. IEEE Trans. Magn. (99), 1–1 (2017)
6. T. Ohkochi, *Gigantic Spin Waves Induced by Ultrashort Laser Pulses* (Research Frontiers, Physical Science, 2017), pp. 44–45
7. S. Kolodny, D. Yudin, I. Iorsh, Resonant spin wave excitation in magnetoplasmonic bilayers using short laser pulses. Nanoscale **11**, 2003–2007 (2019)
8. A.A. Awad, S. Muralidhar, A. Aleman, R. Khymyn, M. Dvornik, D. Lu, D. Hanstorp, J. Akerman, *Stimulated k-vector Selective Magnon Emission in NiFe Films Using Femtosecond Laser Pulse Trains*. arXiv:1908.03388v1 [cond-mat.mes-hall] (2019)
9. R.M.A. Azzam, M. Elshazly-Zaghloul, N.M. Bashara, Combined reflection and transmission thin-film ellipsometry: a unified linear analysis. Appl. Optics **14**(7), 1652–1663 (1975). https://doi.org/10.1364/AO.14.001652
10. I. Ohlidal, E. Schmidt, M. Libezny, Complete unambiguous optical characterization of double layers consisting of two strongly absorbing thin films by combined reflection and transmission ellipsometry. Appl. Optics **29**(4), 593–598 (1990). https://doi.org/10.1364/AO.29.000593
11. G. Bader, P.V. Ashrit, V.-V. Truong, Transmission and reflection ellipsometry of thin films and multilayer systems. Appl. Optics **37**(7), 1146–1151 (1998). https://doi.org/10.1364/AO.37.001146
12. J. Toudert, Spectroscopic ellipsometry for active nano- and meta-materials. Nanotechnol. Rev. **3**(3), 223–245 (2014). https://doi.org/10.1515/ntrev-2013-0043
13. P.B. Johnson, R.W. Christy, Optical constants of transition metals: Ti, V, Cr, Mn, Fe, Co, Ni, and Pd. Phys. Rev. B **9**, 5056 (1973). https://doi.org/10.1103/PhysRevB.9.5056
14. D.A. Rakić, A.B. Djurišić, J.M. Elazar, M.L. Majewski, Optical properties of metallic films for vertical-cavity optoelectronic devices. J. Optical Soc. Am. **37**, 5271–5283 (1998). https://doi.org/10.1364/AO.37.005271
15. H.G. Tompkins, J. Baker, D. Convey, Effect of process parameters on the optical constants of thin metal films. Surface and Interface Analysis **29**, 227–231 (2000)
16. J. Wang, J. Dong, Y. Cheng, Z. Xie, Y. Chen, Visible to near-infrared perfect absorption from alternate silica and chromium layers deposited by magnetron sputtering. Optics Lett. **46**, 4582–485 (2021)
17. T. Grigor'eva, T. Khasanov, Optical constants of Ni nanofilms. Optics Spectrosc. **112**, 796–800 (2012)
18. K.-Y. Chou, C.-L. Wu, C.-C. Shen, J.-K. Sheu, C.-K. Sun, Terahertz photoacoustic generation using ultrathin nickel nanofilms. J. Phys. Chem. C **125**(5), 3134–3142 (2021). https://doi.org/10.1021/acs.jpcc.0c09303
19. P.Y. Yu, M. Cardona, *Fundamentals of Semiconductors: Physics and Materials Properties* (Springer, Berlin, 2001), p. 261. ISBN 978-3-540-25470-6
20. Y. Oberemok, S. Savenkov, X. Chen, Z. Zhao, Z. Sun, A. Sizhuk, O. Prokopenko, Mueller matrix for chromium nanofilms on a glass substrate, in *2021 IEEE 11th International Conference Nanomaterials: Applications Properties (NAP)* (2021), pp. 1–5. https://doi.org/10.1109/NAP51885.2021.9568513

21. Y. Oberemok, S. Savenkov, X. Chen, Z. Zhao, Z. Sun, A. Sizhuk, O. Prokopenko, V. Malyshev, K. Yakimov, T. Rodionova, Polarimetry For Nickel-Chromium two-layer nanofilms and Nickel nanostripe on a glass substrate, in *2022 IEEE 41st International Conference on Electronics and Nanotechnology (ELNANO)* (2022), pp. 224–227. https://doi.org/10.1109/ELNANO54667.2022.9927127
22. F. McCrackin, E. Passaglia, R. Stromberg, H. Steinberg, Measurement of the thickness and refractive index of very thin films and the optical properties of surfaces by ellipsometry. J. Res. Natl. Bureau Stand. A. Phys. Chem. **67A**(4), 363–377 (1963)
23. M. Aouadi, D.M. Mihut, M.L. Kuruppu, S.R. Kirkpatrick, S.L. Rohde, Spectroscopic ellipsometry measurements of chromium nitride coatings. J. Vac. Sci. Technol. A **19**(6), 2800–2804 (2001)
24. J.A. Solovjov, V.A. Pilipenko, Effect of rapid thermal treatment temperature on electrophysical properties of Nickel films on Silicon. Doklady BGUIR **18**(1), 81–88 (2020)
25. T. Ivanova, K.A. Gesheva, A. Szekeres, O. Trofimov, Surface characterization of Chromium Oxide thin films in dependence on CVD growth process parameter. ECS Trans. **2**, 229–236 (2007)

Preconditions and Regularities of Phase Layers Formation in Binary Inhibitive Systems

Yu. P. Vyshnevska, V. M. Ledovskykh, I. V. Brazhnyk, and S. V. Levchenko

Abstract The regularities of protective action and preconditions for synergistic effects in the inhibitive mixtures that are characterized with different mechanism of action of individual components for the protection of steel in aqueous saline and acidic medium have been investigated. A generalized approach for the development of synergistic inhibitive compositions that is based on the concepts of thermodynamics, analysis of the Pourbaix diagram for the Fe-H_2O system and the conditions for the formation of nonmetallic layers on the metal surface in presence of different additives have been discussed. The factors and preconditions that cause the synergistic phenomena in inhibitive mixtures and approaches for optimizing their compositions are considered. The theoretical model that represents joint action of binary inhibitive mixtures based on oxyanions and salt passivators or complexing type inhibitors has been further extended and refined.

1 Introduction

Development of efficient methods for protection of metals and alloys remains an area of high scientific interest and practical significance. Moreover, the materials degradation prevention plays an increasingly important role for stability of nanostructured materials with high specific area in many emerging applications.

Yu. P. Vyshnevska (✉)
National Technical University of Ukraine "Igor Sikorsky Kyiv Polytechnic Institute", 37, Peremohy Ave., Kyiv 03056, Ukraine
e-mail: vishnevsk@ukr.net

V. M. Ledovskykh · S. V. Levchenko
National Aviation University, Liubomyra Huzara Ave, 1, Kyiv 03058, Ukraine

I. V. Brazhnyk
Gimasi SA, Piazzale Roncàa, 4, 6850 Mendrisio, Switzerland

O. Fesenko and L. Yatsenko (eds.), *Nanostructured Surfaces, Nanocomposites and Nanomaterials, and Their Applications*, Springer Proceedings in Physics 296,
https://doi.org/10.1007/978-3-031-42704-6_8

Of particular interest are the corrosion inhibitors that are able to form phase layers on the metal surface that in addition to anticorrosion protection may serve as an advanced metal surface modification technique. Despite extensive studies of the protective mechanism of individual inhibitors of different nature, the regularities of their coaction within an inhibitive mixtures are still understood incompletely.

Synergistic inhibitive mixtures are capable to efficiently suppress the electrochemical corrosion of steel in aqueous saline media [1–7] in some cases providing up to complete protection. Such mixtures may significantly outperform the individual components both in terms of inhibitive efficiency and the condition range where they may be applied. Next-generation inhibitors should combine wide pH range of application, low efficient concentration, and stability of the effect outside optimal parameters that may be achieved by combining inhibitors with different mechanism of action that require solid theoretical representation of the electrochemical processes for designing mixtures in a predictable fashion.

In present work, isomolar series method for designing synergistic inhibitive mixtures and studying regularities of coaction of the mixture components that belong to different types of inhibitors has been demonstrated.

2 Materials and Experimental Methods

Corrosion tests and determination of the inhibitor efficiency were performed gravimetrically according to the standard procedure. Tests were performed on steel 08 kp in the aqueous saline solutions with the composition: 0.3 g/l NaCl; 0.3 g/l Na_2SO_4 and 0.3 g/l $NaHCO_3$, exposure time at 25 °C temperature was 72 h–for passivation type inhibitors; 0.5 M H_2SO_4, exposure time at 25 °C–168 h–for complexing and polymerization type inhibitors.

The inhibition efficiency was determined by the equation $Z = [(K_m - K_{m'})/K_m] \cdot 100\%$, the inhibition coefficient–by the equation $\gamma = K_m/K_{m'}$, where K_m, $K_{m'}$—the corrosion rate of metal in solutions without and with inhibitor ($K_m = \Delta m/(S \cdot \tau)$, where Δm—the loss of the sample weight, g; S—the sample area, m^2; τ—exposure time, h).

Polarization measurements were carried out in potentiostatic regime in a temperature controlled three-electrode cell with separated cathodic and anodic compartments. Carbon steel 08kp was used as the working electrode, platinum–as the counter-electrode and a silver chloride electrode–as the reference one. In this paper, the potential values are given with respect to the normal hydrogen electrode potential.

Investigation of the morphological characteristics and elemental composition of the surface protective films of the inhibitive compositions were carried out with the scanning electron microscope (EVO-50, Zeiss, Germany) equipped with the energy-dispersive detector (INCA PENTA FET×3, Oxford Instruments,. Co., UK) and using Auger microprobe JAMP-9500F in the scanning electron microscopy mode. The energy of the electron beam was 3 keV and the current was 0.5 μA. During the profiling, the surface of the samples was bombarded with argon ions (ion etching) with energy of 4 keV. The pickling rate is 40 Å/min.

3 Theoretical Background

Purposeful design of efficient inhibitive mixtures requires solid theoretical understanding of the protective mechanism of their individual components. Corrosion inhibitors that utilize a protective mechanism of phase layers formation may rely on different reactions, thus, numerous types of such inhibitors may be distinguished. Complexing type inhibitors exploit a complexation process between added ligands and the metal ions resulting in a deposition of slightly soluble complex compounds on the metal surface in a form of dense phase layer [2, 8, 9]. Polymerization type inhibitors utilize the process of polymerization of added monomers under conditions that are established and maintained near the electrode–electrolyte interface [10, 11]. Passivation type inhibitors are able to from phase protective layers by maintaining thermodynamically favorable conditions for a specific corrosion reaction products that are able to from dense protective films (oxidative passivation), or by deposition of slightly soluble salts of added inhibitor and the metal ions (salt passivation) [12, 13].

Discussed types of inhibitors may serve as a basis or building blocks for the development of the synergistic inhibitive mixtures while taking into account individual features and regularities of the protective mechanism aliasing effects that occur as a result of their coaction [13].

Synergistic mixtures provide high protective efficiency in various media, mostly in acidic and neutral solutions [1–7]. In aqueous saline media, the synergistic effects are usually exhibited by the inhibitive mixtures that are capable to form dense passive films on the metal surface that in turn mainly suppresses the anodic reaction of metal ionization. Application of the Pourbaix diagram for Fe-H_2O system allows one to analyze thermodynamic aspects of the inhibitive mechanism of action of the mixtures and preconditions for synergistic phenomena [12, 14].

3.1 Complexing Type Inhibitors

Phase layers formation as a result of complexation process between added organic ligands and metal cations originating from the metal dissolution process has been proposed for wide range of organic ligands of different nature [2, 8]. The main kinetic regularities of the film formation process in such systems have been established showing that polarization resistance associated with phase layers formation obeys to a parabolic law indicating a complexation process with diffusion control. After pre-exposure with inhibitors, a strongly marked aftereffect may be observed for a prolonged time intervals that may be attributed to a high degree of the film continuity. Still, such inhibitors tend to be pH sensitive and applicable predominantly in acidic medium due to presence of metal cations that are mandatory for the complexation process.

3.2 *Polymerization Type Inhibitors*

Polymerization protective mechanism has been proposed for aromatic amines such as aniline derivatives [11, 15] and implies that added monomers undergo oxidative polymerization under conditions that are established and maintained in near-electrode layer. Obtained protective films have been characterized using SEM, AFM, and IR-spectroscopy methods revealing that significant portion of initial monomers is found to be included into protective film setting a precondition for a self-healing capability. The distinctive feature of such conducting polymers is their electroactivity that affects both electrochemical behavior of metal in the case of application as protective coatings as well as may exhibit electrochromic properties that is important from the functional coatings perspective.

3.3 *Passivation Type Inhibitors*

Passivation type inhibitors are most commonly used in neutral aqueous saline medium utilizing thermodynamic stabilization of the corrosion products to obtain dense phase layers in the case of oxidative passivation or deposition of slightly soluble salts of added inhibitor with metal ions in the case of salt passivation enabling them to achieve very high protective efficiency. Possible drawback of individual inhibitors of such type is that they are characterized with limited pH range and in some cases may be prone to promote localized corrosion damage outside of the recommended working concentration range [12].

3.4 *Synergistic Inhibitive Mixtures*

Disadvantages of individual inhibitors may be efficiently addressed using inhibitive mixtures. For this purpose, inhibitive action of each component needs to be splitted onto individual factors that may be positive or negative in regard to electrochemical and corrosion behavior of metal and analyzed in conjunction with other component. Maximum protective effect may be achieved in cases where individual factors of different additives complement one another creating favorable conditions for efficient implementation of protective mechanism of other component resulting in a supplementary, complementary, or tandem protective action [13].

4 Results and Discussion

High protective efficiency may be delivered by the inhibitive mixtures of XO_n^- oxyanions with salt-type passivators, like SiO_3^{2-}, PO_4^{3-}, MoO_4^{2-} [1, 12, 16, 17]. The mechanism of action of passivators usually attributed to their specific interaction with the metal surface involving active electron pair of acidifier, its reduction as oxidizers, and promotion of cathodic process that in turn leads to a potential shift toward a positive direction, where the oxide passivation of surface is thermodynamically possible. These reactions rely on the oxidizing ability of anions as well as the other conditions, primarily the pH value of the medium. According to the Pourbaix diagram, oxide passivation of steel occurs under such metal potentials and pH values, where the Fe-H_2O system is transferred to a thermodynamically stable state of oxyhydroxides of Fe (III)–FeOOH:

$$Fe^{2+} + 3H_2O = Fe(OH)_3 + 3H^+ + \bar{e}$$

coupled with the electron assimilation that appears during this transition. Thus, in case of nitrites, the environment pH value shift to pH ≥ 9 is required for active reduction of adsorbed NO_2^--ions. At the same time, the sole oxide passivation as a protection method has the significant flaws that may lead to a pitting corrosion in the case of partial film destruction. According to the proposed mechanism, this issue may be addressed via introduction of anions that may form slightly soluble salts and complex compounds with Fe^{2+} ions. In addition, silicates, phosphates, and molybdates as salts of weak acids undergo hydrolysis by the anion that leads to an alkalization of the medium that in turn facilitates to transition of the Fe-H_2O system toward a passive state.

As it was previously shown [2, 12], sodium nitrite provides excellent basis for synergistic inhibitive mixtures due to its mechanism of protective action that facilitates establishing of favorable conditions for salt- or complexing type component of the composition. The NO_2^- -ion has the active electron pair on sp^2—hybridized orbital on the central nitrogen atom that result in its adsorption properties. The morphology studies of the protective layer formed in presence of $NaNO_2$ show generally uniform film structure with marked signs of local corrosion damage (Fig. 1c). The ion etching of the surface (pickling rate is 40 Å/min) confirmed the presence of nanoscale protective layer with typical thickness of 15–20 nm, while the surface profiling showed sharp decrease in nitrogen concentration falling to zero levels after 1 min of etching that is also in line with theoretical expectations.

It has been also discovered that the morphology of the support plate after etching of the film formed in the presence of $NaNO_2$ shows marked partially nonlinear relief structure that does not match typical signs of mechanical treatment. Thus, that the presence of NO_2^- ions affects the metal ionization process to such an extent that it leads to the macroscopic changes in the metal surface morphology. According to this representation of the $NaNO_2$ mechanism of action, the second inhibitor in the binary mixture has been selected by its ability to form slightly soluble compounds

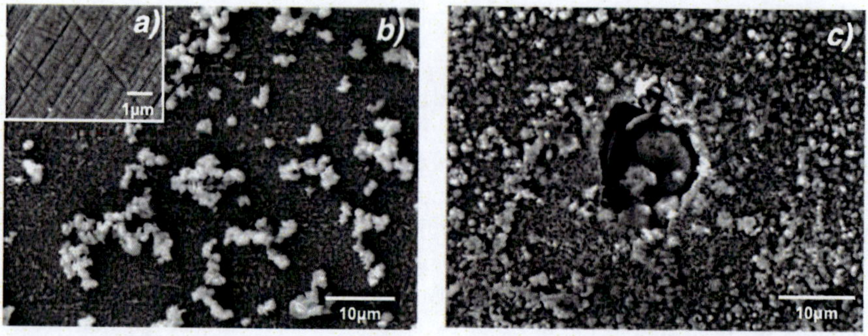

Fig. 1 SEM images of carbon steel after exposition: **a** without exposure, **b** in background aqueous saline solution; **c** in the presence of 30 mmol/l $NaNO_2$. Exposure time: 168 h

with the components present in near-electrode layer. For establishing the synergistic effect regularities, the following binary systems based on sodium nitrite that include both salt- and complexing type inhibitors as second component of the mixture have been studied.

4.1 Passivation Type (Oxidative)–Passivation Type (Salt) Inhibitor

The isomolar series method has been employed for determining the optimal concentration ratio for the $NaNO_2$–Na_2CO_3, $NaNO_2$–Na_2WO_4, and $NaNO_2$–polyhexamethylene guanidine (PHMG) binary mixtures. Sole Na_2CO_3 provide relatively low protective efficiency of 78% and leads to a formation of film with clearly visible discontinuities (Fig. 3a), that is also in line with polarization measurements (Fig. 3a, curve *3, 3′*) (Fig. 2).

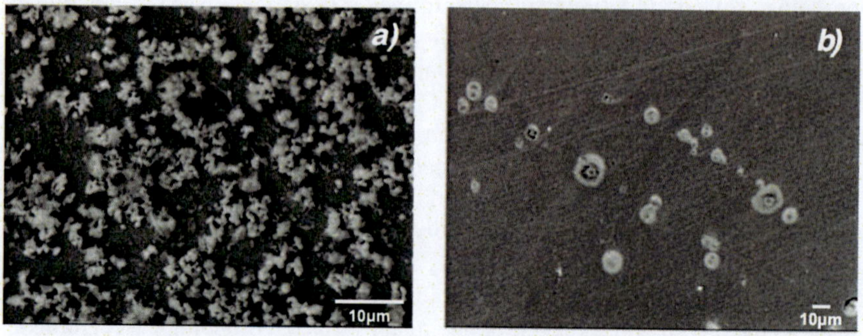

Fig. 2 SEM images of carbon steel after exposition in: **a** the presence of 30 mmol/l Na_2CO_3, **b** the presence of 20 mmol/l $NaNO_2$ + 10 mmol/l Na_2CO_3 Exposure time, 168 h

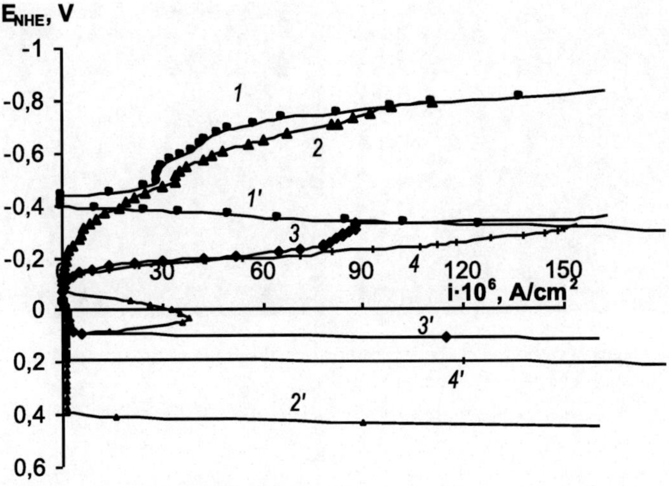

Fig. 3 Cathodic (*1, 2, 3, 4*) and anodic (*1′, 2′, 3′, 4′*) polarization curves for carbon steel in: *1, 1′*—background aqueous saline solution (pH 7.3); *2, 2′*—the presence of 30 mmol/l $NaNO_2$ (pH 7.38); *3, 3′*—the presence of 30 mmol/l Na_2CO_3 (pH 10.9); *4, 4′*—the presence of 20 mmol/l $NaNO_2$ + 10 mmol/l Na_2CO_3 (pH 10.4)

SEM images for the $NaNO_2$–Na_2CO_3 mixture at optimal concentration ratio of 2:1 (Fig. 3b) indicate significantly improved film structure with moderate amount of defects that is also in line with its protective efficiency of 92.3%.

The Na_2WO_4 has been selected for studying the case, where the products are characterized with significantly lower solubility compared to Na_2CO_3 that allows sole Na_2WO_4 to provide protective efficiency of 99.8%. SEM images (Fig. 4a) indicate that despite high protective efficiency, the formed film structure contains defects and irregularities. Switching to a $NaNO_2$ + Na_2WO_4 mixture at optimal concentration as 1:2 is accompanied by a significant change in structure and morphology of the formed film, while the protective efficiency remains at 99.8%.

It should be noted, that the mixture exhibit slight decline in efficiency below level of individual components on the non-optimal concentration ratios that may be attributed to a concurrency between inhibitors. The pH value for the $NaNO_2$ + Na_2WO_4 mixture is found to be pH 8.3 that does not deliver preconditions for reduction of Nitrite ions. Thus, Na_2WO_4 does not ensure a favorable condition for activating the mechanism of synergistic action that is in line with comparative analysis of electrochemical behavior of steel in presence of $NaNO_2$ under different pH values and $NaNO_2$ + Na_2WO_4 (Fig. 5).

Based on obtained results, we may conclude that the specific action of both components contributes to synergistic effects within studied binary mixtures.

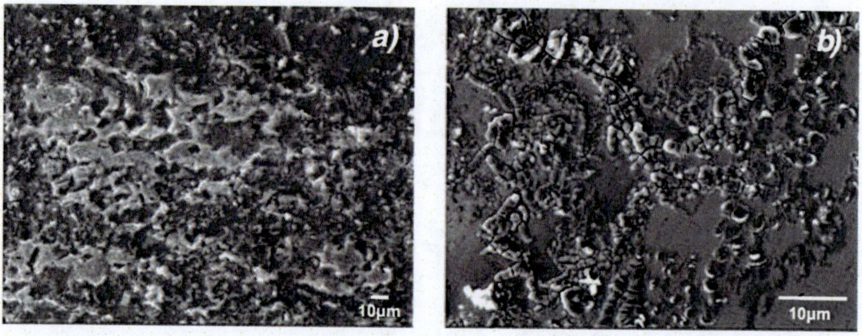

Fig. 4 SEM images of carbon steel after exposition in: **a** the presence of 30 mmol/l Na_2WO_4, **b** the presence of 10 mmol/l $NaNO_2$ + 20 mmol/l Na_2WO_4

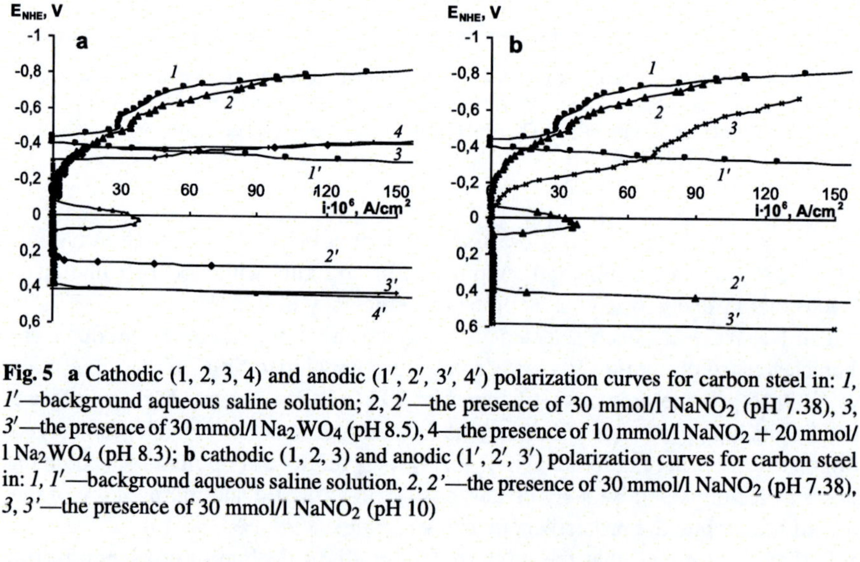

Fig. 5 **a** Cathodic (1, 2, 3, 4) and anodic (1′, 2′, 3′, 4′) polarization curves for carbon steel in: *1, 1′*—background aqueous saline solution; *2, 2′*—the presence of 30 mmol/l $NaNO_2$ (pH 7.38), *3, 3′*—the presence of 30 mmol/l Na_2WO_4 (pH 8.5), 4—the presence of 10 mmol/l $NaNO_2$ + 20 mmol/l Na_2WO_4 (pH 8.3); **b** cathodic (1, 2, 3) and anodic (1′, 2′, 3′) polarization curves for carbon steel in: *1, 1′*—background aqueous saline solution, *2, 2'*—the presence of 30 mmol/l $NaNO_2$ (pH 7.38), *3, 3'*—the presence of 30 mmol/l $NaNO_2$ (pH 10)

4.2 *Passivation Type–Complexing Type Inhibitor*

Sole PHMG in neutral aqueous saline solutions provides relatively low efficiency of 48% that is attributed to a fact that according to a mechanism of action of the complexing type inhibitors, the sufficient amount of metal ions is required for the formation of protective film [2]. In contrast, within $NaNO_2$–PHMG mixture, the protective efficiency reaches the value of 99.8% at optimal concentration ratio as 2:1 that is also in line with the morphology of the surface layer (Fig. 6). It should be noted, that SEM images of the surface layer after 168 h exposition show only slight morphology changes compared to initial condition with marked smoothing of the mechanical treatment traces.

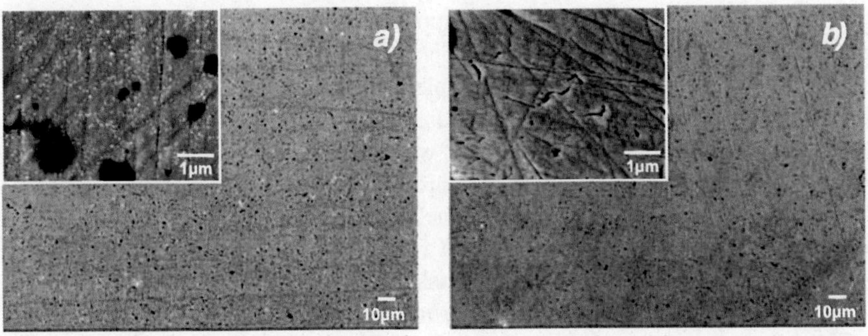

Fig. 6 SEM images of carbon steel after exposition in: **a** the presence of 30 mmol/l PHMG; **b** the presence of 10 mmol/l PHMG + 20 mmol/l $NaNO_2$

The elemental composition of the protective film formed in presence of synergistic mixtures of $NaNO_2$ + PHMG at optimal concentration ratio shows that the nitrogen content increased slightly from 20.6 to 25.5 at.% compared to sole PHMG that adequately corresponds to the presence of NO_2^- ions. The obtained results correspond to the structure of the phase film of slightly soluble PHMG-Fe complexes with some inclusion of iron oxide indicating that the complexation process has a predominant role in the protective mechanism of the mixture that is found to be in a good agreement with previous studies [12].

4.3 Complexing Type–Complexing Type Inhibitor

Synergistic effects within complexing type inhibitors may be illustrated on the serine-cystine binary mixture in acidic medium. Individually, both amino acids demonstrate poor protective performance: serine provide slight reduction in corrosion rate of 28.5%, while cystine tends to promote corrosion for 5.5%. At the same time, within a binary mixture at optimal concentration ratio serine-cystine as 2:1, the protective efficiency reaches value of 47% that may be attributed to a formation of biligand complexes that are characterized with lower solubility than individual ones.

5 Conclusions

Regularities of joint action of inhibitors of different nature have been investigated in the binary mixtures composed according to the following schema: oxidative passivator–salt passivator; oxidative passivator–complexing type inhibitor; and complexing type–complexing type inhibitor.

It is shown, that in studied binary mixtures of oxidative (XO_n^-) and salt passivators at optimal concentration ratio, the synergistic effects may be observed in such cases where specific preconditions are implemented. In particular, to ensure optimal performance of $NaNO_2$-based composition, the second mixture component should facilitate the shift of the value of pH ≥ 9, where active reduction of NO_2^--ions may occur. The mechanism of action of oxyanions is due to their reduction, promoting the cathodic reaction and switching the metal surface toward a thermodynamically stable state owing to a hydrated oxides of Fe(III) in a form of goethite-type phase. Such transition occurs under such potentials and pH values and that correspond to an area of stability of FeOOH on the Pourbaix diagram for the Fe-H_2O system.

The inhibitive mechanism of salt passivators within the mixture is attributed to a modification of oxide phases with a low solubility salts with the Fe(II) ions that improve the inhibitor action against local corrosion. The inhibitive efficiency tends to improve with components that are characterized with lower solubility of the products.

Binary inhibitive mixtures with complexing type inhibitors exploiting the ability of oxyanions to promote cathodic reaction that in turn facilitates establishing of favorable conditions for the complexation process that plays predominant role in the mixture protective mechanism.

References

1. A.T. Truc, N. Pébère, T. Xuan Hang, Y. Hervaud, B. Boutevin, Study of the synergistic effect observed for the corrosion protection of a carbon steel by an association of phosphates. Corros. Sci. **44**, 2055–2071 (2002)
2. V.M. Ledovskykh, Yu.P. Vyshnevska, I.V. Brazhnyk, S.V. Levchenko, Development and optimization of synergistic compositions for the corrosion protection of steel in neutral and acid media. Mater. Sci. **52**, 634–642 (2017). https://doi.org/10.1007/s11003-017-0002-1
3. A.L. Chong, J.I. Mardel, D.R. MacFarlane, M. Forsyth, A.E. Somers, Synergistic corrosion inhibition of mild steel in aqueous chloride solutions by an imidazolinium carboxylate salt. ACS Sustain. Chem. Eng. **4**, 1746–1755 (2016)
4. K.T. Kim, H.W. Kim, H.Y. Chang, B.T. Lim, H.B. Park, Y.S. Kim, Corrosion inhibiting mechanism of nitrite ion on the passivation of carbon steel and ductile cast iron for nuclear power plants, in *Advances in Materials Science and Engineering* (2015), pp. 1–16
5. K.N. Mohana, A.M. Badiea, Effect of sodium nitrite-borax blend on the corrosion rate of low carbon steel in industrial water medium. Corros. Sci. **50**, 2939–2947 (2008)
6. G.D. Eyu, G. Will, W. Dekkers, J. MacLeod, The synergistic effect of iodide and sodium nitrite on the corrosion inhibition of mild steel in bicarbonate-chloride solution. Materials **9**, 868 (2016). https://doi.org/10.3390/ma9110868
7. S. Božović, S. Martinez, V. Grudić, A novel environmentally friendly synergistic mixture for steel corrosion inhibition in 0.51 M NaCl. Acta Chim. Slov. **66**, 1–11 (2019)
8. D.A. Tkalenko, G. Venkatesvaran, Yu.P. Vishnevskaya, S. Keni, M.V. Byk, K. Mate, Inhibitory effect of cysteine in acid media. Prot. Met. Phys. Chem. Surf. **46**, 609–614 (2010). https://doi.org/10.1134/S2070205110050199
9. R.N. El-Tabesh, A. Abdel-Gaber, H.H. Hammud, R. Oweini, Effect of mixed-ligands copper complex on the corrosion inhibition of carbon steel in sulfuric acid solution. J. Bio- and Tribo-Corros. **6**, 29 (2020). https://doi.org/10.1007/s40735-020-0323-8

10. K.J. Croes, A.J. Vreugdenhil, M. Yanb, T.A. Singletona, S. Boraas, V.J. Gelling, An electrochemical study of corrosion protection by in situ oxidative polymerization in phenylenediamine crosslinked sol–gel hybrid coatings. Electrochim. Acta **56**, 7796–7804 (2011)
11. Yu.P. Vyshnevska, I.V. Brazhnyk, S.O. Kudrya, Design and characterization of *in situ* synthesized hybrid polymeric materials, in *Nanoplasmonics, Nano-Optics, Nanocomposites, and Surface Studies*, vol. 167, ed. by O. Fesenko, L. Yatsenko. Springer Proceedings in Physics (2015). https://doi.org/10.1007/978-3-319-18543-9
12. V.M. Ledovskykh, Y.P. Vyshnevska, I.V. Brazhnyk, S.V. Levchenko, Thermodynamic States and transitions diagrams in surface engineering for the material degradation prevention, in *Nanomaterials and Nanocomposites, Nanostructure Surfaces, and Their Applications*, vol. 263 ed. by O. Fesenko, L. Yatsenko. NANO 2020. Springer Proceedings in Physics (Springer, Cham, 2021). https://doi.org/10.1007/978-3-030-74741-1_30
13. V.M. Ledovskykh, Yu.P. Vyshnevska, I.V. Brazhnyk, S.V. Levchenko, Mechanism of coaction of the oxidative and salt passivators in binary inhibiting mixtures. Mater. Sci. **5649**, 678–683 (2021). https://doi.org/10.1007/s11003-021-00482-w
14. M. Pourbaix, *Atlas of Electrochemical Equilibria in Aqueous Solutions* (National Association of Corrosion Engineers, Houston, Texas, 1974), p.551
15. Yu.P. Vyshnevska, I.V. Brazhnyk, The electrochromic feedback methods for obtaining nanoparticles, nanoalloys and core-shell objects in quasi-reversible redox systems, in: *Nanophysics, Nanomaterials, Interface Studies, and Applications*, vol. 195, ed. by O. Fesenko, L. Yatsenko. Springer Proceedings in Physics (Springer, Cham, 2017). https://doi.org/10.1007/978-3-319-56422-7_64
16. K.S. Bokati, C. Dehghanian, S. Yari, Corrosion inhibition of copper, mild steel and galvanically coupled copper- mild steel in artificial sea water in presence of 1H-benzotriazole, sodium molybdate and sodium phosphate. Corros. Sci. **126**, 272–285 (2017)
17. M. Saremi, C. Dehghanian, M.M. Sabet, The effect of molybdate concentration and hydrodynamic effect on mild steel corrosion inhibition in simulated cooling water. Corros. Sci. **48**, 1404–1412 (2006)

Comparative Analysis of Nanostructures Formed by the Chemical Route in the Ferrum–Lanthanum-Containing Systems

O. M. Lavrynenko, O. Yu. Pavlenko, and O. I. Olifan

Abstract Phase formation processes during the chemical co-precipitation of inorganic salts of ferrum and lanthanum in a weakly alkaline medium and subsequent calcination of the precipitates at temperatures of 800 and 1100 °C were studied. It is shown that at the initial stage in the system, hydroxides-oxyhydroxides of ferrum and lanthanum, as well as hydroxosalts of different anionic composition, are formed in the system, which is caused by the presence of chlorides or sulfates of the corresponding metals in the solution. It is shown that the injection of urea species into the system, the dissociation of which, leads to the formation of hydroxocarbonates of ferrum and lanthanum. According to TG-DTA data, dehydroxylation of goethite occurs at $T =$ 250 °C, and lanthanum hydroxide at 330 °C, which causes spatial separation of the process of formation of oxide phases. It is shown that at 800 °C, ferrum oxide is represented by a single phase—hematite, while lanthanum forms a hexagonal-trigonal La_2O_3 phase, with an admixture of cubic body-centered La_2O_3, cubic face-centered LaO, and lanthanum hydroxide. The destruction of carbonate complexes and burning of auxiliary substances occur in the temperature range of 600–700 °C. At the same time, the perovskite structure in the powders at 800 °C is weakly developed, which indicates the need to increase the sample's calcination temperature or the duration of the temperature treatment. It was shown that at a temperature of 1100 °C, two phases were identified in the composition of all nanopowders—lanthanum orthoferrite and hematite with particle sizes of 27–39 nm. It was established that the violation of the stoichiometric ratio of La/Fe in the composition of the precipitate and the formation of two-phase samples is caused by a number of factors, in particular, differences in the pH of precipitation of ferrum and lanthanum hydroxides, the dehydroxylation temperatures of the hydroxide phases, the stability of ferrum and lanthanum oxides in contact with air and less receptive conditions entry of lanthanum into the perovskite structure due to its inherent chemical properties.

O. M. Lavrynenko (✉) · O. Yu. Pavlenko · O. I. Olifan
I. Frantsevich Institute for Problems in Material Science, NAS of Ukraine, Krzhizhanovsky Str., 3, Kyiv 03142, Ukraine
e-mail: alena.lavrynenko@gmail.com

O. Fesenko and L. Yatsenko (eds.), *Nanostructured Surfaces, Nanocomposites and Nanomaterials, and Their Applications*, Springer Proceedings in Physics 296,
https://doi.org/10.1007/978-3-031-42704-6_9

1 Introduction

Nowadays, interest in the preparation and practical application of lanthanum-containing materials is closely related to their high oxidative-catalytic and electrocatalytic activity [1]. At the same time, the combination of the catalytic properties of lanthanum oxides with the magnetic properties of iron oxides and the formation of mixed structures such as ferrites, orthoferrites, and perovskites make it possible to form magnetoelectric and multiferroic materials [2]. In addition, nanocomposites based on rare-earth orthoferrites and iron oxides are expected to be good promising materials for use in biomedicine, for example, for magnetic hyperthermia [3].

At present, obtaining particles of nano-sized perovskite $LaFeO_3$ is carried out using a number of chemical and physical methods, the most widespread of which are: chemical co-precipitation in an alkaline medium [4, 5], sol–gel synthesis [6], citrate-gel approach [7], reversive micelles or microemulsion method [8], hydrothermal synthesis [9, 10], microwave-assisted decomposition of the precursor species [11], sonochemical synthesis [12], the polymerizable complex method [13], thermal decomposition of a heteronuclear complex [14, 15], polyol method [16], combustion synthesis [17, 18], solution combustion synthesis [19], auto combustion synthesis [20, 21], solid-state combustion of a precursor species [22], the solid-state reaction [23], mechanical ball milling [24], template method [25], self-assembly process [26] etc.

Among the variety of composite oxides with a perovskite-type structure, those that contain a combination of rare-earth elements with non-ferrous or heavy metals attract considerable attention of scientists and technologists [19, 27, 28]. The combination of unique magnetic, superconducting, dielectric optical, photocatalytic properties, and electronic structure ensures the wide use of lanthanum orthoferrite particles $LaFeO_3$ in creating a number of technical materials [29]. In particular, particles of composite oxide such as $LaFeO_3$ are widely used in creating composite cathodes for protonic ceramic fuel cells [30], catalysts [31], membranes, and optoelectronic devices. Lanthanum orthoferrite $LaFeO_3$ has been identified as a potential material for the sensors creation for detecting a number of chemicals, such as nitrogen monoxide, xylene, methanol, acetone, formaldehyde, and ethanol [9, 32–34].

At the same time, perovskite-type structures show good catalytic properties and are characterized by lower cost and higher thermal stability in comparison to noble metal catalysts, and therefore, they can be used in the catalysts' manufacture [35]. The catalytic effect's principle of semiconductors in the photo-destruction of organic substances consists of the electron–hole pairs generation in the semiconductor after absorption of visible light with an energy that reaches or exceeds the band gap. Generated electrons (e^-) and holes (h^+) migrate to the surface of the catalyst and enter into a red-ox reaction with adsorbed reagents' molecules. The holes' high oxidation potential on the surface of the catalyst oxidizes toxic organic compounds on the surface and converts them into the form of biologically safe substances [36].

The parameters important for the creation of effective catalysts based on lanthanum orthoferrite are the corresponding crystal structure of the material and the large surface area [24]. In general, the catalytic properties of perovskite are affected by texture, morphology, crystallite size, and specific surface area (SSA), which are determined by the synthesis method and parameters [37], for example, by the pH value of co-precipitation and the aging time of the precipitates [38].

It was shown [39] that nanostructures of perovskite $LaFeO_3$ belonging to semiconductors act as an efficient visible light photocatalyst due to their narrow band gap energy. The influence of particle morphology of the $LaFeO_3$ catalyst, synthesized by a chemical method using as starting materials $La(NO_3)_3 \cdot 6H_2O$ and $K_3[Fe(CN)_6]$, and adding as structure-directing agents urea ($(NH_2)_2CO$) and citric acid ($C_6H_8O_7 \cdot H_2O$), toward the decolorization of Rhodamine B (RhB) in aqueous solution (1 mg/ml) was studied in work [40]. It was found that the $LaFeO_3$ nanospheres exhibit higher photocatalytic activity than the nanorods and nanocubes. At the same time, it was shown that all nanostructure samples $LaFeO_3$ show a higher photocatalytic activity for RhB degradation than Degussa P25 (TiO_2) in the visible light region and form a series in terms of efficiency: Degussa P25 < nanocube < nanorod < nanosphere.

According to the absorption measurements, the band-gap energies of $LaFeO_3$ nanostructures of nanocubes, nanorods, and nanospheres were found to be 2.01, 2.05, and 2.1 eV, respectively [41]. But optical diffuse reflectance spectroscopy studies show that the $LaFeO_3$ sample obtained through mechanical ball milling of lanthanum and iron oxides has semiconductor properties with the band gap energy of 2.67 eV [24]. To improve the properties of catalysts, it is advisable to use the possibilities of defect engineering, a discipline that can be aimed at improving the catalytic efficiency of perovskites by adapting the stoichiometry and incorporating defects into the structure in certain concentrations [42]. In particular, experimental and theoretical studies proved the formation of a mixture of two $LaFO_3$ and Fe_2O_3 phases instead of a change in the La/Fe ratio in a single-phase sample [43], which can positively affect the photocatalytic properties of the $LaFeO_3$ and α-Fe_2O_3 sample. At the same time, it is known that the use of polyphase catalysts helps to increase their activity due to the overlapping of band gap of semiconductors and the expansion of the spectrum of activity of catalysts, for example, for visible and solar light [44].

The purpose of this work is the comparative analysis of nanostructures of lanthanum orthoferrite and hematite formed during the chemical precipitation of starting materials from solutions of lanthanum and ferrum inorganic salts in a slightly alkaline environment and subsequent calcination of the precipitates at 800 and 1100 °C temperatures.

2 Objects and Methods of the Research

The phase's formation was carried out by the method of chemical co-precipitation of inorganic salt's solutions containing ferrum and lanthanum in a weakly alkaline environment in the presence of auxiliary substances. Chemically pure salts $Fe_2(SO_4)_3$, $FeSO_4$, $FeCl_3$, $K_3[Fe(CN)]$, $La_2(SO_4)_3$, $LaCl_3$, $La(NO_3)_3$ with 1:1 ratio of ferrum to lanthanum were used to prepare the initial solutions. Solutions of urea, potassium, sodium, and ammonium hydroxides were introduced into the systems as auxiliary substances. Co-precipitation was carried out at $T = 80$ °C. The precipitates were washed with a water–ethanol solution and lyophilized at $T = 160$ °C. The precipitates were calcined at 800 and 1100 °C for 2 and 5 h. Oxide particles were additionally treated with acidic solution.

The characterization of powders was carried out using X-ray diffraction analysis (XRD), thermogravimetric study (TG-DTA), and scanning electron microscopy method combined with energy-dispersive spectroscopy method (SEM-EDS).

XRD data were obtained using a powder X-ray diffractometer (DRON 3M) equipped with Cu anode tube. The scanning step was 0.05–0.1°, exposure—4 s, range of 2θ angles was taken from 15 to 85°. Samples were taken at standard temperature. The International Powder Standards Committee (JSPDS International Centre for Diffraction Data 1999) database was applied to determine the phase composition. The program Data analysis using Match! according to the Reference database COD-Inorg 2022.06.29, which made it possible to assess the degree of crystallinity of the sample, distribution of phases in the samples, determination of the intensity of reflexes of individual phases, chemical composition of the samples, etc. was applied to obtain more information from XRD data. The calculation of the lattice parameters was received according to standard methods using Debye–Scherrer equation by three planes each mineral phases. A simultaneous study of thermogravimetric and differential thermal properties (TG-DTA) of the composite particles were performed in the static air atmosphere by a thermal analyzer Q-1500D (Hungary). The parameters of the pattern recording were: the samples of 100 mg were heated at the rate of 10 °C/min from 20 to 1000 °C; the sensitivity was 50 mg; TG-500, DTG-500, and DTA-250. To create a homogeneous temperature field, the samples were placed into a corundum crucible and covered by a quartz beaker.

Mira 3 Tescan equipped with EDX module (Oxford INCA x-act) was used for the nanocomposites visualization, morphology study, and performing elemental analysis.

The specific surface area (SSA) of the samples was measured by the chromatographic method of low-temperature argon adsorption at − 196 °C (one-point BET method). Test samples were analyzed under identical conditions with a standard sample with a known and long-term stable specific surface area. In this case, silica (silochrome C-80) with a specific surface area $S_{sp} = 80$ m^2/g was used as a standard. The value of the specific surface of the test sample was calculated according to the formula:

$$S_{sp} = (Sc_{sp} \cdot Fc \cdot mc)/(F \cdot m),$$

where

S_{sp} and Sc_{sp} are the specific surface area of the sample and standard, respectively, m^2/g;
F and Fc are the area of the argon desorption peak of the studied sample and the standard, respectively (chromatographic data);
m and mc are the mass of the sample and the standard, respectively, g.

The permissible measurement error is 20%. The gas mixture contained 5 vol.% Ar (adsorbate gas) and 95 vol.% He (carrier gas).

3 Results and Discussion

The first part of the experimental study consists of considering the processes associated with the formation of individual oxide-hydroxide compounds of lanthanum and their transformation under the influence of temperature. The second part of the experimental study is devoted to the study of the influence of ferrum salt's composition on the formation of structures in a lanthanum-containing environment. At the same time, the characteristics of lyophilized sediments and powders obtained after the temperature treatment of the respective samples at 800 and 1100 °C are given.

4 Characteristics of the Nanostructures Formed via Precipitation of Lanthanum Chloride and Sulfate Solutions

To consider the processes of nanoscale structure's formation in polyphase's systems, it is appropriate to trace the development of individual precursor phases, the interaction of which leads to the formation of lanthanum–ferrum-containing composite particles. Thermal gravimetric analysis of the powder obtained in the pure $LaCl_3$ system (Fig. 1a) shows the first endothermic peak at $T = 270$ °C and shoulder at $T = 300$ °C in the DTA curve accompanied by mass lost at $T = 200$ °C. This may be related to the formation of intermediate phase $LaCl(OH)_2$ and $La(OH)_3$ and their next phase transformation into LaClO and LaO(OH), correspondingly. The second effect of mass loss in the DTG curve points to the dehydroxylation of lanthanum oxy(hydroxide) lattice and the formation of lanthanum oxide phases (endothermic peak at 390 °C). The next shoulder in the DTA curve at the temperature 710–735 °C may be related to the transformation of the crystal lattice of lanthanum oxides (Ia3

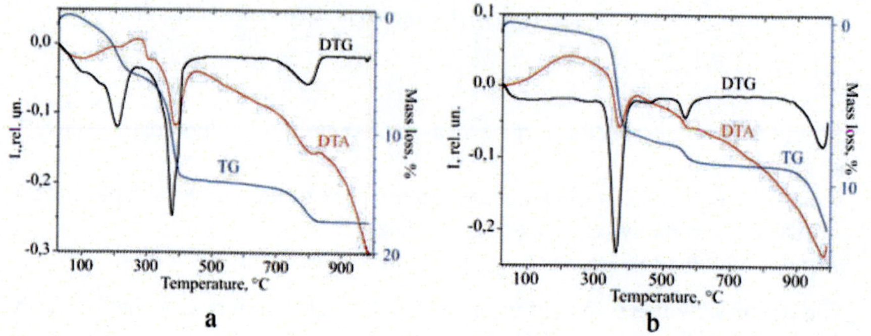

Fig. 1 TG-DTA curves of the precipitates formed in the presence of weak alkaline salt's solutions: **a** $LaCl_3$; **b** $La_2(SO_4)_3$

into P3m1) and oxidation of chlorides. Such mass loss effect is described by DTG in the temperature range 750–860 °C. The last small effect of mass lost in the DTG curve is detected at $T = 950$ °C and may characterize the rearrangement of lanthanum oxide phases. The total mass loss of the sample equals 17.8%.

The analysis of the pure lanthanum sulfate system indicated a deep reflex of mass lost at $T = 360$ °C and an endothermic peak at $T = 370$ °C that may be corresponded to the dehydroxylation of lanthanum hydroxide phase and formation of the primary lanthanum oxide phase. The processes of phase transformation of lanthanum hydroxide into lanthanum oxyhydroxide and the following lanthanum oxyhydroxide transformation into oxide phase were carried out at temperatures 230–420 °C, correspondingly. The next reflexes shown at $T = 600$–610 °C can indicate the destruction sulfate containing components. The total mass loss is 13.7%. It should be noted that the previously performed process of precipitate lyophilization at $T =$ 160 °C led to the absence of the reflexes corresponding to the removal of adsorbed water in the DTG curve (Fig. 1b).

Generally, lanthanum oxides are unstable under standard conditions and they can easily transform into hydroxide–oxyhydroxide phases. Hence, such property significantly complicates achieving the stoichiometric ratio in polymetallic lanthanum-containing systems. Typical XRD patterns of the samples obtained when industrial La_2O_3 powder was contacted with air and lanthanum hydroxide precipitate formed in a low alkaline medium at a temperature of 160 °C are present in Fig. 2b. At the same time, the contact of La_2O_3 powder with air oxygen (Fig. 2a) results in the hydroxylation of lanthanum oxide lattice and the appearance of two-phase's system—$La(OH)_3$ (PDF card No. 06-0558, PDF card No. 36-1481) and LaOOH (PDF card No. 19-0656). According to Match! Analysis of the parameters for $La(OH)_3$ is the following: Space group (S.G.) P 63/m; Crystal system hexagonal; Unit cell $a = 1.58090$ nm, $c = 0.49724$ nm, average particle size—42.9 nm. Degree of crystallinity (DOC) of the sample is 33.87 wt.%. The content of the amorphous components—66.13 wt.%. The

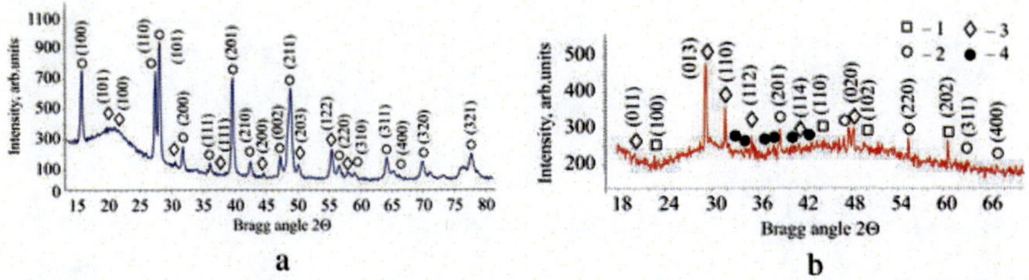

Fig. 2 XRD patterns of the precipitates: **a** industrial La_2O_3 powder that was kept under standard conditions; **b** the precipitate formed in $La(NO_3)_3$ + NH_4OH system (T = 160 °C). Numbers correspond to mineral phases: 1—La_2O_3 (S.G. P3m1(164)); 2—$La(OH)_3$ (S.G. P63/m (176)); 3—LaOOH (S.G. P2 1/m (11)); 4—$La_2O_2CO_3$ (S.G. P63/mmc (194))

lyophilized precipitate (Fig. 2b) contains four phases—LaOOH, a small amount of La_2O_3 (PDF card No.), and $La(OH)_3$, a trace quantity of $La_2O_2CO_3$ (PDF card No. 23-0435). The total degree of crystallinity is 12 wt.%, and the amorphous content is 88 wt.%.

XRD patterns of the lanthanum oxide-oxyhydroxide powders formed in the lanthanum chloride system and calcined in the temperature range from 400 to 1000 °C in 5 h are shown in Fig. 3. According to the X-ray diffraction data, the temperature treatment of the lanthanum-containing precipitates at T = 400 °C is not enough to obtain well crystalline particles (Fig. 3a). The total crystallinity degree of the sample is 10.14 wt.%, whereas the amorphous content reaches 89.96 wt.%. Weak reflexes of lanthanum oxides correspond to trigonal-hexagonal La_2O_3, Space Group (S.G.) P-3m1 (PDF card No. 05-0602; PDF card No. 96-152-3969) and cubic La_2O_3, S.G. Ia3 (PDF card No. 22-0369) with total weight 26.9% can be seen. According to the Reference database COD-Inorg 2022.06.29, the Unit cell of La_2O_3 belonged to S.G. P63/mmc (H) with Entry number (EN) 96-152-3969 is as follows: $a = 0.40570$ nm, and $c = 0.64300$ nm. The calculated density of the lanthanum oxide phase equals 3.386 g/cm^3. The total content of residual lanthanum chloride $LaCl_3 \cdot 7H_2O$ (E.N. 96-154-1972) phase is relatively big and it reaches 73.1%. The lanthanum chloride phase belongs to the triclinic (anorthic) crystal system; unit cell $a = 0.82820$ nm, $b = 0.92370$ nm, $c = 0.80580$ nm, $\alpha = 107.810°$; $\beta = 98.630°$; $\gamma = 71.520°$; calculated density 2.135 g/cm^3. The elemental composition of the powder is La—47.3; O—29.0; Cl—20.9, and H—2.8 wt.%.

Analysis of precipitates formed in pure $LaCl_3$ systems and calcined at 600 °C indicates the formation of a complex three-phase powder—hexagonal-trigonal La_2O_3 (PDF card No. 05-0602), cubic body-centered Ia3 La_2O_3 (PDF card No. 22-0369), and face-centered F LaO (PDF card No. 33-0716) (Fig. 3b). The average size of La_2O_3 particles is 13.5 nm. The lattice parameters of La_2O_3 are the following: $a = 0.3937$ nm, $c = 0.6129$ nm. As a rule, an increase in the calcination temperature leads to an increase in the degree of crystallinity of the sample, that reaches 28.04 wt.%. Therefore, the amorphous part decreases to 71.96 wt.%. It is important to identify

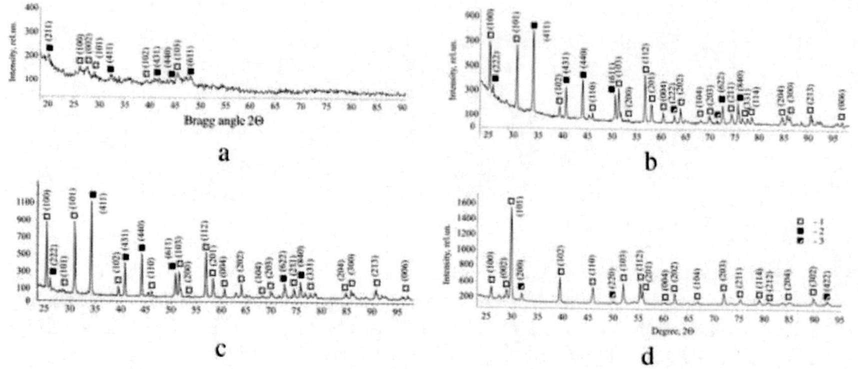

Fig. 3 XRD patterns of lanthanum-containing powders calcined at temperatures, °C: **a** 400; **b** 600; **c** 800; **d** 1000. Numbers correspond to: 1—La_2O_3 (space group P3m1); 2—La_2O_3 (space group Ia3); 3—LaO (space group F)

a new phase—LaOCl (PDF card No. 96-153-9094), using the COD-Inorg database 2022.06.29. The crystal system of LaOCl is tetragonal; S.G. P 4/nm; unit cell $a =$ 0.41090 nm, $c = 0.68650$ nm; estimated density is 5.454 g/cm^3.

The calcination of the sample at $T = 800$ °C in 5 h increases the total degree of crystallinity (DOC) to 60.46 wt.%. Correspondingly, amorphous content decreases to 39.54 wt.% (Fig. 3c). Phase composition of the sample is represented by trigonal-hexagonal La_2O_3, cubic La_2O_3, and tetragonal LaOCl. The elemental composition of the powder is O—7.3 wt.%; Cl—16.2 wt.%, and La—63.5 wt.%. The distribution of phase content for La_2O_3 (H) is 15.03 wt.%, and for La_2O_3 (C) 13.01 wt.%. The average particle size equals 46.7 nm.

The following temperature treatment of the sample at 1000 °C forms the 93% hexagonal-trigonal La_2O_3 (PDF card No. 05-0602) particles (Fig. 3d). Math analysis displays: Elemental composition of the powder is La—85.3 wt.%; O—14.7 wt.%. Space group P 63/mmc, E.N. 96-200-2287; Crystal system hexagonal; Unit cell $a =$ 0.39373 nm, $c = 0.61299$ nm; calculated density 6.574 g/cm^3; CSR 50.8 nm. The last 7 wt.% of the total mass of the sample corresponds to LaO phase. The degree of crystallinity (DOC) of the powder is 32.28%, and it is lower in comparison to the sample obtained at 800 °C. Amorphous content decreases to 67.72 wt.%.

In addition, the calculation of the crystal lattice parameter of body-centered Ia3 La_2O_3 (PDF card No. 22-0369) by three planes (222), (400), (431) shows its invariability for lanthanum chloride and lanthanum sulfate systems: $a = 1.1096$ nm for $La_2(SO_4)_3$ and $a = 1.1322$ nm for $LaCl_3$. A similar situation is typical for the LaO (F) phase (PDF card No. 33-0716), where the calculation of parameters for (111), (220), (311) planes show $a = 0.5147$ nm for $La_2(SO_4)_3$ and $a = 0.5082$ nm for $LaCl_3$. Based on the obtained result, it can be assumed that these oxides are structural elements of trigonal lanthanum oxide, and they are not individual phases.

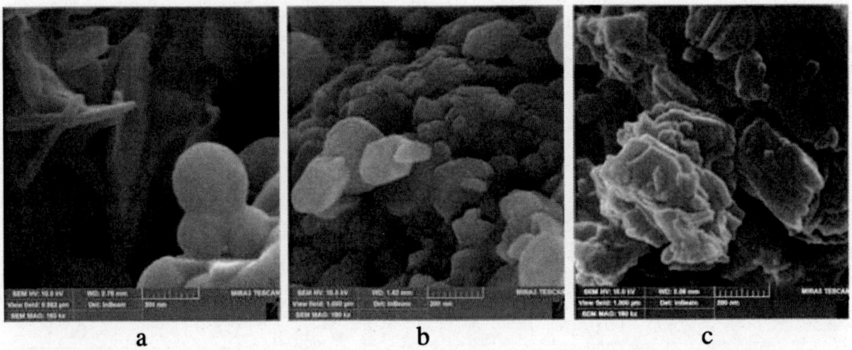

Fig. 4 The powders morphology formed via precipitation in the lanthanum sulfate system and calcined at the temperatures, °C: **a** 600, **b** 800, **c** 1000

The visualization of the samples was performed using scanning electron microscopy. The SEM images of the lyophilized precipitates and heated powders are present in Fig. 4. Analysis of the $La_2(SO_4)_3$ system unambiguously indicates that at a relatively low heating temperature of the samples, the powders are characterized by a heterogeneous composition and imperfect structure (Fig. 4a). The increase in the temperature treatment leads to the structures' formation with smooth edges and a shape tending to spherical (Fig. 4b). Moreover, the structures formed at $T = 800$ °C are more homogeneous in comparison to the structures calcined at $T = 600$ °C. The relatively high-temperature treatment of the samples results in the big aggregates formation (Fig. 4c).

The energy-dispersive spectra of the lanthanum oxide powders are present in Table 1. The obtained data indicate that the samples are relatively homogeneous in chemical composition. Whereas, carrying out the synthesis procedure without introducing additional substances, for example using mechanical synthesis, promotes the structures' formation with containing the main elements: oxygen, and lanthanum. But performing the chemical synthesis in the presence of inorganic salts and the additional species led the obtaining the powders containing anionic part of initial salts: nitrogen (Spectra 1–2), chlorine (Spectra 3–7), sulfur (Spectra 9–10). Moreover, additional elements are included in the composition of the powders: potassium and sodium, which act as precipitant and nucleating agents. In this case, the use of a reducing agent (hydroxylamine hydrochloric acid) leads to an increase in the number and quantity of elements that are included in the powder and influences its quality for subsequent usage (e.g., in biomedicine) consequently [45].

Hence, the EDS data clearly show that carrying out the procedure of washing sediments before their lyophilization does not allow completing the additional components' removal from the sample's composition.

Table 1 Distribution of chemical elements (wt.%) in the composition of lanthanum containing powders lyophilized at $T = 160$ °C and calcined at temperatures of 400 and 600 °C according to EDS data

Spectrum	Element, wt.%							Total
	O	La	Na	K	N	Cl	S	
	$La(NO_3)_3$, T = 400 °C, 5 h							
Spectrum 1	25.71	62.95	25.71	–	25.71	–	–	100.00
Spectrum 2	31.38	55.41	31.38	–	31.38	–	–	100.00
	$LaCl_3$, T = 600 °C, 2 h							
Spectrum 3	7.25	71.66	–	–	–	21.09	–	100.00
Spectrum 4	7.73	70.97	–	–	–	21.30	–	100.00
Spectrum 5	7.52	72.13	–	–	–	20.36	–	100.00
	$LaCl_3$, T = 600 °C, 2 h hydroxylamine hydrochloric acid							
Spectrum 6	5.82	75.72	–	0.77	–	17.68	–	100.00
Spectrum 7	5.90	77.05	–	–	–	17.06	–	100.00
	$La_2(SO_4)_3$, T = 600 °C, 2 h							
Spectrum 8	7.32	71.85	–	1.13	–	18.72	0.97	100.00
Spectrum 9	8.73	69.91	–	0.92	–	19.19	1.25	100.00
Spectrum 10	7.92	73.53	–	1.15	–	14.79	2.61	100.00

5 Thermal Behavior of the Ferrum- and Lanthanum-Containing Precipitates

The analysis of DTA curves makes it possible to analyze the reactions in the powders under the influence of temperature, which are accompanied by a change in their phase composition. For comparison, we selected systems of ferrum and lanthanum chlorides and sulfates, which give the possibility to determine the influence of the anionic component of the initial solutions on the phase formation processes and the crystallization of the corresponding oxides and perovskite.

In particular, in the $La_2(SO_4)_3$–$Fe_2(SO_4)_3$ system (Fig. 5a), the removal of adsorption-bound water takes place in the range of 120–140 °C with a sample mass loss of up to 6%. Further transformation of $Fe(OH)_3$ into goethite was determined at 260 °C with a maximum mass loss at 240 °C (~ 6.5%). Dehydroxylation of lanthanum hydroxide and oxyhydroxide takes place in the temperature range of 340–570 °C at a mass loss of 5.5%. The hematite phase crystallizes at 445 °C. The removal of sulfates and other auxiliary substances is determined in the range of 660–750 °C with a mass loss of ~ 8.8%. The formation of the perovskite phase ends at 960 °C. The total mass loss is 26.8%.

The removal of adsorbed-bonded water in the $La_2(SO_4)_3$–$FeSO_4$ system takes place up to 135 °C with a maximum mass loss (~ 12%) at 120 °C (Fig. 5b). Dehydroxylation of FeOOH with the possible formation of an intermediate phase of magnetite,

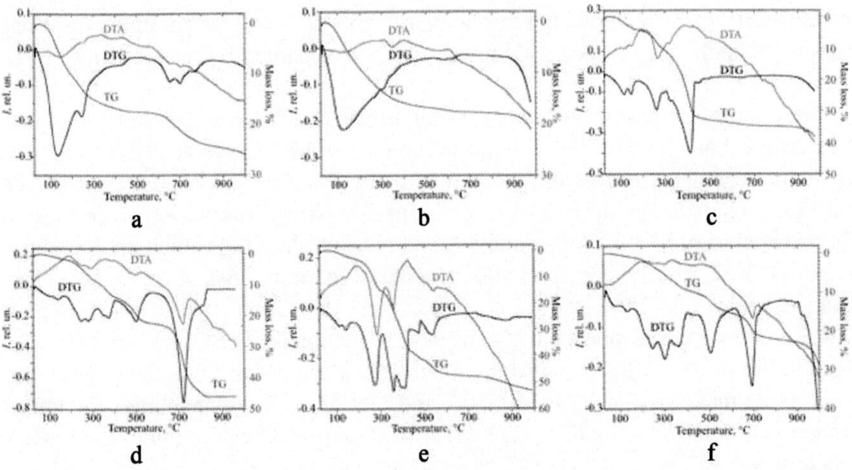

Fig. 5 TG-DTA curves of the precipitates obtained in the systems: **a** $La_2(SO_4)_3$–$Fe_2(SO_4)_3$; **b** $La_2(SO_4)_3$–$FeSO_4$; **c** $FeCl_3$–$La_2(SO_4)_3$; **d** $LaCl_3$–$Fe_2(SO_4)_3$; **e** $LaCl_3$–$FeCl_3$; **f** $LaCl_3$–$FeSO_4$

which is associated with the presence of Fe^{2+} cations in the system, ends at 310 °C. Dehydroxylation of $La(OH)_3$ is indicated by the endo effect at 350 °C. The plateau on the DTA curve at 400–610 °C characterizes the polymorphic transformation of lanthanum oxides and the formation of hematite. At the same time, it is possible that sulfates remain in the composition of powders, for example, in the structure of basic salts or ferric oxy sulfate $Fe_2(SO_4)_2O$. Formation of the perovskite phase is characterized by a bend in the DTA curve at 960 °C. The total mass loss of the sample is 23.2%.

The formation of the goethite phase in the $La_2(SO_4)_3$–$FeCl_3$ system was recorded at 195 °C, and its dehydroxylation took place at 250 °C (Fig. 1c). In general, for all lanthanum chloride systems, in the precipitates, not only the $La(OH)_3$ primary phase transforms into LaOOH at ~ 300 °C but also the additional $La(OH)_2Cl$ phase transforms into the LaOCl phase. Destruction of auxiliary substances occurs at T ~ 725 °C. The formation of the hematite phase is completed at 510–530 °C. Polymorphic transformations of lanthanum oxides and formation of the perovskite structure are carried out at 860–900 °C. Further thermal arrangement of the sample is not accompanied by mass loss, but the total mass loss is 46.1%.

Comparison of the $LaCl_3$–$Fe_2(SO_4)_3$ system (Fig. 1d) to the $La_2(SO_4)_3$–$Fe_2(SO_4)_3$ system (Fig. 1a) gives reason to believe that most of the reflexes on the DTA curve characterize the transformation of chloride compounds. The removal of adsorbed water is determined at 120 °C, and the precipitation of goethite—at 190 °C. The transformation of $La(OH)_3$ into LaOOH occurs at 270 °C, and its dehydroxylation gets at 310 °C. The formation of the La_2O_3 oxide's phase and hematite takes place in the temperature range of 390–425 °C. The removal of the impurity, which is accompanied by the mass loss of the sample, ends at 490 °C. The effects at 740, 860, and 920–930 °C characterize the formation of perovskite and polymorphic transformations of lanthanum oxides and they are not accompanied by mass change. Total mass loss is 39.9%.

The removal of adsorption-bound water for the $LaCl_3$–$FeCl_3$ system (Fig. 1e) is determined at 130–140 °C. The formation of goethite occurs at 190 °C, and the beginning of its transformation into hematite starts at 330 °C. Endo effects at 280 and 360 °C characterize the formation of lanthanum oxyhydroxide and oxide phases. The arrangement of lanthanum oxide's structure takes place in the temperature range of 545–600 °C, which lasts up to 900 °C without the mass loss of the samole. The total mass loss is 54.2%.

Adsorbed water disappears at 120–140 °C in the $LaCl_3$–$FeSO_4$ system (Fig. 1f). Goethite forms at ~ 200 °C, hematite at 340 °C. At the same time, the DTG curve shows three mass loss peaks at 260, 300, and 380 °C, which characterize the transformation of $La(OH)_3$, LaOOH, La(OH) with subsequent destruction of the LaOCl structure at ~ 500 °C.

In general, the TG-DTA analysis indicates that the reactions of the formation of lanthanum- and ferrum-oxygen compounds up to a temperature of ~ 960 °C are spatially separated due to differences in the temperatures of formation and transformation in the series $Fe(OH)_3$ → α-FeOOH → α-Fe_2O_3 and $La(OH)_3$ → LaOOH → La_2O_3. In particular, the crystallization of hematite ends at 425–445 °C, and perovskite at 960 °C, while lanthanum oxygen compounds form a number of unstable phases La_2O_3, LaO, $La(OH)_3$ in the temperature range of 450–960 °C.

Generally, we can see the greater mass loss of the samples corresponded to all chloride systems compared to sulfate systems, which can be explained by the presence of the $La(OH)_2Cl$ phase in the composition of the powders. Dehydroxylation occurs at 520–545 °C. Removal of auxiliary substances in sulfate systems takes place in the range of 660–710 °C, in chloride systems in the range of 700–725 °C.

6 Characteristics of the Ferrum- and Lanthanum-Containing Precipitates

X-ray phase analysis of samples obtained in solutions of inorganic salts of ferrum and lanthanum in a weakly alkaline environment and lyophilized at $T = 160$ °C indicates the presence of the main phases in the sample, the lanthanum hydroxide phase—$La(OH)_3$ (PDF card No. 06-0558) and the ferrum oxyhydroxide phase—goethite α-FeOOH (PDF card No. 17-536).

At the same time, the diffractograms show reflections from the interplanar distances of Fe(II)–Fe(III) layered double hydroxides (LDH), probably Green Rust II of the hydroxy sulfate composition (PDF card No. 3-92) and weak reflections of ferrum and lanthanum oxides. At the same time, the possibility of the formation of LDH compounds of mixed La–Fe composition, which will play the role of the primary structure in the formation of particles of complex oxides, in particular, lanthanum orthoferrites, is not excluded, which requires additional research.

But analysis of XRD spectra was performed using the program Match! and Reference database used COD-Inorg 2022.06.29 indicates the presence in all investigated samples of the non-stoichiometric oxide structure LaFeO Fe_2O_3, which corresponds to the triclinic system (PDF card No. 99-900-0000). The sample obtained in $La_2(SO_4)_3$–$Fe_2(SO_4)_3$ system includes the primary iron oxide particles, whereas the sample of La_2O_3–$Fe_2(SO_4)_3$ system contains the primary perovskite-type structures $FeLaO_3$ (PDF card No. 96-156-1805, S.G. Pnma, Crystal system orthorhombic; Unit cell $a = 0.55627$ nm, $b = 0.78416$ nm, $c = 0.55464$ nm; Calc. density 6.664 g/cm^3) and $FeLaO_3$ (PDF card No. 96-154-2145, S.G. Pm-3m; Crystal system cubic; Unit cell $a = 0.38900$ nm, Calc. density 6.848 g/cm^3). Distribution of chemical elements in the sample La_2O_3–$Fe_2(SO_4)_3$ is the following, wt.%: O—17.3; Fe—45.2; La—37.5; degree of crystallinity (DOC)—13.47 wt.%, the content of amorphous phases—86.53 wt.%. For the sample obtained in the $La(NO_3)_3$–$Fe_2(SO_4)_3$ system, the total value of the degree of crystallinity is 10.28 wt.%, and the amorphous content is 89.72 wt.%. The average particle size is 31 nm. Whereas for the sample formed in the $La_2(SO_4)_3$–$Fe_2(SO_4)_3$ system, the average particle size is 61 nm. DOC equals 9.85 wt.%, and amorphous content is 90.15 wt.%.

Therefore, already at the initial stage of structure formation, during lyophilization of the sediment, differences in the phase composition, degree of crystallinity, and size of the primary particles of the hydroxide samples are observed, depending on the chemical composition of the dispersion medium in which their precipitation was carried out (Fig. 6).

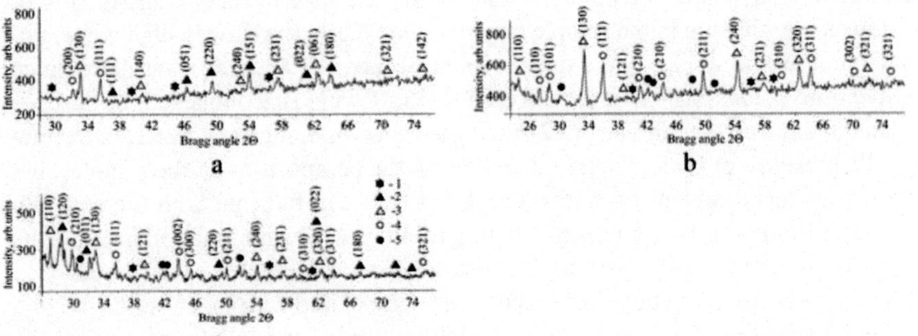

Fig. 6 Diffractograms of the precipitates formed in the systems: **a** $La(NO_3)_3$–$Fe_2(SO_4)_3$; **b** $La_2(SO_4)_3$–$Fe_2(SO_4)_3$; **c** La_2O_3–$Fe_2(SO_4)_3$. Numbers correspond to phases: 1—LDH; 2—$La(OH)_3$; 3—α-FeOOH; 4—La_2O_3; 5—LaO

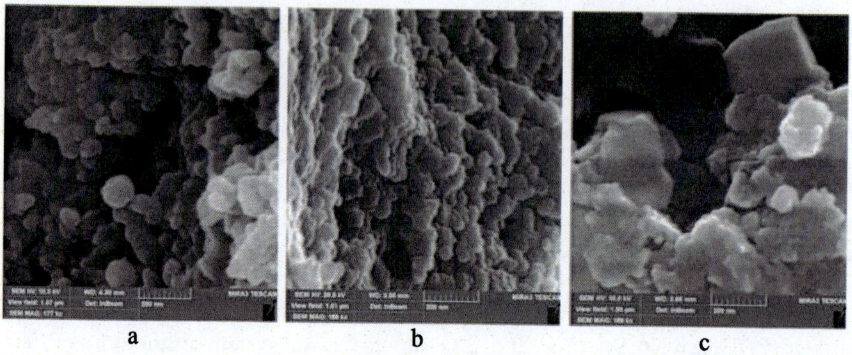

Fig. 7 SEM images of the precipitates lyophilized at $T = 160$ °C that were formed in the systems: **a** $La(NO_3)_3$–$Fe_2(SO_4)_3$; **b** $La_2(SO_4)_3$–$Fe_2(SO_4)_3$; **c** La_2O_3 $Fe_2(SO_4)_3$

The morphology of typical hydroxide samples precipitated in solutions of inorganic salts of ferrum and lanthanum, lyophilized at $T = 160$ °C is presented in Fig. 7. In general, samples of $La(NO_3)_3$–$Fe_2(SO_4)_3$ and $La_2(SO_4)_3$–$Fe_2(SO_4)_3$ systems are characterized by relative uniformity in particle size (Fig. 7a, b). At the same time, for the sample of the La_2O_3–$Fe_2(SO_4)_3$ system, the presence of plates with partially developed faces, probably hydroxides of lanthanum and ferrum, and spherical particles of smaller sizes, which may belong to nuclei of oxides of lanthanum, ferrum and complex oxides based on them, can be traced (Fig. 7c).

EDS spectra of samples correspond to $La_2(SO_4)_3$–$Fe_2(SO_4)_3$ and La_2O_3–$Fe_2(SO_4)_3$ systems lyophilized at $T = 160$ °C (Table 2), indicating that their composition includes the main elements (lanthanum, ferrum, and oxygen) as well as the chemical elements of auxiliary substances, in particular, carbon, nitrogen, sulfur, potassium. Analysis of the obtained data shows the heterogeneous distribution of ferrum and lanthanum in the sample composition. While a relatively high content of lanthanum (30.62 wt.%) with a trace amount of ferrum (0.6 wt.%) was determined in Spectrum 1, localization of ferrum (28.66–55.81 wt.%) accompanied the absence of lanthanum was seen in Spectra 2–4. At the same time, it can be traced in Spectra 5–7 the presence of both metals (La and Fe) in the composition of the samples, but at a higher mass content of lanthanum. Such data can indicate both the spatially separated formation of a phase of (oxy)hydroxides of ferrum and lanthanum (the $La_2(SO_4)_3$–$Fe_2(SO_4)_3$ system) and the precipitation of complex salts (the La_2O_3–$Fe_2(SO_4)_3$ system). The high carbon content is associated with the introduction of a concentrated urea solution into the initial salts solution, which helps to regulate the hydrolysis process. The potassium hydroxide's inclusion into the solutions promotes the carrying out of the nucleation process. The presence of sulfur and nitrogen is explained by the usage of the corresponding lanthanum and ferrum salts.

Table 2 Distribution of chemical elements (wt.%) in the composition of lanthanum–ferrum-containing powders lyophilized at $T = 160$ °C according to EDS data

Spectrum	Elements, wt.%							Total
	C	N	O	S	K	Fe	La	
	$La_2(SO_4)_3$–$Fe_2(SO_4)_3$							
Spectrum 1	18.64	–	33.42	11.34	5.38	0.60	30.62	100.00
Spectrum 2	18.18	–	40.45	5.80	1.11	34.46	–	100.00
Spectrum 3	17.21	–	20.66	5.08	1.24	55.81	–	100.00
Spectrum 4	16.63	–	49.19	4.60	0.92	28.66	–	100.00
	La_2O_3 (+HNO_3)–$Fe_2(SO_4)_3$							
Spectrum 5	31.15	–	32.52	12.91	8.80	4.21	10.41	100.00
Spectrum 6	19.63	–	33.46	10.68	7.72	5.65	22.86	100.00
Spectrum 7	13.03	10.66	39.87	3.38	6.45	7.42	19.19	100.00

7 Characteristic of the Lanthanum- and Iron-Containing Oxide Powders Calcined at $T = 800$ °C

X-ray phase analysis of samples obtained in systems of inorganic ferrum and lanthanum salts and calcined at a temperature of 800 °C indicates the presence of iron oxide phases (hematite) and a few modifications of lanthanum oxides, as well as primary perovskite particles (Fig. 8). At the same time, there are weak reflexes of the lanthanum hydroxide and oxyhydroxide phases on the diffractograms. Graphic interpretation of diffractograms was carried out according to the database of International Powder Standards Committee (JSPDS International Centre for Diffraction Data 1999). Also, among mineral phases, we can see in diffractograms the reflexes corresponding to cubic body-centered Ia3 La_2O_3 (JSPDS No 22-0369), hexagonal-trigonal S.G. P-3m1 La_2O_3 (PDF card No. 05-0602), and face-centered S.G. F LaO (PDF card No. 33-0716), $La(OH)_3$ (PDF card No. 06-0558), LaOOH (PDF card No. 13-0436), and hematite α-Fe_2O_3; S.G. R-3c (PDF card No. 13-534).

Data analysis using Match! According to the Reference database, COD-Inorg 2022.06.29 makes it is possible to clarify the characteristics of powders. That, the application of this software gives the possibility to identify the following phase composition of the powders by the number and intensity of reflexes on the diffractograms:

- complicate oxide phase LaFeO Fe_2O_3, crystal system triclinic (*anorthic*) (PDF card No. 99-900-000);
- $FeLaO_3$ Crystal system cubic, Space group *Pm-3m* (PDF card No. 96-154-2145); Unit cell $a = 0.38900$ nm; Calc. density 6.848 g/cm^3;
- $LaFeO_3$ Crystal system orthorhombic, Space group *Pbnm* (PDF card No. 96-152-6451); Unit cell $a = 0.55520$ nm, $b = 0.55630$ nm, $c = 0.78430$ nm; Calc. density 6.656 g/cm^3;

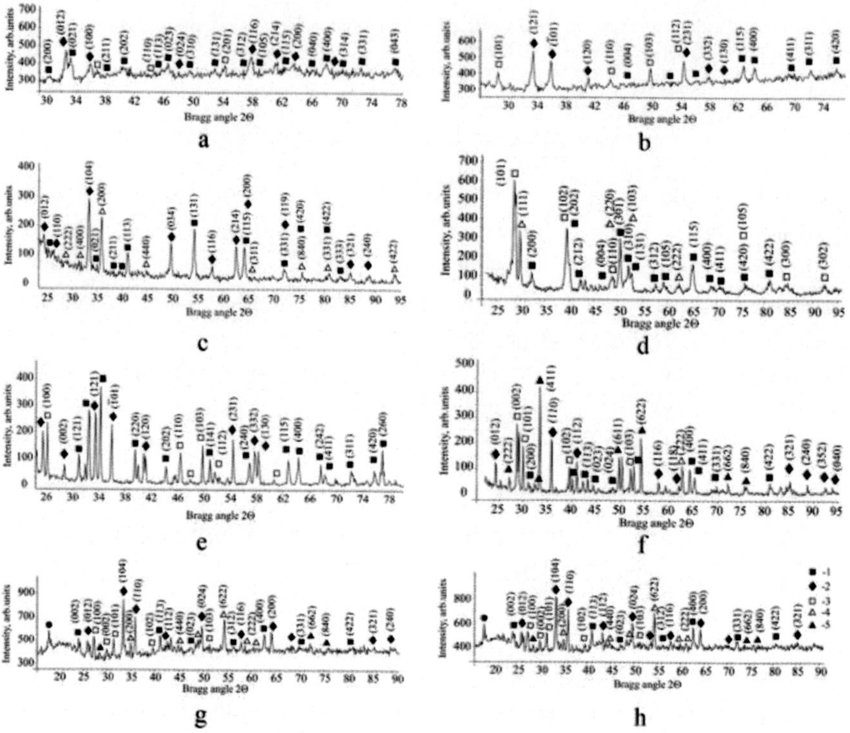

Fig. 8 XRD patterns of the powders formed at $T = 800$ °C in the systems: **a** $La(NO_3)_3$–$Fe_2(SO_4)_3$; **b** La_2O_3–$Fe_2(SO_4)_3$; **c** $La_2(SO_4)_3$–$FeSO_4$; **d** $La_2(SO_4)_3$–$K_3[Fe(CN)]$; **e** $LaCl_3$–$FeCl_3$; **f** $La_2(SO_4)_3$–$FeCl_3$; **g** LaCl3–$FeSO_4$; **h** $LaCl_3$–$Fe_2(SO_4)_3$. Numbers correspond to: 1—$LaFeO_3$; 2—α-Fe_2O_3; 3—La_2O_3 (P-3m1); 4—LaO (F); 5—La_2O_3 (Ia3)

- hematite α-Fe_2O_3, Crystal system rhombohedral, Space group *R-3c* (PDF card No. 96-101-1241); Unit cell $a = 0.54300$ nm, $\alpha = 55.280°$; Meas. density 5.260 g/cm^3; Calc. density 5.261 g/cm^3.

The Peak data Counts Amount option makes it possible to determine the contribution of specified compounds to the phase composition of sediments, the degree of powders crystallinity, the average size of crystallites, the elemental composition of samples, etc.

Table 3 shows the interpretation of the data of diffractograms using Match! Phase Analysis summarizing the results of the X-ray phase analysis of samples obtained in systems based on lanthanum and ferrum oxides and calcined at a temperature of 800 °C. It should be noted that the degree of crystallinity (DOC) of the powders varies from 4.88% for the $La(NO_3)_3$–$Fe_2(SO_4)_3$ system to 86.37% for the $LaCl_3$–$FeCl_3$ system. The size of the primary particles (CSR) varies from 60 nm for the $La(NO_3)_3$–$Fe_2(SO_4)_3$ system to 112 for the system $La_2(SO_4)_3$–$FeSO_4$.

Table 3 Match! Phase Analysis using diffraction data of the La–Fe containing oxide powders calcined at 800 °C

Element amount (weight %)	Degree of crystallinity analysis		Integrated profile areas			
	DOC, %	AC, %	Selected phases (SP) (Match!), peak area (%*), CSR, nm	Peak intensity SP, %	Other phases (JSPDS)	Peak intensity OP, %
$La(NO_3)_3$–$Fe_2(SO_4)_3$						
–	4.88	95.12	–	–	$LaFeO_3$ La_2O_3 (P-3m1) α-Fe_2O_3	–
La_2O_3–$Fe_2(SO_4)_3$						
–	11.01	88.99	α-Fe_2O_3, LaFeO Fe_2O_3 112 nm	29.14	$LaFeO_3$ La_2O_3 (P-3m1)	70.86
$La_2(SO_4)_3$–$FeSO_4$						
Fe—68.7% O—29.8% La—1.6%	47.49	52.51	$FeLaO_3$ (2.7%)* Fe_2O_3 (97.3%)* 60 nm	23.72	$LaFeO_3$ La_2O_3 (P-3m1) LaO	76.18
$La_2(SO_4)_3$–$K_3[Fe(CN)]$						
–	55.09	44.91	$FeLaO_3$ 69 nm	34.76	La_2O_3 (Ia3) La_2O_3 (P-3m1) α-Fe_2O_3 $La(Fe(CN)_6)$	65.24
$LaCl_3$–$FeCl_3$						
Fe—45.2% O—17.3% La—37.5%	86.37	13.63	LaFeO Fe_2O_3 (4.57%) 60 nm	43.62	La_2O_3 (P-3m1) α-Fe_2O_3	56.38
$La_2(SO_4)_3$–$FeCl_3$						
Fe—68.5% O—29.7% La—1.8%	69.21	30.79	α-Fe_2O_3 (10.09%)* 84 nm $FeLaO_3$ (1.66%)* LaFeO Fe_2O_3 (4.24%)*, 90 nm	85.21	La_2O_3 (Ia3) La_2O_3 (P-3m1)	14.79
$LaCl_3$–$FeSO_4$						
La—57.2% Fe—23.0% O—19.8%	10.70	89.30	α-Fe_2O_3, LaFeO Fe_2O_3 85 nm $LaFeO_3$	30.06	La_2O_3 (Ia3) La_2O_3 (P-3m1)	69.94
$LaCl_3$–$Fe_2(SO_4)_3$						
Fe—45.2% La—37.5% O—17.3%	10.76	89.24	LaFeO Fe_2O_3 97 nm	5.98	La_2O_3 (Ia3) La_2O_3 (P-3m1) α-Fe_2O_3	94.02

The presence of an additional phase $C_5Fe_2N_6O_3$ (17.2%) was detected in the $LaCl_3$–$FeCl_3$ system. Its PDF card No. 96-156-1261; S.G. Pnma, C.S. orthorhombic; Unit cell $a = 1.39674$ nm, $b = 0.74341$ nm, $c = 1.04630$ nm; Calc. density 1.857 g/cm^3; elemental composition, %: Fe—36.8; N—27.7; C—19.8; O—15.8.

It is not excluded the presence of an admixture of phases $LaK(Fe(CN)_6)\cdot 4H_2O$ (PDF card No. 96-210-64830), $La(Fe(CN)_6)\cdot 4H_2O$ (PDF card No. 96-210-6581) or $La(Fe(CN)_6)\cdot 5H_2O$ (PDF card No. 96-210-6831) in the powders corresponds to the $La_2(SO_4)_3$–$K_3[Fe(CN)]$ system.

Typical morphology of the structures formed in the iron–lanthanum-containing oxide systems at $T = 800$ °C is present on Fig. 9. Generally, particles are characterized as heterogeneous components of the powders. Relatively, big plate-like structures correspond to hexagonal (trigonal) lanthanum oxide (Fig. 9a, b, f), and spherical-shaped small structures may be related to hematite particles (Fig. 9c, d, e). Probably, orthorhombic perovskite $LaFeO_3$ and admixture cubic ferrite $FeLaO_3$ are the smallest spherical particles. At the same time, taking into account the results of Match! Analysis, the 60–112 nm aggregates may be estimated as mixed triclinic (anorthic) structure LaFeO Fe_2O_3. Hematite and lanthanum ferrite particles have less than 85 nm of size.

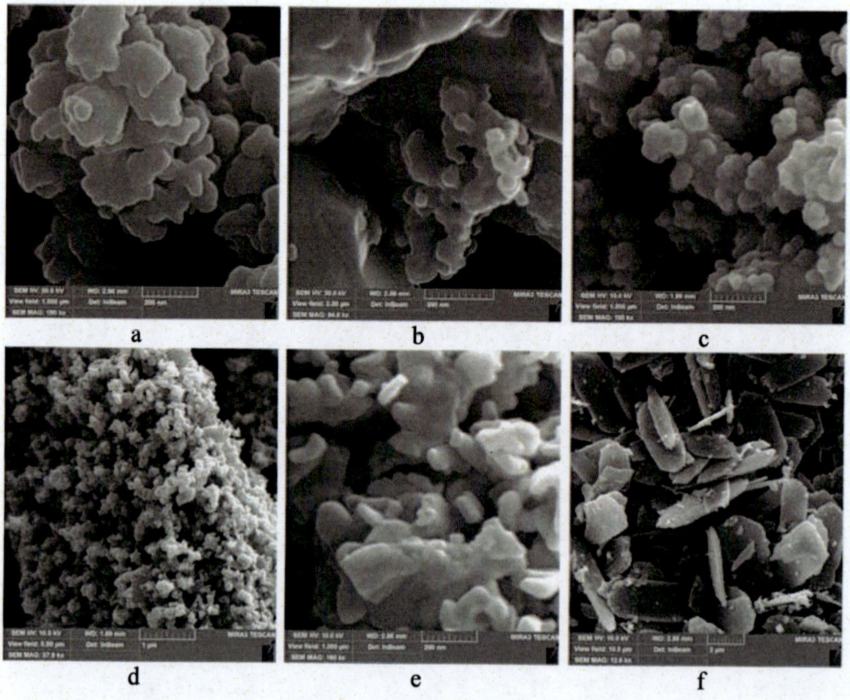

Fig. 9 SEM images of the structures formed at $T = 800$ °C in the systems: **a** $La(NO_3)_3$–$Fe_2(SO_4)_3$; **b** La_2O_3–$Fe_2(SO_4)_3$; **c** $La_2(SO_4)_3$–$Fe_2(SO_4)_3$ nano; **d** $La_2(SO_4)_3$–$Fe_2(SO_4)_3$ mkm; **e** $La_2(SO_4)_3$–$K_3[Fe(CN)]$ nano; **f** $La_2(SO_4)_3$–$K_3[Fe(CN)]$ mkm

Table 4 Distribution of chemical elements (wt.%) in the composition of lanthanum–ferrum-containing oxide powders calcined at $T = 800$ °C according to EDS data

Spectrum	Elements, wt.%						Total
	C	O	Fe	La	S	K	
	La(NO₃)₃–Fe₂(SO₄)₃						
Spectrum 1	19.79	26.41	44.42	9.38	–	–	100.00
Spectrum 2	20.06	26.24	44.76	8.94	–	–	100.00
	La₂O₃–Fe₂(SO₄)₃						
Spectrum 3	–	20.87	76.13	1.94	0.23	0.83	100.00
Spectrum 4	–	16.87	80.67	1.86	0.00	0.59	100.00
Spectrum 5	–	24.12	5.91	40.98	14.84	14.14	100.00
Spectrum 6	–	26.29	2.40	39.82	19.39	12.11	100.00
	La₂(SO₄)₃–Fe₂(SO₄)₃						
Spectrum 7	–	24.26	9.03	48.85	17.86	–	100.00
Spectrum 8	–	25.36	47.10	20.09	7.45	–	100.00
Spectrum 9	–	31.57	41.31	19.98	7.14	–	100.00
	La₂(SO₄)₃–K₃[Fe(CN)]						
Spectrum 10	–	22.58	1.03	70.20	2.66	3.53	100.00
Spectrum 11	–	22.48	1.93	65.94	8.45	1.20	100.00
Spectrum 12	–	17.67	1.74	72.77	6.44	1.37	100.00

Whereas Element Amount (weight %) calculated using XRD data gives information about the average distribution of the chemical elements in the powders (Table 3). the EDS analysis of the powders confirms the formation of a few kinds of structures and it shows the incorporation of the residual elements into the individual mineral phases.

Thus, we can see the spectra of relatively pure hematite (Spectra 3, 4), mixed lanthanum–ferrum structures, probably perovskites (Spectra 1, 2, 8, 9), lanthanum oxides (Spectra 10–12), mixed LaFeO Fe_2O_3 structure (Spectrum 7), and non-stoichiometric structures containing La–Fe (Spectra 5, 6) (Table 4).

8 Characteristic of the Lanthanum- and Iron-Containing Oxide Powders Calcined at $T = 1100$ °C

X-ray phase analysis data (using the data processing Match! program) shows that the composition of the initial solutions affects not only the composition, structure, properties, and morphology of the primary precipitates but also the listed parameters of the calcined samples. It is found that the main phases include the composition of the lanthanum–ferrum-containing powders obtained via thermal treatment at the *T*

= 1100 °C. There are hematite S.G. *R-3c*, a compound of mixed composition LaFeO Fe_2O_3, crystal system triclinic (anorthic), two lanthanum orthoferrites $LaFeO_3$ S.G. *Pbnm* and $FeLaO_3$ S.G. *Pm-3m.*

At the same time, lanthanum oxide La_2O_3 (S.G. *P-3m1*), as well as complex salts (for example, $Fe_2(SO_4)_2O\ 7H_2O$), can be identified among additional phases. Table 5 shows the data on Element Amount (weight %), Degree of crystallinity analysis, Peak area (%), and Peak intensity of selected (SP) and other (OP) phases, and CSR (nm) of several phases.

In general, the difference in Element Amount (weight %) of samples formed in different systems draws attention. In particular, for the systems $La_2(SO_4)_3$–$FeSO_4$, $La_2(SO_4)_3$–$K_3[Fe(CN)]$, $LaCl_3$–$FeCl_3$, $LaCl_3$–$Fe_2(SO_4)_3$ the ratio of lanthanum and ferrum is close to each other, while as for the $La(NO_3)_3$–$Fe_2(SO_4)_3$ and $LaCl_3$–$FeSO_4$ systems, the content of lanthanum exceeds the content of ferrum by 7 and 19 times, and in the system $La_2(SO_4)_3$–$FeCl_3$, on the contrary, the content of lanthanum is 2.5 times higher than the content of ferrum.

The degree of crystallinity (DOC) of the samples is higher on comparison to the similar samples calcined at T = 800 °C and varies within 62–74%. The intensity of the peaks of the identified phases ranges from 94 to 99%, with the exception of the sample $La_2(SO_4)_3$–$FeCl_3$, in which almost 36% of the peak intensity probably belongs to lanthanum oxide, as evidenced by the mass distribution of elements.

Figure 10 shows the diffractograms of lanthanum–ferrum-containing oxide samples calcined at a temperature of 1100 °C. The main defined phases are the structure of perovskite $FeLaO_3$ (PDF card No. 37-1493) and hematite α-Fe_2O_3 (PDF card No. 13-534). The main differences between the samples lie in the distribution of the specified phases, which is manifested in the variation of the intensity of reflexes on the diffractograms, and the degree of crystallinity, which is confirmed by the Match! Analysis (Table 5). Calculation of the primary particle size (CSR) of perovskite and hematite using the Debye–Scherrer formula indicates nanometer particle sizes and a narrow range in their divergence (Table 6). In particular, for the perovskite structure, the CSR index calculated on the (200), (202), and (004) planes ranges from 27.1 nm (La_2O_3–$Fe_2(SO_4)$ system) to 35.9 nm ($LaCl_3$–$Fe_2(SO_4)_3$ system). For the hematite structure, the CSR index calculated on the (104), (110), and (018) planes varies from 29.7 nm ($LaCl_3$–$FeCl_3$ system) to 37.3 nm (La_2O_3–$Fe_2(SO_4)_3$ system).

The SEM images of the samples calcined at T = 1100 °C in 5 h are presented in Fig. 11. The morphology of the powders includes two types of observed structures: well-crystallized crystals of the hexagonal and pseudo-hexagonal shape of perovskite-type particles (up to 200 nm) and smaller spherical particles of hematite (up to 60 nm) that are localized in the surface of perovskite's faces as a separate phase (Fig. 11a, b). The resulting structure is characterized by high porosity and 3d shape as aggregates. Perovskite particles formed in Fig. 11c are well crystallized and they do not form aggregates. The particles obtained in systems from Fig. 11d–f are relatively homogeneous and their shape is characterized by smoothed corners and edges of crystals.

Table 5 Match! Phase Analysis using diffraction data of the La–Fe containing powders calcined at 1100 °C

Element amount (weight %)	Degree of crystallinity analysis		Integrated profile areas			
	DOC, %	AC, %	Selected phases (SP) (Match!), peak area (%), CSR, nm	Peak intensity SP, %	Other phases (PC PDF)	Peak intensity OP, %
$La(NO_3)_3$–$Fe_2(SO_4)_3$ П1						
Fe—62.7% O—28.5% La—8.8%	70.23	29.77	$LaFeO_3$ (15.3%) α-Fe_2O_3, (84.7%) 440 nm	94.47	–	5.53
$La_2(SO_4)_3$–$FeSO_4$ П 10						
Fe—34.4% O—22.3% La—43.3%	67.63	32.37	LaFeO Fe_2O_3 (25.64%)* α-Fe_2O_3 (16.35%)	94.95	–	5.05
$La_2(SO_4)_3$–$K_3[Fe(CN)]$ П 13						
Fe—43.3% O—17.3% La—37.5	74.38	25.62	$LaFeO_3$ (75.7%) α-Fe_2O_3 (24.3%)	96.53	–	3.47
$LaCl_3$–$FeCl_3$ П15						
Fe—45.2% O—22.3% La—43.3%	74.27	25.7	$FeLaO_3$ + $LaFeO_3$ (75.7%); 169 nm α-Fe_2O_3 (24.3%)*	96.53	–	3.47
$La_2(SO_4)_3$–$FeCl_3$ П16						
Fe—23.0% O—19.8% La—57.2%	65.90	34.10	$LaFeO_3$ (80.6%)*, 526 nm $Fe_2(SO_4)_2O$ $7H_2O$	80.28	LaFeO Fe_2O_3 α-Fe_2O_3	19.72
$LaCl_3$–$FeSO_4$ П 17						
La—3.5% Fe—67.1% O—29.4%	69.09	30.91	α-Fe_2O_3 (93.9%) $FeLaO_3$ (6.1%) 915 nm	96.45	–	3.55
$LaCl_3$–$Fe_2(SO_4)_3$ П18						
Fe—45.2% La—37.5% O—17.3%	62.75	37.25	LaFeO Fe_2O_3 (32.97%) 820 nm $LaFeO_3$ (0.33%), α-Fe_2O_3	99.55	$Fe_2(SO_4)_2O$ $7H_2O$ (0.63%)	0.45

EDS analysis of the powders calcined at $T = 1100$ °C shows the heterogeneous compositions of the samples including a few mineral phases. Analysis of the spectra indicates the presence of pure hematite particles (Spectra 1, 10, 13, 16), lanthanum-doped hematite (Spectra 3–5, 18, 19), $LaFeO_3$ perovskite-type structures (Spectra 15, 17), lanthanum ferrite $FeLaO_3$ (Spectra 11, 12), non-stoichiometric compounds of the type LaFeO Fe_2O_3 (Spectra 2, 6, 7, 9, 20), lanthanum oxide doped with ferrum (Spectrum 14).

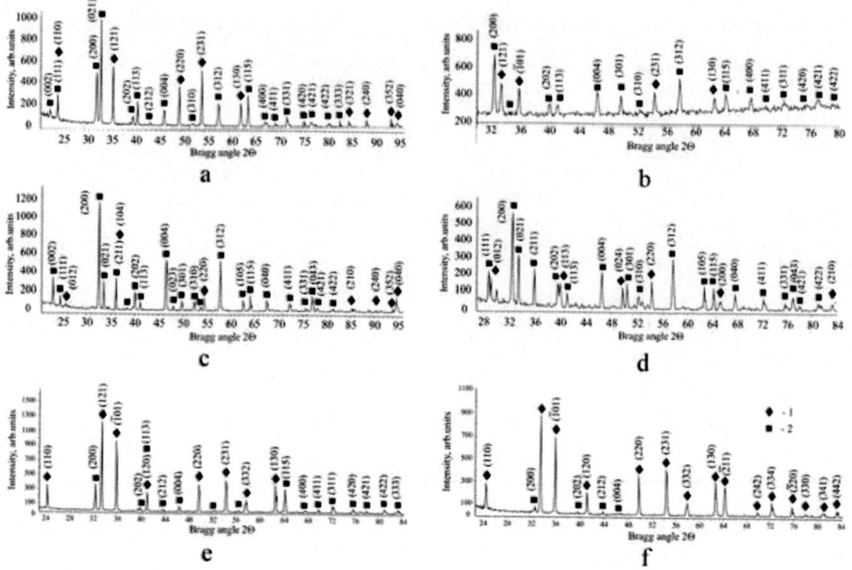

Fig. 10 XRD patterns of the powders formed at T = 1100 °C in the systems: **a** $La(NO_3)_3$–$Fe_2(SO_4)_3$; **b** La_2O_3–$Fe_2(SO_4)_3$; **c** $LaCl_3$–$FeCl_3$; **d** $La_2(SO_4)_3$–$FeCl_3$; **e** $LaCl_3$–$FeSO_4$; **f** $LaCl_3$–$Fe_2(SO_4)_3$. Numbers correspond to: 1—α-Fe_2O_3; 2—$LaFeO_3$

Table 6 The primary particle's size (CSR) of perovskite and hematite (T = 1100 °C), calculated according to the Debye–Scherrer equation

Characteristics of the system	Average particle size, nm	
	Perovskite	Hematite
$La(NO_3)_3$–$Fe_2(SO_4)_3$	28.9	30.3
La_2O_3–$Fe_2(SO_4)_3$	27.1	37.3
$La_2(SO_4)_3$–$FeSO_4$	29.3	31.5
$LaCl_3$–$FeCl_3$	30.7	29.7
$La_2(SO_4)_3$–$FeCl_3$	31.7	33.5
$LaCl_3$–$FeSO_4$	33.9	32.6
$LaCl_3$–$Fe_2(SO_4)_3$	35.9	30.8

At the same time, attention is drawn to the presence of auxiliary elements in the composition of the structure, including Spectrum 17, in which there is an admixture of sulfur can be associated, for example, with the oxo salt structure of $Fe_2(SO_4)_2O$ $7H_2O$, which was identified using X-ray diffraction in the $LaCl_3$–$Fe_2(SO_4)_3$ system (see Table 5). In Spectrum 12 ($LaCl_3$–$FeCl_3$ system), an admixture of chlorine (0.16 wt.%) was determined.

Therefore, temperature treatment of the ferrum–lanthanum oxide powders at 1100 °C leads to the formation of heterogeneous structures, that chemical compositions contain first of all ferrum, oxygen, and lanthanum. At the same time, the elements that are introduced into the systems during the synthesis of the primary

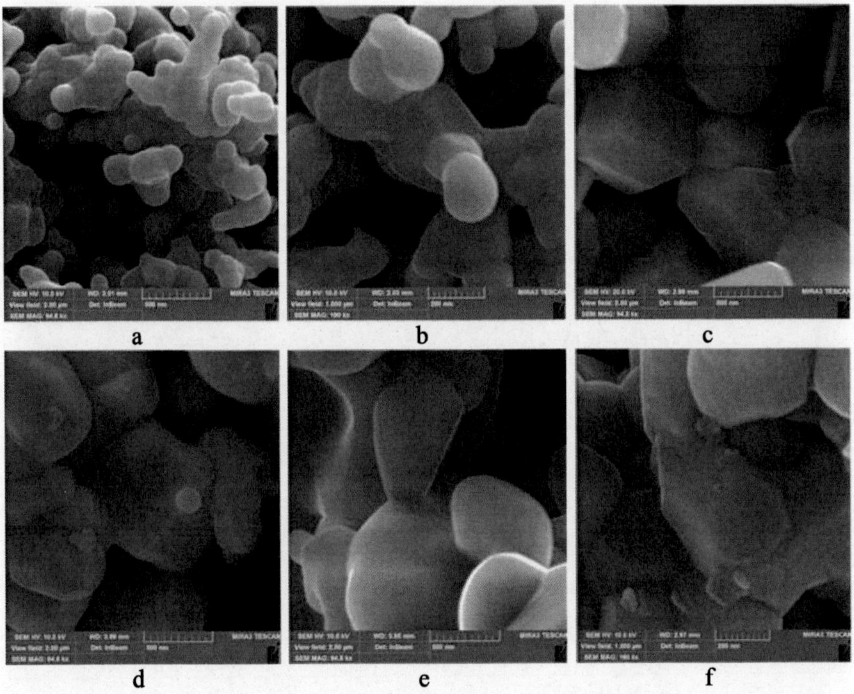

Fig. 11 SEM images of the powders obtained via temperature treatment at $T = 1100$ °C in the systems: **a** $La(NO_3)_3$–$Fe_2(SO_4)_3$, micro-sized aggregates; **b** $La(NO_3)_3$–$Fe_2(SO_4)_3$, nanostructures; **c** La_2O_3–$Fe_2(SO_4)_3$; **d** $LaCl_3$–$FeCl_3$; **e** $La_2(SO_4)_3$–$FeCl_3$; **f** $LaCl_3$–$FeSO_4$

precipitates, as part of auxiliary substances, are removed during the calcination of the samples by destruction, burning or gas formation (Table 7).

Typical EDS spectra of hematite, perovskite $LaFeO_3$, mixed LaFeO Fe_2O_3, and La_2O_3 structure are present in Fig. 12. We can see the main elements Fe, O (Fig. 12a) and, correspondingly, La, Fe, O with La to Fe ratio close to 1:1 (Fig. 12b) for perovskite as well as two phases aggregates with non-stoichiometric ratio La to Fe (Fig. 12c).

9 Measurement of Specific Surface Area

For the use of powders, for example, for photocatalysis, the composition of which includes particles of lanthanum orthoferrite and hematite, an important characteristic is the specific surface area. Table 8 presents the results of measuring samples of two systems $LaCl_3$–$FeCl_3$ and $LaCl_3$–$FeSO_4$. In both cases, we can see the decrease in the Specific surface area (m^2/g), when the sample firing temperature is increased. At the same time, it is likely that the presence of sulfates in the system contributes to the

Table 7 Distribution of chemical elements (wt.%) in the composition of the lanthanum–ferrum oxide powders (calcined at $T = 1100$ °C) according to EDS data

Spectrum	Elements, wt.%					Total
	O	Fe	La	Cl	S	
	$La(NO_3)_3$–$Fe_2(SO_4)_3$					
Spectrum 1	31.40	68.60	0.00	–	–	100.00
Spectrum 2	24.60	58.80	16.60	–	–	100.00
Spectrum 3	25.17	67.86	6.96	–	–	100.00
Spectrum 4	30.70	61.85	7.45	–	–	100.00
	La_2O_3–$Fe_2(SO_4)_3$					
Spectrum 5	19.26	77.93	2.80	–	–	100.00
Spectrum 6	26.98	59.95	13.08	–	–	100.00
Spectrum 7	22.09	56.66	21.25	–	–	100.00
Spectrum 8	6.77	39.68	53.56	–	–	100.00
Spectrum 9	24.46	56.56	18.98	–	–	100.00
	$LaCl_3$–$FeCl_3$					
Spectrum 10	26.93	73.07	–	–	–	100.00
Spectrum 11	17.62	25.43	56.95	–	–	100.00
Spectrum 12	13.72	25.70	60.43	0.16	–	100.00
	$La_2(SO_4)_3$–$FeCl_3$					
Spectrum 13	21.84	78.16	–	–	–	100.00
Spectrum 14	20.31	9.06	65.68	–	4.96	100.00
Spectrum 15	26.48	47.93	25.59	–	–	100.00
	$LaCl_3$–$FeSO_4$					
Spectrum 16	27.93	72.07	0.00	–	–	100.00
Spectrum 17	23.16	42.66	34.18	–	–	100.00
Spectrum 18	23.79	74.17	2.05	–	–	100.00
Spectrum 19	25.07	72.22	2.71	–	–	100.00
Spectrum 20	24.56	57.44	18.00	–	–	100.00

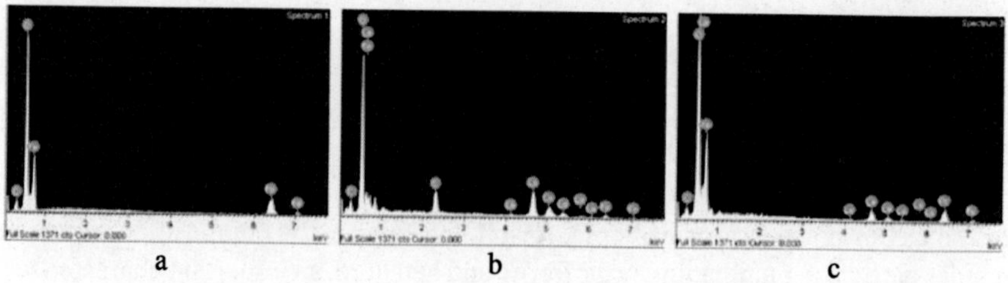

Fig. 12 EDS spectra of: **a** α-Fe_2O_3, **b** $LaFeO_3$, **c** lanthanum oxide doped with ferrum

Table 8 The value of specific surface area of the calcined powders

Temperature	Hinge, mg	Specific surface area, m^2/g
	$LaCl_3$–$FeCl_3$	
160	197.0	1.26
800	438.3	0.52
1100	591.5	0.39
	$LaCl_3$–$FeSO_4$	
160	226.6	3.72
800	233.2	2.31
1100	160.2	2.02

increase of SSA compared to the chloride system. Probably, for carrying out photocatalysis, it is advisable to use samples calcined at 800 °C, which are characterized by a polyphase composition that can be effective for overlapping the band gap of semiconductors, lower solubility compared to lyophilized samples (160°) and the absence of sintered aggregates (1100 °C). Generally, the obtained data are well correlated with the SSA measurement of the $LaFeO_3$ powders formed via microwave-assisted synthesis, where it is 2.60 m^2/g [41].

10 Discussion

Thus, the analysis of the obtained data gives the reason to draw a conclusion about the colloidal-chemical mechanism of the formation of nanoscale structures of orthoferrite and lanthanum hematite during the chemical precipitation of hydroxide precipitates in an alkaline environment, their lyophilization and heat treatment at 800 and 1100 °C temperatures.

The study of the formation process of lanthanum oxide particles was carried out by a chemical method in the systems of lanthanum sulfate and chloride in the presence of potassium and ammonium hydroxides. According to RFA, lanthanum hydroxide $La(OH)_3$ or lanthanum hydrochloride $LaCl(OH)_2$ precipitates in the suspension, because of the presence of chlorides in the initial solution. Thermal treatment of hydroxide precipitates (250–300 °C) leads to the phase transformations of the specified compounds into, respectively, the LaOOH and LaClO phases accompanied by the removal of structurally bound water and the loss of sample mass according to TG-DTA data.

The transformation of oxyhydroxide phases into hexagonal-trigonal lanthanum oxide begins at 390 °C temperature and continues up to 710–735 °C temperatures. Probably, the combustion of impurity components, in particular, chlorides, ends at $T = 950$ °C, which is accompanied by a loss of sample mass. The lanthanum sulfate system is also characterized by the transformation of lanthanum hydroxide into the corresponding oxyhydroxide, and then the oxide. At the same time, sulfur can become part of unstable hydroxide compounds and be removed from the composition of the precipitates at temperatures above 600 °C. The obtained results are in good agreement with those proposed in the work [46], the double-step mechanism of the dehydration process from lanthanum hydroxide to lanthanum hydroxide oxide. It consists in the transformation of $La(OH)_3$ into LaOOH, which crystallizes in a monoclinic crystal structure with the space group P21/m (no. 11) (endo effect at ~ 330 °C). The process of dehydroxylation of LaOOH with the formation of La_2O_3 ends with the endo effect at 490 °C temperature. At the same time, the phase formation process by a chemical method is significantly affected by the introduction of urea into the initial solution, which dissociates into CO_3^{2-}, OH^-, and NH_4^+ ions in the solution and leads to an increase in the pH value and the homogeneous precipitation of lanthanum carbonate particles [12]. The study of the process of the formation of the La_2O_3 particles in situ in the presence of carbonates under the influence of temperature indicates the existence of $La(OH)_3$ particles up to 200 °C, the coexistence of structures of tetragonal $La_2O_2CO_3$ and hexagonal La_2O_3 at $T = 400$ °C, removal of CO_2 after 700 °C and formation of homogeneous La_2O_3 phase at 800 °C [1]. At the same time, unstable compounds LaOOH and $La_2(OH)_4CO_3$ were found as intermediate phases at the low-temperature stage of phase formation.

The introduction of a ferric sulfate solution into a solution of lanthanum salts changes the lyophilized precipitates' composition at $T = 160$ °C. In particular, in addition to the reflexes of lanthanum hydroxide, the diffractograms revealed the reflexes of iron-containing phases—goethite and, presumably, Fe(II)–Fe(III) layered double hydroxides (Green Rust). Thus, it is possible to make an assumption about the spatial separation of the development of lanthanum-containing and ferrum-containing hydroxide compounds. Numerous studies of structure formation in aqueous solutions of ferric and ferrous iron salts testify to the formation of primary phases of ferrihydrite or Green Rust, the anionic component of which determines the type of crystal structure of LDH [47]. In particular, hydroxy sulfate Green Rust was formed in the presence of sulfates, the transformation of which was carried out through the formation of iron oxyhydroxides—lepidocrocite and goethite [48]. Dehydroxylation of the crystal lattice of ferrum oxyhydroxides at a 250–280 °C temperature leads to the formation of unstable phases of magnetite (in the presence of ferrous cations) or maghemite (in the presence of ferric cations only). Further maghemite polymorphic transformation into hematite takes place at T ~ 490–545 °C.

It is important that the diffractograms of the lyophilized sediments contain reflections that correspond to the phases of several lanthanum oxides, including hexagonal-trigonal La_2O_3, body-centered (Ia3) La_2O_3, and face-centered (F) LaO phases. In our opinion, the presence of additional chemical reagents in the system can lead to their entry into the crystal lattice or adsorption on individual faces of lanthanum oxide crystals. In addition, the course of the red-ox process, for example, the oxidation of Fe(II) to Fe(III) and, accordingly, the reduction of La(III) to La(II) is not excluded. The question about the existence of layered double hydroxides in the Fe(II)–La(III) suspension remains open [49]. They can play the role of a precursor phase for the formation of lanthanum orthoferrite. So, in fact, the systems are characterized by heterogeneity up to 800 °C temperature. The powders include individual particles of iron oxide, hematite and of lanthanum oxide La_2O_3, mainly hexagonal (trigonal) structure, but their solid-state interaction with the formation of a homogeneous lanthanum orthoferrite phase does not occur.

The diffractograms show weak reflections of the $LaFeO_3$ perovskite or the structure identified by Match! Analysis as $LaFeO{\cdot}Fe_2O_3$, or, taking into account stoichiometry, $LaFeO_3$ and Fe_2O_3. Thus, several mineral phases coexist in the composition of sediments calcined at $T = 800$ °C—hematite; hematite doped with lanthanum cations; lanthanum oxide doped with ferrum; two mixed oxides $LaFeO_3$ (orthorhombic) and $FeLaO_3$ (cubic), as well as non-stoichiometric complex oxides of the type $LaFeO{\cdot}Fe_2O_3$. Whereas, the powders heat-treated at a temperature of 1100 °C include only two phases—hematite and lanthanum orthoferrite.

Analysis of the mechanisms of formation of lanthanum orthoferrite with a perovskite-type structure indicates that during solid-phase synthesis, homogeneous perovskite can be formed only in a narrow range of the molar ratio of metals, but in another case, binary or ternary systems are present in the powders [50]. Modeling of phase formation processes in ternary systems La–Fe–O taking into account the oxygen non-stoichiometry of the structure of lanthanum orthoferrite $La_{1\pm x}Fe_{1\pm y}O_3{}^{2-}$ was carried out using the compound energy formalism, and the model is submitted to a defect chemical analysis [51].

At the same time, the presence of auxiliary substances in the system greatly complicates the process of phase formation and leads to the formation of a significant number of heterogeneous unstable compounds. In general, the reason for the formation of double systems $LaFeO_3$ and Fe_2O_3 is the lack of lanthanum, which is inherent in powders obtained by wet-chemical synthesis routes from nitrates [43]. The formation of metastable compounds or large surface fractions of oxide phases may lead to different surface configurations, at which locally a lower La stoichiometry may be realized.

To prevent this and ensure the physical and chemical conditions, under which the formation of a single phase of the perovskite $LaFeO_3$ takes place in systems with a stoichiometric ratio of lanthanum and ferrum metals, for example, citric acid is used. It leads to the formation of stable metal–citric acid complexes $HCit^{2-}$ and Cit^{3-} ions at an appropriate pH above pH 6, the decomposition of which takes place through the formation of $LaFeO_3$ particles and the simultaneous evolution of CO_2 and water vapor at 700–750 °C [35]. During chemical synthesis, urea is added to aqueous solutions, which converts lanthanum and ferrum hydroxides into carbonate complexes [52]. During hydrothermal synthesis, the presence of urea in the system, which induced the crystal growth process via the interaction of unreacted La^{3+}/Fe^{3+} to form $La_2(CO_3)_3/Fe_2(CO_3)_3$, produces $LaFeO_3$ with the elimination of CO_2 upon calcination [53]. An alternative is to use a single precursor compound that contains both cations that will make up the perovskite structure, e.g., $Ln[Fe(CN)_6]{\cdot}nH_2O$ (Ln = La, Gd, Lu) [15].

In our opinion, the problem of forming a homogeneous perovskite phase by a chemical method lies in the difference in the properties of the oxygen compounds of ferrum and lanthanum. In particular, with a gradual increase in the pH of the solution in the suspension, ferric hydroxide first precipitates in the pH range ~ 3.5–4.1, but lanthanum is in an ionized state and it may interact with $Fe(OH)_3$ by the sorption mechanism only. Precipitation of the lanthanum hydroxide phase takes place at the pH value of the ferrum hydroxide precipitate's existence. At the initial stage of the phase formation process, the formation of primary particles of ferrum and lanthanum hydroxides is spatially separated. Dehydroxylation of hydroxides to form ferric and lanthanum oxides also occurs at different temperatures of 250 and 330 °C, respectively, which supports receptive conditions for the spatially separated formation of hematite and lanthanum oxide particles. At the same time, lanthanum oxide partially changes into the hydroxide phase upon contact with air, which further disrupts the stoichiometry of the compounds. Therefore, the formation of the orthoferrite phase occurs under the conditions of the existence of a stable hematite phase and an unstable oxide-hydroxide mixture of lanthanum compounds (Fig. 1a), which leads to a violation of the stoichiometric ratio in the powders and, probably, becomes the cause of the deficiency of lanthanum in the final product of the reaction (Table 5, Fig. 10). The exception is the samples formed in $FeCl_3$–$LaCl_3$, in which the lanthanum content exceeds the ferrum content (Spectra 10–12, Table 7). Perhaps, the reason is the formation of an intermediate phase of LaOCl and a change in the interaction conditions of the components in the aqueous environment.

11 Conclusion

Using the methods of X-ray phase analysis, thermogravimetric analysis, and scanning electron microscopy, a study of phase formation processes during the chemical co-precipitation of inorganic salts' solutions of ferrum and lanthanum in a weakly alkaline medium, subsequent precipitate's lyophilization and powder's calcination at temperatures 800 and 1100 °C was carried out. It is shown that the precipitation of ferrum and lanthanum oxides-oxyhydroxides at the initial stage of the process is spatially separated due to the difference in the values of the hydrogen index at which the formation of particles of each of the compounds takes place. Anions of the dispersion medium are part of the hydroxide phases and form unstable intermediate compounds, for example, lanthanum oxychloride or iron and lanthanum carbonates.

Calcination of the samples at 800 °C does not lead to their homogenization, which is explained by the differences in the dehydroxylation temperatures of ferrum and lanthanum compounds at 250–280 and 330 °C, respectively. The formation of lanthanum orthoferrite particles takes place under the coexistence of a stable hematite phase and a less stable lanthanum oxides-hydroxides mixture in the system, which leads to a violation of the stoichiometric ratio in the sediment structure. Therefore, the final reaction products at 1100 °C temperature are a mixture of hematite and lanthanum orthoferrite with particle sizes in the range from 27 to 39 nm.

References

1. X. Zhou, E.I. Vovk, Y. Liu, C. Guan, Y. Yang, An in situ temperature-dependent study of La_2O_3 reactivation process. Front. Chem. **9**, 694559 (2021). https://doi.org/10.3389/fchem.2021.694559
2. S. Katba, S. Jethva, M. Vagadia, A. Ravalia, D.G. Kuberkar, Effect of La-substitution on magnetic properties of $ErFeO_3$ orthoferrites. J. Magn. Magn. Mater. **514**, 167–170 (2020). https://doi.org/10.1016/j.jmmm.2020.167170
3. A. Oliveira, M.L. Hneda, L.E. Fernandez-Outon, E.M.B. de Sousa, J.D. Ardisson, Synthesis and characterization of nanocomposites based on rare-earth orthoferrites and iron oxides for magnetic hyperthermia applications. Ceram. Int. **45**, 17920–17929 (2019). https://doi.org/10.1016/j.ceramint.2019.06.009
4. A.D. Jadhav, A.B. Gaikwad, V. Samuel, V. Ravi, A low temperature route to prepare $LaFeO_3$ and $LaCoO_3$. Mater. Lett. **61**, 2030–2032 (2007). https://doi.org/10.1016/j.matlet.2006.08.009
5. F.A. Fabian, P.P. Pedra, J.L.S. Filho et al., Synthesis and characterization of La(Cr, Fe, Mn)O_3 nanoparticles obtained by co-precipitation method. J. Magn. Magn. Mater. **379**, 80–83 (2015). https://doi.org/10.1016/j.jmmm.2014.12.004
6. C. Vázquez-Vázquez, P. Kögerler, M.A. López-Quintela, R.D. Sánchez, J. Rivas, Preparation of $LaFeO_3$ particles by sol-gel technology. J. Mater. Res. **13**(2), 451–456 (1998). https://doi.org/10.1557/JMR.1998.0058
7. G. Pecchi, P. Reyes, R. Zamora, C. Campos, L.E. Cadus, B.P. Barbero, Effect of the preparation method on the catalytic activity of $La_{1-x}Ca_xFeO_3$ perovskite-type oxides. Catal. Today **133**, 420–427 (2008). https://doi.org/10.1016/j.cattod.2007.11.011
8. A.E. Giannakas, A.K. Ladavos, P.J. Pomonis, Preparation, characterization and investigation of catalytic activity for NO + CO reaction of $LaMnO_3$ and $LaFeO_3$ perovskites prepared via microemulsion method. Appl. Catal. B **49**, 147–158 (2004). https://doi.org/10.1016/j.apcatb.2003.12.002
9. F. Tong, Y. Zhao, M.-H. Wang, Cube-like $LaFeO_3$ microstructures synthesised by a hydrothermal method and their optical properties. Micro Nano Lett. **14**(3), 259–262 (2019). https://doi.org/10.1049/mnl.2018.5366
10. W. Zheng, R. Liu, D. Peng, G. Meng, Hydrothermal synthesis of $LaFeO_3$ under carbonate-containing medium. Mater. Lett. **43**, 19–22 (2000). https://doi.org/10.1016/S0167-577X(99)00223-2
11. S. Farhadi, Z. Momeni, M. Taherimehr, Rapid synthesis of perovskite-type $LaFeO_3$ nanoparticles by microwave-assisted decomposition of bimetallic $La[Fe(CN)_6]\cdot 5H_2O$ compound. J. Alloys Compd. **471**, L5–L8 (2009). https://doi.org/10.1016/j.jallcom.2008.03.113
12. M. Sivakumar, A. Gedanken, W. Zhong, Y.H. Jiang, Y.W. Du, I. Brukental, D. Bhattacharya, Y. Yeshurun, I. Nowik, Sonochemical synthesis of nanocrystalline $LaFeO_3$. J. Mater. Chem. **14**, 764–769 (2004). https://doi.org/10.1039/B310110J
13. M. Popa, J. Frantti, M. Kakihana, Lanthanum ferrite $LaFeO_3$ + d nanopowders obtained by the polymerizable complex method. Solid State Ionics **154**, 437–445 (2002). https://doi.org/10.1016/S0167-2738(02)00480-0
14. E. Traversa, P. Nunziante, M. Sakamoto, Y. Sadaoka, M.C. Carotta, G. Martinelliand, Thermal evolution of the microstructure of nanosized $LaFeO_3$ powders from the thermal decomposition of a heteronuclear complex, $La[Fe(CN)_6]\cdot 5H_2O$. J. Mater. Res. **13**(5), 1335–1344 (1998). https://doi.org/10.1557/JMR.1998.0189
15. N. Kondo, H. Itoh, M. Kurihara, M. Sakamoto, H. Aono, Y. Sadaoka, New high-yield preparation procedure of $Ln[Fe(CN)_6]\cdot nH_2O$ (Ln = La, Gd, and Lu) and their thermal decomposition into perovskite-type oxides. J. Alloys Compd. **408**, 1026–1029 (2006). https://doi.org/10.1016/j.jallcom.2004.11.095
16. M. Siemons, A. Leifert, U. Simon, Preparation and gas sensing characteristics of nanoparticulate p-type semiconducting $LnFeO_3$ and $LnCrO_3$ materials. Adv. Funct. Mater. **17**, 2189–2197 (2007). https://doi.org/10.1002/adfm.200600454

17. S. Matei, D. Berger, S. Stoleriu, F. Papa, V. Fruth, Synthesis of lanthanum ferrite nanopowder by combustion method. J. Optoelectron. Adv. Mater. **9**, 1793–1796 (2007)
18. H. Shen, G. Cheng, A. Wu, J. Xu, J. Zhao, Combustion synthesis and characterization of nanocrystalline $LaFeO_3$ powder. Phys. Stat. Sol. A **206**, 1420–1424 (2009). https://doi.org/10.1002/pssa.200824266
19. C. Zhu, A. Nobuta, I. Nakatsugawa, T. Akiyama, Solution combustion synthesis of $LaMO_3$ (M = Fe Co, Mn) perovskite nanoparticles and the measurement of their electrocatalytic properties for air cathode. Int. J. Hydrogen Energy **38**, 13238–13248 (2013). https://doi.org/10.1016/j.ijhydene.2013.07.113
20. X. Qi, J. Zhou, Z. Yue, Z. Gui, L. Li, Auto-combustion synthesis of nanocrystalline $LaFeO_3$. Mater. Chem. Phys. **78**, 25–29 (2002). https://doi.org/10.1016/S0254-0584(02)00341-3
21. M.A. Ahmed, A.A. Azab, E.H. El-Khawas, Structural, magnetic and electrical properties of Bi doped $LaFeO_3$ nano-crystals, synthesized by auto-combustion method. J. Mater. Sci. Mater. Electron. **26**, 8765–8773 (2015). https://doi.org/10.1007/s10854-015-3556-4
22. D. Sánchez-Rodríguez, H. Wada, S. Yamaguchi, J. Farjas, H. Yahiro, Synthesis of $LaFeO_3$ perovskite-type oxide via solid-state combustion of a cyano complex precursor: the effect of oxygen diffusion. Ceram. Int. **43**, 3156–3165 (2017). https://doi.org/10.1016/j.ceramint.2016.11.134
23. S.W. Ding, Y.Y. Zhang, C.Y. Feng et al., Synthesis and mechanism research of nano-perovskite materials by solid-state reaction. Acta Chim. Sin. **69**, 752–756 (2011)
24. M. Sorescu, T. Xu, J.D. Burnett, J.A. Aitken, Investigation of $LaFeO_3$ perovskite growth mechanism through mechanical ball milling of lanthanum and iron oxides. J. Mater. Sci. **46**, 6709–6717 (2011). https://doi.org/10.1007/s10853-011-5625-2
25. J.H. Su, L.Q. Jing, K. Shi, H.G. Fu, Synthesis of large surface area $LaFeO_3$ nanoparticles by SBA-16 template method as high active visible photocatalysts. J. Nanopart. Res. **12**, 967–974 (2010). https://doi.org/10.1007/s11051-009-9647-5
26. S. Thirumalairajan, K. Girija, I. Ganesh et al., Controlled synthesis of perovskite $LaFeO_3$ microsphere composed of nanoparticles via self-assembly process and their associated photocatalytic activity. Chem. Eng. J. **209**, 420–428 (2012). https://doi.org/10.1016/j.cej.2012.08.012
27. W. Kaituo, W. Xuehang, W. Wenwei, L. Yongni, L. Sen, Synthesis of perovskite $LaCoO_3$ by thermal decomposition of oxalates: phase evolution and kinetics of the thermal transformation of the precursor. Ceram. Int. **40**, 5997–6004 (2014). https://doi.org/10.1016/j.ceramint.2013.11.048
28. H. Xu, X. Hu, L. Zhang, Generalized low-temperature synthesis of nanocrystalline rare-earth orthoferrites $LnFeO_3$ (Ln = La, Pr, Nd, Sm, Eu, Gd). Cryst. Growth Des. **8**(7), 2061–2065 (2008). https://doi.org/10.1021/cg800014
29. P. Kanhere, J. Nisar, Y. Tang, B. Pathak, R. Ahuja, J. Zheng, Z. Chen, Electronic structure, optical properties, and photocatalytic activities of $LaFeO_3$–$NaTaO_3$ solid solution. J. Phys. Chem. C **116**, 22767–22773 (2012). https://doi.org/10.1021/jp307857h
30. J. Choi, B. Kim, S.H. Song, J.S. Park, A composite cathode with undoped $LaFeO_3$ for protonic ceramic fuel cells. Int. J. Hydrogen Energy **41**(22), 9619–9626 (2016). https://doi.org/10.1016/j.ijhydene.2016.03.115
31. B.P. Barbero, J.A. Gamboa, L.E. Cadús, Synthesis and characterisation of $La_{1-x}Ca_xFeO_3$ perovskite-type oxide catalysts for total oxidation of volatile organic compounds. Appl. Catal. B Environ. **65**, 21–30 (2006). https://doi.org/10.1016/j.apcatb.2005.11.018
32. Y. Teraoka, H. Fukuda, S. Kagawa, Catalytic activity of perovskite-type oxides for the direct decomposition of nitrogen monoxide. Chem. Lett. **19**, 1–4 (1990). https://doi.org/10.1246/cl.1990.1
33. Y. Matuura, S. Matsushima, M. Sakamoto, Y. Sadaoka, NO_2-sensitive $LaFeO_3$ film prepared by thermal decomposition of the heteronuclear complex, $La[Fe(CN)_6]\cdot 5H_2O$. J. Mater. Chem. **3**, 767–769 (1993). https://doi.org/10.1039/JM9930300767
34. X.P. Dai, Q. Wu, R.J. Li, C.C. Yu, Z.P. Hao, Hydrogen production from a combination of the water−gas shift and redox cycle process of methane partial oxidation via lattice oxygen over LaFeO3 perovskite catalyst. J. Phys. Chem. B **110**, 25856–25862 (2006). https://doi.org/10.1021/jp0654664

35. B. Kucharczyk, J. Winiarski, I. Szczygieł, K. Adamska, Physicochemical properties of $LaFeO_3$ perovskite prepared by various methods and its activity in the oxidation of hydrocarbons. Ind. Eng. Chem. Res. **59**, 16603–16613 (2020). https://doi.org/10.1021/acs.iecr.0c03035
36. J. Xia, S. Yin, H. Li, H. Xu, L. Xu, Y. Xu, Improved visible light photocatalytic activity of sphere-like BiOBr hollow and porous structures synthesized via a reactable ionic liquid. Dalton Trans. **40**, 5249–5258 (2011). https://doi.org/10.1039/C0DT01511C
37. S. Keav, S.K. Matam, D. Ferri, A. Weidenkaff, Structured perovskite-based catalysts and their application as three-way catalytic converters—a review. Catalysts **4**, 226–255 (2014). https://doi.org/10.3390/catal4030226
38. G. Pecchi, P. Reyes, R. Zamora, L.E. Cadus, B.P. Barbero, Catalytic combustion of methane over $LaFeO_3$ perovskites: the influence of coprecipitation pH and ageing time. J. Chil. Chem. Soc. **51**, 1001–1005 (2006). https://doi.org/10.4067/S0717-97072006000400001
39. K.M. Parida, K.H. Reddy, S. Martha, D.P. Das, N. Biswal, Fabrication of nanocrystalline $LaFeO_3$: an efficient sol–gel auto-combustion assisted visible light responsive photocatalyst for water decomposition. Int. J. Hydrogen Energy **35**, 12161–12168 (2010). https://doi.org/10.1016/j.ijhydene.2010.08.029
40. S. Thirumalairajan, K. Girija, N.Y. Hebalkar, D. Mangalaraj, C. Viswanathana, N. Ponpandian, Shape evolution of perovskite $LaFeO_3$ nanostructures: a systematic investigation of growth mechanism, properties and morphology dependent photocatalytic activities. RSC Adv. **3**, 7549–7561 (2013). https://doi.org/10.1039/C3RA00006K
41. J. Ding, X. Lu, H. Shu, J. Xie, H. Zhang, Microwave-assisted synthesis of perovskite $ReFeO_3$ (Re: La, Sm, Eu, Gd) photocatalyst. Mater. Sci. Eng. B **171**, 31–34 (2010). https://doi.org/10.1016/j.mseb.2010.03.050
42. X. Li, H. Zhao, J. Liang, Y. Luo, G. Chen, X. Shi, S. Lu, S. Gao, J. Hu, Q. Liu, X. Sun, A-site perovskite oxides: an emerging functional material for electrocatalysis and photocatalysis. J. Mater. Chem. A **9**, 6650–6670 (2021). https://doi.org/10.1039/d0ta09756j
43. D. Mutter, R. Schierholz, D.F. Urban, S.A. Heuer, T. Ohlerth, H. Kungl, C. Elsässer, R.-A. Eichel, Defects and phase formation in non-stoichiometric $LaFeO_3$: a combined theoretical and experimental study. Chem. Mater. **33**(24), 9473–9485 (2021). https://doi.org/10.1021/acs.chemmater.1c02106
44. P. Garcia-Muñoz, F. Fresno, J. Ivanez, D. Robert, N. Keller, Activity enhancement pathways in $LaFeO_3$@TiO_2 heterojunction photocatalysts for visible and solar light driven degradation of myclobutanil pesticide in water. J. Hazard. Mater. **400**, 123099 (2020). https://doi.org/10.1016/j.jhazmat.2020.123099
45. O.M. Lavrynenko, M.M. Zahornyi, V.V. Vember, O.Y. Pavlenko, T.F. Lobunets, O.F. Kolomys, O.Y. Povnitsa, L.O. Artiukh, K.S. Naumenko, S.D. Zahorodnia, I.L. Garmasheva, Nanocomposites based on cerium, lanthanum, and titanium oxides doped with silver for biomedical application. Condens. Matter **7**, 45 (2022). https://doi.org/10.3390/condmat7030045
46. A. Neumann, D. Walter, The thermal transformation from lanthanum hydroxide to lanthanum hydroxide oxide. Thermochim. Acta **445**, 200–204 (2006). https://doi.org/10.1016/j.tca.2005.06.013
47. X. Duan, D.G. Evans, *Layered Double Hydroxides: Structure and Bonding*, vol. 119 (Springer, 2006), 234 p.
48. O.M. Lavrynenko, Y.S. Shchukin, Development of the hydroxysulfate green rust on the steel surface contacting with water ferric and ferrous salt solutions. Mineral. J. (Ukr.) **37**(2), 23–36 (2015)
49. E.S. Seliverstov, S.N. Golovin, O.E. Lebedeva, Layered double hydroxides containing rare earth cations: synthesis and applications. Front. Chem. Eng. **4**, 867615 (2022). https://doi.org/10.3389/fceng.2022.867615
50. V.L. Moruzzi, M.W. Shafer, Phase equilibria in the system La_2O_3-iron oxide in air. J. Am. Ceram. Soc. **43**(7), 367–372 (1960)
51. E. Povoden-Karadeniz, A.N. Grundy, M. Chen, T. Ivas, L.J. Gauckler, Thermodynamic assessment of the La–Fe–O system. J. Phase Equilib. Diffus. **30**, 351–366 (2009). https://doi.org/10.1007/s11669-009-9501-6

52. E.A. Tugova, O.N. Karpov, Nanocrystalline perovskite-like oxides formation in Ln_2O_3–Fe_2O_3–H_2O (Ln = La, Gd) systems. Nanosyst. Phys. Chem. Math. **5**(6), 854–860 (2014)
53. P.K. Thiruppathi, D. Nataraj, Precisely tailored $LaFeO_3$ dendrites using urea and piperazine hexahydrate for the highly selective and sensitive detection of trace level acetone. Mater. Adv. **1**, 2971 (2020). https://doi.org/10.1039/d0ma00602c

Metal-Based Nanocrystalline Materials Condensed from the Vapor Phase

V. Grechanyuk, I. Grechanyuk, A. Kozyrev, A. Matsenko, V. Chornovol, and Yulia Kovalchuk

Abstract The research work is aimed at produced of metal-based nanocrystalline materials condensed from the vapor phase in the form of thin (up to 5 microns) and thick (more than 5 microns) films for use them as contacts in radio engineering, microelectronics, etc. Composite copper-based materials with nanocrystalline components of the second phase were produced on an electron beam installation by the evaporation–condensation method. Doping of the copper matrix with zirconium and yttrium with a total content of up to 0.1% in the composite material was carried out by evaporating copper from the cuprum, zirconium and yttrium alloy through an intermediary bath. It has been show that materials with copper matrix doped with molybdenum, tungsten or chromium have characteristics, which make them promising for widespread industrial applications. The source for such properties is the formation of supersaturated solid solutions at the submicron level due to deposition from the vapor phase, which, as a result of them decomposition, leads to the formation of a layered structure with a hierarchy of layers at the macro-, micro-, and submicron levels.

1 Introduction

It is impossible to achieve significant progress in any of the most important areas of science and technology without the creation of new materials with high characteristics. One of the most promising areas for creating of fundamentally new materials with predetermined properties is high-speed electron beam evaporation followed by condensation of metallic and non-metallic materials in a vacuum [1–3].

V. Grechanyuk · I. Grechanyuk · A. Kozyrev (✉) · A. Matsenko · V. Chornovol · Y. Kovalchuk
Department of Chemistry, Kyiv National University of Construction and Architecture, Povitroflotsky Av., 31, Kyiv 03037, Ukraine
e-mail: artem.v.kozyrev@gmail.com

O. Fesenko and L. Yatsenko (eds.), *Nanostructured Surfaces, Nanocomposites and Nanomaterials, and Their Applications*, Springer Proceedings in Physics 296,
https://doi.org/10.1007/978-3-031-42704-6_10

Such evaporation–condensation processes make it possible form at the atomic–molecular level thin films (up to 5 microns) for radio engineering, microelectronics, computer technology, etc., as well as thick films (more than 5 microns) used as effective protective and wear-resistant coatings, and also as massive composite materials isolated from the substrates. Vapor flows do not have solubility limits, therefore, it is possible to regulate the structure and dispersion of the created materials over a wide range, by mixing their components according to a definite intention and controlling their subsequent condensation. In this way, it is possible to create compositions of a dispersed or layered type, even if this is not possible by other methods.

2 Research Materials and Methods

Composite copper-based materials with nanocrystalline components of the second phase were produced on a universal electron beam installation L-2 by the evaporation–condensation method [4]. The initial materials were evaporated from two independent sources onto a substrate made of steel St3 (GOST), which was stationary or rotating, depending on the tasks. To obtain gradient condensates with different concentrations of the reinforcing phase, a stationary substrate 700 × 400 × 15 mm was used. The scheme of this technological experiment is shown in Fig. 1.

The deposition of condensates with the required concentration of components was carried out on a substrate with a diameter of 800 mm and a thickness of 25–30 mm, rotating at a speed of 36 rpm (Fig. 2).

In this work, alloying a copper matrix with zirconium and yttrium with a total content of up to 0.1% in the composite material was carried out by evaporating copper from the Cu–Zr–Y alloy through an intermediary bath, which resulted in an increase in the copper evaporation rate by 2–3 times.

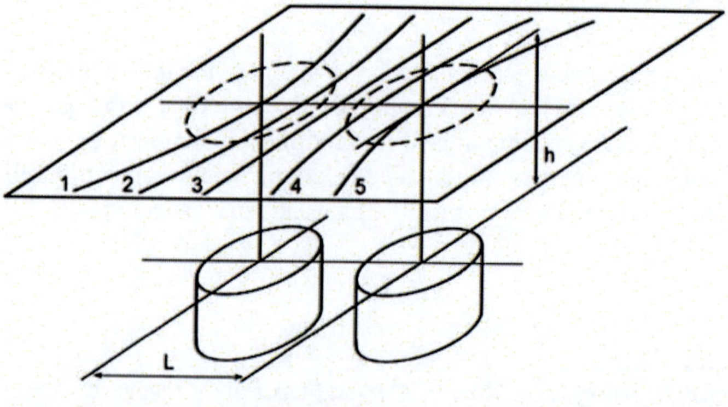

Fig. 1 The scheme of obtaining condensed composite materials with varying concentration along the length of the substrate by evaporation–condensation of components from two independent sources: 1, 2, 3, 4, 5—lines of constant concentrations of copper and molybdenum [5]

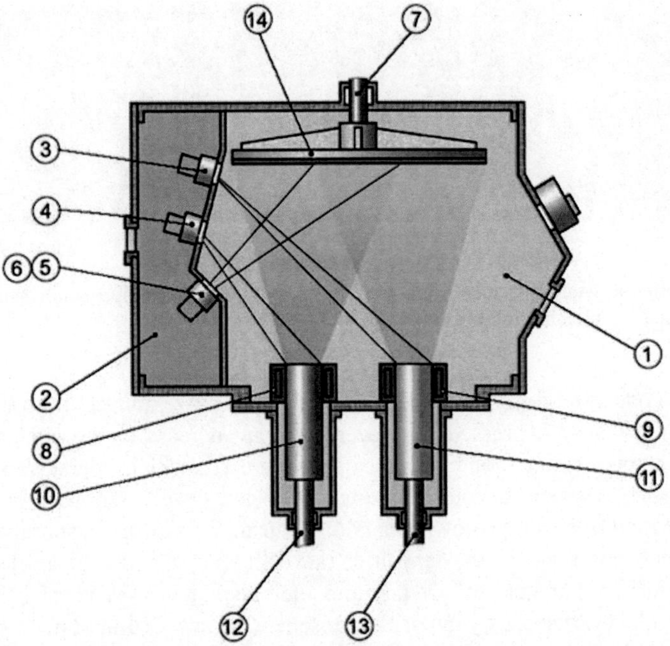

Fig. 2 The scheme of obtaining condensed composite materials on a rotating substrate by deposition of copper and molybdenum from two independent sources without separation of steam flows: 1—operational space, 2—space of equipment, 3–6—electron beam heaters, 7—substrate rotation mechanism, 8, 9—copper water-cooled crucibles, 10, 11—initial (evaporated) materials, 12, 13—feeding materials mechanisms, 14—substrate

This research was aimed at alloying a copper matrix with zirconium and yttrium with a total content of up to 0.1% in the composite material, by evaporating copper from the Cu–Zr–Y alloy through an intermediary bath, which resulted in an increase in the copper evaporation rate by 2–3 times. To separate the condensate from the substrate, it was covered with a separating layer of calcium fluoride. The resulting condensate was used to prepare test samples with the required dimensions.

3 Results and Discussions

The structure, physicochemical mechanical and operational properties of Cu–Mo, Cu–W, Cu–Cr, (Cu–Zr–Y)–Mo composite materials in the concentration range of refractory components up to 50% were comprehensively studied. The result of the research was finding the formation of supersaturated solid solutions in composite materials Cu–W, Cu–Mo, Cu–Cr at the submicron level. As a result of their decomposition, this leads to the formation of a layered structure with a hierarchy of layers at the macro-, micro- and submicron levels as it shown in Fig. 3.

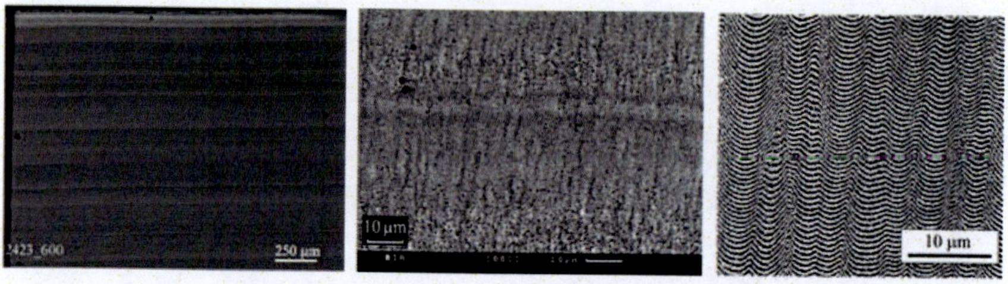

a b c

Fig. 3 Layered structure of a composite material based on copper and molybdenum on macro- (**a**), micro- (**b**) and submicron level (**c**)

At low concentrations of molybdenum (up to 7–8 wt%), tungsten (up to 4 wt%) layering is not very expressed. However, with an increase in the content of the refractory component, the contrast of the image increases, which indicates an increase in layering due to various factors. Layering at the macro-level is probably the result of the of electrical micro-breakdowns that occur during high-speed evaporation of the technical pure components (the deposition rate of copper on a rotating steel substrate with 1000 mm in diameter is 60–70 µm/min, molybdenum – 6–8 µm/min). Layering at the micro-level is caused by impurities present in the initial (evaporated) materials. The formation of layers at the submicron level is associated with the formation of supersaturated solid solutions, which, decomposing, form the corresponding micro-layers [6].

It has been experimentally confirmed that layered condensed composite materials based on low-alloy copper and molybdenum alloys up to 6 mm thick, obtained on a rotating substrate heated to 700 ± 30 °C, are bulk nanocrystalline materials. The grain size in them does not exceed 96 nm, and the size of dispersed inclusions of the reinforcing phase is 18 nm.

Transmission electron microscopic studies of layered composite materials (Cu-0.1 wt% Zr, Y)—Mo (8–12 wt%) showed that the average grain size of copper ranges from 58 to 96 nm, and the grain size of molybdenum is from 46 to 62 nm. Inside the grains, dispersed particles of reinforcing phases are evenly distributed. Their size ranges from 10 to 18 nm and depends on the nature of the reinforcing phases (oxides, metals) [7]. A typical electron microscopic image of the structure is shown in Fig. 4.

As can be seen in Fig. 4, the microelectron diffraction pattern of the composite has a characteristic annular shape inherent in nanocrystalline structures.

The composition and basic physical and mechanical properties of these materials are given in the Table 1.

Composite materials produced by the electron beam evaporation–condensation method are widely used in various industries due to their unique properties. Such materials of the (Cu–Zn-Y)–Mo system have mainly found industrial application as electrical contacts [8]. Switching tests have shown that in such a gradient layered nanomaterials, a change in the chemical composition of the layers significantly limits the zone of the thermal effect of the discharge. In some types of switching devices,

Fig. 4 Microelectron diffraction pattern of a composite material (Cu-0.1 wt% Zr, Y)—8–12 wt% Mo

Table 1 Chemical composition and physical and mechanical properties of (Cu–Zr–Y)–Mo composite materials containing Cu, 0.05–0.1 wt % Zr and Y, and various amounts of molybdenum (σ_B is the ultimate tensile strength, δ is the relative elongation)

Chemical composition (wt.% Mo)	Density (g/cm^3)	Electrical resistivity (μOhm × m)	Hardness H_V (MPa)	Mechanical characteristics			
				Before annealing		After annealing	
				σ_B (MPa)	δ (%)	σ_B (MPa)	δ (%)
3–5	8.9–9.0	0.021–0.022	1000–1500	300–430	10.3–7.3	295–420	17.6–9.3
5.1–8	9.0–9.05	0.022–0.024	1500–1650	440–630	7.25–3.4	425–600	9,45–4.9
8.1–12	9.05–9.1	0.024–0.028	1650–1800	635–785	3.25–1.8	605–730	4.85–3.9

when using these materials, there are smaller changes in the working layer of contacts and electrodes and an increase in erosion resistance compared to analogues obtained by powder metallurgy methods.

Composite materials (Cu–Zr, Y)–Cr with 0.05–0.1 wt % of Zr and Y, containing 35–50 wt % of chromium, are used for produced connectors in vacuum arc chutes [9]. Usability of condensed composite materials of this system is due to the features of the chemical composition and morphology of the "secondary" structure, which is formed on the surface of the contacts. In a non-equilibrium state of an arc discharge, the mutual solubility of copper and chromium increases and the decomposition of solid solutions take place with the formation of a dispersed structure. Condensates (Cu–Zr–Y)–Cr with such a chromium content have a layered structure at the macro-, micro-, and submicron levels. The layering of the last two levels is due to the anisotropy of normal grain expansion. This contributes to the formation of "pillars" within several layers of condensate. In such a structure, under the temperature and time, in the section of the layer perpendicular to the pillars, a structure is formed,

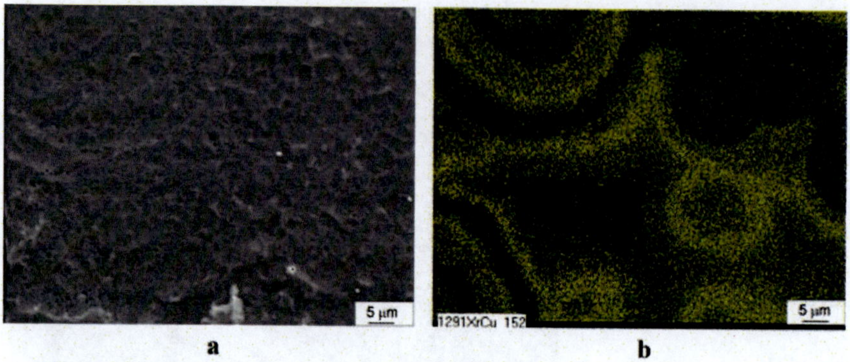

Fig. 5 Microstructure of a composite material (Cu-Zr,Y)–Cr with a chromium content of 35–50 wt % in secondary electrons (**a**) and in x-rays of copper (**b**)

which is characterized by a polygonal shape of grains (Fig. 5a), and contains signs of separation of the solid solution (Fig. 5b).

Hardness (by Vickers) varies with chromium content linearly. In the concentration range of 35–50% Cr, the hardness of the material is within 2069–2503 MPa. In a tensile test, the ultimate tensile strength increases to 550 MPa; however, the composite material in this case has zero plasticity.

Composite (Cu–Zr–Y)–Cr materials condensed from the vapor phase have many advantages: they are obtained in one technological cycle, they are cheaper than analogues obtained by powder metallurgy methods (by 1.5–1.7 times) and significantly (by 4 times) cheaper than the materials of silver-containing contacts.

Composite (Cu–Zr–Y)–Cr materials condensed from the vapor phase have several advantages, due to they are produced in one technological cycle. They are cheaper than analogues produced by powder metallurgy (by 1.5–1.7 times) and significantly cheaper (by 4 times) than the materials of silver-containing contacts. In terms of operational durability, condensed composite materials are no concede to materials based on silver-containing compositions. They are well processed by cutting, grinding, drilling, easily soldered by any of the known soldering methods using standard silver-containing and silver-free solders.

4 Conclusion

The studies have shown that composite materials with a metallic copper matrix doped with molybdenum, tungsten or chromium, produced by method condensation from the vapor phase, have a nanoscale structure that lend them characteristics, which make them promising for widespread industrial applications.

References

1. B.A. Movchan, V.A. Kuz'menko, Y.F. Lugovskoi, V.A. Osokin, N.I. Grechanyuk, Elasticity and fatigue strength of condensed microlayered materials Cu-Y-Mo. Powder Metall. Met. Ceram. **34**, 631–633 (1996). https://doi.org/10.1007/BF00559490
2. J. Safarian, T.A. Engh, Vacuum evaporation of pure metals. Metall. Mater. Trans. A. **44**, 747–753 (2013). https://doi.org/10.1007/s11661-012-1464-2
3. V. Bukhanovsky, M. Rudnytsky, M. Grechanyuk, R. Minakova, Ch. Zhang, Vapour-phase condensed composite materials based on copper and carbon. Mater. Technol. **50**(4), 523–530 (2016). https://doi.org/10.17222/mit.2015.057
4. N.I. Grechanyuk, G.A. Baglyuk, P.P. Kucherenko, A.G. Melnik, I.N. Grechanyuk, V.G. Grechanyuk, Y.A. Smashnyuk, Laboratory electron-beam multipurpose installation L-2 for producing alloys, composites, coatings, and powders. Powder Metall. Met. Ceram. **56**(1), 113–121 (2017). https://doi.org/10.1007/s11106-017-9878-4
5. V.G. Grechanyuk, N.I. Grechanyuk, V.O. Chornovol, A.V. Kozyrev, V.I. Gots, A.V. Matsenko, V.A. Kulichenko, T.D. Grabina, I.K. Yu, Copper and molybdenum-based nanocrystalline materials. Metallofiz. Noveishie Tekhnol **44**(7), 927–942 (2022). https://doi.org/10.15407/mfint.44.07.0927
6. V.G. Grechanyuk, Physico-mechanical bases for forming condensed from the vapor phase composite materials based on copper. Dissertation, Frantsevich Institute for Problems of Materials Science, National academy of science of Ukraine, 2013.
7. N.I. Grechanyuk, V.G. Grechanyuk, Precipitation-strengthened and microlayered bulk copper- and molybdenum-based nanocrystalline materials produced by high-speed electron-beam evaporation-condensation in vacuum: structure and phase composition. Powder Metall. Met. Ceram. **56**(11–12), 633–646 (2018). https://doi.org/10.1007/s11106-018-9938-4
8. M.I. Hrechaniuk, I.N. Hrechaniuk, V.O. Denysenko, V.G. Hrechaniuk, Patent No. 86434 UA Composite material for electric contacts and electrodes and method for its production (2009). https://uapatents.com/4-86434-kompozicijjnijj-material-dlya-elektrichnikh-kontaktiv-i-elektrodiv-ta-sposib-jjogo-otrimannya.html. Accessed 29 Nov. 2022
9. M.I. Hrechaniuk, M.M. Plashchenko, A.V. Zvarich, V.O. Osokin, Patent No. 76737 UA Contact system of vacuum arc chute (2006). https://uapatents.com/4-76737-kontaktna-sistema-vakuumno-dugogasno-kameri.html. Accessed 29 Nov. 2022.

Polymer–Magnet Nanosystems

O. I. Aksimentyeva and Yu. Yu. Horbenko

Abstract Macroscopic magnetism in organic materials is a very complex but promising scientific problem. The need to create an organic magnet is due to several expected advantages, such as lightness, transparency, flexibility, ability to switch under the influence of light (magneto-optics), sensitivity to external factors (sensors), absorption of radiation, etc. Studying the state of a spin glass, biomagnetism, and the mechanism of paramagnetic probe action in living tissues can help to understand the mechanism of human thinking and DNA functioning. This chapter is devoted to molecular magnetism in organic polymer–magnet nanosystems. The ways of synthesis and magnetic properties of materials based on conjugated polymers and their complexes with transition metals are considered. Particular attention is paid to the hybrid polymer–magnetic nanosystems formation and application. The study of magnetic properties of polymer–magnet nanosystems is helpful for understanding the nature of "organic magnetism" and identifying new approaches to the fabrication of functional hybrid nanomaterials.

1 Introduction

A new field of scientific research related to obtaining organic and molecular magnets is developing rapidly [1–20]. Physico-chemistry of molecular and organic magnetism is described in only a few reviews [1–5] but many pioneering works [6–14].

One of the main tasks of modern physical and chemical research is the development of scientific foundations and new approaches to the creation of polymer and composite materials with specified functional properties—electrical, optical, and magnetic, the ability to accumulate and transform charge, catalyze some reactions, sensors, etc. [6–14]. Research and development of highly dispersed composite materials based on polymers doped or filled with organic and inorganic components, in particular compounds of transition metals, is an urgent problem of science and

O. I. Aksimentyeva (✉) · Yu. Yu. Horbenko
Ivan Franko National University of Lviv, Kyryla and Mefodia str., 6/8, Lviv 79005, Ukraine
e-mail: olena.aksimentyeva@lnu.edu.ua

O. Fesenko and L. Yatsenko (eds.), *Nanostructured Surfaces, Nanocomposites and Nanomaterials, and Their Applications*, Springer Proceedings in Physics 296,
https://doi.org/10.1007/978-3-031-42704-6_11

technology. Due to the nanostructure of composites (the particle size ranges from a few units to tens of nm), they have unique magnetic, spectral, and electrochemical properties.

Providing macroscopic magnetism in organic materials is an arduous but promising scientific task. The need to create an organic magnet is due to several expected advantages, such as lightness, transparency, flexibility, the ability to switch under the influence of light (magneto-optics) or compounds (sensors), the creation of modern toners for digital printing, materials for chemical current sources, etc. [9–13]. Investigation of the spin glass state, biomagnetism, and the paramagnetic probe action in living tissues will help in understanding the mechanism of biological processes, in particular, human thinking and the functioning of DNA [14–19].

One of the ways to solve the problem of creating organic magnets is obtaining crystals of organometallic complexes—molecular magnets [1–3, 15], doping organic matrices with transition metal ions—polymeric organometallic magnets [7, 16–19], synthesis of high-spin macromolecules based on organic compounds [4, 8, 20], filling polymer matrices with colloidal metals and oxides of transition metals—magnetic dispersions [6, 11–14]. Studying the magnetic properties of new organometallic materials can help to understand the nature of "organic magnetism" and develop new approaches to organic magnet creation.

Forms of magnetism, namely diamagnetism, paramagnetism, ferro-, antiferro- and ferrimagnetism, and spin glasses, are ultimately related to the electron magnetic moment. The difference between the forms is the types of interaction of moments. Thus, molecular magnets based on materials in which unpaired electrons are associated with discrete molecules would provide possibilities for controlling the interaction of elementary moments. As well as achievements in the molecular synthesis make it possible to place a molecule in a defined position.

All known ferromagnets are built of metal atoms with unfilled *d*- or *f*-electron orbital, i.e., based on the spin magnetism of internal *d*- and *f*-electrons. The magnetic interaction is positive and arranges the electron spins of nearby atoms in parallel—ferromagnetically. In contrast, the interatomic/intermolecular interaction is negative for atoms with *s*- or *p*- electrons, and their spins orient antiparallel—antiferromagnetically.

Diamagnetic materials contain only spin-paired electrons, and the density of field lines decreases in a magnetic field. Paramagnetism is due to the unpaired electrons, and the magnetic field strength increases in material that kind. The magnetic susceptibility in the last case is described by the Curie law or the Curie–Weiss law:

$$\chi = C/(T - \Theta)$$

where C—Curie constant (temperature), Θ—Neel temperature.

Depending on the sign and value of the parameter Θ, the following types of magnetic interaction are distinguished:

$\Theta > 0$ is a ferromagnet; $\Theta < 0$—antiferromagnet; $\Theta = 0$—paramagnet.

Living systems mainly consist of diamagnets, and as a result, the magnetic susceptibility of the biomass of organic substances turns out to be negative. However, the degree of diamagnetism in different cells, organs, and tissues has not been comparatively studied. The known studies determine the value of the cells' magnetic susceptibility by the ratio of dia- and paramagnetic compounds, special magnetosomes, the electromagnetic field generated by the cell, the nature and intensity of energy exchange. It was noted that it is different in living and dead objects, reflects the degree of cell viability, and is a kind of barometer of dissociative shifts in bacteria when the geomagnetic field changes. Therefore, magnetic susceptibility is related to many vital functions of organisms. Perhaps, it even provides their ability to navigate in space and migrate along the lines of force of the geomagnetic field.

There are few natural magnetic materials, mainly transition metals and their oxides. The increase in the number of magnetic materials is achieved mainly due to the production of alloys, double oxides, and intermetallic compounds. Research in the design of molecular ferromagnets and polymer nanocomposites significantly expands the range of magnetically active compounds and magnetic materials. In addition, using conventional synthetic approaches, it is possible to create new magnetic materials by methods alternative to the known energy-intensive metallurgical technologies.

2 Basic Approaches to Creating "Organic Magnets"

An organic ferromagnet is a magnetic material built from organic paramagnetic molecules [3–5, 20]. This is a partial case of a molecular ferromagnet consisting of paramagnetic molecules that can contain metal atoms (e.g., organometallic molecules). The ferromagnetic properties of known metallic magnets are determined by the orderly arrangement of d- and f-electron spins in a definite "domain". In contrast, research in molecular magnetism is related to obtaining substances whose solid phases are formed from individual molecules or ions that contain paramagnetic centers.

A classical ferromagnet is a macro-object, for example, a piece of iron, nickel, cobalt, or their alloys, but in no case is an individual molecule. By its nature, the magnetic phase transition to the ferromagnetic state is a cooperative phenomenon.

The basis of cooperative physicochemical phenomena characteristic of substances, such as electrical conductivity, magnetic susceptibility, and optical activity, is the interaction between electric and magnetic charges that make up the structure of substances. Quasiparticles of the soliton or polaron type (radical ions) with charge $q = 1$ and spin $s = ½$ can be elementary charged particles in the case of conjugated polyarenes and their composites with organic and inorganic compounds [21–23]. The interaction of charges in a substance determines the electrical conductivity, magnetization, the ability to convert electric and light energy, absorption, and emission of light in the visible part of the spectrum, which ensures the connection between the electrical, magnetic, and optical properties of materials.

The word "molecular" reflects the methodology of the approach—individual molecules or ions form a spatial structure that contributes to the realization of a magnetic phase transition when a solid phase is formed. The problem can be solved by various chemical methods united by the term "chemical design of molecular ferromagnets". The layered or framework structure formation in the solid phase is necessary for the magnetic phase transition into a ferromagnetic state. However, this condition is not sufficient. Paramagnetic centers in polymers must be connected by a definite chain of atoms, serving as an "exchange channel" so that the polymer can "magnetize" in a magnetic field [3–5].

The problem of the synthesis of a molecular ferromagnet can be solved by organizing paramagnetic molecules with a high density of oriented spins into a definite supramolecular structure, providing the necessary energy of intermolecular interaction and thus ferromagnetic ordering of spins [4, 24].

The class of organometallic polymers is based on multi-spin organometallic macromolecules in which spins belong to paramagnetic ions. There are polycomplexes of inorganic ions with organic ligands synthesized by inorganic chemistry methods. Methods of synthesis of ferromagnets based on organometallic polymers with paramagnetic metal atoms, covalently bound in the main polymer chain, are also being developed. It is necessary to ensure exchange interaction between metal atoms for turning such polymers into ferromagnets. In terms of dynamic properties, these high-temperature ferromagnets are rigid—magnetic susceptibility is close to unity, and the magnetic loss tangent is extremely small in the wave range of 0.8–30 cm. This direction in the chemical design of ferromagnets overlaps with the modern materials science of metal-filled polymer composites.

An attempt to obtain a purely organic magnet using high-spin molecules of electrically conductive polymers as a "building material" was made by us in cooperation with scientists from the Institute of Physics of the Polish Academy of Sciences [25–27]. The idea was to use supramolecular compounds as alloying agents to create "exchange channels" in polymer formations. The paramagnetic–antiferromagnetic phase transition is possible upon the formation of so-called polymer blends—molecular mixtures of conjugated high-spin molecules—and a dielectric polymer matrix [20, 28, 29]. In this case, for the first time, we established several interesting electrical and magnetic phenomena, the interpretation of which is quite complicated.

3 Organic Magnets Based on Polymer Nanosystems

The nature of magnetism in polymers, as in other substances, is determined by the total contribution of diamagnetic and paramagnetic components. The paramagnetic susceptibility of polymers is caused by the unpaired spins due to the breaking of chemical bonds, stable free radicals, macroradicals, ions with uncompensated spins, and temperature-independent spin paramagnetism of conduction electrons (Pauli paramagnetism).

Conducting polymers can be considered as "synthetic nanometals" with a particle diameter of 10–20 nm and unique electronic, optical, electrochemical, and catalytic properties, the ability to absorb radioactive rays [30]. Delocalization of *p*-electrons is due to conjugated double bonds in the polymer backbone. The conductivity of conjugated polymers can be very high in doped states (*p*-doping or *n*-doping). Charge transfer proceeds according to a hopping mechanism along the polymer chain with π-conjugated bonds [21–23].

The widely used conjugated polymer—polyaniline (PAn), and its copolymers with organic sulfonic amino acids were studied as a model object with a stable paramagnetic response.

Polymer films were obtained on the plane Pt electrode by electrochemical polymerization of 0.1 M aniline solution in 0.5 M sulfuric acid at current density $i =$ 0.1–1.0 mA/cm^2. The structure of the film was investigated using transmission electron microscopy (UEMB-100K), scanning electron microscopy (ISI-DS-130), electronography (EG-100K), and X-ray diffraction analysis (DRON-2, FeKα-radiation). A statistical method of correlative functions, namely radial distribution of atomic density, was used to describe a polymer design in a thin layer [31]. The radial distribution of intensity (I) as a function of scattering vector $S = 4\pi \sin\theta/\lambda$ was obtained according to electronography at low angles of scattering and photometry of the electronograph.

It is found that polymer film thinner than 80 nm is amorphous with a close order in the disposition of structure elements (Fig. 1). Wide diffusion maximums at $S =$ 1.8 Å^{-1}, 2.9 Å^{-1}, and additional low angle maximum at $S = 1.35$ Å^{-1} are observed at the scattering intensity distribution for thin polymer films (Fig. 1b).

Thin films up to 60–80 nm in thickness are formed from ordering islands (domains) with a diameter of 100–1000 nm (Fig. 1a). The domain is an ensemble of spherical globules (grains) with a diameter of 10–30 nm.

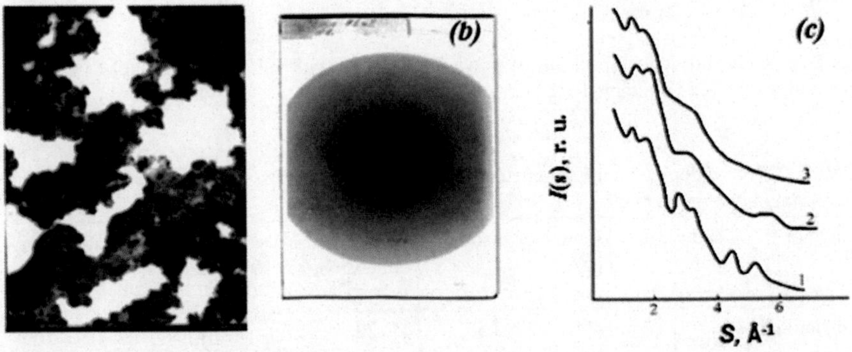

Fig. 1 **a** TEM image of PAn layer on Pt surface (×10,000), film thickness 60 nm; **b** electron diffraction pattern and **c** distribution of scattering intensity for PAn films obtained at current density: 0.5 (*1*); 0.7 (2); 1.0 mA/cm^2 (*3*), film thickness 60–80 nm

Based on the radial distribution of intensity as a function of scattering vector $S = 4\pi \sin\theta/\lambda$, parameters of polyaniline short-range ordered structure preserved the coplanar orientation of aromatic fragments have been calculated (Table 1).

Due to X-ray diffraction spectra (Fig. 2a), crystallinity level of PAn increases from 25% to 41–44% because of acid doping. Size of crystallites increases in the range of 15–30 Å. The degree of crystallinity decreases for the iodine-doped polymers (Table 2). The estimated parameters of short-ordered structure and the localization radius of electronic states in metallic grains are grain sized and the scaled of repetitive polymer units.

Table 1 Parameters of short-range ordered structure for polyaniline

Period of identity, *d*, Å	Length of conjugation, Å	Number of monomer links
4.6	25–30	5–6

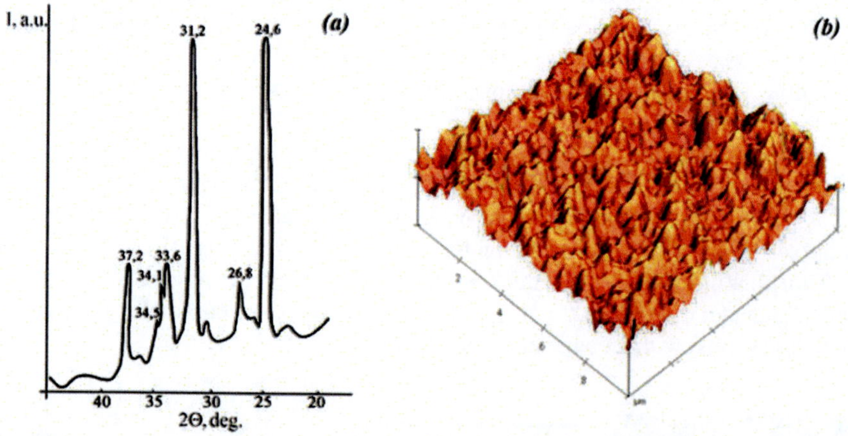

Fig. 2 **a** X-ray diffraction spectrum of polyaniline doped with sulfate acid; **b** AFM image of polyaniline film. Size of grains 10–30 nm

Table 2 Effect of doping on the parameters of X-ray diffraction, size of crystallites and degree of the crystallinity of polyaniline

Polymer	Doping agent	Interplane distance, d, Å	Crystallite dimension, l ± 10%, Å	Level of crystallinity, I ± 2%
Polyaniline	–	4.47; 4.23; 3.58	15–20	25
	H_2SO_4	4.55; 4.37; 3.49	25–30	41
	HCl	4.66; 4.31; 3.57	20–30	43
	I_2	4.46; 4.27; 3.54	15–25	18

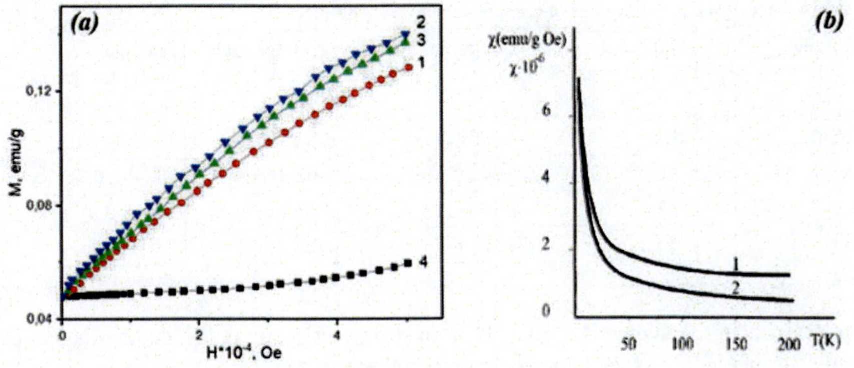

Fig. 3 **a** Field dependence of magnetization at $T = 5$ K for PAn (*1*), PAn-I_2 (2), PAn-H_2SO_4 (*3*), and PAn at $T = 300$ K (*4*); **b** Temperature dependence of magnetic susceptibility of PAn doped with I_2 in the field of 0.1 T (*1*) and 5 T (*2*)

The magnetization of undoped PAn and doped with sulfuric acid (50 mol %) or iodine (2.6 mol %) was studied in the temperature range 5–300 K. It was found that the diamagnetic contribution predominates at room temperature, while the paramagnetic—at lower temperatures. The "degree" of paramagnetism depends on the type of doping agent (Fig. 3).

It was found that the magnetic susceptibility of polymers in weak fields can be described by an equation that includes a temperature-independent component (χ_o) and a temperature-dependent component according to the Curie–Weiss law

$$\chi = \chi_o + C(T - \Theta)$$

Due to this dependence and the minimum of the root-mean-square deviation of the experimental susceptibility data at $H = 5$ T, the values of the paramagnetic Weiss constant (Θ), the Curie constant (C), and the parameter χ_o were obtained (Table 3). The concentration of unpaired spins N_s was calculated from the found values of C [25].

It was established that the magnetic susceptibility is almost independent of the temperature at $T > 200$ K [25, 29]. Pauli paramagnetism ($\chi_o > 0$) was found for PAn doped with iodine. For acid-doped PAn $\chi_o > \chi_{diam}$ indicates the presence of a paramagnetic contribution. Thus, there is a superposition of diamagnetic and paramagnetic

Table 3 Parameters of magnetic susceptibility of polyaniline $\chi = \chi_o + C(T - \Theta)$

Polymer	$C \times 10^5$, K/gE	$\chi_o \times 10^7$, 1/gOe	Θ, K	$N_s \times 10^{-19}$, 1/g
PAn base undoped	2.9	− 3.5	− 0.89	4.7
PAn-H_2SO_4	3.4	− 0.7	− 0.94	5.5
PAn-I_2	3.9	+ 1.3	− 1.19	6.2

Table 4 Results of X-ray studies of composites Fe_3O_4–polymer

Sample	Magnetite content, mass %	Lattice parameter, *a*, Å	Average grain size, Å	Average maximum stress, %
Fe_3O_4	100	8.3479 (17)	75.5 (± 7.3)	13.1 (± 8.3)
Fe_3O_4–polymer	10	8.345 (2)	102.0 (± 9.5)	10.0 (± 6.7)
	12.4	8.3491 (12)	111.1 (± 1.8)	8.8 (± 2.8)
	13.3	8.351 (3)	98.7 (± 2.8)	9.9 (± 3.1)

components in the magnetization of PAn. In strong magnetic fields, the diamagnetic contribution (M_{diam}) is well approximated by the dependence:

$$M_{\text{diam}} = M_0 + \chi_{\text{diam}} \cdot H$$

where χ_{diam}—the diamagnetic susceptibility of the polymer matrix, H—the magnetic field strength, M_0—a constant associated with hardware corrections. The values found at $T = 300$ K are shown in Table 3. To estimate the real diamagnetic contribution with correction for paramagnetism, the obtained values of χ_{diam} must be supplemented with the paramagnetic contribution C/T at $T = 300$ K, where C—the Curie constant.

The polymer matrix diamagnetism does not depend on temperature, so all the obtained dependences were corrected. Polyaniline doped by iodine demonstrates a paramagnetism without magnetic metal ions (Table 3). The study of injection of the doping magnetic ions on the magnetic behavior of the conjugated polymers caused growing interest.

4 Magnetic Properties of Polymers Doped with Magnetic Ions. Temperature Dynamics of EPR Magnetic Centers in Polymer–Magnetic Nanosystems

The influence of doping on the magnetic properties of polyaniline (PAn) and polyparaphenylene (PPP) was studied. As noted earlier, the undoped matrix of polyaniline and PAn doped with sulfuric acid have mainly diamagnetic properties [25]. Doping with HCl or iodine causes paramagnetic behavior, which can be described by the Curie–Weiss law at $T < 120$ K. All investigated samples of PAn, its derivatives, and copolymers give a time-stable EPR signal already at room temperature with a g-factor of 2.0003–2.0027 [26, 29]. However, the effect of doping with magnetic ions on the polymer paramagnetic properties is unstudied, and the influence of the polymer matrix on the magnetic center's closest surroundings on the magnetic behavior of polymer composites too. To solve this problem, we studied the temperature dependence of EPR for polyaniline composite matrices doped with the complex compound $K_3[Fe(CN)_6]$.

The samples were synthesized by oxidative polymerization followed by treatment with an ammonia solution to remove inorganic impurities and obtain a polymer in the form of an emeraldine base. The dedoped sample was kept for 24 h in a 0.02 M potassium hexacyanoferrate in 0.05 M sulfuric acid to obtain the hybrid material based on PAn by the ion exchange method. After washing and drying, the samples were stored in quartz ampoules. The content of doping impurities, determined by X-ray microanalysis, was 0.3% $K_3[Fe(CN)_6]$.

The spectrum of the obtained doped polymer is a superposition of the EPR spectrum of free radicals of the PAn polymer matrix and Fe(III) ions. The EPR spectrum of pure PAn in the temperature range $T = 4.2$–295 K has one resonance line with $g = 2.000 \pm 0.001$, the width of which (ΔH_{pp}) decreases with increasing temperature [26].

The value of the g-factor does not depend on the temperature. The resonance line observed in undoped PAn is caused by the radicals stabilized by the π-conjugation system [22, 32].

The EPR spectrum of the sample doped with $K_3[Fe(CN)_6]$ consists of 3 resonance lines (Fig. 4a). The first and second lines, probably due to Fe(III) ions, appear after doping. The third resonance line, according to the value of the g-factor and the nature of the temperature behavior, refers to the free radicals of the PAn.

The value of the g-factor of line 1 at 4.2 K is equal to $g_1 = 4.22 \pm 0.03$ and is almost independent of temperature. For line 2 $g_2 = 2.13 \pm 0.05$ at $T = 295$ K. While the temperature decreases, the intensity of line 1 decreases until it completely disappears at $T = 290$ K (Fig. 4). At the same time, the width of the line increases by no more than 30% (Fig. 4b). In contrast, width decreases with increasing temperature for undoped PAn (Fig. 4c).

The revealed features of EPR spectra should appear in other disordered polymer systems doped with magnetic ions. Therefore, polyparaphenylene (PPP) matrices obtained by chemical synthesis and characterized by structural and spectral methods [33] were studied.

A dispersed fine-crystalline sample was kept for 24 h in a solution of anhydrous ferric chloride in nitromethane to introduce magnetic ions into the PPP matrix, washed with a solvent, and dried. According to the spectral analysis, the Fe content in this material was 2.3% by mass.

One (300 K) or two (4.2–100 K) resonance lines can be observed in the EPR spectrum of PPP doped with ferric chloride, depending on the temperature. Low-intensity line with $g_1 = 4.21 \pm 0.05$ and a high-intensity line with $g \approx 2$ appears at low temperatures. The second one is a superposition of two lines, one of which refers to the impurity ion Fe^{3+} (Fig. 5). The EPR line for undoped PPP at $T = 4.2$ K has the value $g = 2.12 \pm 0.03$ and the width $\Delta H_{pp} = 0.38$ kE, while for resonance line 2 the width $\Delta H_{pp} = 0, 65$ kE and the g-factor $g_2 = 2.00 \pm 0.09$. The contribution of the PPP to the total intensity of line 2 is 4% due to the computer simulation method.

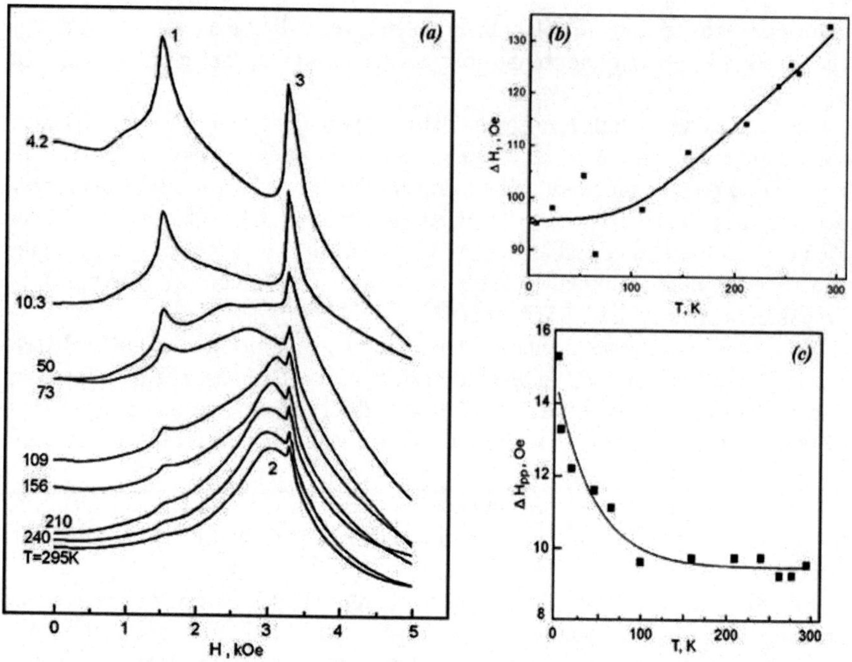

Fig. 4 **a** Temperature-induced changes in the resulting absorption line in the EPR spectrum of polyaniline doped with $K_3[Fe(CN)_6]$; **b** temperature dependence of the width of the first resonance line ΔH_1; **c** temperature dependence of the resonance linewidth ΔH_{pp} for pure polyaniline

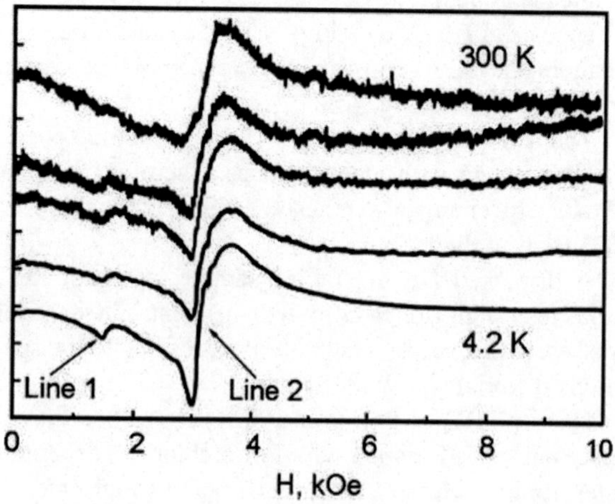

Fig. 5 EPR spectrum of polyparaphenylene doped with $FeCl_3$ at $T = 4.2$, 25, 49, 100, and 300 K

Therefore, in some organic magnets—complexes of iron with nitrosonaphthol [27], disordered polymer matrices—polyaniline [7, 26], polyparaphenylene [33], polymethoxyaniline [29, 34], polyaminothiazole [29, 35, 36], doped with $FeCl_3$ or $K_3[Fe(CN)_6]$, interesting temperature dependence of the EPR spectra was revealed. In these substances, the EPR spectrum is a superposition of several lines, two of which refer to the magnetic center—Fe(III). When the temperature decreases, the intensity of one-line increases, and the other contra. So, the first line has the maximum intensity at low (helium) temperatures and the other at room temperatures. The first line is the low-temperature component of the EPR spectrum, and the second is the high-temperature component. A change in temperature leads to a redistribution of absorption intensity between its. The EPR spectrum indicates the unusual dynamics of the molecules surrounding the Fe^{3+} ion. These dynamics can significantly influence various properties of substances. Based on the temperature dependence of EPR spectra and the crystal field theory [32, 37], it can be assumed that the temperature changes of the EPR spectrum are not caused by movements of Fe^{3+} ions from one non-equivalent position to another but rather the transition of the magnetic center from one state to another [29, 34].

Therefore, the Fe^{3+} ion can act as a paramagnetic probe in the polymer complexes revealing temperature dynamics different from the polymer matrix. The magnetic center dynamic depends on the iron ion external molecular (ligand) environment and the structural non-equivalence of its positions in conjugated polymer matrices. This phenomenon can be used in biology and medicine to monitor and predict the behavior of nerve cells and other biological objects [37–41].

5 Micro- and Nanocomposites Polymer–Magnetite

Nowadays, with intensive studies of molecular polymer magnets [3–5, 8], much attention is paid to hybrid polymer–magnetic composites [6, 9–14, 42]. Most of the studies are devoted to the methods for obtaining systems, their structure, morphology, and some physical properties, namely electrical conductivity, magnetization, and thermal stability. At the same time, the chain structure of polymer matrix macromolecules and the fluctuation network determine the mechanical properties of polymer composites depending on the action of external factors—temperature, pressure, and load. The relative value of elastic deformation in polymers tends to be tens of percent, so composites of polymers with magnetic particles will likely exhibit useful properties, primarily as plastic magnetic materials of small specific gravity, as well as toner materials for digital printing [13].

The enhanced functionality of hybrid nanocomposites is achieved by adding various functions—magnetic, conductive, fluorescent, etc. [9, 43–47]. Functionalized magnetic polymer capsules are promising for this aim [9]. Marking capsules with fluorescent substances, such as semiconductor nanocrystals, makes it possible to trace their way into the tissues and capillaries. Loading capsules with magnetic nanoparticles simultaneously allows controlling their movement by changing the gradient of the external magnetic field. An intermediate polymer layer between the magnetic particles and fluorescent labels is necessary to reduce the adverse effects of quenching the nanocrystal luminescence. It is achieved by the polymer shell encapsulation of magnetic particles [9]. According to magnetometry and Mossbauer spectroscopy, encapsulated magnetite particles preserve superparamagnetic properties [44].

Improvement of the properties of the composites is also possible by providing electrical functions using conducting polymers [6]. Such modification allows controlling the nanoparticle movement by both electric and magnetic fields and tracking it in different mediums [9, 43]. Conducting polymer–magnetic composites provide magneto-switchable charge transport and bioelectrocatalysis [43, 48]. However, there were reported only a few attempts to obtain hybrid magnetic composites with both luminescence and conductivity. In addition, despite intensive research on hybrid systems, the influence of the polymer matrix on the properties of fluorescent substances remains to be fully elucidated [49, 50].

We have studied the conditions of preparation, structure, luminescence, and electrical properties of composites based on magnetite nanoparticles encapsulated with polymer shells and modified by barium zirconate and polyaniline [14, 51].

Highly dispersed magnetite Fe_3O_4 was synthesized by alkaline hydrolysis of iron II and iron III salts with sodium oleate as a stabilizer. For better compatibility of the magnetic part with the epoxy matrix, the magnetite nanoparticles were modified with a polystyrene (PS) shell [42]. Spherical particles (granules) of dispersed magnetite with a size of about 1–2 μm were separated from the suspension by magnetic decanting.

According to the diffractogram of the magnetite (Fig. 6), the peaks denoted with hkl indicate the cubic phase Fe_3O_4 with spinel structure and cell parameter a = 8.3490(3) Å. The average size of the Fe_3O_4 granules, determined by microstructure studies, is 75.5 (± 7.3) Å. The amorphous halo, observed for magnetite with polymer shells, refers to polystyrene and diffraction reflections to magnetite (Fig. 6b). At the same time, diffraction peaks of cubic Fe_3O_4 phase with low intensity for powder diffraction are almost vanishing.

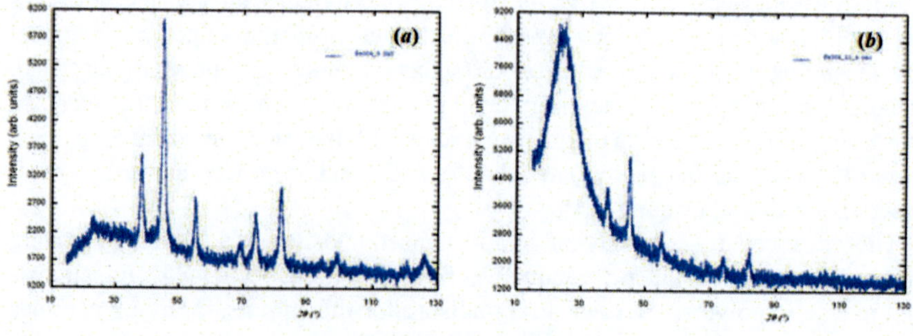

Fig. 6 X-ray diffractogram of Fe_3O_4 (**a**) and polymer/magnetite composite (**b**)

The cell parameter a is almost the same for both samples within standard deviation. But it is a bit larger $a = 8.3491$ (12) Å for the composite. It is important to note that the average size of the composite granules exceeds one for Fe_3O_4 without polymer (Table 4). It confirms the formation of a polymer shell around magnetite particles.

The oxidative polymerization of aniline in the aqueous dispersion of polymer–magnetite capsules leads to additional microsphere encapsulation by conducting polyaniline shells. The adsorption of luminescent nanocrystals $BaZrO_3$ (BZO) was used for surface modification of polymer–magnet microcapsules.

EDX analysis confirmed the adsorption of BZO nanocrystals on the surface of polystyrene capsulated magnetite. According to the elements distribution profile on the surface of the composite, the intensive peaks from Zr, Ba, and Fe are observed (Fig. 7). The number of adsorbed nanocrystals grows with the increase of $BaZrO_3$ content in the initial reaction mixture from 1 to 20%. Experimental diffraction patterns (Fig. 8) indicate an amorphous halo from polymer (PS), semi-amorphous reflections of magnetite [42, 52], and formed peaks from $BaZrO_3$ [53].

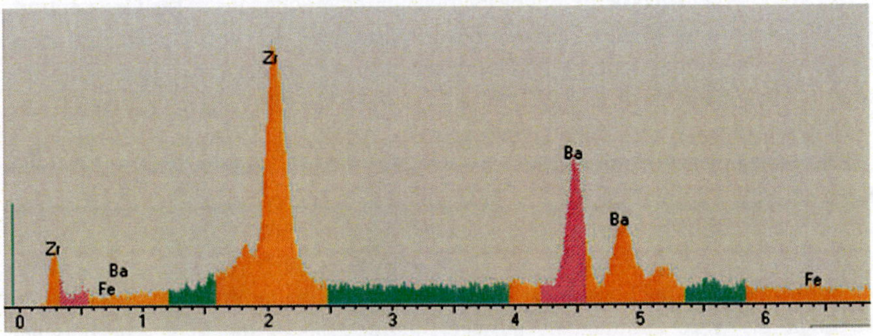

Fig. 7 Profile of distribution of elements on the surface of polymer–magnetite composite, modified by nanocrystals at $BaZrO_3$ at its 10% content in initial dispersion

Fig. 8 XPD pattern of Mt-PS-BZO-PAn composites obtained at 1 (*1*), 5 (*2*), 10 (*3*) and 20% (*4*) content of BZO in the initial dispersion

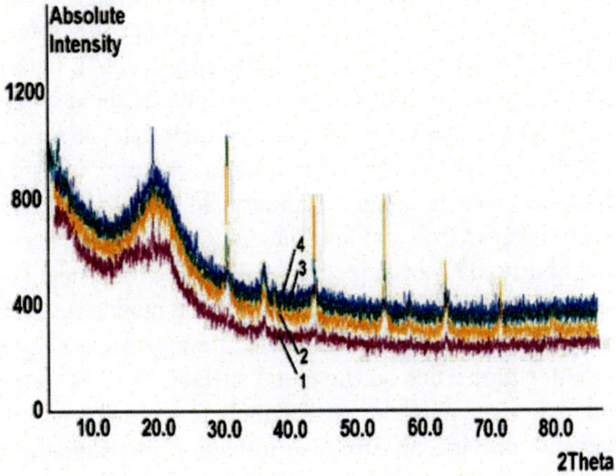

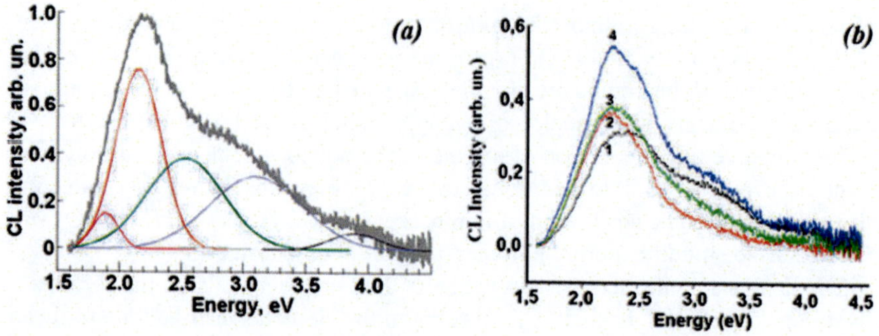

Fig. 9 **a** CL spectrum of BZO-PS composites at BZO content 5%; **b** CL spectrum of Mt–PS–BZO–PAn hybrid composites. Content of BZO, %: 1—1; 2—5; 3—10; 4—20

Light radiation of polymer–magnetite nanocapsules was studied by cathode luminescence (CL) spectra of the pellet samples. Adsorption of BZO nanocrystals on the surface of polymer–magnetite grains causes significant changes in the CL spectrum [49, 50]. According to the Gaussian decomposition of CL curves, the additional lowest and the highest energy bands appear at a significant reduction of the CL intensity. New bands at $E = 1.9$, 2.15, 2.45, 3.0, and 3.96 eV were observed in the PS-BZO composites spectra (Fig. 9a).

The composites with polyaniline preserve the ability of radiation. The CL intensity increases with the increase of the BZO content in the composite (Fig. 9b). The polyaniline modification of nanocapsules causes a decrease in the specific volume resistance ρ ($T = 293$) from 4×10^8 Om $*$ m to 18.2 ± 0.5 Om $*$ m.

For the first time, the hybrid composites with magnetic, luminescent, and conductive functions were obtained by surface modification of the magnetite-polymer nanoparticles by $BaZrO_3$ and polyaniline [14, 51]. The influence of the polymer matrix on the cathodoluminescence spectrum of $BaZrO_3$ was found. CL spectra modification in composites is probably due to the change in the nanocrystal substructure caused by the PS matrix. The lattice parameter decreases from 4.19083 (6) for nano-BZO to 4.1879 (2) Å in BZO-PS composite. Interaction between the PS matrix and BZO was confirmed by FTIR spectra [14]. The proposed method of surface modification may be used in sensors and medical diagnostics [54, 55].

Composite materials containing dielectric or semiconductor nanoparticles of a particular shape or micro- and nanostructures capable of spatial redistribution of scattered electromagnetic radiation in desired directions are perspectives as anti-radar screens. Iron oxides, magnetite, or Fe_3O_4 are especially interesting due to their ability to absorb electromagnetic waves of near IR and microwave ranges [56]. Carbon threads or nanotube inclusions in magnetite-based composites may provide a synergetic effect, i.e., enhance electromagnetic energy absorption and improve anti-corrosive properties on the metal surface [57, 58]. We suggest filling the composite with magnetite particles with a diameter commensurate with the microwave's length that may provide an effective reflection and radiation scattering [59]. The hybrid composites with stabilized by a polymer shell magnetite microparticles were prepared as described above. Instead of expensive and scarce nanotubes, we use conducting polymer polyaniline as an organic additive [60]. Variation in the size of Fe_3O_4–PS particles does not affect the period of the magnetite elementary cell. In contrast, it influences the magnetite–polymer composite magnetic susceptibility and microhardness (Table 5). Partial aggregation of magnetite grains and formation of polymer shell on their surface results in the appearance of spherical core–shell particles with an average size of 1.5–2 μm.

The granules Fe_3O_4–PS of ≈ 2 μm size were used as magnetic fillers to obtain thermosetting polymer composites with antiradar and protective functions [59]. We propose to use polyaniline doped with TSA as a component with intrinsic electron conductivity. The polymer has high specific conductivity of $(3.0 \pm 0.3)10^{-2}$ S/cm. So, the formation of conducting channels inside the composite structure is possible. It is necessary to enhance the absorption and scattering of electromagnetic radiation [56, 57].

Magnetic and electric properties of hybrid nanocomposite materials depend on the nature of the initial components, the amount of ferromagnetic and conductive fillers, polyaniline dopants, and the preparation procedure. Composite films with conducting polymer prepared in the presence of ferromagnetic nanoparticles can be alternatives to metal-containing systems due to improved properties and lower prices.

The polymer composite was spilled onto a flat Teflon substrate and peeled off the surface after film formation. Thus, a free polymer composite film of 0.2 mm thickness was obtained. Details of synthesis can be found in [59].

It is worth noting that the encapsulation of magnetite into a polymer shell weakly affects the magnetic susceptibility of magnetic filler [51]. Thus, composites can absorb IR and microwave electromagnetic radiation effectively [58].

The microstructures of the composite polymer coatings of the thermosetting mixtures based on epoxy resin, amine hardener, and different fillers are noticeably different (Fig. 10a, b). The uniform microstructure is observed for the composite containing a mixture of fillers of 10% and a weight ratio of 1:1 (Fig. 10b). In this case, there is a better interaction of components inside of composite, probably due to the compatibility of magnetite microgranules surrounded by polystyrene shell and PAn doped with TSA (plasticizer) with hydrophobic epoxy matrix.

Table 5 Physical properties of magnetite–polystyrene granules

Granule size of Fe_3O_4–PS composite, μm	Magnetic susceptibility, $\chi \times 10^{-6}$, cm^3/g	Period of the elementary cell, Å	Microhardness, $F_\infty \times 10^{-8}$, N/m^2
1.5–1.7	1797	8.36 (2)	2.45
1.7–2.0	1969	8.357 (9)	2.47
2.0–2.5	2822	8.333 (9)	3.03
2.2–2.5	2589	8.35 (2)	3.89
2.5–3.0	2123	8.36 (2)	2.78

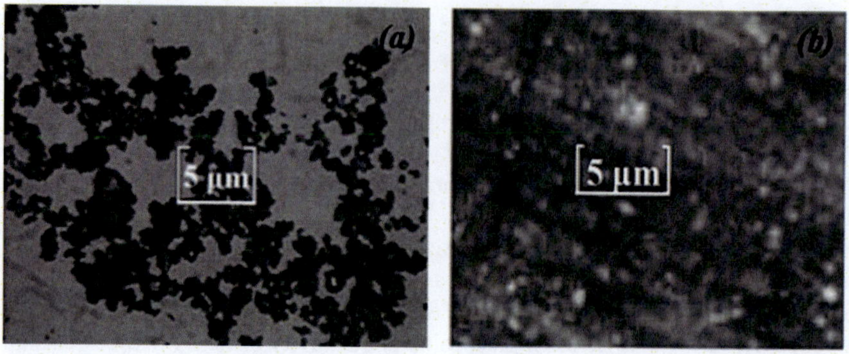

Fig. 10 Micrograph of magnetite Fe_3O_4 particles encapsulated into polystyrene shell (**a**); micrograph of the composite (5% magnetite + 5% PAn/TSA powder) taken by "Olympus" optical microscope (**b**). A long-focus lens with a magnification of 50× was used

The interaction leads to a significant increase in the microhardness F_∞ of the filled composites—up to 18.7×10^9 N/m^2 compared to the unfilled epoxy composite with $F_\infty = 1.25 \times 10^9$ N/m^2 [59].

Synthesized magnetite microgranules with polystyrene shells and TSA-doped polyaniline were used to form composite coatings with antiradar functions. The highest microhardness and corrosion resistance of coatings are observed for the composite with 10% filler. It is achieved by combining magnetic and conducting components due to the synergistic effect.

The composite, containing magnetic microparticles and polyaniline doped with toluene sulfonic acid at a 1:1 ratio, reveals the strongest IR absorption coefficient $A = 35.2$ cm^{-1} and low reflectivity $R = 0.022$ (Table 6). At the same time, this composite has excellent anti-corrosive properties and high microhardness.

The composite containing only the conductive PAn-TSA component has a low level of both reflectivity and absorption (Table 6). The composites doped with 10% of magnetite reveal high absorption and reflectivity undesired for protective coatings. Doping with both PAn and magnetite fillers leads to the expected properties—high absorption and low reflectivity of composite.

Table 6 The influence of fillers on IR absorption and reflectivity

Magnetite, w.%	PAn-TSA, w.%	Reflectivity, R (at $\lambda =$ 1200 nm)	Absorption coefficient, A, cm^{-1}
0	0	0.082	3.5
5	5	0.022	35.2
10	0	0.056	31.7
0	10	0.014	14.8

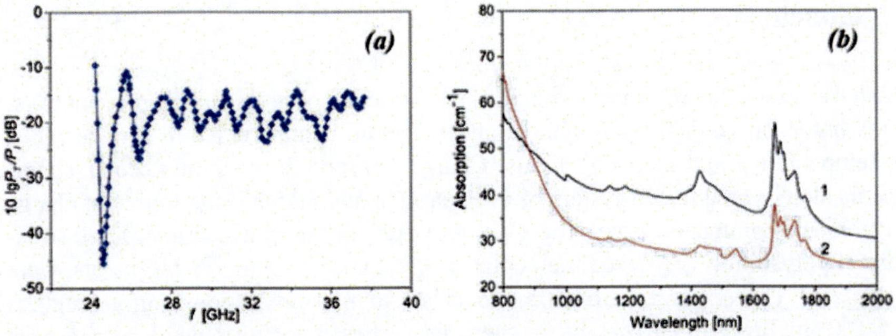

Fig. 11 **a** RF radiation power absorption spectrum measured by HP 8722C spectrum analyzer for ED20/PEPA composites with 10% content of fillers (5% Fe_3O_4 + 5% PAn/TSA). Film thickness 0.2 μm; **b** IR absorption spectra of film samples of epoxy composition with different content of fillers: 1—5% magnetite + 5% PAn/TSA; 2—10% magnetite. Film thickness 0.2 μm

The microwave absorption of the composite was studied with HP 8722C spectrum analyzer. The sample of 0.2 mm thickness was placed across the waveguide perpendicularly to the direction of electromagnetic radiation (EMR) propagation. The composite film was cut to fit the internal cavity of the waveguide. The composite reveals high attenuation of EMR in K band (18–26.5 GHz) and Ka-band (26.5–40 GHz) at the level of – 25 dB with a peak value of – 47 dB at 23 GHz (Fig. 11a). That makes the material perspective for electromagnetic radiation shielding and anti-radar purposes [60]. Shielding properties of similar composite materials based on epoxy resins with graphene nanoplates in the microwave frequency range show only 10 dB at 27 GHz [61].

The film samples of different compositions were prepared to study the absorption of the composites in the near IR range. The IR absorption spectra were recorded using an MDR-23 monochromator with a halogen lamp as a light source. The absorption spectra in the near IR (NIR) range of samples with different filler content are shown in Fig. 11b.

The composite containing magnetic microparticles and TSA-doped polyaniline at a 1:1 ratio reveals stronger IR absorption than the composite filled only with magnetite (Fig. 11b).

In conclusion, we have synthesized composite material with high absorption of electromagnetic radiation in microwave and NIR spectral ranges. The composite comprises polyaniline doped with TSA and magnetite microparticles stabilized by polystyrene shells. This coating may find dual military and common-purpose applications. It can be used for electromagnetic shielding, antiradar, and light harvesting in solar thermal energy applications.

6 Conclusion

We have studied the magnetic properties and structure of polymer–magnet nanosystems based on conjugated polymers and magnetic nanoparticles in a wide range of temperatures and magnetic fields. Charge transport in macromolecular chains occurs along a conjugated system of π-electron bonds, which is the source of charge carriers and paramagnetic centers. The parameters of the short-range ordered structure of polyaniline, preserved the coplanar orientation of aromatic fragments, were calculated. The size of repetitive fragments is 0.46 nm, and the conjugation length is 2.5–3.0 nm. It was found that the magnetic properties of conjugated polymers are the superposition of dia- and paramagnetic contributions depending on the type of doping and temperature. The interesting new phenomena in the temperature dynamic of paramagnetic centers in a series of conjugated polymers doped with magnetic ions were revealed by EPR spectroscopy. Ion Fe^{3+} in the formed polymer complexes can act as a paramagnetic probe demonstrating its own temperature dynamic, different from the polymer matrix. The peculiarities of the behavior of paramagnetic "probe"—the dynamics of magnetic centers depend on the external molecular (ligand) environment of the Fe atom and the presence of structural non-equivalence of its positions in conjugated polymer matrices. This phenomenon can be used in biology and medicine to monitor and predict the behavior of nerve cells and other biological objects.

Particular attention is given to hybrid polymer–magnetic micro- and nanocomposites from the point of view of their practical use as materials sensitive to magnetic and electric fields, modern absorbers of radiation, etc.

For the first time, we obtained hybrid composites with magnetic, luminescent, and conductive functions simultaneously by surface modification of the magnetite-polymer nanoparticles with luminescent $BaZrO_3$ nanocrystals and polyaniline as conducting substance. The proposed surface modification method may be used to develop sensors and functional materials for medical diagnostics.

Also, we have synthesized a composite material with high absorption of electromagnetic radiation in microwave and NIR spectral ranges. This coating can have military and common purposes—for electromagnetic shielding and antiradar, as well as solar thermal energy applications. Composite films of conducting polymer with ferromagnetic nanoparticles can be alternatives to metal-containing systems due to improved properties and lower prices.

References

1. J.S. Miller, Organic- and molecule-based magnets. Mater. Today **17**(5), 225–235 (2014). https://doi.org/10.1016/j.mattod.2014.04.023
2. Y. Aoki, Y. Orimoto, A. Imamura, *Quantum Chemical Approach for Organic Ferromagnetic Material Design* (Springer, Cham, 2017). https://doi.org/10.1007/978-3-319-49829-4
3. E. Coronado, Molecular magnetism: from chemical design to spin control in molecules, materials and devices. Nat. Rev. Mater. **5**, 87–104 (2020). https://doi.org/10.1038/s41578-019-0146-8

4. A. Rajca, From high-spin organic molecules to organic polymers with magnetic ordering. Chem. Eur. **98**, 4834–4841 (2002). https://doi.org/10.1002/1521-3765(20021104)8:21%3c4834::AID-CHEM4834%3e3.0.CO;2-E
5. A. Rajca, J. Wongsriratanakul, S. Rajca, Magnetic ordering in an organic polymer. Science **294**, 1503–1505 (2001). https://doi.org/10.1126/science.1065477
6. Z. Zhang, M. Wan, Nanostructures of polyaniline composites containing nano-magnet. Synth. Met. **132**, 205–212 (2003). https://doi.org/10.1016/S0379-6779(02)00447-2
7. V.N. Vasyukov, V.P. Dyakonov, V.A. Shapovalov, E.I. Aksimentyeva, H. Szymczak, S. Piehota, Temperature-induced change in the ESR spectrum of the Fe^{3+} ion in polyaniline. Low Temp. Phys. **26**(4), 265–269 (2000). https://doi.org/10.1063/1.593896
8. M.M. Matsushita, H. Kawakami, T. Sugawara, M. Ogata, Molecule-based system with coexisting conduct9ivity and magnetism and without magnetic inorganic ions. Phys. Rev. (2008)https://doi.org/10.1103/PhysRevB.77.195208
9. H. Sawada, H. Yoshioka, T. Kawase, K. Ueno, K. Hamazaki, Preparation of magnetic nanoparticles by the use of self-assembled fluorinated oligomeric aggregates. A new approach to the dispersion of magnetic particles on poly(methyl methacrylate) film surface. J. Fluorine Chem. **126**, 914–917 (2005). https://doi.org/10.1016/j.jfluchem.2005.04.015
10. C. Janaky, C. Visy, O. Berkesi, E. Tomba, Conducting polymer-based electrode with magnetic behavior: electrochemical synthesis of poly(3-thiophene-acetic-acid)/magnetite nanocomposite thin layers. J. Phys. Chem. C **113**, 1352–1358 (2009). https://doi.org/10.1021/jp809345b
11. I. Opainych, O. Aksimentyeva, V. Dyakonov et al., Structure and thermodeformation properties of polymer-magnetite hybrid composites. Mater. Sci. **48**, 95–100 (2012). https://doi.org/10.1007/s11003-012-9477-y
12. B. Zebli, A.S. Susha, G.B. Sukhorukov, A.L. Rogach, W.J. Parak, Magnetic targeting and cellular uptake of polymer microcapsules simultaneously functionalized with magnetic and luminescent nanocrystals. Langmuir **21**, 4262–4265 (2005). https://doi.org/10.1021/la0502286
13. P. Tiberto, G. Barrera, F. Celegato et al., Magnetic properties of jet-printer inks containing dispersed magnetite nanoparticles. Eur. Phys. J. (2013).https://doi.org/10.1140/epjb/e2013-30983-8
14. O.I. Aksimentyeva, V.P. Savchyn, V.P. Dyakonov et al., Modification of polymer-magnetic nanoparticles by luminescent and conducting substances. Mol. Cryst. Liq. Cryst. **590**, 35–42 (2014). https://doi.org/10.1080/15421406.2013.873646
15. L.S. Hegedus, *Transition Metals in the Synthesis of Complex Organic Molecules* (University Science Books, US, 1994)
16. Yu. Horbenko, O. Aksimentyeva, Structure and physicochemical properties of poly-ortho-anisidine doped with ferric (III) chloride. Visnyk Lviv Univ. Ser. Khimia **54**, 353–357 (2013)
17. C.G.C.M. Netto, H.E. Toma, L.H. Andrade, Superparamagnetic nanoparticles as versatile carriers and supporting materials for enzymes. J. Mol. Catal. **71**, 85–86 (2013). https://doi.org/10.1016/j.molcatb.2012.08.010
18. L.H.M. Fonseca, A.W. Rinaldi, A.F. Rubira et al., Structural, magnetic, and electrochemical properties of poly(o-anisidine)/maghemite thin films. Mater. Chem. Phys. **97**, 252–255 (2006). https://doi.org/10.1016/j.matchemphys.2005.08.007
19. V.A. Shapovalov, V.V. Shapovalov, M. Rafailovich, S. Piechota, A. Dmitruk, E. Aksimentyeva, A. Mazur, Dynamic characteristic of molecular structure of poly-ortho-methoxyaniline with magnetic probes. J. Phys. Chem. **117**, 7830–7834 (2013). https://doi.org/10.1021/jp311456a
20. O.I. Aksimentyeva, V.P. Dyakonov, Effect of aminonaphthalene sulfonic acid nature on the structure and physical properties of their copolymers with aniline, in *Functional Polymer Blends and Nanocomposites. A Practical Engineering Approaches*. ed. by G.E. Zaikov, L.I. Bazylak, A.K. Haghi (Apple Academic Press Ink, Toronto, 2014), pp. 217–231
21. A.J. Heeger, Semiconducting and metallic polymers: the fourth generation of polymeric materials. Synth. Met. **123**, 23–42 (2002). https://doi.org/10.1021/jp011611
22. A. MacDiarmid, "'Synthetic metals'": a novel role for organic polymers. Curr. Appl. Phys. **1**, 269–279 (2001)

23. V.N. Prigodin, A.J. Epstein, Nature of insulator-metal transition and novel mechanism of charge transport in the metallic state of highly doped electronic polymers. Synth. Met. **125**, 43–53 (2002)
24. A.L. Buchachenko, Organic and molecular ferromagnets. Achievements and successes. Adv. Chem. **59**(4), 529–550 (1990)
25. E.I. Aksimentyeva, M. Baran, V.P. Dyakonov, S. Pehota, R. Shimchak, Magnetic properties of doped polyaniline. Solid State Phys. **38**(7), 2277–2285 (1996)
26. V.N. Vasyukov, V.P. Dyakonov, V.A. Shapovalov, E.I. Aksimentyeva, H. Szymczak, S. Piehota, Temperature-induced change in the ESR spectrum of the Fe^{3+} ion in polyaniline. Low Temp. Phys. **26**(4), 265–269 (2000)
27. V.P. Dyakonov, E. Zubov, E. Aksimentyeva, S. Piehota, H. Szymczak, Low-temperature magnetic behavior of the organic-based magnet $Na[FeO_6(C_{10}H_6N)_3]$. Low Temp. Phys. **40**(7), 835–841 (2014)
28. G.V. Martynyuk, O.I. Aksimentyeva, Features of charge transport in polymer composites polymethylmethacrylate–polyaniline. Phys. Chem. Solid State **21**(2), 319–324 (2020). https://doi.org/10.15330/pcss.21.2.319-324
29. O.I. Aksimentyeva, O.I. Konopelnyk, Yu.Yu. Horbenko, Hybrid nanosystems based on conjugated polyaminoarenes doped by ferrum-containing compounds, in *Computational and Experimental Analysis of Functional Materials*. ed. by O.V. Reshetnyak, G.E. Zaikov (Apple Academic Press Ink, Toronto, 2017), pp. 473–496. https://doi.org/10.1201/9781315366357
30. F. Tavakolinia, M. Yousef, S. Salman, S. Afghahi, S. Baghshahi, S. Samadi, Effect of polyaniline on magnetic and microwave absorption properties in $SrFe_{12}O_{19}/Zn_{0.4}Co_{0.2}Ni_{0.4}Fe_2O_4$ ferrite nanocomposites. J. Inorg. Organomet. Polym Mater. **30**, 4014–4026 (2020). https://doi.org/10.1007/s10904-020-01547-0
31. O.I. Aksimentyeva, O.I. Konopelnyk, Structure of near order in conducting polyaniline films. Mol. Cryst. Liq. Cryst. **427**(1), 117–125 (2005). https://doi.org/10.1080/15421400590892109
32. Y.G. Klyava, *EPR Spectroscopy of Disordered Solids* (Zinatne, Riga, Latvia, 1988)
33. A.F. Dmitruk, E.I. Aksimentyeva, V.P. Dyakonov et al., Investigation of structure of Fe^{3+} magnetic center in polyparaphenylene. Intern. J. Quant. Chem. **88**, 525–529 (2002)
34. V. Shapovalov, M. Rafailovich, S. Piechota, A. Dmitruk, E. Aksimentyeva, A. Mazur, Dynamic characteristic of molecular structure of poly-ortho-methoxyaniline with magnetic probes. J. Phys. Chem. **117**, 7830–7834 (2013). https://doi.org/10.1021/jp311456a
35. Yu. Horbenko, O. Aksimentyeva, Electron paramagnetic resonance of polyaminothiazole complexes doped with ferrum (III) chloride. Visnyk Lviv Univ. Ser. Chem. **56**(2), 334–338 (2015)
36. Y.Y. Horbenko, O.I. Aksimentyeva, V.A. Shapovalov, ESR study of model magnetic centers in metalloproteins, in *Abstracts of the II International Conference on "Nanobiophysics: Fundamental and Applied Aspects"*, Kyiv, Ukraine, 6–9 Oct 2011
37. V.A. Shapovalov, E.I. Aksimentyeva, V.V. Shapovalov, Commonality of the properties of compounds characterized by different symmetry of the structure. Mol. Cryst. Liq. Cryst. **671**(1), 175–182 (2018). https://doi.org/10.1080/15421406.2018.1542100
38. Y. Lai, Y. Kuo, Y. Chiang, Identifying protein conformational dynamics using spin-label ESR. Chem. Asian J. **14**(22), 3981–3991 (2019). https://doi.org/10.1002/asia.201900855
39. B. Dobosz, R. Krzyminiewski, Electron spin resonance (ESR) study of human blood and its interaction with magnetite nanoparticles. Electron. Spin Resonance Spectrosc. Med. **2018**, 1–22 (2018). https://doi.org/10.1007/978-981-13-2230-3_1
40. R.G. Sajfutdinov, L.I. Larina, T.I. Vakul'skaya, M.G. Voronkov, *Electron Paramagnetic Resonance in Biochemistry and Medicine* (Springer, New York, 2002)
41. A. Zarnani, M. Nejadmoghaddam, M.M. Moghaddam, F. Mohammadi, M. Eskandari, R. Ghahremanzadeh, Bioimaging based on antibody-conjugated amphiphilic polymer-core@shell quantum dots. Emerg. Mater. Res. **7**(4), 209–217 (2018). https://doi.org/10.1680/jemmr.17.00067
42. O.I. Aksimentyeva, I.Ye. Opaynych, V.P. Dyakonov, S. Piechota, V.P. Zakordonskyi, P.Yu. Demchenko, H. Szymczak, Structure and thermal stability of the polymer—magnet nanocomposites. Phys. Chem. Solid State **13**(2), 438–442 (2012)

43. M. Riskin, B. Basnar, Ye. Huang, I. Willner, Agnetoswitchable charge transport and bioelectrocatalysis using maghemite-Au core-shell nanoparticle/polyaniline composites. Adv. Mater. **19**, 2691–2695 (2007)
44. D. Wang, J. He, N. Rosenzweig, Z. Rosenzweig, Superparamagnetic Fe_2O_3 beads—CdSe/ZnS quantum dots core—shell nanocomposite particles for cell separation. Nanoletters **4**, 409–413 (2004). https://doi.org/10.1021/nl035010n
45. K.D. Mahajan, G. Ruan, G. Vieira, T. Porter, J.J. Chalmers, R. Sooryakumar, J.O. Winter, Biomolecular detection, tracking, and manipulation using a magnetic nanoparticle-quantum dot platform. J. Mater. Chem. B **8**(16), 3534–3541 (2020). https://doi.org/10.1039/C9TB02481F
46. J. Lee, M. Senna, Preparation of monodispersed polystyrene microspheres uniformly coated by magnetite via heterogeneous polymerization. Colloid Polym. Sci. **273**, 76–82 (1995)
47. S.E. Jacobo, J.C. Aphesteguy, R.L. Anton, N.N. Schegoleva, G. Kurlyandskaya, Influence of the preparation procedure on the properties of polyaniline based magnetic composites. Eur. Polym. J. **43**(4), 1333–1346 (2007). https://doi.org/10.1016/j.eurpolymj.2007.01.024
48. S. Ansari, M. Karimi, Recent configurations and progressive uses of magnetic molecularly imprinted polymers for drug analysis. Talanta **167**, 470–485 (2017). https://doi.org/10.1016/j.talanta.2017.02.049
49. O. Aksimentyeva, V. Savchyn, I. Opaynych, P. Demchenko, Yu. Horbenko, V. Pankratov, A.I. Popov, Effect of polymer matrix on the structure and luminescence properties of barium zirconate nanocrystals. Chem. Metals Alloys **6**, 72–76 (2013)
50. V.P. Savchyn, A.I. Popov, O.I. Aksimentyeva, H. Klym, Yu.Yu. Horbenko, V. Serga, A. Moskina, I. Karbovnyk, Cathodoluminescence characterization of polystyrene-$BaZrO_3$ hybrid composites. Low Temp. Phys. **42**(7), 597–600 (2016). https://doi.org/10.1063/1.4959020
51. Y.Y. Horbenko, L.Y. Kit, O.I. Aksimentyeva, *Method of Surface Modification of Magnetite Nanoparticles*. UA Patent 115956, 10 Jan 2018
52. M.E. Fleet, The structure of magnetite. Acta Crystallogr. B **37**(4), 917–920 (1981). https://doi.org/10.1107/S0567740881004597
53. L.S. Cavalcante, J.C. Sczancoski, V.M. Longo et al., Intense violet–blue photoluminescence in $BaZrO_3$ powders: a theoretical and experimental investigation of structural order–disorder. J. Alloys Comp. Optics Commun. (14), 3715–3720 (2008). https://doi.org/10.1016/j.optcom.2008.03.069
54. A.A. Belanova, N. Gavalas, Y.M. Makarenko, M.M. Belousova, A.V. Soldatov, P.V. Zolotukhin, Physicochemical properties of magnetic nanoparticles: Implications for biomedical applications in vitro and in vivo. Oncol. Res. Treat **41**(3), 139–143 (2018). https://doi.org/10.1016/10.1159/000485020
55. R.C. Popescu, E. Andronescu, B.S. Vasile, Recent advances in magnetite nanoparticle functionalization for nanomedicine. Nanomaterials **9**(12) (2019). https://doi.org/10.1016/10.3390/nano9121791.
56. D. Micheli, M. Marchetti, Mitigation of human exposure to electromagnetic fields using carbon foam and carbon nanotubes. Engineering **4**, 928–943 (2012). https://doi.org/10.4236/eng.2012.412A118
57. P. Mathur, S. Raman, Electromagnetic interference (EMI): measurement and reduction techniques. J. Electron. Mater. **49**(5), 2975–2998 (2020). https://doi.org/10.1007/s11664-020-07979-1
58. National Counterintelligence and Security Center, *Technical Specifications for Construction and Management of Sensitive Compartmented Information Facilities*, Washington (2020). https://www.dni.gov/files/Governance/IC-Tech-Specs-for-Const-and-Mgmt-of-SCIFs-v15.pdf
59. O.I. Aksimentyeva, I.B. Chepkov, R.V. Filipsonov, S.Z. Malynych, R.V. Gamernyk, G.V. Martyniuk, YuYu. Horbenko, Hybrid composites with low reflection of IR radiation. Phys. Chem. Solid State **21**(4), 764–770 (2020)
60. O. Aksimentyeva, S. Malynych, R. Filipsonov, R. Gamernyk, Shielding broadband electromagnetic radiation with conducting polymer composites, in *Abstracts of the Fourth International*

Conference "Actual problems of fundamental science", Lutsk–Svityaz', Ukraine, 01–05 June 2021.
61. O. Aksimentyeva, S. Malynych, R. Filipsonov, R. Gamernyk, Broadband electromagnetic radiation absorber based on bifunctional polymer-magnetite composite. Acta Phys. Pol. A (144), 356–360 (2022). https://doi.org/10.12693/APhysPolA.141.356

Influence of Pyrogenic SiO_2 Nanoparticles on the Photodegradation of Polymer Coatings Under UV Irradiation

Oleksiy Myronyuk, Denys Baklan, and Anna Bilousova

Abstract The study is focused on the determination of the effect of functionalized silica nanoparticles on the process of styrene acrylic polymer-based coatings photo destruction. The thin layer coatings UV degradation in accelerated aging conditions is characterized by infrared spectroscopy, ablation speed and water contact angle measurements. The HALS amine loaded coatings are used as the reference. It is shown that the process of photodegradation of silica nanoparticles-polymer composite thin film coatings is determined by two factors. The first one is the screening of the UV radiation, that becomes noticeable at nanoparticles loads more than 15 wt.% and is evident from the decrease of the thickness ablation speed from 1.3 for the neat polymer to 1.2 μm/h. The second one is increased polymer hydrophilization speed due to the reduced thickness of it films and increased fraction of polymer under exposition that leads to a sharp decrease of contact angle values at the first 30 min of exposition from initial 89° to as low as 40°.

1 Background

Polymer-based coatings are widely used for protection of various substrates in a number of aggressive environments and also as functional materials, e.g., with increased water repellency due to the texture features.

It is known that polymer-mineral compositions are unstable under the solar ultraviolet irradiation mainly due to photodestruction of the organic part, which includes complex photo-oxidation processes with the participation of air oxygen. This leads

O. Myronyuk · D. Baklan (✉) · A. Bilousova
Department of Chemical Technology of Composite Materials, Chemical Technology Faculty, Igor Sikorsky Kyiv Polytechnic Institute, Peremohy Ave. 37, Kyiv 03056, Ukraine
e-mail: d.baklan@kpi.ua

O. Myronyuk
e-mail: o.myronyuk@kpi.ua

O. Fesenko and L. Yatsenko (eds.), *Nanostructured Surfaces, Nanocomposites and Nanomaterials, and Their Applications*, Springer Proceedings in Physics 296,
https://doi.org/10.1007/978-3-031-42704-6_12

to a decrease in a number of important performance properties: primarily due to the appearance of polar surface groups the total surface energy of such systems and their hydrophilicity increases [1], mechanical properties decrease [2], additional porosity appears due to the growth of contractual cracks. It is known that the effect of UV radiation is mainly concentrated in the top surface layers and extends to hundreds of micrometers in thickness of the coating.

There are a number of approaches to decrease the photodestruction of polymers, which can be roughly divided into: (1) the use of light absorbers that convert the light energy into the heat [3]; (2) the use of neutralizers of photo-oxidation intermediates (excited-state quenchers [4, 5], peroxide decomposers [6, 7], free radical scavengers [8, 9]) and (3) spatial shielding of the polymer from UV light screeners [10]. Approaches number 1 and 2 are quite widely implemented in practice, e.g., there are several thousand commercial products of the type [11]. The advantage of the third approach is expressed in the relative inertness of prospective shielding additives and their weak effect on the chemical composition of the polymer matrix [12]. At the same time, the shielding additive particles should be the smallest in size to be most effective, which can be achieved by using, e.g.,, pyrogenic silica nanoparticles.

The purpose of this work is to determine the effect of nanoparticles on the rate and nature of the photodestruction process of styrene acrylic polymer-based coatings.

In order to achieve this goal, the tasks of determining the photoablation rate of neat polymer, the change of surface energy and the chemical composition of this substance were solved. The comparison of process speed depending on nanoparticles content in the film was carried out. The efficiency of deceleration of photodegradation by nanoparticles is compared with that of the UV adsorber.

2 Materials and Methods

NeoCryl B880 (DSM N.V.), referred to as AC, was used as a polymeric binder. The choice of styrene acrylic copolymer is based on good film forming ability, high temperature resistance and corrosion resistance [13, 14]. TINUVIN 5151 (BASF), referred to as 5151, which is 2-(2-hydroxyphenyl)-benzotriazole has been used as a UV absorber in a basic hindered amine light stabilizer (HALS). Aerosil R972 (Evonik), which is dimethyldichlorosilane treated silica, designated R972, was used as a filler as an agent for blocking UV radiation.

Coatings were applied to polished aluminum substrates (6063 alloy) 30 × 30 mm in size and on KBr windows 40 mm in diameter, 3 mm thick. The application was performed by air spraying, o-xylene (Merck) was used as a carrier. Coating thickness value was calculated from a measurement of substrate area, applied coating weight and density.

UV exposure was conducted based on the ASTM D 4329 standard procedure. A 400 W fluorescent lamp, 340 UV-A, in the medium to short UV wavelength range, was used for the test. According to ASTM 4329, an 8-h UV exposure at 60 ± 5 °C was used.

The static water contact angle value was determined by the sessile drop technique (using a Delta Optical HCDE-50 digital camera and ScopeTek View software). Every value of the contact angle was measured with 5 drops of water, and then the average value was calculated. A TESCAN MIRA3 scanning electron microscope was used to evaluate the surface of the samples. To obtain IR spectra, a Nicolet 4700 IR Fourier spectrophotometer was used. The IR spectrum was obtained in the range of 4000–400 cm^{-1} by the transmission method.

The carbonyl index (CI) was used to characterize the degree of polymer degradation. There are various methods for selecting indicative bands, which have different accuracy of results [15]. CI was calculated as the ratio of the carbonyl band absorbance at 1780–1750 cm^{-1} and the reference band at 700 cm^{-1} absorbance based on the approaches described in [16–19].

3 Results and Discussion

The IR spectrum (Fig. 1a) of copolymer AC corresponds to the set of bonds in the styrene-buthyl metacrylate. The absorption bands at 3052 and 3025 cm^{-1} correspond to aromatic stretching vibrations of C–H bond of styrene [20], at the same time, there are bands of stretching vibrations of symmetric and asymmetric C–H vibrations at wave numbers 2930 and 2869 cm^{-1}. Bands at wavenumbers of 1598 and 1495 cm^{-1} indicate vibrations of C=C bonds in aromatic groups [21]. Bands at 760 and 700 cm^{-1} correspond to out-of-plane bending vibrations of C-H groups. The spectrum has a high intensity band at 1720 cm^{-1}, which corresponds to the C=O bond stretching of acrylic acid. A broad band at 1175 cm^{-1} indicates a C–O bond [22]. Bonds at 1457 and 1380 cm^{-1} correspond to bending vibrations of C–H groups.

The main noticeable effect that can be seen on the spectra on Fig. 1 is the appearance of the shoulder at 1742 cm^{-1}, that is attributed to stretching vibrations of C=O bonds and indicates the formation of additional carbonyl groups due to the oxidation processes [23]. It is worth to mention that this band is hindered by the main band of carbonyl C=O in the ester unit of the buthyl methacrylate side group. The carbonyl index of the film increases gradually: after the first appearance of the said shoulder, the CI value remains constant up to 4 h of the test and then rapidly increases almost 3 times. After 8 h of irradiation, the spectrum lines fade. This can be explained by the full ablation of the substance on the surface of the glass. Considering the initial average thickness of the layer 10.3 μm, it gives the ablation rate 1.3 μm/h in the experimental conditions.

The addition of HALS type UV absorber provides the change in kinetics of CI evolution (Fig. 2). First of all, the increase happens smoother with even slope, but, surprisingly, values of CI after 2 h are higher than in the case of a neat polymer (Fig. 1).

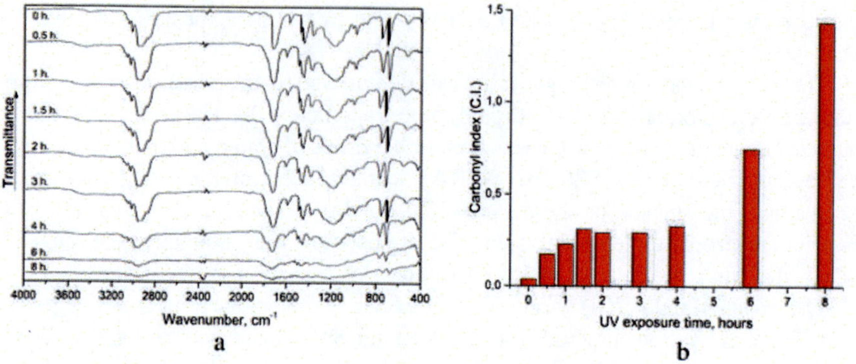

Fig. 1 Photooxidation of AC polymer under UV exposure: **a**-FTIR spectra of thin films; **b**-carbonyl index of thin films

However, the type of FTIR spectra change of the degradation products remains close to that of the neat polymer and limited, in general, by the appearance of the shoulder at 1740 cm^{-1}. This may be explained with the accumulation of inactive carboxyl containing products of radicals deactivation, as it described in [24].

The addition of silica nanoparticles provides the surface porosity of thin coatings (Fig. 3) by forming arrays of submicron scale asperities and air pockets distributed between them. It may be seen that the primary particle diameter of the hydrophobized pyrogenic silica is 25–30 nm. And after UV irradiation, the polymer phase that plays a role of film forming agent is notably consumed by the ablation that may cause the detachment of the separate particles or their arrays during the artificial aging.

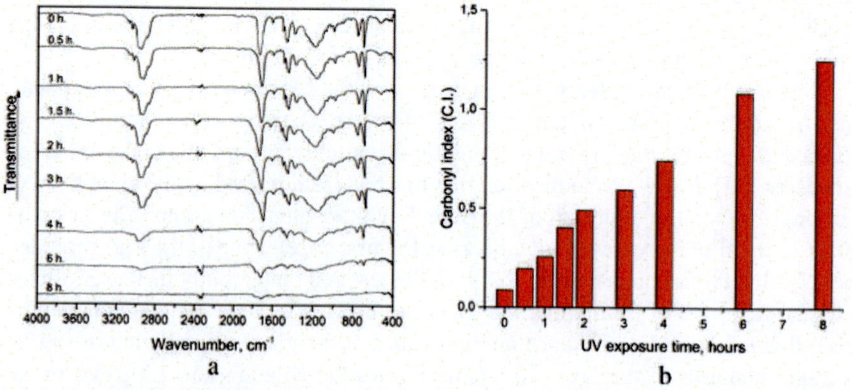

Fig. 2 Photooxidation of AC polymer + 6 wt.% 5151 under UV exposure: **a**-FTIR spectra of thin films; **b**-carbonyl index of thin films

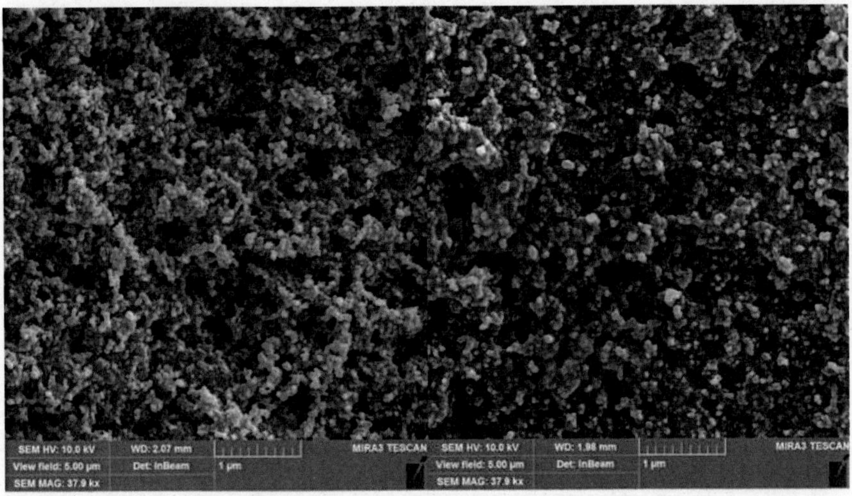

a b

Fig. 3 SEM images of AC + 15 wt.% Aerosil R972 (before **a** and after **b** UV irradiation)

Alongside with it, we can assume the certain fraction of the AC polymer is screened by the particles layer.

The polymer ablation causes the change of the weight ratio between the inorganic silica particles (with characteristic Si–O band at 1100 cm^{-1}) and organic acrylics with above mentioned bands (Fig. 4). It may be noted that the spectrum of 7 wt.% filled composite (Fig. 4a) becomes equal to 15 wt.% filled (Fig. 4b) after 0.5 h of irradiation. The ablation of the 7 wt.% 22 μm layer that results in the formation of residue consisted almost solely of the silica nanoparticles. The speed of the process in this case is even higher than of the neat polymer that may be explained by the entrainment of non-bonded particles. However, in the case of 15 wt.% load, the ablation speed is reduced.

The CI of both thin films change pattern is closer to that of 5151 additive provides than of the neat polymer and it is evident that the increase of nanoparticles content to 15 wt.% lowers the CI values almost by 30%.

The ablation speed is used as the characteristic parameter of the photodestruction resistance of the material that can be utilized for the comparison of effectiveness between the experimental samples (Table 1). As it can be seen from Fig. 5, the thickness reduction of the sample may be approximated as a straight line with a certain slope that is equal to the ablation speed. The deviation of the line shape in the final stages of destruction, especially in the case of thinner coatings, is probably caused by the surface residue influence.

It may be seen (Table 1) that the most effective concentration of HALS type UV absorber additive 5151 is 6 wt.%, that provides almost a 1.7 fold reduction of the ablation rate. The nanoparticles in concentration of 7 wt.% even increase the

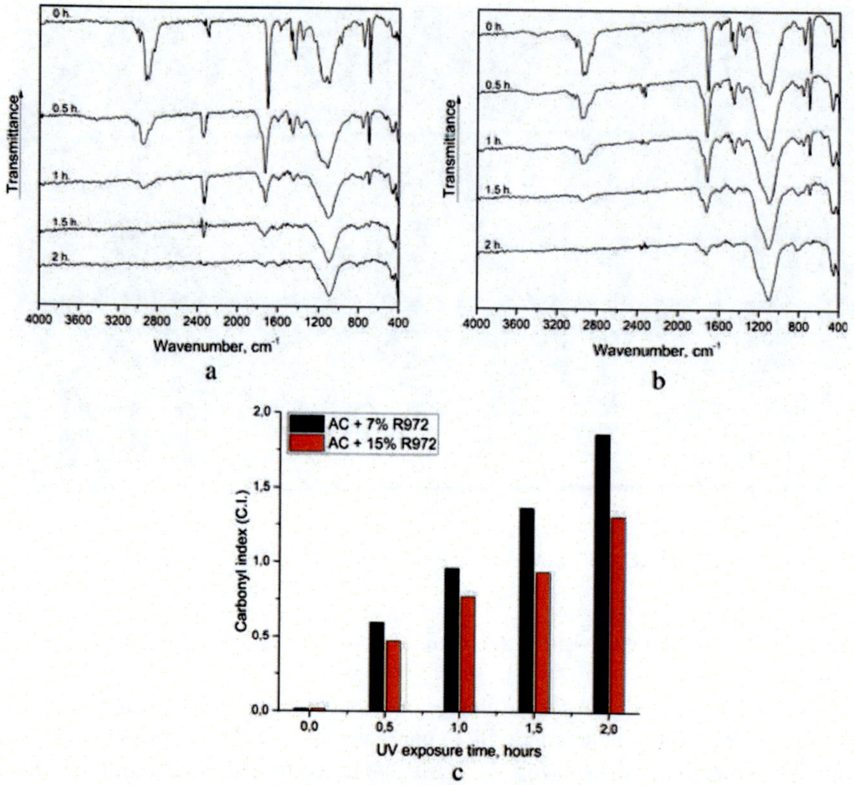

Fig. 4 Photooxidation of AC polymer loaded with Aerosil R972 under UV exposure: **a**-FTIR spectra of thin films with 7 wt.% of nanoparticles; **b**-FTIR spectra of thin films with 7 wt.% of nanoparticles; **c**-carbonyl index of thin films

Table 1 Sample thickness ablation rate

Sample	Ablation rate (μm/h)
AC	1.29
AC + 2% 5151	1.10
AC + 4% 5151	1.25
AC + 6% 5151	0.77
AC + 7% R972	1.74
AC + 15% R972	1.20

thickness loose speed, but at loads 15 wt.% their addition provides the UV screening effect.

The contact angle value is considered to be more sensitive to the oxidation than the FTIR measurements [25]. It is clearly seen (Table 2) that any level of HALS additive inhibits the photodestruction. It is a sharp decrease of the water contact angle value

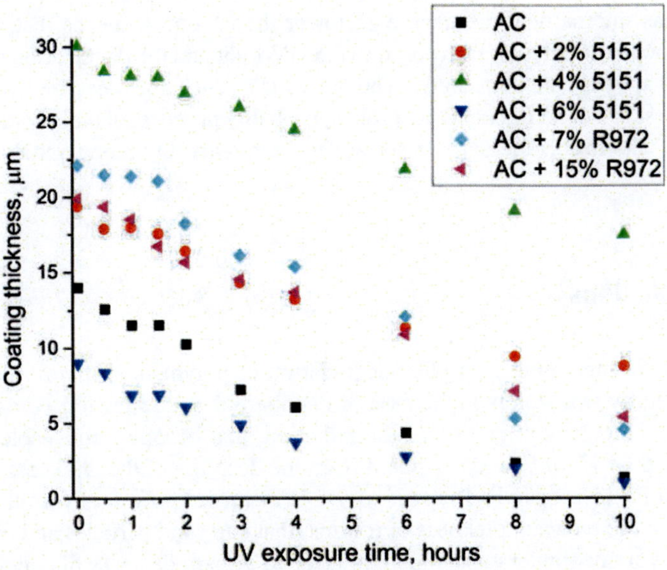

Fig. 5 Thickness change of AC composites in conditions of UV irradiation

after 1 h of exposition of a neat AC to 40°, but all films with 5151 remain at levels above 75°. In a contrast, the contact angle decrease in case of AC-R972 compositions is even sharper than in case of neat polymer and starts even after 0.5 h of exposition.

It can be assumed that the decrease of UV resistance in this case may be explained with the texturing of a surface in case nanoparticles are present. Air pockets of the structure play a role of capillaries with hydrophilized inner walls that causes the positive capillary tension and sucking of a probe liquid due to a Washburn effect [26, 27]. In the other hand, the layer, that is subjected to the oxidation in case of the particle presence, is much more tiny than in case of the bulk coatings and may be roughly

Table 2 Water contact angles of thin films after UV exposition

Sample	Water contact angle (°)								
	0	0.5	1	1.5	2	3	4	6	8
AC	88	85	40	35	26	10	21	12	0
AC + 2% 5151	78	87	82	67	6	44	20	10	0
AC + 4% 5151	87	86	76	72	30	42	29	18	0
AC + 6% 5151	86	81	86	77	42	38	26	9	15
AC + 7% R972	89	48	41	28	5	0	0	0	0
AC + 15% R972	82	71.5	43	16	0	0	0	0	0

estimated as interparticle distance, e.g., lower than 5 nm. It means that a greater specific amount of polymer is exposed to the UV-light and the degradation of such a thin layer happens almost instantly. The products of such a process are hydrophilic due to a presence of additional C=O groups, probably products of buthyl side groups oxidation. This makes the particle surface to be remarkably more wettable by the water.

4 Conclusions

It is shown that the process of photodegradation of polymers in thin film coatings is influenced by two factors in the case of nanoparticles introduction. The first one is the spatial blocking a fraction of the radiation, that becomes noticeable at their loads more than 15 wt.% and is evident from the decrease of the thickness ablation speed from 1.3 to 1.2 μm/h. The second one is increased polymer hydrophilization speed due to the reduced thickness of it films, that surround particles and increased fraction of polymer under exposition that leads to a sharp decrease of contact angle values at the first 30 min of exposition from initial 89° to as low as 40°.

Acknowledgements The authors are grateful to the 10th International Conference "Nanotechnologies and Nanomaterials" NANO-2022 committee for creating an ambience of friendship and for giving the opportunity to present our research.

Competing Interests The authors declare that they have no competing interests.

Author's Contributions OM and DB provide guidance regarding research; AB and DB obtained series of superhydrophobic coatings for research; OM, DB and AB have made substantial contributions to the analysis and interpretation of data and contributed to the draft of the manuscript. All authors read and approved the final manuscript.

References

1. A. Benke, J. Sonnenberg, K. Oelschlägel et al., Wettability after artificial and natural weathering of polyethylene terephthalate. Environments **9**, 134 (2022). https://doi.org/10.3390/environments9110134
2. C. Signoret, M. Edo, D. Lafon et al., Degradation of styrenic plastics during recycling: impact of reprocessing photodegraded material on aspect and mechanical properties. J. Polym. Environ. **28**, 2055–2077 (2020). https://doi.org/10.1007/s10924-020-01741-8
3. J.U. Kim, S. Lee, S.J. Kang, K. T-il, Materials and design of nanostructured broadband light absorbers for advanced light-to-heat conversion. Nanoscale **10**, 21555–21574 (2018). https://doi.org/10.1039/c8nr06024j
4. J.S. Hargreaves, D. Phillips, Photo-oxidation poly(butadiene); quenching of excited states by nitroxides. Eur. Polymer J. **15**, 119–124 (1979). https://doi.org/10.1016/0014-3057(79)90195-2
5. S. Santabarbara, Limited sensitivity of pigment photo-oxidation in isolated thylakoids to singlet excited state quenching in photosystem II antenna. Arch. Biochem. Biophys. **455**, 77–88 (2006). https://doi.org/10.1016/j.abb.2006.08.017

6. C. Sato, I. Kim, H. Tanaka, Effect of H2O2 on UV photo-oxidation of pharmaceuticals and personal care products in wastewater. J. Environ. Eng. (2016). https://doi.org/10.1061/(asce)ee.1943-7870.0001132
7. L.C. Santos, A.L. Poli, C.C. Cavalheiro, M.G. Neumann, The UV/H_2O_2—photodegradation of poly(ethyleneglycol) and model compounds. J. Braz. Chem. Soc. **20**, 1467–1472 (2009). https://doi.org/10.1590/s0103-50532009000800012
8. C. Maier, T. Calafut. Additives. Polypropylene 27–47 (1998). https://doi.org/10.1016/b978-188420758-7.50008-4
9. P. Gijsman, Polymer stabilization. In: *Handbook of Environmental Degradation of Materials*, pp. 673–714 (2012). https://doi.org/10.1016/b978-1-4377-3455-3.00023-7
10. E. Yousif, R. Haddad, Photodegradation and photostabilization of polymers, especially polystyrene: review. Springerplus (2013). https://doi.org/10.1186/2193-1801-2-398
11. G. Wypych, *Handbook of UV Degradation and Stabilization*. ChemTec Publishing, Toronto (2020). ISBN 1927885582
12. D.G. Goodwin, S.-J. Shen, Y. Lyu et al., Graphene/polymer nanocomposite degradation by ultraviolet light: the effects of graphene nanofillers and their potential for release. Polym. Degrad. Stab. **182**, 109365 (2020). https://doi.org/10.1016/j.polymdegradstab.2020.109365
13. H. Mogami, H. Mori, Enhanced heat resistance and adhesive strength of styrene–acrylic Triblock copolymer elastomers by incorporating acrylic acid. ACS Omega **5**, 3678–3688 (2020). https://doi.org/10.1021/acsomega.9b04282
14. J.J. Gallagher, M.A. Hillmyer, T.M. Reineke, Acrylic triblock copolymers incorporating isosorbide for pressure sensitive adhesives. ACS Sustain. Chem. Eng. **4**, 3379–3387 (2016). https://doi.org/10.1021/acssuschemeng.6b00455
15. J. Almond, P. Sugumaar, M.N. Wenzel, et al., Determination of the carbonyl index of polyethylene and polypropylene using specified area under band methodology with ATR-FTIR spectroscopy. e-Polymers **20**, 369–381 (2020). https://doi.org/10.1515/epoly-2020-0041
16. R. Shanti, A.N. Hadi, Y.S. Salim et al., Degradation of ultra-high molecular weight poly(methyl methacrylate-co-butyl acrylate-co-acrylic acid) under ultra violet irradiation. RSC Adv. **7**, 112–120 (2017). https://doi.org/10.1039/c6ra25313j
17. A. Benítez, J.J. Sánchez, M.L. Arnal et al., Abiotic degradation of LDPE and LLDPE formulated with a pro-oxidant additive. Polym. Degrad. Stab. **98**, 490–501 (2013). https://doi.org/10.1016/j.polymdegradstab.2012.12.011
18. W.W. Focke, R.P. Mashele, N.S. Nhlapo, Stabilization of low-density polyethylene films containing metal stearates as photodegradants. J. Vinyl Add. Tech. **17**, 21–27 (2011). https://doi.org/10.1002/vnl.20248
19. M.C. Antunes, J.A.M. Agnelli, A.S. Babetto et al., Abiotic thermo-oxidative degradation of high density polyethylene: effect of manganese stearate concentration. Polym. Degrad. Stab. **143**, 95–103 (2017). https://doi.org/10.1016/j.polymdegradstab.2017.06.012
20. A. Sakr, A. Naser, H. Abd El-Wahab et al., Preparation and characterization of modified reclaimed asphalt by using styrene—butyl acrylate nanoemulsion copolymer. Egypt. J. Chem. **61**, 280–290 (2018). https://doi.org/10.21608/ejchem.2018.2956.1245
21. J.F. Fang, Y.M. Xuan, Q. Li, Preparation of polystyrene spheres in different particle sizes and assembly of the PS colloidal crystals. Sci. Chin. Technol. Sci. **53**, 3088–3093 (2010). https://doi.org/10.1007/s11431-010-4110-5
22. O.A. Mohamed, A.B. Moustafa, M.A. Mehawed, N.H. El-Sayed, Styrene and butyl methacrylate copolymers and their application in leather finishing. J. Appl. Polym. Sci. **111**, 1488–1495 (2009). https://doi.org/10.1002/app.29022
23. O. Chiantore, L. Trossarelli, M. Lazzari, Photooxidative degradation of acrylic and methacrylic polymers. Polymer **41**, 1657–1668 (2000). https://doi.org/10.1016/s0032-3861(99)00349-3
24. C.W. Klampfl, M. Himmelsbach, Advances in the determination of hindered amine light stabilizers—a review. Anal. Chim. Acta **933**, 10–22 (2016). https://doi.org/10.1016/j.aca.2016.06.001
25. O. Myronyuk, D. Baklan, Z. Yong, A.M. Rodin, Complex destruction of textured water-repellent coatings under the influence of UV and water flow. Mater. Today Commun. **33**, 104509 (2022). https://doi.org/10.1016/j.mtcomm.2022.104509

26. S. Kirdponpattara, M. Phisalaphong, Z. Newby B-min, Applicability of washburn capillary rise for determining contact angles of powders/porous materials. J. Colloid Interface Sci. **397**, 169–176 (2013). https://doi.org/10.1016/j.jcis.2013.01.033
27. L. Galet, S. Patry, J. Dodds, Determination of the wettability of powders by the Washburn capillary rise method with bed preparation by a centrifugal packing technique. J. Colloid Interface Sci. **346**, 470–475 (2010). https://doi.org/10.1016/j.jcis.2010.02.051

Nanocomposites and Nanomaterials

Effect of Low-Molecular Weight Arenes Additives on the Interphase Layer Structure of SAN-PS Systems and Dielectric Behavior of Such Compositions

M. Anoshenko, T. Bazylyuk, N. Iukhymenko, and V. Budzinska

Abstract The influence of low-molecular weight arena's (LMA) on the dielectric behavior of a binary system with limited compatibility between components based on styrene-acrylonitrile (SAN)-polystyrene (PS) copolymers studied. Benzene found to selectively interact with SAN and PS, blocking the polymer–polymer interaction in interphase layer (IPL). The restructuring of the IPL structure in the presence of toluene revealed, as well as the increase in the stabilization of the lyophilic fluctuation network that forms the IPL. Styrene can be used as a compatibilizer of the studied polymer–polymer system. Benzonitrile causes weakening of intermolecular interaction in IPL due to selectivity, as well as dipole–dipole interaction of nitrile groups of the copolymer and benzonitrile. It was concluded that the possibility of the formation of a rigid lyophobic network, as well as the heterogeneity of mixtures of mixtures associated with this process, is controlled, in general, by the content of SAN in the compositions.

1 Introduction

Stable complexes of low-molecular arene-arene systems are known from the literature. These complexes are stabilized by π–π electron conjugation [1–6]. Such an interaction requires a co-planar arrangement of the phenyls of the corresponding components, which in this case is easily achieved.

M. Anoshenko (✉) · T. Bazylyuk · N. Iukhymenko
Faculty of Chemistry, Macromolecular Chemistry Department, Taras Shevchenko National University of Kyiv, 60 Volodymyrska Street, Kyiv 01033, Ukraine
e-mail: drim-story@ukr.net

V. Budzinska
Institute of Macromolecular Chemistry of National Academy of Sciences of Ukraine, 48 Kharkivs'ke Shoseose, Kyiv 02160, Ukraine

O. Fesenko and L. Yatsenko (eds.), *Nanostructured Surfaces, Nanocomposites and Nanomaterials, and Their Applications*, Springer Proceedings in Physics 296,
https://doi.org/10.1007/978-3-031-42704-6_13

On the other hand, the effects of a similar interaction can be expected for systems polymer–polymer with limited compatibility of components. However, in this case, the intermolecular interaction (IMI) process "run into" a significant complication, which is that the phenyls associated with the macromolecule cannot freely to get closer with the corresponding groups of another macromolecule, moreover, cannot show clear coplanarity. In our opinion, there is a possibility of coplanarity correction due to the third component—the low-molecular weight arena (LMWA). In this case, in our opinion, we can expect the formation of a sandwich polymer *A*—LMWA—polymer *B*, if both polymers contain phenyl core in their composition.

Good models for studying the processes of redistribution of IMI forces in the interphase layer (IPL) of polymer–polymer compositions with limited compatibility are systems of copolymer AB-homopolymer B, if component B is non-polar. Compositions whose components are non-polar can form advanced IPL. It is known that mixtures in which a polystyrene component is present can form structures, in which the IPL content is ~ 20%. On the other hand, the interphase layers formed by non-polar polymers are not strong. There is a problem of increasing IMI in these layers. Such are mixtures based on a copolymer of styrene and acrylonitrile and polystyrene (SAN-PS).

On the other hand, the mixtures described above show in the experiment the anomalies of many structurally sensitive characteristics. Such characteristics include, in particular, the dielectric behavior of the corresponding compositions.

The aim of this study was to study the influence of low-molecular weight arenas on the dielectric behavior of a binary system of copolymers of styrene and acrylonitrile-polystyrene.

2 Experimental

The industrial thermoplastic polymers of SAN and PS: SAN of azeotropic composition (the ratio of styrene and acrylonitrile units by gas chromatographic analysis was 62:38) and industrial suspension PS were used in the work.

The viscometric average molecular weight (Mη) was for SAN-54560 (DMF, 20 °C), for PS-237775 (chloroform, 25 °C). The glass transition temperature (T_g) of the SAN was 108 °C, and of the PS was 98 °C (inverse gas chromatography (IGC) method).

Mixing of SAN with PS carried out by extrusion method at temperature 20–30 °C higher than SAN and PS flow temperatures. After the mixing, prepared mixtures were slowly cooled down to the laboratory temperature and crumbled and, for better mixing, re-extrusion performed. The product of double extrusion crumbled again at a liquid nitrogen temperature (to prevent degradation). Samples for the experiment in the form of films prepared by compression at a temperature of 150 °C and a pressure of 100 atm. in fast and slow cooling mode.

As low-molecular weight components benzene, toluene, styrene, benzonitrile (BN) used. LMWA incorporated into a polymer system by sorption of the corresponding liquid under saturated vapor (desiccator method). The amount of LMWA incorporated into polymers was ~ 0.5% wt.% relative to the polymer matrix.

To compare of the relaxation processes in the initial polymers and in the corresponding compositions, the dielectric relaxation method used. The following dielectric parameters, such as the tangent of the dielectric loss angle (tg δ) and dielectric permittivity (ε'), were estimated. Conclusions about the mechanisms of intermolecular interaction changing in the IPL made based on the infrared spectroscopy data (Nexus-475). IR spectroscopic investigations carried out in the range 400–4000 cm^{-1}.

3 Results and Discussion

Analysis of the electrical characteristics in the region of α-relaxation, and especially in the temperature range of crankshaft oscillations (molecular motions of the type "crankshaft"), gives an opportunity to make conclusions about the relative level and redistribution of IMI forces in systems with different ratio of components.

If the manifestation of molecular crankshaft type motions is established in the nature of the PS, then with regarding styrene copolymers (e.g., SAN), such rearrangements can only occur when the copolymers contain styrene oligoads. It was calculated [8] and NMR C13 was used to prove experimentally [9] that SANs (obtained by initiating radical copolymerization at azeotropic ratio of styrene and acrylonitrile) include styrene oligoads, at least dyads and triads. We have analyzed our SAN using IR spectroscopy. The absorption band at 547 cm^{-1} is clearly registered in the spectrum (Fig. 1). According to [9], a copolymer containing styrene (C) sequences of CC and CCC absorb at 550–555 cm^{-1} if the number of styrene sequences is $n =$ 4–5 then at 545 cm^{-1}. Thus, it can be argued that our SAN sample contains styrene dyads, triads, and probably tetrads, and this, in turn, is the basis for the manifestation of molecular motions such as crankshaft oscillations of the crankshaft type.

Figure 2 shows the dependence of tg δ (tangent of dielectric loss angle) on the temperature of the original SAN and the SAN-PS compositions at a frequency of 1 kHz.

We see that in the temperature range 20–60 °C, there is a dipole-group relaxation process, which, in view of [7], we attribute to the manifestation of independent crankshaft oscillations in the styrene oligoads of the SAN and PS.

On the other hand, a study of the basic relaxation process showed that the mixtures are characterized by a single T_{α}, which, when increasing the PS content, moves to lower temperatures by 3–4 °C. Considering that the glass transition temperatures determined by the inverse gas chromatography method are equal to 108 °C (SAN) and 96 °C (PS), it is possible to state that there is a sufficient level of IMI in IPL of these partially compatible systems, at least in the temperature range 100–140 °C.

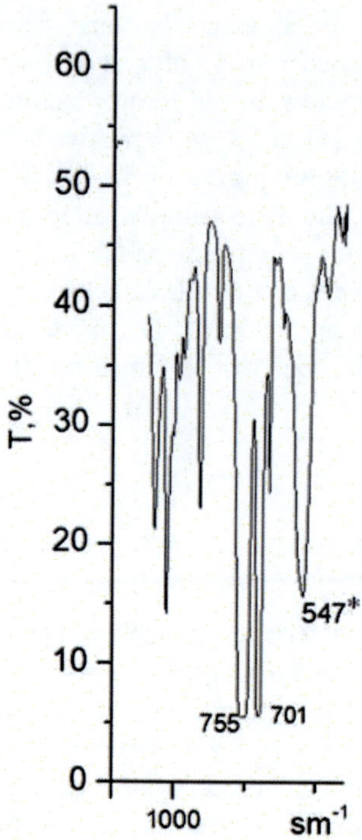

Fig. 1 Fragment of the SAN IR spectrum

Figure 3 shows similar dependencies for SAN-PS mixtures in the presence of LMWA: benzene, toluene, and styrene.

The completely different effect of benzene and its substituted homologues on the polymer–polymer interaction seen.

Thus, in the presence of benzene, active relaxation behavior observed at temperatures in the region of crankshaft oscillations. Therefore, it should concluded that in the temperature range of 35–50 °C, benzene not incorporated into the structure of the polymer–polymer sandwich.

The α-relaxation region in the SAN-benzene system (Fig. 4a) and in the microheterogeneous SAN-PS-benzene mixtures (PS content does not exceed 5%) (Fig. 4b) is located at 118–119 °C and this is lower than T_α SAN = 123 °C, which indicates a decrease of IMI in IPL polymer–polymer compositions in the presence of this arenas, and also it indicates that benzene is not played as a plasticizer. The last conclusion supported by the dielectric behavior of a macro heterogenic mixture containing 10% PS Fig. 5.

There is a harsh redistribution of the dipole-segmental losses intensity in behalf of dipole-group losses, and T_α of this mixture is also reduced to 92 °C.

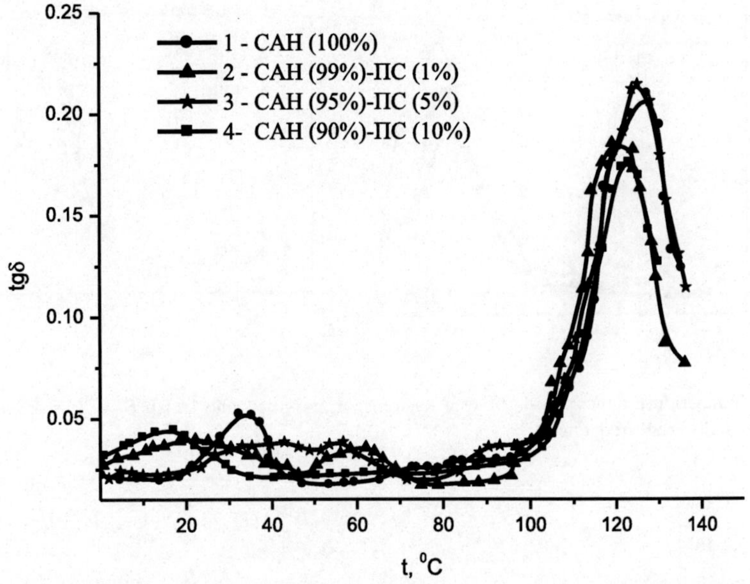

Fig. 2 Temperature dependence of the tangent of dielectric loss angle (tg δ) of SAN and SAN-PS compositions

Fig. 3 Temperature dependence of the dielectric loss angle tangent (tg δ) of SAN and systems: SAN-benzene, SAN-toluene, SAN-styrene at a frequency of 1 kHz

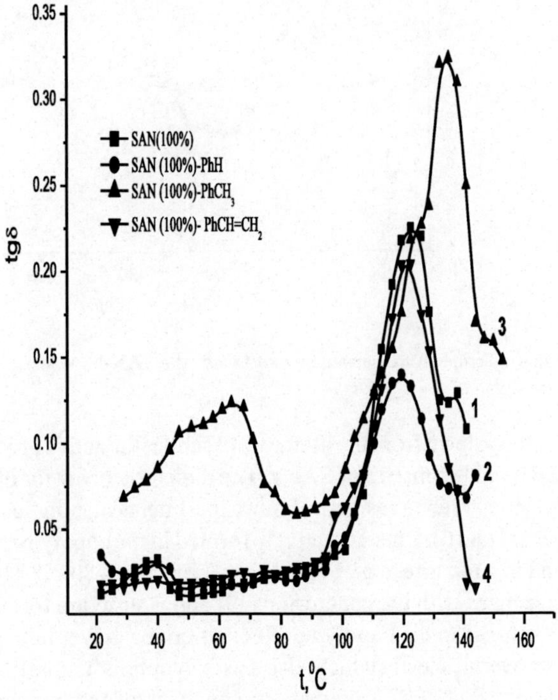

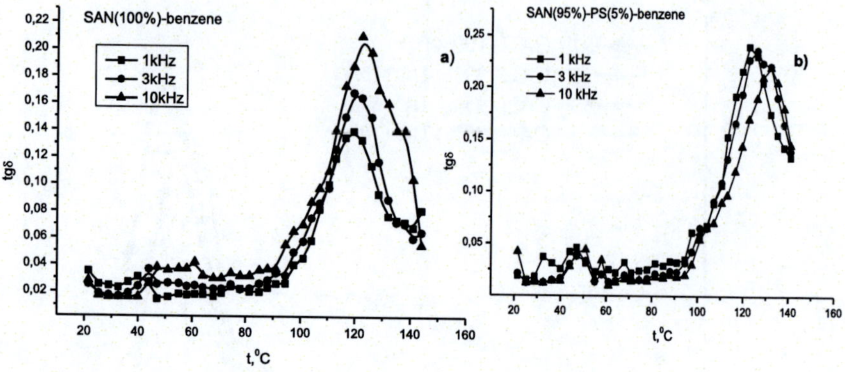

Fig. 4 Temperature dependences of tg δ systems: **a** SAN-benzene and **b** SAN (95)-PS (5%)-benzene at different frequencies

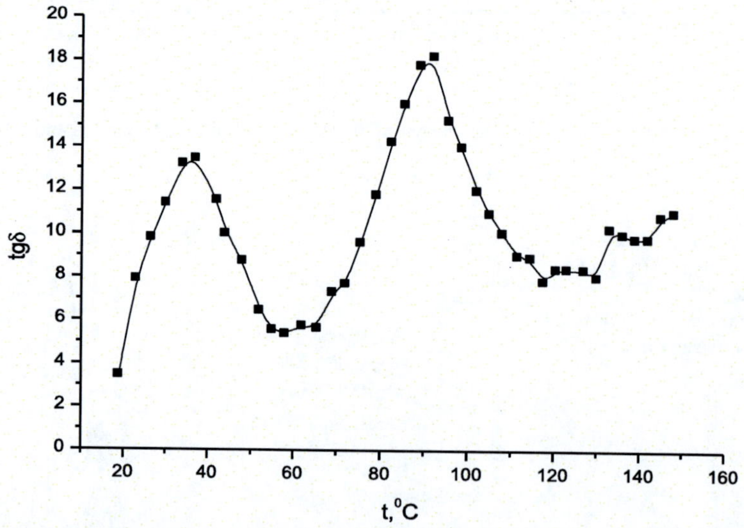

Fig. 5 Temperature dependence of tg δ of the SAN-PS system (10%) at a frequency of 1 kHz

It is known from the literature [9] about the active processes of toluene clustering in the lyophilic matrix of SANs. The dielectric behavior of SAN-toluene and SAN-PS-toluene systems (as part of our experiment) also indicates this possibility: (curve 3, Figs. 3 and 6). Clusters actively formed in the copolymer at temperatures of 40–75 °C, and in a mixture, e.g., SAN: PS = 95: 5-at 60–80 °C. Then, we made conclusion that correlates with the conclusions [9] about structure formation caused by the presence of toluene in the copolymer fluctuation mesh lyophilic component, and probably, in the overall mesh (which also has a lyophilic nature), which is realized in the IPL system SAN-PS. Moreover, the analysis of the temperature position tg δ_{max} leads

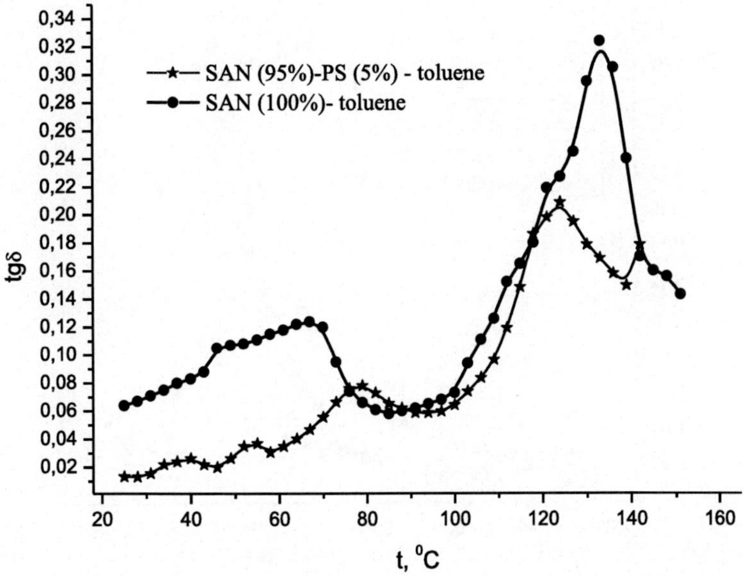

Fig. 6 Temperature dependence of tg δ of the SAN-PS-toluene and SAN-toluene system at 1 kHz

to the conclusion that the IMI in the SAN-toluene-PS system is greater than in the SAN-toluene system.

The presence of styrene (in the amount of ~ 0.5% relative to the polymeric matrix of SAN) completely freezes the dipole-group losses, which are characteristic of molecular motions of the crankshaft type in the SAN and PS. This fact can testify to the formation of overall fluctuation meshes of the lyophilic nature in both the binary SAN-styrene system and the triple SAN-styrene-PS systems. Moreover, the IMI in such meshes is larger than in the matrix copolymer, as evidenced by the temperature position of tg δ_{max}. This temperature does not drift downwards, even for a macro heterogenic mixture containing 10% PS.

Figure 7 shows that the behavior of SAN-PS-BN mixtures in the range of crankshaft oscillations indicates the selective interaction of phenyls of the corresponding polymers with LMWA; at the same time, the interaction of phenyls of SAN and PS in IPL getting weak. The α-relaxation region for compositions containing benzonitrile changing significantly with any change in the ratio of components in the system. A complex change in the symmetry of the maxima is visible.

The basic maximum tg δ (α-relaxation process) in the binary composition of SAN-BN is observed at ~ 121 °C, and in the triple mixtures of SAN-PS-BN (if PS content does not exceed 5%) at slightly higher (5 and 7 °C) temperatures (Fig. 7). It can be attributed to the fact that in the region of temperatures 100–140 °C increasing the role of rigid lyophobic mesh that compete with lyophilic mesh. Dipole–dipole interaction involving the nitrile groups of SAN and BN is responsible for lyophobic mesh formation.

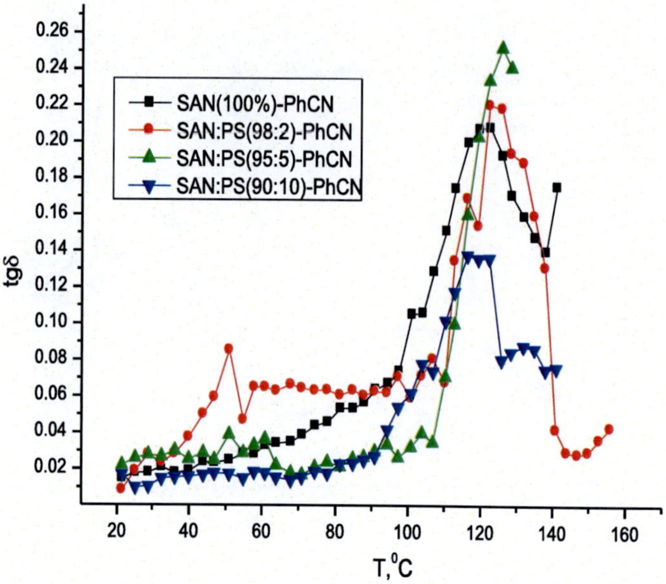

Fig. 7 Temperature dependence of tg δ (1 kHz) for composition: SAN-BN-PS

The SAN-PS system (90:10)-BN in the α-relaxation region already shows two maxima tg δ (similar to the SAN-PS composition (90:10)-AN), but with a more clear separation effect. The half-width of the peaks increases significantly in the case of this mixture. This indicates a wide distribution of relaxation times and, therefore, of molecular motions responsible for the range of relaxation processes. (Fig. 7.)

It can be said that in the presence of BN, there is a very complex redistribution of IMI forces in the whole spectrum of interphase layers, which is formed by the initial lyophilic mesh of the matrix copolymer and the elements of the lyophobic gmesh formed in the presence of BN. The analysis of temperature and concentration dependences of dielectric constant has not added anything new to our conclusions, so not stated in the paper.

4 Conclusions

Thus, as a study result of the styrene and acrylonitrile copolymer that obtained by the initiated radical copolymerization in mass (azeotropic ratio of the original monomers) by IR spectroscopy proved the presence of styrene dyads, triads and tetrads in SAN macromolecules. Such a structure of macromolecules opens the possibility for the active participation of the styrene component of the copolymer in the redistribution of the IMI forces.

Analysis of the dielectric behavior of the samples showed that under the conditions of our experiment, the corresponding molecular moves (crankshaft oscillations) manifested in the dipole-group loss region (temperature range of 20–60 °C). In addition, the evaluation of the behavior of the polymers studied (in the absence and in the presence of LMWA) in the temperature regions of dipole-group-(crankshaft molecular rearrangements) and dipole-segmental losses showed that in the presence of toluene and styrene the systems behave as the only complexes of SAN-LMWA-PS. This effect manifested despite the different nature of the substituents on the benzene ring: the main thing is that the aromatic compound has an asymmetric distribution of the electron density in the ring. In the first case, the possibility of forming a stable complex is associated with an increase in the electrical charge of the ring of the toluene molecule (compared to the benzene molecule).

The presence of such a strong electron-withdrawing group as vinyl (and the π-double bonds of this group are in conjunction with the phenyl p-electrons) leads to high polarization and delocalization of the electron cloud. Judging by the dielectric behavior of the SAN-styrene-PS mixtures studied by us, this arena may serve as a compatibilizer for them.

Benzene selectively interacts with SAN and PS, blocking interaction in the IPL.

In the case of polymer–polymer, systems containing benzonitrile, the decrease in intermolecular interaction in IPL is obviously associated not only with similar selectivity but also with the possibility of dipole–dipole interaction due to the nitrile groups of the copolymer and benzonitrile. As a result, the complex structure of IPL-s formed by competing (lyophilic and lyophobic) fluctuating meshes. The possibility of forming a rigid lyophobic mesh and the related heterogeneity of the mixtures generally controlled by the content of SAN in the compositions.

References

1. L. Fieser, M. Fieser, *Reagents for Organic Synthesis* (USA, 1967)
2. R. Bruckner, *Advanced Organic Chemistry: Reaction Mechanisms* (Academic Pres, USA, 2001)
3. T. Schaller, U.P. Büchele, F.G. Klärner, D. Bläser, R. Boese, S.P. Brown, H.W. Spiess, F. Koziol, J. Kussmann, C. Ochsenfeld, Structure of molecular tweezer complexes in the solid state: NMR experiments, X-ray investigations, and quantum chemical calculations (2006)
4. S. Liu, J. Shi, E.W. Forsythe, S.M. Blomquist, D. Chiu, Polymer charge-transfer complexes for opto-electronic applications (2009)
5. Pure & Appl, Chern. **63**(2), 255 (1991)
6. M. Anoshenko, T. Bazylyuk, N. Iukhymenko, S. Studzinsky, Possibilities for intermolecular layer strengthening inside microgeterogeneous systems of styrene-acrylonitrile copolymer-polystyrene. Mol. Cryst. Liq Cryst. **670**, 50–59 (2018). https://doi.org/10.1080/15421406.2018.1542064
7. Р. Бойер, *Transition and Relaxation Phenomena of polymer*. M (1968); R.H. Boyd, G.D. Smith, *Polymer Dynamics and Relaxation* (CUP, 2007), p. 255.
8. Майо, *Chemistry and Technology Polymer*, №5. C.3 (1967); F.R. Mayo, Ch. Walling, Chem. Revs. **46**, 191 (1950)
9. L.M. Kalyuzhnaya, S.I. Ganicheva, V.M. Denisov, P.N. Yu, F.S. Ya, Polym. Sci. **1**(1), 114 (1985)

Electrostatics of a Nanowire Radial Heterostructure *p–i–n* Diode

Vitalii L. Borblik

Abstract Spatial (along the radius) distribution of built-in electric field in a nanowire radial heterostructure *p–i–n* diode and its capacitance is calculated in the depletion approximation. The cases of the symmetric structure (when the doped layers are made of the same material) and the asymmetric one (when the doped layers are made of different materials) are considered. It is shown that in advance-given profile of the built-in electric field can be achieved by means of corresponding combination of the different materials. It is shown also that extension of the electric field region of a *p–n* junction at the expense of introducing an *i*–layer into it is achieved not always. Furthermore, in contrast to a planar heterostructure *p–i–n* diode, in a nanowire radial heterostructure *p–i–n* diode, sequence of the materials used for doped layers has significant meaning.

1 Introduction

At the present time, semiconductor nanowires are used mostly in the form of multilayer structures, in which the material and/or doping type changes either in radial or axial direction. In particular, nanowires with *p–n* junctions have found broad application as radiation detectors [1, 2], solar cells [3, 4], light emitting devices [5, 6] and devices for energy harvesting [7]. Numerous investigations have shown that the radial core–shell nanowires have certain advances compared with axial ones because shell structure of a nanowire provides passivation of the internal layers.

In many cases, *p–i–n* junctions prove to be more effective than simple *p–n* junctions because they extend a region of the built-in electric field. This concerns both photovoltaic applications [8] and light emitting devices [9]. Detailed analysis [10] of electrostatics of a nanowire radial homo *p–i–n* junction has shown that the electric field in it falls from the nanowire centre and its maximum value can be both lower and higher than electric field in the equivalent planar *p–i–n* diode.

V. L. Borblik (✉)
V. Lashkaryov Institute of Semiconductor Physics, Kyiv, Ukraine
e-mail: borblik@isp.kiev.ua

O. Fesenko and L. Yatsenko (eds.), *Nanostructured Surfaces, Nanocomposites and Nanomaterials, and Their Applications*, Springer Proceedings in Physics 296,
https://doi.org/10.1007/978-3-031-42704-6_14

Herewith not homo but hetero p–i–n junctions are used often in the applications when it is necessary for the certain layer to have other material properties. It is known that in nanostructures, restrictions connected with lattice mismatch are eliminated in significant degree because stress in them relaxes and therefore does not create interface defects. This gives great freedom in choosing of materials for the constructions. Furthermore, the i-shell with refraction index smaller than that of the p– and n-layers can create useful waveguide effect. And selective etching of the individual layers simplifies creation of the contacts to them.

Therefore, it is advisable to consider in details electronic properties of a radial heterostructure p–i–n diode.

2 Theory

Schematic view of the structure under investigation is given by Fig. 1, where r_p is the depletion region boundary in the p-core, r_n is the depletion region boundary in the n-shell, and i-layer is located between r_1 and r_2.

2.1 Solution of the Poisson's Equation

In the depletion approximation, Poisson's equation has a following form

$$\frac{1}{r}\frac{d}{\mathrm{d}r}(rE) = \frac{qN_{A1}}{\varepsilon_1}, r_p \le r \le r_1 \quad (1a)$$

$$\frac{1}{r}\frac{d}{\mathrm{d}r}(rE) = 0, r_1 \le r \le r_2, \quad (1b)$$

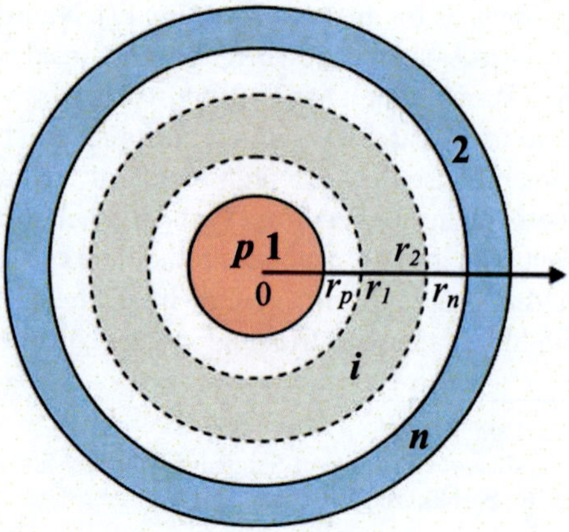

Fig. 1 Schematic view of a radial heterostructure p–i–n diode

$$\frac{1}{r}\frac{d}{\mathrm{dr}}(rE) = -\frac{qN_{D2}}{\varepsilon_2}, r_2 \le r \le r_n, \tag{1c}$$

where q is electron charge, ε_1 and ε_2 are dielectric constants of the core and shell, respectively, N_{A1} and N_{D2} are concentrations of acceptors and donors in them. Solution of these equations gives the electric field distribution in the structure

$$E = -\frac{qN_{A1}}{2\varepsilon_1}\frac{r^2 - r_p^2}{r}, r_p \le r \le r_1, \tag{2a}$$

$$E = \frac{A}{r}, r_1 \le r \le r_2, \tag{2b}$$

$$E = \frac{qN_{D2}}{2\varepsilon_2}\frac{r^2 - r_n^2}{r}, r_2 \le r \le r_n, \tag{2c}$$

where A is the integration constant to be determined. By stitching electric induction $D = \varepsilon E$ at r_1 and r_2, we obtain

$$A = -\frac{qN_{A1}}{2\varepsilon_i}\left(r_1^2 - r_p^2\right) = \frac{qN_{D2}}{2\varepsilon_i}\left(r_2^2 - r_n^2\right) \tag{3}$$

where ε_i is dielectric constant of the *i*-layer. It follows also from (3) that

$$N_{A1}\left(r_1^2 - r_p^2\right) = N_{D2}\left(r_n^2 - r_2^2\right). \tag{4}$$

Second integration of (2) gives the potential distribution

$$V(r) = \frac{qN_{A1}}{2\varepsilon_1}\left(\frac{r^2 - r_p^2}{2} + r_p^2 \ln\left(\frac{r_p}{r}\right)\right), r_p \le r \le r_1, \tag{5a}$$

$$V(r) = -A(\ln(r) + B), r_1 \le r \le r_2, \tag{5b}$$

$$V(r) = -\frac{qN_{D2}}{2\varepsilon_2}\left(\frac{r^2 - r_n^2}{2} + r_n^2 \ln\left(\frac{r_n}{r}\right)\right) + V_{bi}, r_2 \le r \le r_n, \tag{5c}$$

where B is the integration constant and following boundary conditions are used

$$V(r_p) = 0, V(r_n) = V_{bi} - U, \tag{6}$$

Here, V_{bi} is the built-in potential of the junction and U is the applied voltage. Stitching of the potentials at $r = r_1$ and $r = r_2$ allows us to exclude B and obtain transcendent equation

$$\frac{qN_{A1}}{2\varepsilon_1}\left(\frac{r_1^2 - r_p^2}{2} + r_p^2 \ln\left(\frac{r_p}{r_1}\right)\right) + \frac{qN_{D2}}{2\varepsilon_2}\left(\frac{r_2^2 - r_n^2}{2} + r_n^2 \ln\left(\frac{r_n}{r_2}\right)\right)$$
$$- A\ln\left(\frac{r_2}{r_1}\right) = V_{bi} - U \tag{7}$$

Equations (4) and (7) have to be solved jointly in order to obtain r_p and r_n. All the rest quantities are expressed through them. This can be made only numerically.

2.2 *Calculation of the Built-In Potential*

Availability in the structure of 3 different materials gives a large number of possible their combinations. Let us consider firstly *p–i–n* structures symmetrical on materials of the doped layers, i.e. the structures where *p*- and *n*-layers are made of the same semiconductor and *i*-layer–of other material. Then, (7) simplifies

$$\frac{qN_{A1}}{2\varepsilon_{12}} r_p^2 \ln\left(\frac{r_p}{r_1}\right) + \frac{qN_{D2}}{2\varepsilon_{12}} r_n^2 \ln\left(\frac{r_n}{r_2}\right) - A\ln\left(\frac{r_2}{r_1}\right) = V_{bi} - U \tag{8}$$

where $\varepsilon_1 = \varepsilon_2 = \varepsilon_{12}$.

Two possible band structures of such type are shown in Fig. 2 (before bringing three materials into contact) where we consider abrupt anisotype *p-n* heterojunctions of the type I separated by *i*-layer.

In the case of the non-generated semiconductors, we have

$$E_{F1} - E_{v1} = kT \ln \frac{N_{v1}}{N_{A1}}, \tag{9}$$

$$E_{c2} - E_{F2} = kT \ln \frac{N_{c2}}{N_{D2}}, \tag{10}$$

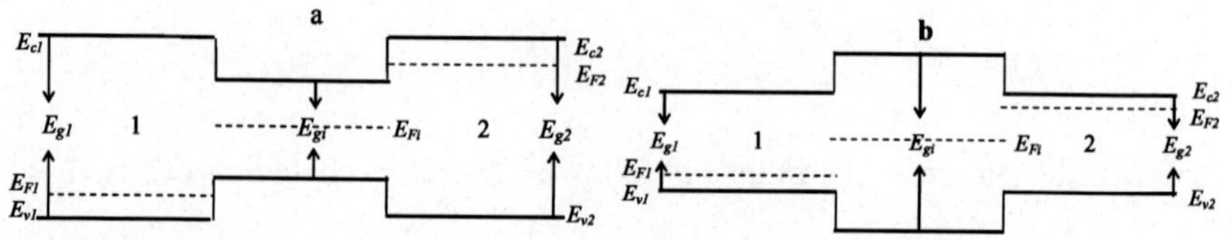

Fig. 2 Sketches of symmetrical band structures, where band gap of the *i*-layer is smaller (case a) and larger (case b) than band gap of the doped layers; here E_{c1} and E_{c2} are bottoms of the conductivity bands, E_{v1} and E_{v2} are tops of the valence bands, ΔE_c and ΔE_v are the energy bands discontinuities, E_{g1}, E_{gi} and E_{g2} are the band gaps, E_{F1} and E_{F2} are the Fermi levels

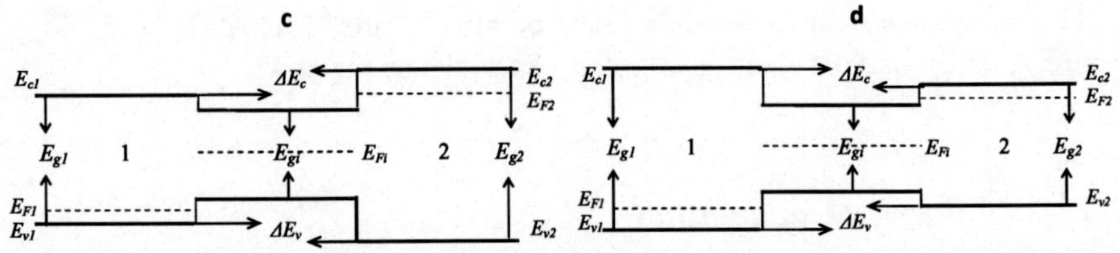

Fig. 3 Sketches of asymmetrical band structures where band gap of the core is smaller (case c) and larger (case d) than band gap of the doped shell; here E_{c1} and E_{c2} are bottoms of the conductivity bands, E_{v1} and E_{v2} are tops of the valence bands, ΔE_c and ΔE_v are the energy bands discontinuities, E_{g1}, E_{gi} and E_{g2} are the band gaps, E_{F1} and E_{F2} are the Fermi levels.

where N_{v1} and N_{c2} are the effective densities of states in the valence and conduction bands, respectively. Adding (9) and (10), we obtain

$$E_{F2} - E_{F1} \equiv qV_{bi} = E_{c2} - E_{v1} - kT\ln\left(\frac{N_{v1}N_{c2}}{N_{A1}N_{D2}}\right). \tag{11}$$

In both cases (a and b in Fig. 2), $E_{g1} = E_{g2} \equiv E_{g12}$, therefore

$$qV_{bi} = E_{g12} + kT\ln\left(\frac{N_{A1}N_{D2}}{N_{v1}N_{c2}}\right) \tag{12}$$

i.e. band gap of the *i*-layer has not any role (as well as the band discontinuities between *i*-layer and doped layers).

More general cases when $E_{g1} \neq E_{g2}$ (Fig. 3) we will call asymmetric (relatively to materials of doped layers). Herewith, the band discontinuities for doped layers acquire important meaning.

Then, we have for the structure of c-type

$$\begin{aligned} qV_{bi} &= E_{c2} - E_{v1} - kT\ln\left(\frac{N_{v1}N_{c2}}{N_{A1}N_{D2}}\right) \\ &= E_{g1} + \Delta E_c + kT\ln\left(\frac{N_{A1}N_{D2}}{N_{v1}N_{c2}}\right), \end{aligned} \tag{13}$$

and for the structure of d-type

$$\begin{aligned} qV_{bi} &= E_{c2} - E_{v1} - kT\ln\left(\frac{N_{v1}N_{c2}}{N_{A1}N_{D2}}\right) \\ &= E_{g1} - \Delta E_c + kT\ln\left(\frac{N_{A1}N_{D2}}{N_{v1}N_{c2}}\right), \end{aligned} \tag{14}$$

i.e. band discontinuity results in either increasing or decreasing built-in potential of the junction.

In concordance with Anderson's "electron affinity rule" [11], $\Delta E_c = \chi_1 - \chi_2$ where χ_1 and χ_2 are the electron affinities of two doped materials.

2.3 *The Barrier Capacitance*

The barrier capacitance $C = \frac{dQ_p}{dU}$, where Q_p is the electric charge concentrated in the depleted, e.g. *p*-region of the junction. This charge is given by

$$Q_p = qN_A\pi\left(r_1^2 - r_p^2\right)L \tag{15}$$

where r_p is voltage-dependent, L is length of the nanowire. Inasmuch as

$$\frac{dr_p}{dU} = \frac{2\varepsilon_2}{qN_A r_p} \frac{1}{\frac{\varepsilon_2}{\varepsilon_1} - 1 + \ln\left(r_n/r_p\right)^2 + \left(\frac{\varepsilon_2}{\varepsilon_i} - 1\right)\ln\left(r_2/r_1\right)^2}, \tag{16}$$

capacitance per unit area of the *p–i–n* junction is

$$C = \frac{1}{r_1} \frac{\varepsilon_2}{\frac{1}{2}\left(\frac{\varepsilon_2}{\varepsilon_1} - 1\right) + \ln\left(r_n/r_p\right) + \left(\frac{\varepsilon_2}{\varepsilon_i} - 1\right)\ln(r_2/r_1)}. \tag{17}$$

3 Results of the Numerical Calculations

The numerical calculations have been performed for radial *p–i–n* structures on the base of Ge and GaAs because this heteropair has good lattice matching; therefore, no appreciable density of interface states can be associated with this heterojunction.

In accordance with data from [12], in this case ΔE_c = 4.13 eV – 4.07 eV = 0.06 eV. Corresponding value of $\Delta E_v = E_{g2} - E_{g1} - \Delta E_c$ = (1.42–0.66) eV – 0.06 eV = 0.7 eV. The densities of the band states for Ge and GaAs (as well as values of their energy gaps and dielectric constants) have been taken from [13].

3.1 Symmetrical *p–i–n* Structures

Some examples of the electric field profile in radial symmetrical *p–i–n* heterostructures are presented in Fig. 4 for structure *p*-GaAs/*i*-Ge/*n*-GaAs (a-type, left) and for structure *p*-Ge/*i*-GaAs/*n*-Ge (b-type, right).

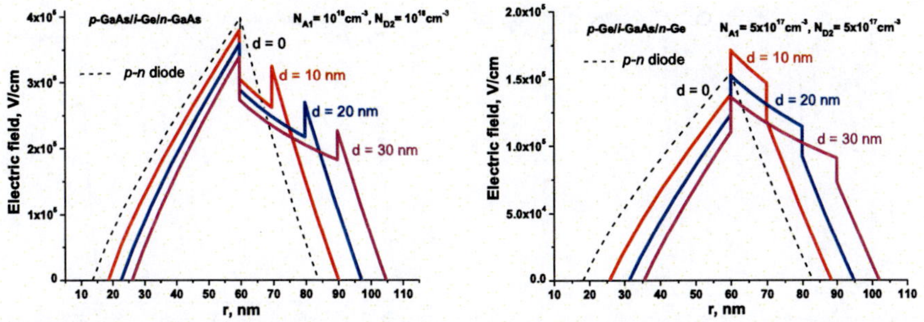

Fig. 4 Radial electric field distribution in symmetrical *p–i–n* heterostructures of a-type (left) and *b*-type (right); $d = r_2 - r_1$

As follows from these figures, introducing into *p–n* junction of the *i*-layer with larger dielectric constant than that of the doped layers lowers the electric field in it, and introducing of *i*-layer with smaller dielectric constant raises the field. The reason is continuity of electric induction at the heterointerfaces.

It should be noted that introducing *i*-layer into *p–n* junction extends the electric field region not always. This fact is illustrated by Figs. 5 and 6, where depletion region width in the *p*-core $w_p = r_1 - r_p$, depletion region width in the *n*-shell $w_n = r_n - r_2$ and total extent of the electric field region $w = w_p + w_n + d$ ($d = r_2 - r_1$) are shown as functions of radius r_1. It is seen that at high enough doping levels (when depletion widths in the *n*- and *p*-layers have small extent), increasing w at the expense of *i*-layer is really achieved. But at middle doping level in one of doped layers, the depletion widths decrease under introducing *i*-layer so strongly that total extent of the built-in electric field region does not increase practically. This takes place both in structure of a-type (Fig. 5) and particularly in structure of b-type (Fig. 6).

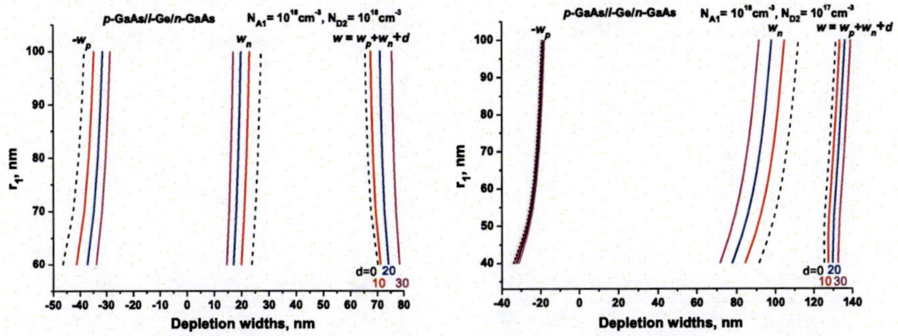

Fig. 5 Depletion widths w_p in the core and w_n in the shell, and also total extent w of electric field region in *a*-type *p–i–n* structure *p*-GaAs/*i*-Ge/*n*-GaAs at high (left) and middle (right) doping levels as functions of radius r_1 (an abscissa equal to zero corresponds to r_1); numbers near the curves are values of *d* in nms

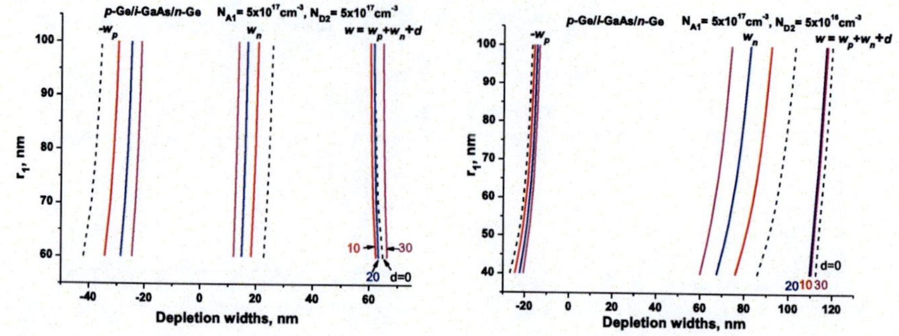

Fig. 6 Depletion widths w_p in the core and w_n in the shell, and total extent w of electric field region in *b*-type *p*–*i*–*n* structure *p*-Ge/*i*-GaAs/*n*-Ge at high (left) and middle (right) doping levels as functions of radius r_1 (an abscissa equal to zero corresponds to r_1); numbers near the curves are values of d in nms

The barrier capacitance equal (in the symmetrical case) to

$$C = \frac{1}{r_1} \frac{\varepsilon_{12}}{\ln(r_n/r_p) + \left(\frac{\varepsilon_{12}}{\varepsilon_i} - 1\right) \ln(r_2/r_1)}, \tag{18}$$

behaves in complete consistence with above presented results, i.e. it grows noticeably with increasing *i*-layer thickness at high doping levels of the both doped regions (Fig. 7, left and Fig. 8, left) but does not change practically at middle doping of one of them (Fig. 7, right and Fig. 8, right).

Herewith in a-type structure, capacitance grows with increasing *i*-layer thickness but in b-type structure, wise versa, it decreases. And in both cases, capacitance goes to more weak dependence on applied voltage as it should be in planar case.

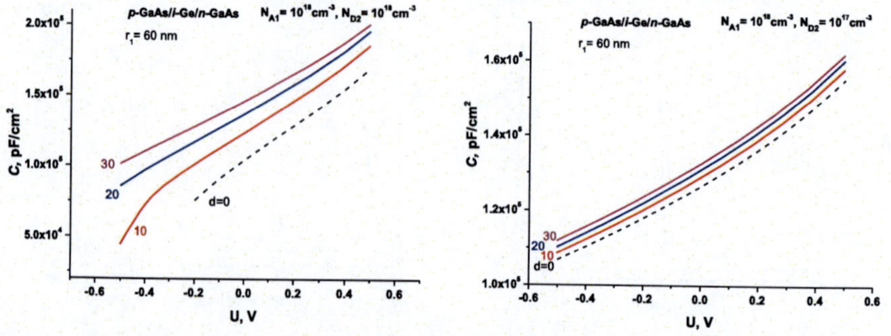

Fig. 7 Capacitance of structure *p*-GaAs/*i*-Ge/*n*-GaAs (a-type) at high (left) and middle (right) doping levels; numbers near the curves are values of d in nms

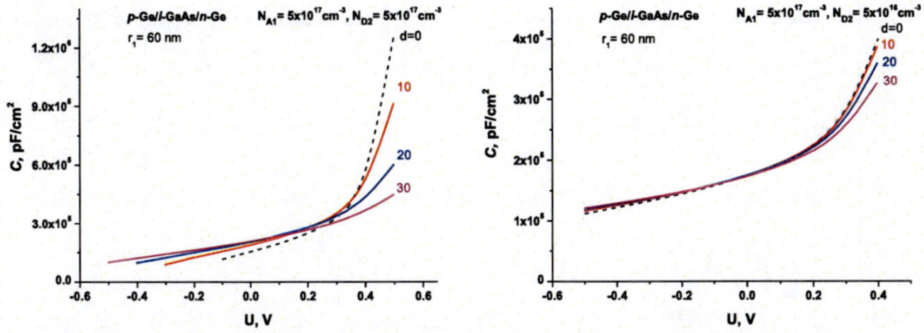

Fig. 8 Capacitance of structure *p*-Ge/*i*-GaAs/*n*-Ge (b-type) at high (left) and middle (right) doping levels; numbers near the curves are values of *d* in nms

3.2 Asymmetrical *p–i–n* Structures

Though *i*-layer in asymmetric *p–i–n* structure can be made of the third material, we consider here only the cases when it is made of either GaAs or Ge in order to not deal with possible (in general case) surface states.

Figures 9 and 10 represent the radial distributions of electric field in asymmetrical *p–i–n* structures of c-type and d-type, respectively, where *i*-layer is made of either GaAs (left) or Ge (right).

As follows from these figures, structure of d-type provides significantly higher built-in electric field than c-type structure does. And as well as in the symmetric structures, smaller dielectric constant of the *i*-layer works towards enhancement of electric field in it and larger dielectric constant of the *i*-layer works towards its lowering.

Note also that in planar case, electrostatics of *p–i–n* heterostructure c- and d-types does not differ.

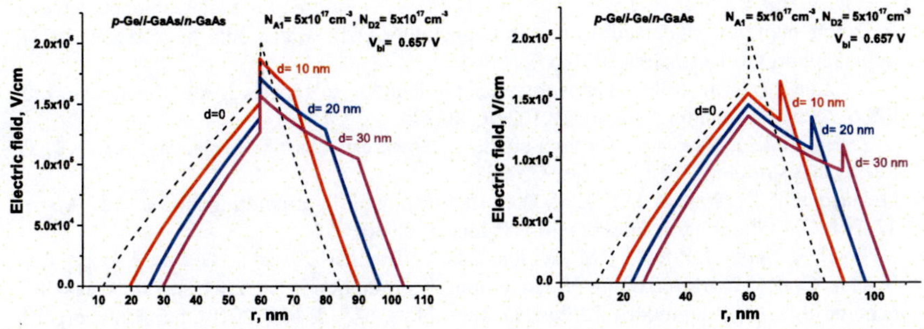

Fig. 9 Radial electric field distribution in asymmetrical c-type *p–i–n* structures *p*-Ge/*i*-GaAs/*n*-GaAs (left) and *p*-Ge/*i*-Ge/*n*-GaAs (right) as a function of the *i*- layer thickness *d*

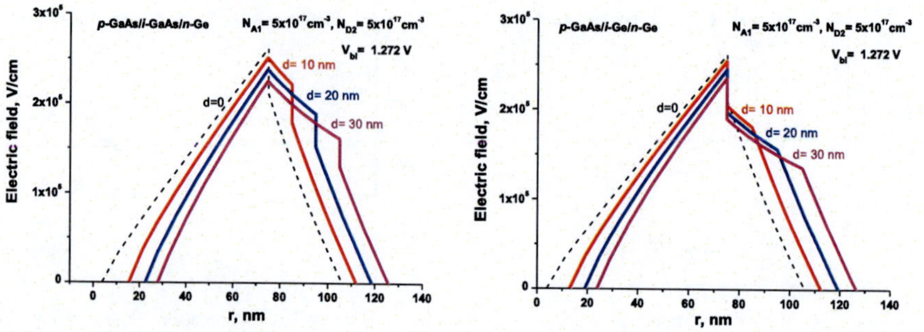

Fig. 10 Radial electric field distribution in asymmetrical *d*-type *p–i–n* structures *p*-GaAs/*i*-GaAs/*n*-Ge (left) and *p*-GaAs/*i*-Ge/*n*-Ge (right) as a function of the *i*- layer thickness *d*

4 Conclusion

In summary,

(1) heterostructure nature of the *p–i–n* diode allows us to create in advance-given profile of the built-in electric field,
(2) possibilities to extend the region of built-in electric field in a *p-n* diode at the expense of introducing *i*-layer turn out rather limited,
(3) in contrast to planar heterostructure *p–i–n* diode, sequence of the doped layers in radial heterostructure *p–i–n* diode has significant meaning.

Acknowledgements This work was supported by the National Academy of Sciences of Ukraine.

References

1. C. Soci, A. Zhang, X.-Y. Bao, H. Kim, Y. Lo, D. Wang, Nanowire photodetectors. J. Nanosci. Nanotechnol. **10**, 1430–1449 (2010). https://doi.org/10.1166/jnn.2010.2157
2. Y. Zhu, V. Raj, Z. Li, H.H. Tan, C. Jagadish, L. Fu, Self-powered InP nanowire photodetector for single photon level detection at room temperature. Adv. Mater. **33**(49), 2105729 (2021). https://doi.org/10.1002/adma.202105729
3. E.C. Garnett, P. Yang, Silicon nanowire radial p-n junction solar cells. J. Am. Chem. Soc. **130**, 9224–9225 (2008). https://doi.org/10.1021/ja8032907
4. Y. Zhang, H. Liu, Nanowires for high-efficiency, low-cost solar photovoltaics. Crystals **9**, 87 (2019). https://doi.org/10.3390/cryst9020087
5. O. Hayden, A.B. Greytak, D.C. Bell, Core-shell nanowire light-emitting diodes. Adv. Mater. **17**, 701–704 (2005). https://doi.org/10.1002/adma.200401235
6. I. Girgel, A. Šatka, J. Priesol, P.M. Coulon, E.D. Le Boulbar, T. Batten, D.W.E. Allsopp, P.A. Shields, Optical characterization of magnesium incorporation in p-GaN layers for core–shell nanorod light-emitting diodes. J. Phys. D Appl. Phys. **51**, 155103 (2018). https://doi.org/10.1088/1361-6463/aab16b
7. N.I. Goktas, P. Wilson, A. Ghukasyan, D. Wagner, S. McNamee, R.R. LaPierre, Nanowires for energy: a review. Appl. Phys. Rev. **5**, 041305 (2018). https://doi.org/10.1063/1.5054842

8. J. Yoo, S.A. Dayeh, W. Tang, S.T. Picraux, Epitaxial growth of radial Si p-i-n junctions for photovoltaic applications. Appl. Phys. Lett. **102**, 093113 (2013). https://doi.org/10.1063/1.4794541
9. F. Qian, Y. Li, S. Gradečak, D. Wang, C.J. Barrelet, C.M. Lieber, Gallium nitride-based nanowire radial heterostructures for nanophotonics. Nano Lett. **4**, 1975–1979 (2004). https://doi.org/10.1021/nl0487774
10. V.L. Borblik, Electrostatics of nanowire radial *p–i–n* diode, Semicond. Phys. Quant. Electron. Optoelectron. **22**, 201–205 (2019). https://doi.org/10.15407/spqeo22.02.201
11. R.L. Anderson, Experiments on Ge-GaAs heterojunctions. Solid-State Electron. **5**, 341–351 (1962). https://doi.org/10.1016/0038-1101(62)90115-6
12. B.L. Sharma, R.K. Purohit, in *Semiconductor Heterojunctions* (Pergamon Press, Oxford–New York–Toronto–Sydney, 1974)
13. S.M. Sze, *Physics of Semiconductor Devices,* 2nd edn. (John Wiley & Sons, New York–Chichester–Brisbane–Toronto–Singapore, 1981)

Vapor-Condensed Composite Materials $Ni–Al_2O_3$, $NiCr–Al_2O_3$ with Oxide Nanophase

V. Grechanyuk, M. Grechanyuk, V. Chornovol, A. Kozyrev, and Kovalchuk Yu

Abstract The structure and mechanical properties of $Ni–Al_2O_3$ and $NiCr–Al_2O_3$ dispersion-strengthened materials condensed from the vapor phase were investigated. It was shown that nanoparticles are uniformly distributed throughout the volume of the condensates. The factor affecting the structure is the contact interaction at the particle–matrix interphase boundary. The studied mechanical properties at temperatures of 700 and 1000 ± 20 °C show that small concentrations of dispersed Al_2O_3 particles (0.25–0.4%) in $Ni–Al_2O_3$ condensates and up to 1% Al_2O_3 in $NiCr–Al_2O_3$ condensates lead to an increase in strength characteristics and non-monotonic decrease in plasticity. Based on the obtained data, the optimal concentration of the strengthening Al_2O_3 nanophase in condensed dispersion-strengthened $Ni–Al_2O_3$ and $NiCr–Al_2O_3$ materials was determined, which ensures a high level of strength and plasticity.

1 Introduction

Dispersion-reinforced materials (DRM) were obtained exclusively by powder metallurgy methods until recently. The authors of [1] conducted detailed studies of a wide range of properties of these materials. Many of them have found practical application. Obtaining DRM is a multi-stage technological process that requires strict adherence to technological discipline at all stages of their production.

V. Grechanyuk (✉) · V. Chornovol · A. Kozyrev · K. Yu
Department of Chemistry, Kyiv National University of Construction and Architecture, Povitroflotsky Av., 31, Kyiv 03037, Ukraine
e-mail: missvik@ukr.net

M. Grechanyuk
Department of Abrasion- and Corrosion-Resistant Powder Construction, Materials Institute for Problems of Material Sciences, NAS of Ukraine, St. Krzhizhanovsky 3, Kyiv 03142, Ukraine

O. Fesenko and L. Yatsenko (eds.), *Nanostructured Surfaces, Nanocomposites and Nanomaterials, and Their Applications*, Springer Proceedings in Physics 296,
https://doi.org/10.1007/978-3-031-42704-6_15

High-speed electron beam evaporation–condensation as a new technological process has found wide application for the deposition of various protective coatings on products, primarily on the blades of gas turbines. The rate of vapor flow deposition on substrates of various configurations can reach 150 μm/min, which allows to deposit quite a lot of material on the substrates. Therefore, it is of scientific and practical interest to use the specified technological process, controlled at the atomic-molecular level, to obtain massive (separated from the substrate) composite materials.

Fundamental studies to establish the basic physical and mechanical laws of the formation of thick (0.01–2 mm) condensates were carried out at the Paton's Institute of Electric Welding under the leadership of Academician Movchan [2, 3]. The main physico-chemical regularities of the formation of thick vacuum condensates of some pure metals, alloys, oxides, carbides, borides were determined, and their physico-mechanical characteristics were studied depending on the composition and condensation parameters.

Composite dispersion-reinforced porous and layered (microlayer) materials should be included in the new materials obtained by vapor deposition in a vacuum.

Currently, intensive research is being conducted on new composite materials condensed from the vapor phase with a reinforcing nanophase (oxides, carbides, borides, refractory metals). Dispersion-strengthened composite materials that condense from the vapor phase (condensates) consist of a polycrystalline metal or ceramic matrix with nanodisperse particles of the second phase uniformly distributed by volume. By varying the substrate temperature and cooling rate, the average crystallite size of the matrix can be varied from several hundred microns to several hundred nanometers, and the particle size of the master phase can be varied from several nanometers to several microns. As a result of the influence on the morphology, dispersion and nature of the distribution of the strengthening phase, it is possible to obtain in dispersion-strengthened materials a combination of properties that are unattainable in ordinary alloys [4, 5].

The use of stable refractory compounds as strengthening phases, e.g., oxides that do not actively interact with the base metal and do not dissolve in it up to its melting temperature, ensures the preservation of the microheterogeneous structure and dislocation substructure up to pre-melting temperatures. This allows you to preserve long-term operational characteristics of materials (0.9–0.95 T_{mel}).

2 Results and Discussions

Condensed from the vapor phase dispersion-strengthened materials Ni–Al_2O_3 (KDSM) were obtained on laboratory and industrial equipment manufactured at the IES named after Paton of the National Academy of Sciences of Ukraine. Sheet rectangular condensates (220 × 320 × 0.8–2) mm with a concentration gradient of dispersed oxide nanophase were obtained for research.

A similar technological technique makes it possible to obtain a significant number of samples of different composition. Alternating vapor flow deposition was carried out on a substrate with Art. 3, processed to a purity class of 0.63 at two condensation temperatures of 700 and 1000 ± 20 °C. For easy separation of condensates from the substrate, a separating layer of calcium fluoride (CaF_2) with a thickness of 20–40 microns was previously applied to the surface on which condensation was carried out. For industrial applying, condensates were formed in the form of cylindrical sheet blanks with a thickness of 1–4 mm and a diameter of 800 mm.

In the work, the structure, chemical, phase composition and mechanical properties were investigated according to known standard methods [4].

According to the Fig. 1 of the Ni–Al_2O_3 KDSM structure that dispersed aluminum oxide nanoparticles are evenly distributed throughout the volume of the condensate. The determining factor that affects the structure and, as a result, the mechanical properties of KDSM is the contact interaction at the particle–matrix interphase boundary. The quantitative criterion of such contact interaction is the wetting angle. It is largely influenced by the environment in which the crystallizing liquid phase interacts with the solid oxide particle, the purity of the metal and oxide phases, the condensation temperature, and other factors. Currently, a sufficient number of studies have been conducted, which are based on their contact interaction when obtaining new materials in metal (alloy)—MeO systems in a vacuum [6]. KDSM Ni–Al_2O_3 with an acceptable set of mechanical characteristics can be obtained in a narrow interval of oxide phase concentration (up to 0.6 wt.%).

Using the methods of electron microscopic and X-ray phase analysis, it was established that only nickel and aluminum oxide are present in the material.

Figure 2a, b shows the dependences of strength limits (σ), yield strength ($\sigma_{0,2}$) and relative elongation (δ) of Ni–Al_2O_3 condensates obtained at substrate temperatures (T_s) of 700 and 1000 ± 20 °C.

The analysis of obtained data shows that small concentrations of dispersed Al_2O_3 particles lead to an increase in strength characteristics and a non-monotonic decrease in plasticity in a relatively narrow range of Al_2O_3 concentrations (0.25–0.4 wt.%).

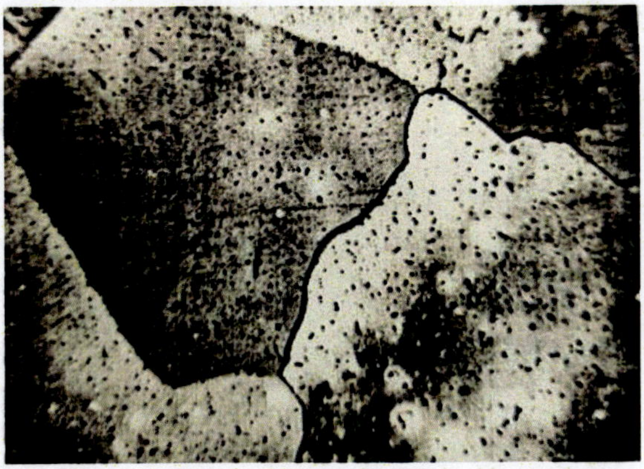

Fig. 1 Structure of KDSM Ni–Al_2O_3 (×30,000)

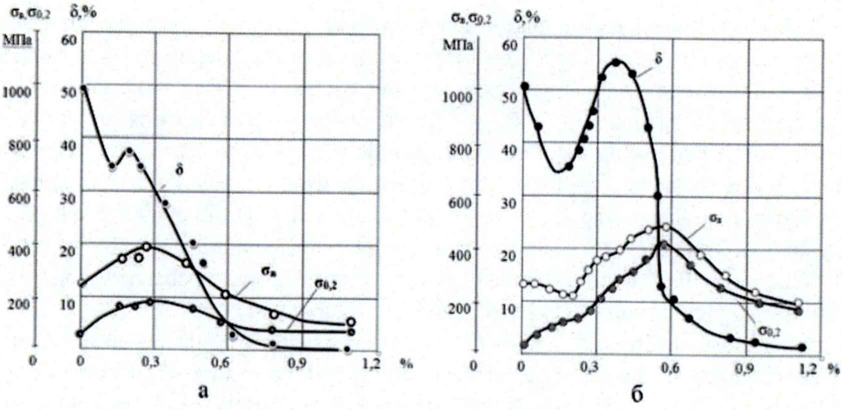

Fig. 2 Dependencies of strength, yield strength and relative elongation on the content of Al_2O_3 in Ni–Al_2O_3 KDSM obtained at temperatures: **a**—700 °C; **b**—1000 °C

The increase in plasticity is explained by the fulfillment of the structural condition: the average grain size of the metal matrix (D_3) is equal to the average distance between dispersed particles of the strengthening phase (Λ) [2].

It should be emphasized that the maximum of the curves of the dependence of plasticity on the content of Al_2O_3 shifts toward a higher content of aluminum oxide with an increase in the temperature of the substrate. The absolute values of the plasticity of two-phase Ni–Al_2O_3 materials with an optimal content of dispersed particles increase with an increase in the condensation temperature. For example, at $T_s = 1000$ °C, KDSM Ni–(0.35–0.4) mass % Al_2O_3 has a value of relative elongation greater than that of pure nickel (Fig. 2b).

In terms of strength characteristics, Ni–Al_2O_3 KDSM is not inferior to industrial VDU-2 DUM (98% Ni + 2% HfO_2), obtained by powder metallurgy methods.

In more complex two-phase condensed systems that form solid solutions, e.g., NiCr-Al_2O_3, qualitatively similar changes in mechanical characteristics are observed (Fig. 3a, b). The strength limit and yield strength increase in a wider range of Al_2O_3 concentrations (up to 1%). However, with this content of dispersed refractory particles, condensates have low plasticity. A similar change of the mechanical characteristics in the KDSM is caused by the almost complete absence of particle–matrix interphase interface interaction. The marginal wetting angle of Al_2O_3 with liquid nickel ranges from 150 to 1150. Due to the lack of mutual action in the condensates, pores are formed, which leads to a loss of strength and plasticity. Improvement of interphase interaction in the NiCr–Al_2O_3 system (up to 85o) leads to some increase in strength and plasticity in a wide range of Al_2O_3 concentrations compared to Ni–Al_2O_3 composites.

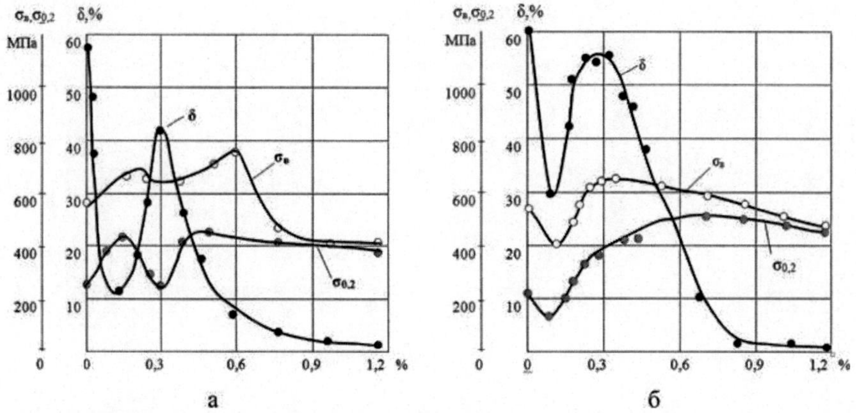

Fig. 3 Dependencies of strength, yield strength and relative elongation on the content of Al_2O_3 in NiCr–Al_2O_3 KDSM obtained at temperatures: **a**—700 °C; **b**—1000 °C

3 Conclusion

Thus, as a result of the research of the mechanical properties of condensed dispersion-strengthened materials (KDSM) Ni–Al_2O_3, NiCr–Al_2O_3, the optimal concentration of the reinforcing nanophase Al_2O_3 in KDSM Ni–Al_2O_3 and NiCr–Al_2O_3 was determined, which ensures a high level of strength and plasticity.

References

1. E. Khomenko, N. Grechanyuk, V. Zatovskiy, Modern composite materials for switching and welding equipment. Message 1. Powder composite materials. Autom. Weld. **10**, 38–44 2015. (In Russian)
2. B. Movchan, K. Yakovchuk, Electron-beam installations for the deposition of inorganic materials and coatings. Spec. Electrometallurgy **2**, 10–15 (2004)
3. B. Movchan, S. Firsov, Yu. Lugovskoy, in *Structure, Strength and Resistance to Fatigue of Microcrystalline and Microlayer Materials* (Naukova Dumka, Kyiv, 2016)
4. I. Grechanyuk, Scientific and technological principles of creating new composite materials based on nickel, cobalt, titanium and copper and the latest industrial radiation equipment for their production. Abstract of the dissertation of Ph.D. Kyiv, 2021, 42 p.
5. N. Grecyanyuk, G. Baglyuk, P. Kucherenko, A. Melnik, I. Grechanyuk, V. Grechanyuk, Y. Smashnyuk, Laboratory electron-beam multipurpose installation L-2 for producing alloys, composites, coatings, and powders. Powder Metall. Metal Ceram. **56**(1–2), 113–121 (2017)
6. N. Grechanyuk, V. Konoval, V. Grechanyuk, G. Bagliuk, D. Myroniuk, Properties of Cu–Mo materials produced by physical vapor deposition for electrical contacts. Powder Metall. Met. Ceram. **60**(3–4), 183–190 (2021)

Transmission of the Ultrarelativistic Quasi-particles in the Nanostructure with the Step-Like Barrier Based on the Dice Latice

A. M. Korol, N. V. Medvid, and O. Shevchenko

1 Introduction

Nanostructures based on the dice lattice belong to the so-called Dirac materials. They have been intensively studied in recent years, due to the fact that they have interesting physical properties and also because of good prospects for practical application (see [1] and references therein). The state and behavior of charge carriers in these structures are described by equations similar to the Dirac relativistic equation. The transport characteristics of the dice lattice-based nanostructures were studied in a number of recent publications (see e.g. [2] and references therein). In them, it was shown in particular that such phenomena as Klein tunneling, the supertunneling can be observed in the considered structures. At the same time, it is known that an important characteristic of the dispersion relation in these materials is the Fermi velocity [1]. Many physical processes, including transport ones, depend significantly on this value. If we talk about the most known and most studied Dirac material, namely graphene, then as it was shown in the works [3–13], that the transmission coefficient of quasi-electrons in the graphene-based structures is very sensitive to the Fermi velocity. It is obvious that in nanostructures based on the dice lattice, the Fermi velocity also plays an important role. In particular, this concerns the transport characteristics of these objects. Being motivated by the above, in this work, we study the transmission process in one of the structures based on the dice lattice, namely the structure with a step-like barrier. At the same time, it is considered that the barrier has a rectangular shape and a dual nature: it is created, on the one hand, by the

A. M. Korol (✉) · N. V. Medvid · O. Shevchenko
National University for Food Technologies, Volodymyrska Str., 68, Kyiv, Ukraine
e-mail: korolam@ukr.net

A. M. Korol
Laboratory on Quantum Theory in Linkoping, International Society for Independent ReSearch (ISIR), Box 8017, 58080 Linkoping, Sweden

O. Fesenko and L. Yatsenko (eds.), *Nanostructured Surfaces, Nanocomposites and Nanomaterials, and Their Applications*, Springer Proceedings in Physics 296,
https://doi.org/10.1007/978-3-031-42704-6_16

electrostatic potential, and on the other hand, by the difference in the values of the Fermi velocity in the barrier (v_2) and out-of-barrier (v_1) regions of the considered contact. We show that the presence of a flat energy band has a strong influence on the transmission of quasi-electrons.

2 Model and Formulae

The dice lattice can be schematically represented as follows:

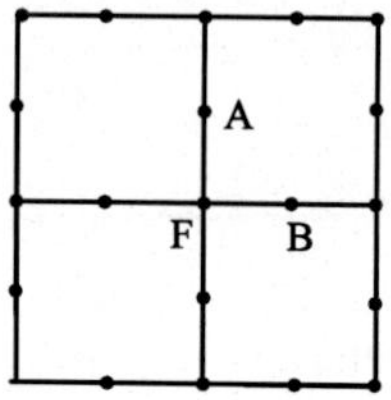

The elementary cell contains three atoms designated as *A*, *B*, *F*. There are three sublattices associated with these atoms. For small energies, the Schrödinger equation for the system considered can be deduced from a nearest neighbor tight-binding model Hamiltonian and reads:

$$(-i\hat{S} \cdot \nabla + \Delta I + U I_0)\Psi = E\Psi, \tag{1}$$

where we accepted the units of measurement in which $\hbar = v_F = 1$, v_F the Fermi velocity (which is associated with the hopping (h) and the lattice (a) constants as $v_F = ha$), $\hat{S}$ the pseudospin-1 operator with the components:

$$S_x = \begin{pmatrix} 0\,1\,0 \\ 1\,0\,0 \\ 0\,0\,0 \end{pmatrix}; S_y = \begin{pmatrix} 0\,0\,0 \\ 0\,0\,1 \\ 0\,1\,0 \end{pmatrix}; \quad S_z = \begin{pmatrix} 0\;0\,-i \\ 0\;0\;\;0 \\ i\;0\;\;0 \end{pmatrix}; \tag{2}$$

The second term on the left-hand side of formula (1) corresponds to the site-energy on the three sublattices, and

$$I = \begin{pmatrix} 1 & 0 & 0 \\ 0 & -1 & 0 \\ 0 & 0 & 1 \end{pmatrix} \tag{3}$$

There are three branches of the dispersion law for the structure considered. For the massless particles ($\Delta = 0$), they are represented by the two cones contiguous in a Dirac point and giving a linear dependence of energy on the quasi-momentum;

the third branch forms a non-dispersive planar band, also passing through the Dirac point. In the case of massive particles, there is an energy gap between the cones of the value of 2Δ, and the mass of particles in the adopted units is equal to Δ. The third term in the left side of the Schrödinger Eq. (1) represents the external potential U (I_0 unit matrix).

The eigenfunctions for Eq. (1) can be represented in the spinor form; for the out-of-barrier region

$$\begin{pmatrix} \psi_A(x) \\ \psi_F(x) \\ \psi_B(x) \end{pmatrix}_I = \begin{pmatrix} \cos\varphi \\ 1 \\ -\sin\varphi \end{pmatrix} e^{-ik_x x} + r \begin{pmatrix} -\cos\varphi \\ 1 \\ \sin\varphi \end{pmatrix} e^{ik_x x} \tag{4}$$

for the barrier region

$$\begin{pmatrix} \psi_A(x) \\ \psi_F(x) \\ \psi_B(x) \end{pmatrix}_{II} = t \begin{pmatrix} \cos\varphi \\ 1 \\ -\sin\varphi \end{pmatrix} e^{-iq_x x} \tag{5}$$

where k_x, k_y are the quasi-momentum components in the out-of-barrier region, q_x—in the barrier region, φ the incidence angle, θ the refraction angle, $\alpha = v_2/v_1$,

$$k_x = \sqrt{E - k_y^2}; \quad q_x = \sqrt{\frac{(E-U)^2}{\alpha^2} - \frac{\Delta^2}{\alpha^2} - k_y^2}; \theta = \mathrm{arctg}\frac{k_y}{q_x}. \tag{6}$$

Using the relevant boundary conditions [2], we can deduce the expression for the transmission rates:

$$T = \frac{B^2 - A^2}{B^2 + C^2} \tag{7}$$

$$\delta = \sqrt{1 + \frac{\Delta^2}{E^2}}; \quad \gamma = \sqrt{1 + \frac{\Delta^2\alpha^2}{(E-U)^2}}; A = s \cdot \cos\varphi - \frac{\delta s' \cos\varphi}{\gamma}$$

$$B = s \cdot \cos\varphi + \frac{\delta s' \cos\varphi}{\gamma}; C = \frac{\Delta \sin\varphi \cdot}{|E-U|} - \frac{\Delta s' \delta \sin\theta}{|E-U|\gamma};$$

and $s = \mathrm{sign}E, \quad s' = \mathrm{sign}(E-U)$.

3 Results and Discussion

Figure 1 shows the dependence of the transmission coefficient T on the angle of the quasi-electron incidence on the potential barrier φ.

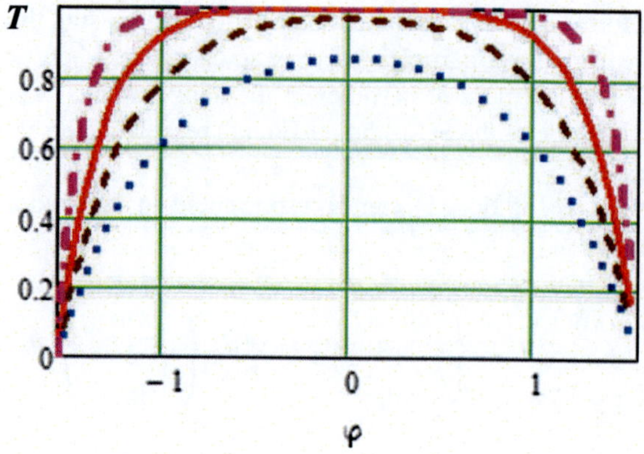

Fig. 1 Dependence of the transmission coefficient T on the incidence angle φ. The parameter values are as follows: $E = 1$, $U = 4$, $\alpha = 0.2$, but different Δ values: $\Delta = 0,\ 1,\ 2$ for the solid, dashed, dotted lines, respectively and there is an additional curve with the values $\alpha = 2.7$, $\Delta = 3$—the dashed-dotted line

It can be seen that the transmission spectra have a pronounced angular dependence, that is, different values of φ correspond to values of T that differ significantly from each other; the largest values of T are formed for the normal incidence of quasi-electrons on the barrier, and the values of T gradually decrease with increasing of φ, so that in the end, T becomes equal to zero at $\varphi = 0.5\pi$. The spectra are symmetrical with respect to the value $\varphi = 0$, i.e. T has the same values for the same values of φ taken with different signs.

It is important that the presence of an energy gap ($\Delta \neq 0$) leads to a noticeable decrease in T, and T decreases in proportion to Δ. The dashed-dotted curve (magenta) shows that for a certain combination of parameter values, in particular, for a certain ratio between the values of Δ and α, the value of T becomes equal to one at the point $\varphi = 0$, that is, in the case of a normal incidence of particles on the barrier. In other words, this means that it is possible to observe the Klein tunneling (Klein paradox) in this structure. At the same time, at arbitrary parameter values, the Klein tunneling is suppressed due to the Δ energy gap. The solid (red) curve refers to the case of the absence of an energy gap—for which the Klein tunneling takes place.

Figure 2, unlike Fig. 1, shows the dependence of $T(\varphi)$ for the same values of $\Delta = 3$, but for different values of $\alpha = 0.2,\ 1,\ 2$ for the solid (red), dotted (blue), dashed (brown) respectively $E = 1$, $U = 4$. Here, first of all, the dash-dotted line (magenta) attracts attention, for which $T = 1$ for all incidence angles. It describes the phenomenon of the supertunneling, which consists in the fact that $T = 1$ regardless of the incidence angle. It follows from the above formulas that the phenomenon of the supertunneling occurs when such a relationship between the parameters of the problem is fulfilled

$$E_{\pm} = \frac{U}{1 \pm \alpha}$$

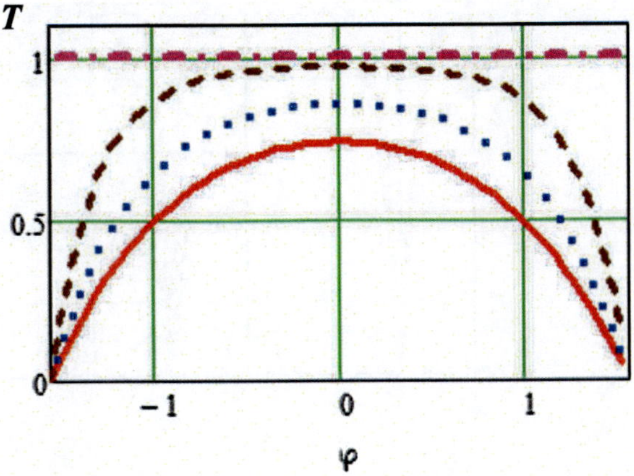

Fig. 2 T versus φ dependence for the following set of the parameters: $E = 1$, $U = 4$, $\Delta = 3$, and different values of $\alpha = 0.2,\ 1,\ 2$ for the solid (red), dotted (blue), dashed (brown), respectively

For the minus sign in this formula, we can obtain the negative values for the supertunneling energies.

In the absence of the velocity barrier (($v_1 = v_2, \alpha = 1$), the energy at which the supertunneling is observed is equal to half the height of the electrostatic barrier ($U/2$), as in [2].

We emphasize that the transmission spectra shown in Figs. 1 and 2 strongly depend on the value of α. And more specifically, the value of the transmission coefficient T decreases with increasing in α.

The similarity of the dotted and dashed curves in Figs. 1 and 2 draws attention. It means that it is convenient to flexibly adjust the analyzed spectra with the help of α and Δ parameters, in particular, we can see that a change in one of these parameters can be compensated by changing another one of them (in a wide range of incidence angle values).

It is interesting that for certain values of the parameters, there may be a wide plateau of incidence angles for which T is close to unity. This phenomenon can be called the limited or reduced supertunneling.

Figures 3 and 4 show the dependence of the transmission coefficient on the energy of quasi-electrons; Fig. 3—for the same α and different Δ, and Fig. 4—for the same Δ and different α. Among all the energies in Figs. 3 and 4, there are those that correspond to the phenomenon of the supertunneling. In particular, for the solid line, the values of these energies are $E_+ = 2$ and $E_- = 6$.

Figures 3 and 4 correspond to the case of a normal incidence, $\varphi = 0$. The effect of α and Δ parameters on the transmission spectra is similar over a wide range of energy values. For most values of E, an increase in α or Δ leads to a decrease in T.

However, in the area of small energies, we have that the influence of α on T is practically the same for different α, while the Δ on T depends on the value of Δ.

Figures 5 and 6 demonstrate how strongly the $T(E)$ spectra depend on the angle of incidence φ; $\varphi = 0.5$ for these figures. In particular, by comparing them to the previous figures (for which $\varphi = 0$), one can see how rapidly the transmission gap expands with the growth of φ.

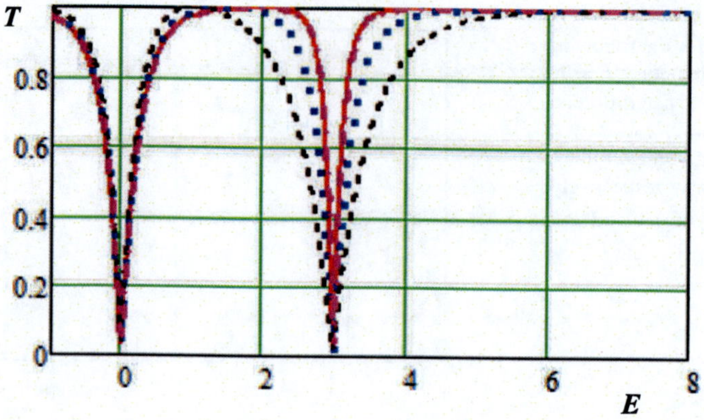

Fig. 3 Dependence of the transmission coefficient T on the quasi-electron energy E. The parameter values are as follows: $\varphi = 0$, $U = 3$, $\Delta = 1$, but different α values: $\alpha = 0.5, 1, 2$ for the solid, dotted, dashed lines, respectively

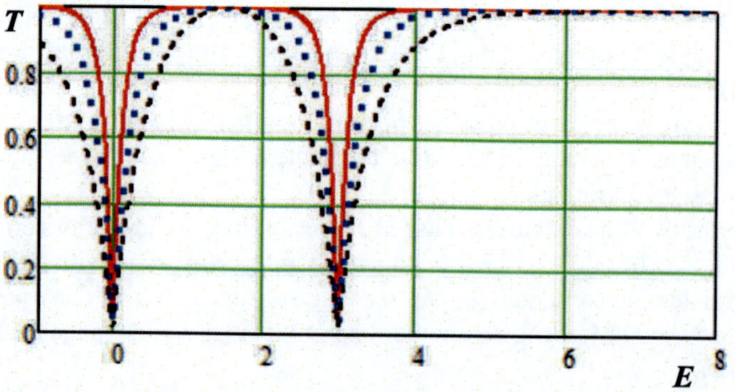

Fig. 4 Dependence of the transmission coefficient T on the quasi-electron energy E. The parameter values are as follows: $\varphi = 0$, $U = 3$, $\alpha = 1$ but different Δ values: $\Delta = 0.5, 1, 2$ for the solid, dotted, dashed lines, respectively

This statement is also illustrated with the help of Fig. 7, which shows the dependences of $T(E)$ for different values of φ and Δ. For Fig. 8, which presents the dependence of T on α, we took parameter values that are significantly different from the values in other figures to emphasize that the results analyzed in this chapter are valid in a wide range of parameter values of the considered structure.

Figures 1, 2, 3, 4, 5, 6, 7 and 8 presented in this chapter show how diverse the range of transmission spectra can be, and the spectra are very sensitive to even small changes in the values of the parameters of the considered structure. This circumstance makes it possible to flexibly adjust the spectra with the help of the parameters in a wide range of these parameters values.

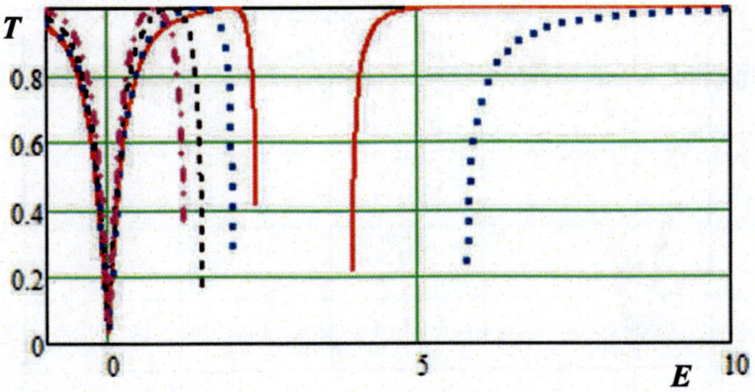

Fig. 5 Dependence of the transmission coefficient T on the quasi-electron energy E. The parameter values are as follows: $\varphi = 0.5$, $U = 3$, $\alpha = 1$, but different α values: $\alpha = 0.5, 1, 2, 3$ for the solid, dotted, dashed, dashed-dotted lines, respectively

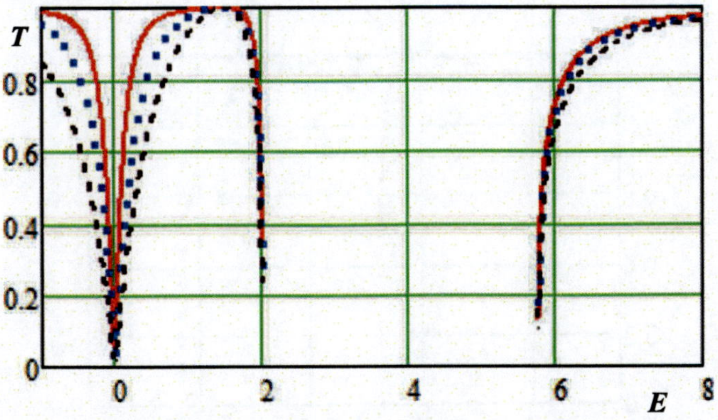

Fig. 6 Dependence of the transmission coefficient T on the quasi-electron energy E. The parameter values are as follows: $\varphi = 0.5$, $U = 3$, $\alpha = 1$ but different Δ values: $\Delta = 0.5, 1, 2$ for the solid, dotted, dashed lines, respectively

4 Conclusions

The ballistic motion of the ultrarelativistic quasi-electrons in a structure based on the dice lattice is considered and studied. It is assumed that the step-like barrier has a rectangular shape and is created by the external electrostatic potential. The transmission coefficient T is calculated in the continuous model by matching of the eigenfunctions at the interface of the barrier and out-of-barrier areas. The eigenfunctions, in turn, are found as a solution of the Dirac-type equation with the Hamiltonian, which takes into account the presence of a flat energy band. The difference in the values of the Fermi velocity in the barrier and out-of-barrier regions (v_1 and v_2 respectively) is taken into account. It is also believed that the flat band is located halfway between the valence

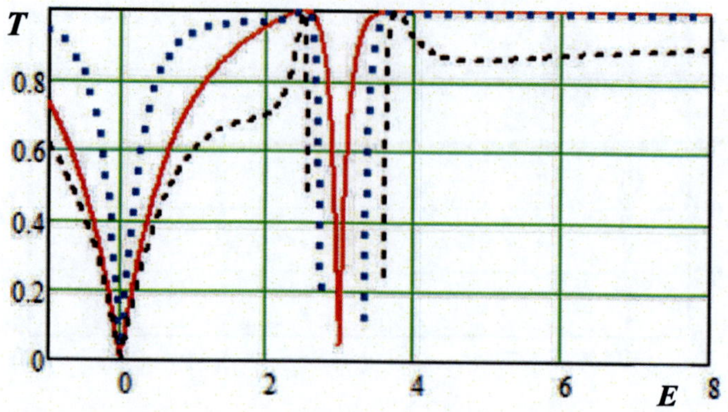

Fig. 7 Dependence of the transmission coefficient T on the quasi-electron energy E for different values of the parameters involved. The parameter values are as follows: $\alpha = 0.2$, $U = 3$ for all curves; $\varphi = 0$, $\Delta = 3$ for the solid line, $\varphi = 0.5$, $\Delta = 1$ for the dotted line, $\varphi = 1$, $\Delta = 2$ for the dashed line

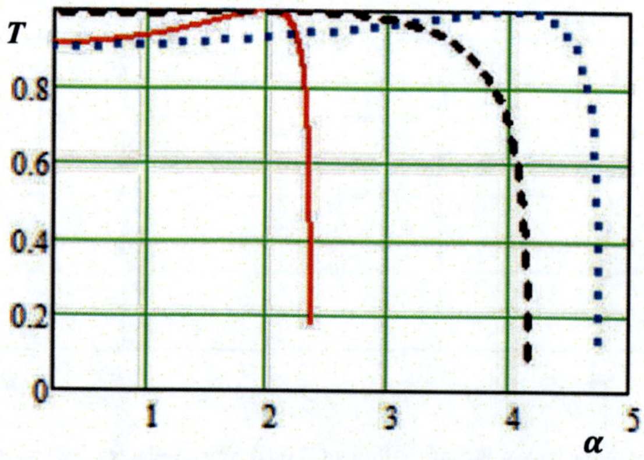

Fig. 8 Dependence of the transmission coefficient T on the parameter α. The other parameter values are as follows: $U = 30$ for all curves, $E = 10$, $\varphi = 1$, $\Delta = 1$ for the solid line, $E = -10$, $\varphi = 1$, $\Delta = 2$ for the dotted line, $\varphi = 0.5$, $\Delta = 3$ for the dashed line

band and the conduction band and is separated from each of them by the energy gap (Δ The detailed analysis of the tunneling spectra (dependences of the transmission coefficient T on the quasi-electron energy, incidence angle, Fermi velocity, electrostatic potential, energy gap Δ) is provided. Depending on the values of the parameters of the considered structure, it is possible to implement a large number of different spectra. In particular, it is shown the following. Tunneling spectra have a pronounced angular dependence, i.e. the transmission coefficient T varies greatly when the angle of incidence of quasi-electrons on the structure changes. Spectra are highly sensitive to changes in the Fermi velocity (in the parameter $\alpha = v_2/v_1$). At certain combinations of the parameter values, the phenomenon of the supertunneling is observed, which is that the transmission coefficient is equal to unity, regardless of the angle of incidence of quasi-electrons on the structure. The Klein tunneling can also be manifested in this structure. For many parameter values, the Klein tunneling can be suppressed. For certain values of the parameters, there may be a wide plateau of incidence angles for which T is close to unity. This phenomenon can be called the limited supertunneling.

References

1. T.O. Wehling, A.M. Black-Schaffer, A.V. Balatsky, Adv. In Phys. **63**(1), 1–76 (2014). https://doi.org/10.1080/00018732.2014.927109
2. D.F. Urban, D. Bercioux, M. Wimmer, W. Hausler, Phys. Rev. B **84**, 115136 (2011). https://doi.org/10.1103/PhysRevB.84.115136
3. L. Liu, Y. Li, J.J. Liu, Phys. Lett. A **376**(45), 3342 (2012). https://doi.org/10.1016/j.physleta.2012.08.047
4. Y. Wang, Y. Liu, B. Wang, Physica E **53**, 186 (2013). https://doi.org/10.1016/j.physe.2013.05.010
5. L.-F. Sun, Fang C., Liang T.-X. Chin. Phys. Lett. **30**(4), 047201 (2013). https://doi.org/10.1088/0256-307X/30/4/047201
6. A. Raoux, M. Polini, R. Asgari, A.R. Hamilton, R. Fasio, A.H. MacDonald, Phys. Rev. B **81**, 073407 (2010). https://doi.org/10.1103/PhysRevB.81.073407
7. A. Concha, Z. Tešanović, Phys. Rev. B **82**, 033413 (2010). https://doi.org/10.1103/PhysRevB.82.033413
8. J.H. Yuan, J.J. Zhang, Q.J. Zeng, J.P. Zhang, Z. Cheng, Physica B Phys. Condens. Matter **406**(22), 4214 (2011). https://doi.org/10.1016/j.physb.2011.08.010
9. P.M. Krstajic, P. Vasilopoulos, J. Phys. Condens. Matter **23**(13), 135302 (2011). https://doi.org/10.1088/0953-8984/23/13/135302
10. A.M. Korol, N.V. Medvid, A.I. Sokolenko, Physica Status Solidi (B) Basic Res. **255**(9), 1800046 (2018). https://doi.org/10.1002/pssb.201800046
11. A.M. Korol, N.V. Medvid, Low Temp. Phys. **45**(10), 1117 (2019). https://doi.org/10.1063/1.5125914
12. A.M. Korol, Low Temp. Phys. **45**(5), 493 (2019). https://doi.org/10.1063/1.5097357
13. A.M. Korol, Fiz Nizk Temp. **47**(2)**,** 147 (2021) [Low Temp. Phys. **47**, 000000 (2021). https://doi.org/10.1063/10.0003174

Study of the Effects of MgO Additive and Sintering Temperature on Mechanical Behavior of Fine-Grained ZrO_2–MgO Ceramics

V. V. Kulyk, Z. A. Duriagina, B. D. Vasyliv, V. I. Vavrukh, P. Ya. Lyutyy, T. M. Kovbasiuk, V. V. Vira, and V. A. Vynar

Abstract Zirconia ceramics partially stabilized with various amounts of MgO have been studied. The aim of this work was to investigate changes in mechanical behavior of the ceramics depending on the sintering temperature and percentage of MgO additive. Conventional sintering of series of ceramic specimens in a temperature range of 1510–1850 °C with and without annealing was performed. A set of mechanical tests were carried out: (i) strength test of beam specimens under three-point bending; (ii) Vickers microhardness test; (iii) fracture toughness test by Vickers indentation (VI method); (iv) single-edge notch beam test under three-point bending (SENB method). Among the materials studied, ZrO_2–2.5 mol% MgO ceramics sintered at 1850 °C showed the best crack growth resistance behavior for both VI and SENB methods. In contrast to other ceramic variants containing about 12–15 wt% c-ZrO_2 and 85–88 wt% m-ZrO_2, this material contained 58 wt% c-ZrO_2 and 42 wt% m-ZrO_2. Such a comparatively high percentage of c-ZrO_2 in this material variant caused a distinct difference in its mechanical behavior. Based on the obtained results of mechanical tests and the analysis of fracture surface morphology, it was concluded that

V. V. Kulyk (✉) · Z. A. Duriagina · V. I. Vavrukh · P. Ya. Lyutyy · T. M. Kovbasiuk
Department of Materials Science and Engineering, Lviv Polytechnic National University, 12 S. Bandera Street, Lviv 79013, Ukraine
e-mail: kulykvolodymyrvolodymyrovych@gmail.com

Z. A. Duriagina
Department of Materials Engineering, The John Paul II Catholic University of Lublin, 14 Racławickie Al, 20-950 Lublin, Poland

B. D. Vasyliv
Department of Hydrogen Technologies and Alternative Energy Materials, Karpenko Physico-Mechanical Institute of the NAS of Ukraine, 5 Naukova Street, Lviv 79060, Ukraine

V. V. Vira
Department of Strength of Materials and Structural Mechanics, Lviv Polytechnic National University, 12 S. Bandera Street, Lviv 79013, Ukraine

V. A. Vynar
Department of Corrosion and Corrosion Protection, Karpenko Physico-Mechanical Institute of the NAS of Ukraine, 5 Naukova Street, Lviv 79060, Ukraine

O. Fesenko and L. Yatsenko (eds.), *Nanostructured Surfaces, Nanocomposites and Nanomaterials, and Their Applications*, Springer Proceedings in Physics 296,
https://doi.org/10.1007/978-3-031-42704-6_17

the fracture toughness of optimally fabricated ZrO_2–MgO ceramics is attributed to particle-crack interactions, which result in a transformation zone around the crack tip. An increase in crack growth resistance of this material variant was accompanied by a slight increase in its strength and microhardness.

1 Introduction

Zirconia ceramics (ZrO_2) partially stabilized with oxides MgO [1], Y_2O_3 [2–5], CaO [6], or CeO_2 [7–9] are being widely used for producing various structural and functional components in many industry branches [10–26]. By adding such oxides, stabilization of the metastable tetragonal ZrO_2 phase at room temperature is achieved that allows reaching improved mechanical properties. Besides, the transformation of the tetragonal (t) phase into the monoclinic (m) one known as transformation toughening [27–31] is implemented in these ceramics. Therefore, the tetragonal ZrO_2 phase should be ensured in microstructure of the ceramics to increase their fracture toughness and strength. Zirconia ceramics can additionally be modified by other oxides [27, 32–42] to attain specific physical properties.

The authors of the works [43–47] studied the influence of sintering parameters (temperature, holding time, etc.) on the microstructure, morphology of grains and their size as well as mechanical properties of zirconia ceramics. The grain growth in zirconia-based ceramics was studied in detail in [43]. For 8 mol% Y_2O_3–ZrO_2 ceramic sintered at 1700 °C for 20 h, a maximum grain size was 20 μm, whereas for 4 mol% Y_2O_3–ZrO_2 ceramic sintered at 1800 °C for 100 h, it was more than 11 μm. In the work [45] doped zirconia annealed at 1500 °C was studied. The evolution of grain growth was exhibited in terms of the maximum grain size. It was shown that for 12Ti–ZrO_2 ceramic such conditions are reached after sintering at 1400 °C for 70 h. In several works [46], zirconia ceramics sintered at 1500 °C for a few hours was studied. For such a condition, formation of microstructure with the average grain size of about 1 μm was revealed. In the work [47], it was found that the average grain size for zirconia sintered at 1600 °C for 2 h is 0.35 μm.

The correlations between the porosity, grain size, and mechanical properties of zirconia depending on the sintering temperature have been shown in the works [48–50]. A single m-ZrO_2 phase microstructure had been revealed in zirconia specimens without Y_2O_3 additive already after pressing [49]. However, a single t-ZrO_2 phase microstructure was formed in zirconia-based ceramics containing 1.5 mol% Y_2O_3 after sintering at 1100 °C. For zirconia containing a comparatively large Y_2O_3 amount (8 mol%), a single c-ZrO_2 phase microstructure was revealed after pressing. It was found that the maximum pore diameter in all the compositions is about 12.5 nm after pressing. When the sintering temperature increased to 600 °C, an increase in the pore diameter was revealed in the material without Y_2O_3 additive. With a further increase in sintering temperature to 1100 °C, the lowering of porosity along with intense grain growth was observed. Materials with 1.5 and 8 mol% of Y_2O_3 additives showed intense pore growth with a gradual increase in the sintering temperature to 600 °C. With an increase in the temperature up to 1100 °C, densification of the ceramics without a significant increase in the grain size occurred.

Spark plasma sintering is reported as one of the promising techniques for preparing yttria-stabilized zirconia [51, 52], ceria-stabilized zirconia [53–55], and other ceramics [56–60].

The authors of the work [61] investigated the influence of a zirconia additive on the fracture toughness and strength of Al_2O_3–ZrO_2 ceramics. Both the monoclinic and tetragonal ZrO_2 phases as well as the α-Al_2O_3 phase were revealed in these materials. The fine-grained Al_2O_3–ZrO_2 microstructure was formed due to adding 10–20% ZrO_2. In such material, improved mechanical strength and fracture toughness were reached. It was shown [62] that both the fracture toughness and flexural strength of ceramic materials should be determined while optimizing their sintering mode.

YSZ ceramics with 3–5 mol% Y_2O_3 additives were studied in the works [63, 64]. The correlations between the chemical and phase compositions of the ceramics as well as between their mechanical properties were revealed. The highest fracture toughness was found for 5YSZ ceramic sintered at 1450 °C for 2 h. This was revealed to relate to a high tetragonal ZrO_2 phase fraction [63]. Similar correlations between the sintering mode, phase composition, and mechanical properties of YSZ ceramics with 6–8 mol% Y_2O_3 additives were obtained [64]. The highest fracture toughness was found for 7YSZ ceramic sintered at 1600 °C for 2 h, similarly to the results published in [65, 66].

In this work, the effects of MgO percentage and sintering temperature on the strength, microhardness, and fracture toughness of fine-grained ZrO_2–MgO ceramics were studied in relation to their microstructure and fracture micromechanisms.

2 Materials and Methods

In this work, fine-grained ZrO_2–MgO ceramics obtained in various sintering modes have been studied.

Five series of bar specimens of zirconia partially stabilized with MgO (MgO–PSZ) of size 2.7 × 2.7 × 40 mm^3 were produced by conventional sintering in argon. The sintering modes are given in Table 1. The sintering temperature was in the range of 1510–1850 °C. A grinding/polishing machine was used for preparing specimen surface to avoid phase transformations and provide a high-quality surface.

Microhardness of the ceramics was measured on a microhardness tester NOVOTEST TC-MKB1. We used the following set of the indentation load: 0.25, 0.49, 0.98, 1.96, 2.94, 4.91, and 9.81 N. At least 10 indentations under each load were made according to the relevant standards [67, 68], and resulting geometry of imprints was determined.

Table 1 Marking of the investigated ceramics, their chemical composition, and sintering mode

Series	System and chemical composition, mol%	Sintering mode		Annealing mode	
		Temperature, °C	Time, h	Temperature, °C	Time, h
1	$97.5ZrO_2$–2.5MgO	1850	2	–	–
2	$97.5ZrO_2$–2.5MgO	1850	2	1100	1
3	$97.5ZrO_2$–2.5MgO	1850	2	1100	2
4	$97ZrO_2$–3MgO	1510	2	–	–
5	$97ZrO_2$–3MgO	1590	2	–	–

There are plenty of formulas for calculating the microhardness of brittle materials [69–78] developed by various scientist teams around the world.

In this work, we calculated Vickers microhardness {in GPa} by the formula [68]:

$$H = 0.0018544\left(\frac{P}{d^2}\right) \tag{1}$$

where P is the applied load (N), and d is the average length of the imprint diagonals (mm).

An optical microscope Neophot-21 was utilized for measuring the lengths of the diagonals of imprints and cracks.

To evaluate resistance of the ceramics to brittle fracture, the critical stress intensity factor (SIF), K_{Ic} can be determined as a measure of their fracture toughness [59, 63]. A Vickers indentation method of evaluating the fracture toughness of brittle materials [79–81] is used most widely. In the various formulas developed for the K_{Ic} value determination, physical and mechanical parameters and experimental coefficients are used. In the works [82–84], it was found for the ZrO_2–Y_2O_3 ceramics that the formula given in [80] provides the best fitting to the values of fracture toughness obtained by conventional methods of fracture mechanics. So the formula proposed in [80] was used in our work:

$$K_{Ic} = 0.016\left(\frac{E}{H}\right)^{1/2}\left(\frac{P}{c^{3/2}}\right) \tag{2}$$

where E is Young's modulus (GPa), H is microhardness (GPa), P is the indentation load (N), and c is the radial crack length (m).

Flexural strength of the ceramics was estimated by performing the three-point bend test of ceramic specimens at 20 °C in air. The fracture stress was calculated using specimen geometry parameters and a value of fracture load [85, 86].

Along with the Vickers indentation (VI) method, another method of determining the fracture toughness of the ceramics, namely a single-edge notch beam (SENB) method [87–89] was used. We utilized a three-point bend scheme for testing the

SENB specimens in air at 20 °C with determining corresponding values of fracture load [83, 84, 90–98]. The critical SIF K_{Ic} was calculated using formulas given in [87–89].

The average K_{Ic} value for five measurements was calculated for each series.

The microstructure study with analyzing fracture images was performed with a scanning electron microscope (SEM) Carl Zeiss EVO-40XVP. An INCA Energy 350 system was utilized for an energy-dispersive X-ray (EDX) microanalysis. X-ray diffraction (XRD) study of the materials was carried out with a DRON-4.07 M diffractometer. The refinement of data followed by the calculation of fractions of zirconia phases was made using the WinCSD program.

3 Results and Discussion

The XRD analysis of the studied ceramics was performed to obtain quantitative data of the phase fractions.

Based on the XRD patterns, we revealed peaks of the m-ZrO_2 and c-ZrO_2 phases in all the studied ceramics (Fig. 1). However, such a two-phase structure of the materials does not mean the same phase balance. No significant difference in the peak heights for the various ceramic series was revealed except for the peaks corresponding to smaller angles. For the last case, the peaks of the c-ZrO_2 phase are higher for series 1 compared to those for all other series. And vice versa, the peaks of the m-ZrO_2 phase are lower for series 1.

Based on the obtained XRD patterns, the phase balances in the studied ceramics were estimated (Fig. 2). For series 1 sintered at 1850 °C without annealing, cubic and monoclinic phase fractions were 58.1% and 41.9%, respectively, whereas for series 2 sintered at 1850 °C and annealed at 1100 °C for 1 h, the corresponding fractions were 12.7 and 87.3%. Similar results were found for series 3 sintered at 1850 °C and annealed at 1100 °C for 2 h, for which the corresponding fractions were 13.8% and 86.2%. For series sintered at lower temperatures (1510 °C and 1590 °C) without annealing, determined fractions were close to those for the last one. Namely, for series 4 sintered at 1510 °C, cubic and monoclinic phase fractions were 13.6% and 86.4%, respectively, whereas for series 5 sintered at 1590 °C, the corresponding fractions were 14.1% and 85.9%. Therefore, no discernible correlation between the sintering temperature for the material series and the monoclinic and cubic ZrO_2 phase fractions was found.

Strength and microhardness of zirconia in relation to its microstructure have been thoroughly investigated in a number of works [82, 90, 95, 99–103]. In particular, the results for ZrO_2–8 mol% Y_2O_3 ceramics presented in the works [82, 84] showed a decrease in microhardness of the material while increasing the indentation load in the range of 0.49–9.81 N. Such lowering of microhardness of ZrO_2–8 mol% Y_2O_3 ceramics known widely as the indentation size effect [104] allows reaching

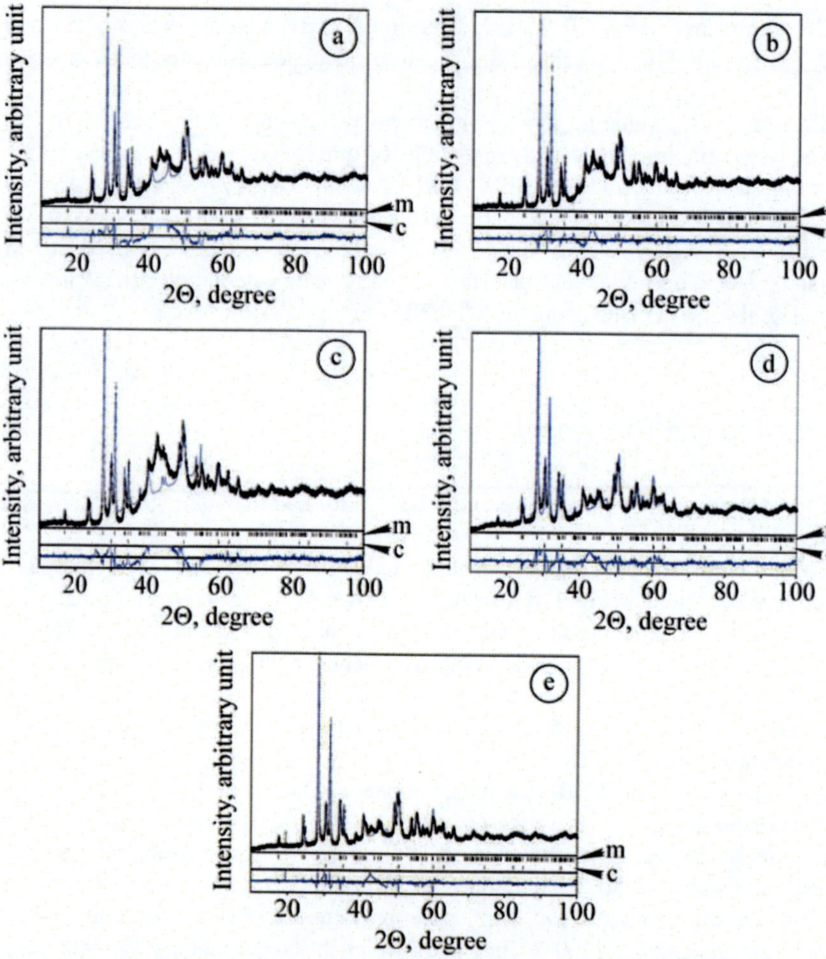

Fig. 1 XRD patterns of the studied ceramics of **a** series 1, **b** 2, **c** 3, **d** 4, and **e** 5 (Table 1). Notation: *m* is monoclinic and c is cubic ZrO_2

the threshold, namely yielding on the plateau at the loads in a range of 4.91–9.81 N. In this loading range, we can obtain invariant values of both the microhardness and fracture toughness.

To validate invariant data for the studied series, we applied a set of the indentation loads (0.25, 0.49, 0.98, 1.96, 2.94, 4.91, and 9.81 N) to the corresponding samples and calculated microhardness values (Fig. 3a).

The load-dependent changes in microhardness of the studied series (Fig. 3a) do not display a well-known indentation size effect [104, 105] typical of yttria-stabilized zirconia [63]. In the range 0.25–2.94 N, we can observe more or less steep increase in microhardness with an increase in the load for all series (Fig. 3a). In the range of 2.94–9.81 N, inverse dependences exhibiting a slight indentation size effect were

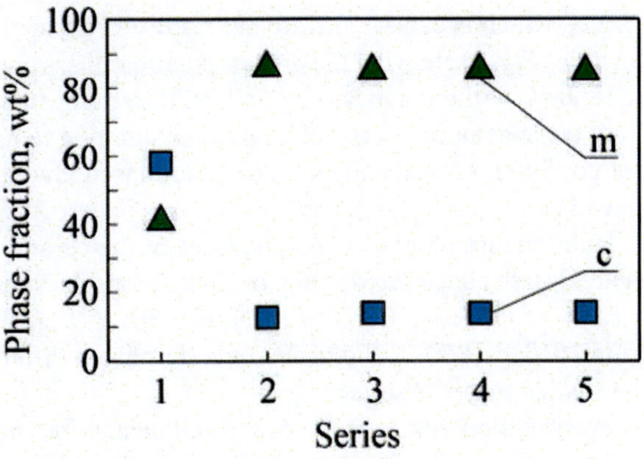

Fig. 2 Phase balance in the studied ceramics (Table 1). Notation: *m* is monoclinic and *c* is cubic ZrO_2

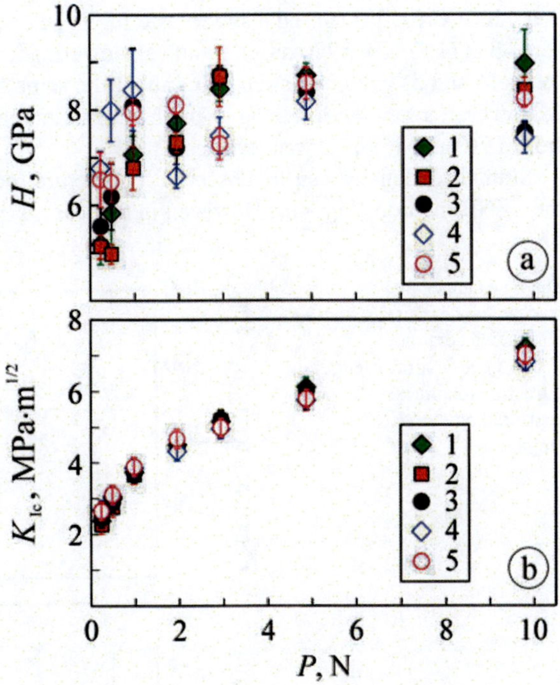

Fig. 3 Changes in mechanical characteristics of the studied ceramics depending on the indentation load: **a** Vickers microhardness; **b** fracture toughness measured by the Vickers method. The symbol marking corresponds to the series marking given in Table 1

obtained for series 2, 3, and 4, whereas the average values of microhardness for series 1 and 5 yield on the plateau in the range of 4.91–9.81 N. It can be assumed that for indentation loads of 4.91–9.81 N invariant values of microhardness can be obtained for all the studied materials. In the case of applying the maximal load (9.81 N), a significant difference in microhardness values for various series was found (Fig. 3a). Namely, they showed an increase in the following series orders: 4–3–5–2–1. According to literature data, the microhardness of m-ZrO_2 is lower than c-ZrO_2

[3, 31], which is consistent with our results for series 1 with the highest cubic phase fraction (Fig. 2) showing the highest microhardness under a load of 9.81 N. Sintering at 1850 °C without annealing provided the strong stabilization of the t-ZrO_2 phase. Both the annealing at 1100 °C and lower sintering temperatures (1510 and 1590 °C) not providing such stabilization (Fig. 2) caused lowering in ceramic microhardness (Fig. 3a).

The fracture toughness dependences of the ceramics by the Vickers indentation method were constructed using for calculations the values of the material microhardness under the load of 0.25–9.81 N (Fig. 3b). A change in the sintering temperature from 1510 to 1850 °C did not cause a significant difference in fracture toughness of the studied material series.

We revealed for series 1–5 a significant difference in strength related to the sintering mode (Fig. 4a). Namely, for series 1 sintered at 1850 °C without annealing, a comparatively high level of strength was achieved, whereas the annealing of series 2 at 1100 °C for 1 h caused a slight lowering in strength (by about 5%) and this of series 3 at 1100 °C for 2 h caused significant lowering in strength (by 50%). Two other series (4 and 5) exhibited strength of about the same level as those of series 1 and 2. However, a trend to lowering strength with an increase in a sintering temperature from 1510 and 1590 °C was revealed.

Similar tendencies can be observed for fracture toughness values measured by the SENB method (Fig. 4b). However, in this case, significant lowering in fracture

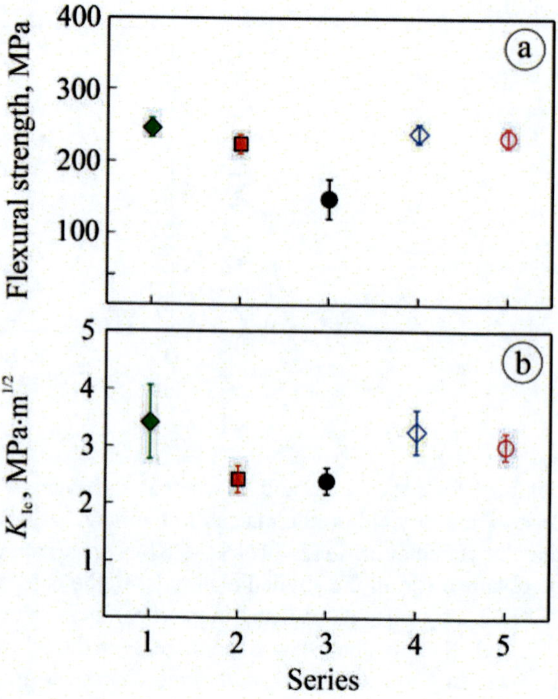

Fig. 4 Changes in mechanical characteristics of the studied ceramics (Table 1): **a** flexural strength; **b** fracture toughness measured by the SENB method

toughness was detected for both series 2 and 3 (by 25% and 26%, respectively). As compared to the trend in strength, a more significant lowering in fracture toughness was revealed for series 5 in relation to series 4 (by about 9%).

Therefore, the ceramic sintered at 1850 °C without annealing exhibited both the highest strength and fracture toughness measured by the SENB method (Fig. 4). It is suggested that a sintering temperature of 1850 °C provides formation of a two-phase crystal structure (c-ZrO_2 and m-ZrO_2) with the domination of the cubic zirconia phase (Fig. 5a). The small grains of the c-ZrO_2 phase (about 1–2 μm in size) are connected to each other without visible boundaries confirming thus high bond strength between cubic zirconia grains (Fig. 6a). Such crystal structure is optimal for the selected chemical composition and allows reaching comparatively high fracture toughness due to the implementation of the mixed fracture micromechanism (Figs. 5b and 6b) [106]. This mechanism comprises both the fracture through the microagglomerates of c-ZrO_2 grains and cleavage fracture along the appropriate planes alternating with bridging micromechanism in large m-ZrO_2 agglomerates.

For the same sintering temperature followed by annealing for 1 h and 2 h, corresponding variants of material exhibited microstructure enriched with agglomerates of the m-ZrO_2 phase (series 2 and 3 in Fig. 5c, e, respectively). These ceramics contain large m-ZrO_2 agglomerates surrounded with a comparatively small amount of the c-ZrO_2 phase (Fig. 6c, e, respectively). The mixed fracture micromechanism is implemented here (Figs. 5d, f and 6d, f) which comprises intergranular fracture along the boundaries of small c-ZrO_2 grains as well as large cleavage areas corresponding to the m-ZrO_2 phase. Such fracture patterns explain lower fracture toughness for series 2 and 3 compared to that for series 1 (Fig. 4b).

Quite different microstructure comprising fine m-ZrO_2 and c-ZrO_2 grains was revealed in series 4 and 5 (Fig. 5g, i, respectively). A detailed analysis showed that both the materials contain fine c-ZrO_2 grains (of about 2 μm in size) uniformly distributed among grains of the m-ZrO_2 phase (Fig. 6g, i, respectively). For both series, intergranular fracture micromechanism is observed (Figs. 5h, j and 6h, j). However, in these cases, an advancing crack grows along the boundaries of the c-ZrO_2 phase grains and occasionally along appropriate cleavage planes of the m-ZrO_2 phase grains. Therefore, a comparatively high fracture toughness for series 4 and 5 in relation to that for series 2 and 3 was found (Fig. 4b).

In summary, with the sintering temperature decreasing, we revealed ambiguous changes in microstructure, phase composition, and mechanical properties (microhardness, flexural strength, fracture toughness) for ZrO_2–MgO ceramics. A slightly higher percentage of MgO additive in the ceramic series 4 sintered at a comparatively low temperature (1510 °C) allowed reaching almost the same levels of strength and fracture toughness as it was reached in this ceramic with lower percentage of MgO additive sintered at much higher temperature (1850 °C). This is mainly due to the implementation of intergranular fracture micromechanism in formed fine two-phase microstructure of zirconia. This micromechanism comprises fracture along the boundaries of the c-ZrO_2 phase grains as well as along appropriate cleavage planes of the m-ZrO_2 phase grains.

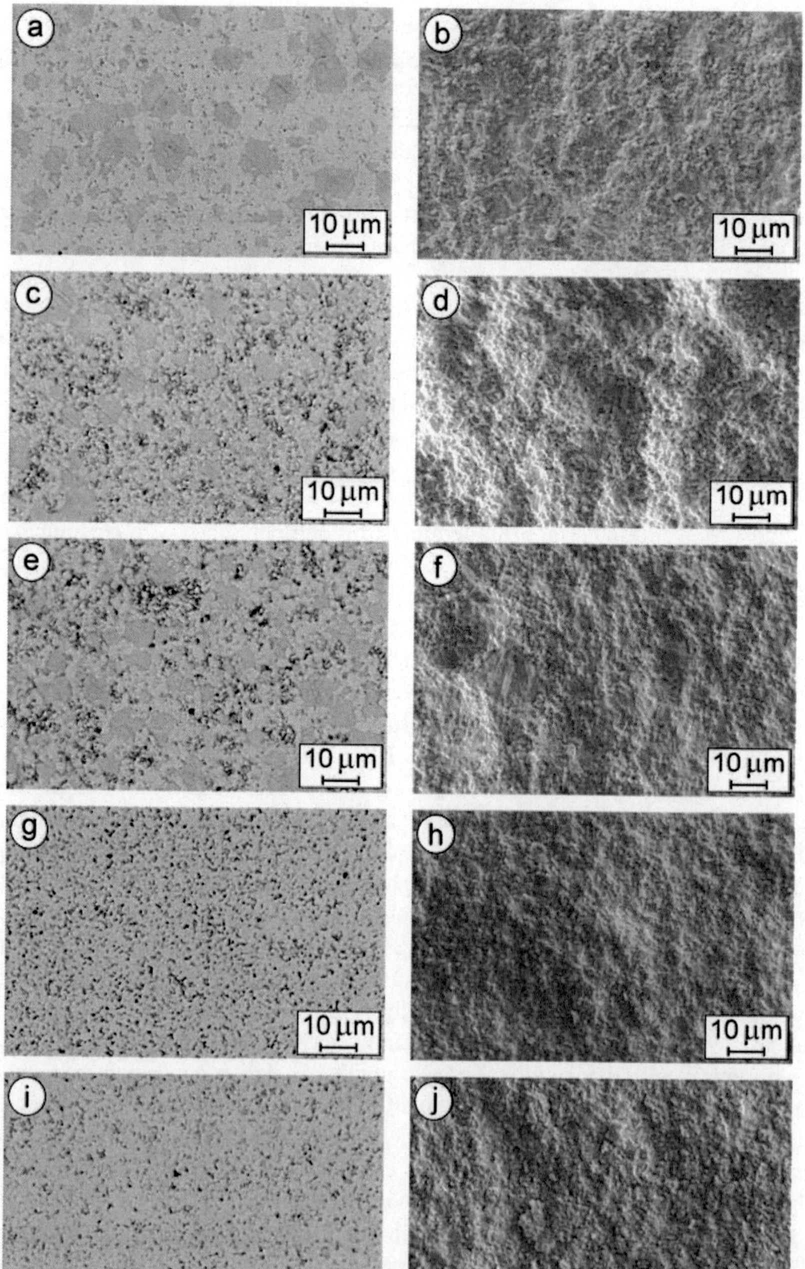

Fig. 5 SEM (**a, c, e, g, i**) microstructures (BSD images at low magnifications) and (**b, d, f, h, j**) fractography (SE images at low magnifications) of (**a, b**) specimen series 1, (**c, d**) 2, (**e, f**) 3, (**g, h**) 4, and (**i, j**) 5 (Table 1)

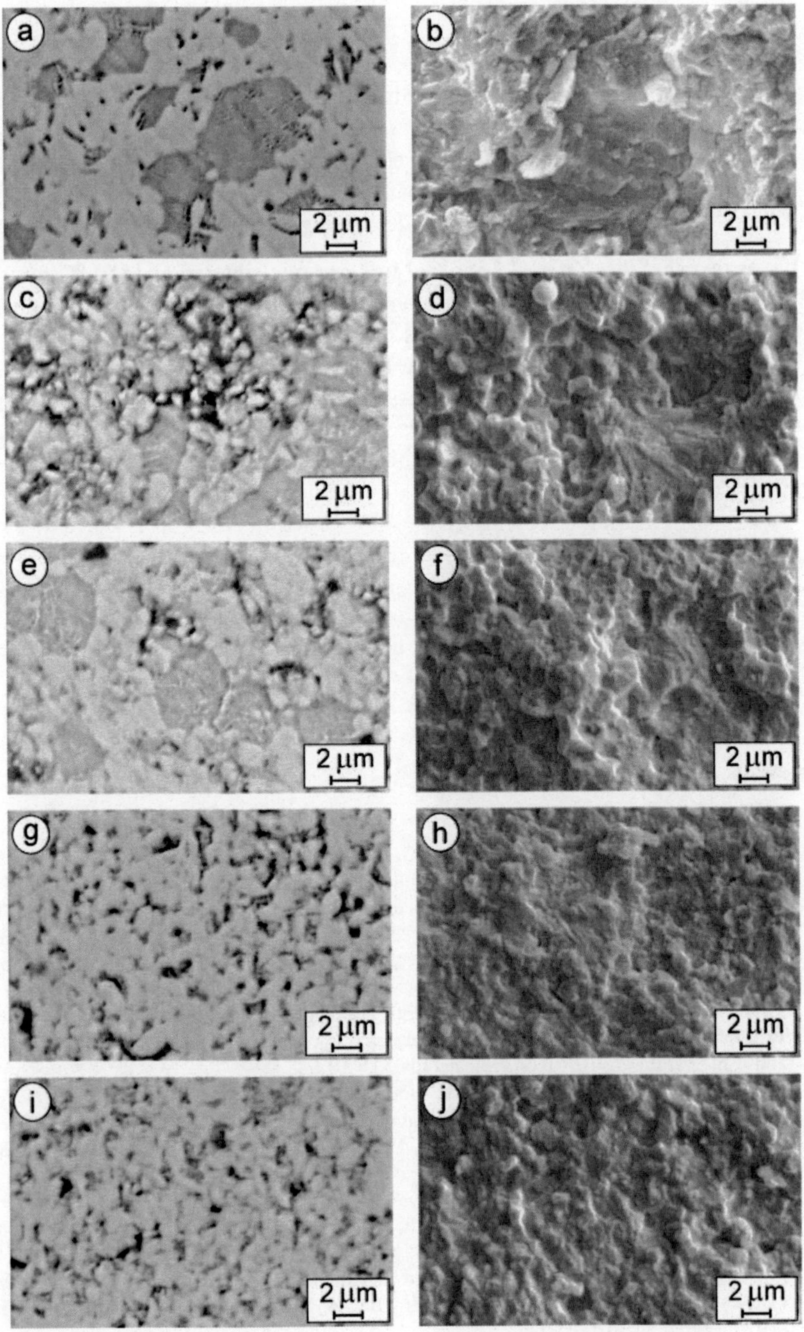

Fig. 6 SEM (**a, c, e, g, i**) microstructures (BSD images at high magnifications) and (**b, d, f, h, j**) fractography (SE images at high magnifications) of (**a, b**) specimen series 1, (**c, d**) 2, (**e, f**) 3, (**g, h**) 4, and (**i, j**) 5 (Table 1)

4 Conclusions

In this study, the influence of MgO additive and sintering temperature on the microstructure, microhardness, strength, and fracture toughness of ZrO_2–MgO ceramics has been estimated. Ambiguous changes in microstructure, phase composition, and mechanical properties (microhardness, flexural strength, fracture toughness) were found for these ceramics with decreasing the sintering temperature:

1. It was revealed that ZrO_2–2.5 mol% MgO ceramic sintered at 1850 °C for 2 h without annealing has the highest microhardness, strength, and fracture toughness as compared to other studied materials.
2. ZrO_2–3 mol% MgO ceramic sintered at 1510 °C for 2 h without annealing exhibited strength and fracture toughness close to that sintered at 1850 °C. This is due to the implementation of intergranular fracture micromechanism comprising fracture along the boundaries of the c-ZrO_2 phase grains and cleavage of grains of the m-ZrO_2 phase.

Acknowledgements The authors are thankful to the staff of the Scientific Equipment Collective Use Center "Laboratory of Advanced Technologies, Creation and Physicochemical Analysis of a New Substances and Functional Materials" at Lviv Polytechnic National University (https://lpnu.ua/ckkno) for their kind help in performing X-ray diffraction studies.

References

1. G. Chen, Q. Li, Y. Ling, H. Zheng, J. Chen, Q. Jiang, K. Li, J. Peng, M. Omran, L. Gao, Phase stability and microstructure morphology of microwave-sintered magnesia-partially stabilised zirconia. Ceram. Int. **47**, 4076–4082 (2021). https://doi.org/10.1016/j.ceramint.2020.09.281
2. A.R. Studart, F. Filser, P. Kocher et al., Fatigue of zirconia under cycling loading in water and its implications for the design of dental bridges. Dent. Mater. **23**, 106–114 (2007). https://doi.org/10.1016/j.dental.2005.12.008
3. J. Chevalier, S. Deville, E. Munch et al., Critical effect of cubic phase on aging in 3 mol% yttria-stabilized zirconia ceramics for hip replacement prosthesis. Biomaterials **25**, 5539–5545 (2004). https://doi.org/10.1016/j.biomaterials.2004.01.002
4. Y.V. Kharchenko, Z.Y. Blikharskyy, V.V. Vira et al., Study of structural changes in a nickel oxide containing anode material during reduction and oxidation at 600 °C. Nanocomposites Nanostruct. Appl. Springer Proc. Phys. **221**, 595–604 (2019). https://doi.org/10.1007/978-3-030-17759-1_42
5. S. Deville, L. Gremillard, J. Chevalier et al., A critical comparison of methods for the determination of the aging sensitivity in biomedical grade yttria-stabilized zirconia. J. Biomed. Mater. Res. B Appl. Biomater. **72B**, 239–245 (2005). https://doi.org/10.1002/jbm.b.30123
6. G. Chen, Y. Ling, Q. Li et al., Crystal structure and thermomechanical properties of CaO-PSZ ceramics synthesised from fused ZrO_2. Ceram. Int. **46**, 15357–15363 (2020). https://doi.org/10.1016/j.ceramint.2020.03.079
7. G.D. Girolamo, C. Blasi, M. Schioppa et al., Structure and thermal properties of heat treated plasma sprayed ceria–yttria co-stabilized zirconia coatings. Ceram. Int. **36**, 961–968 (2010). https://doi.org/10.1016/j.ceramint.2009.10.020

8. M. Alfano, G.D. Girolamo, L. Pagnotta et al., Processing, microstructure and mechanical properties of air plasma-sprayed ceria–yttria co-stabilized zirconia coatings. Strain **46**, 409–418 (2010). https://doi.org/10.1111/j.1475-1305.2009.00659.x
9. B.D. Vasyliv, V.Y. Podhurs'ka, O.P. Ostash, et al., Influence of reducing and oxidizing media on the physicomechanical properties of ScCeSZ–NiO and YSZ–NiO ceramics. Mater. Sci. **49**(2), 135–144 (2013). https://doi.org/10.1007/s11003-013-9593-3
10. N. Juntavee, S. Attashu, Effect of different sintering process on flexural strength of translucency monolithic zirconia. J. Clin. Exp. Dent. **10**, e821–e830 (2018). https://doi.org/10.4317/jced.54749
11. D. Li, Y. Liu, Y. Zhong et al., Dense and strong ZrO_2 ceramics fully densified in <15 min. Adv. Appl. Ceram. **118**, 23–29 (2019). https://doi.org/10.1080/17436753.2018.1449580
12. A.D. Bona, O.E. Pecho, R. Alessandretti, Zirconia as a dental biomaterial. Materials **8**, 4978–4991 (2015). https://doi.org/10.3390/ma8084978
13. M. Zhao, W. Pan, C. Wan et al., Defect engineering in development of low thermal conductivity materials: a review. J. Eur. Ceram. Soc. **37**, 1–13 (2016). https://doi.org/10.1016/j.jeurceramsoc.2016.07.036
14. M.J. Kim, J.S. Ahn, J.H. Kim et al., Effects of the sintering conditions of dental zirconia ceramics on the grain size and translucency. J. Adv. Prosthodont. **5**, 161–166 (2013). https://doi.org/10.4047/jap.2013.5.2.161
15. J. Zou, Y. Zhong, M. Eriksson et al., Tougher zirconia nanoceramics with less yttria. Adv. Appl. Ceram. **118**, 9–15 (2018). https://doi.org/10.1080/17436753.2018.1445464
16. R. Chintapalli, A. Mestra, F.G. Marro et al., Stability of nanocrystalline spark plasma sintered 3Y-TZP. Materials **3**, 800–814 (2010). https://doi.org/10.3390/ma3020800
17. B. Vasyliv, J. Milewski, V. Podhurska et al., Study of the degradation of a fine-grained YSZ–NiO anode material during reduction in hydrogen and reoxidation in air. Appl. Nanosci. **12**, 965–975 (2022). https://doi.org/10.1007/s13204-021-01768-w
18. J. Gui, Z. Xie, Phase transformation and slow crack growth study of Y-TZP dental ceramic. Mater. Sci. Eng. A **676**, 531–535 (2016). https://doi.org/10.1016/j.msea.2016.09.026
19. M. Amaral, P.F. Cesar, M.A. Bottino et al., Fatigue behavior of Y-TZP ceramic after surface treatments. J. Mech. Behav. Biomed. Mater. **57**, 149–156 (2016). https://doi.org/10.1016/j.jmbbm.2015.11.042
20. A.J. Santos, S. Garcia-Segura, S. Dosta et al., A ceramic electrode of ZrO_2–Y_2O_3 for the generation of oxidant species in anodic oxidation. Assessment of the treatment of Acid Blue 29 dye in sulfate and chloride media. Sep. Purif. Technol. **228**, 115747 (2019). https://doi.org/10.1016/j.seppur.2019.115747
21. C. Gautam, J. Joyner, A. Gautam et al., Zirconia based dental ceramics: structure, mechanical properties, biocompatibility and applications. Dalton. Trans. **45**, 19194–19215 (2016). https://doi.org/10.1039/c6dt03484e
22. B. Stawarczyk, M. Özcan, L. Hallmann et al., The effect of zirconia sintering temperature on flexural strength, grain size, and contrast ratio. Clin. Oral. Invest. **17**, 269–274 (2013). https://doi.org/10.1007/s00784-012-0692-6
23. Y. Yan, Z. Ma, J. Sun et al., Surface microstructure-controlled ZrO_2 for highly sensitive room-temperature NO_2 sensors. Nano. Mater. Sci. **3**, 268–275 (2021). https://doi.org/10.1016/j.nanoms.2021.02.001
24. J. Chevalier, What future for zirconia as a biomaterial? Biomaterials **27**, 535–543 (2006). https://doi.org/10.1016/j.biomaterials.2005.07.034
25. M. Guazzato, M. Albakry, S.P. Ringer et al., Strength, fracture toughness and microstructure of a selection of all-ceramic materials. Part II Zirconia-based dental ceramics. Dent. Mater. **20**, 449–456 (2004). https://doi.org/10.1016/j.dental.2003.05.003
26. T.A. Schaedler, R.M. Leckie, S. Kramer et al., Toughening of nontransformable t'-YSZ by addition of titania. J. Am. Ceram. Soc. **90**, 3896–3901 (2007). https://doi.org/10.1111/j.1551-2916.2007.01990.x
27. J. Chevalier, L. Gremillard, A.V. Virkar et al., The tetragonal-monoclinic transformation in zirconia: lessons learned and future trends. J. Am. Ceram. Soc. **92**, 1901–1920 (2009). https://doi.org/10.1111/j.1551-2916.2009.03278.x

28. H.G. Scott, Phase relationships in the zirconia–yttria system. J. Mater. Sci. **10**, 1527–1535 (1975). https://doi.org/10.1007/BF01031853
29. T.K. Gupta, J.H. Bechtold, R.C. Kuznicki et al., Stabilization of tetragonal phase in polycrystalline zirconia. J. Mater. Sci. **12**, 2421–2426 (1977). https://doi.org/10.1007/bf00553928
30. A.D. Ivasyshyn, B.D. Vasyliv, Effect of the size and form of specimens on the diagram of growth rates of fatigue cracks. Mater. Sci. **37**(6), 1002–1004 (2001). https://doi.org/10.1023/A:1015669913601
31. A. Kumar, P. Kumar, Dhaliwal AS (2021) Structural studies of zirconia and yttria doped zirconia for analysing it phase stabilization criteria. IOP Conf. Ser. Mater. Sci. Eng. **1033**, 012052 (2021). https://doi.org/10.1088/1757-899X/1033/1/012052
32. C. Piconi, S. Sprio, Zirconia implants: is there a future? Curr. Oral. Health Rep. **5**, 186–193 (2018). https://doi.org/10.1007/s40496-018-0187-x
33. V. Podhurska, B. Vasyliv, O. Ostash et al., Influence of treatment temperature on microstructure and properties of YSZ–NiO anode materials. Nanoscale. Res. Lett. **11**, 93 (2016). https://doi.org/10.1186/s11671-016-1306-z
34. M.F.R.P. Alves, S. Ribeiro, P.A. Suzuki et al., Effect of Fe_2O_3 addition and sintering temperature on mechanical properties and translucence of zirconia dental ceramics with different Y_2O_3 content. Mater. Res. **24**(2), e20200402 (2021). https://doi.org/10.1590/1980-5373-MR-2020-0402
35. B. Vasyliv, V. Podhurska, O. Ostash, Preconditioning of the YSZ–NiO fuel cell anode in hydrogenous atmospheres containing water vapor. Nanoscale Res. Lett. **12**, 265 (2017). https://doi.org/10.1186/s11671-017-2038-4
36. P. Khajavi, P.V. Hendriksen, J. Chevalier et al., Improving the fracture toughness of stabilized zirconia-based solid oxide cells fuel electrode supports: effects of type and concentration of stabilizer(s). J. Eur. Ceram. Soc. **40**, 5670–5682 (2020). https://doi.org/10.1016/j.jeurceramsoc.2020.05.042
37. T.A. Prikhna, O.P. Ostash, A.S. Kuprin et al., A new MAX phases-based electroconductive coating for high-temperature oxidizing environment. Compos. Struct. **277**, 114649 (2021). https://doi.org/10.1016/j.compstruct.2021.114649
38. H.A. Shabri, M.H.D. Othman, M.A. Mohamed et al., Recent progress in metal-ceramic anode of solid oxide fuel cell for direct hydrocarbon fuel utilization: a review. Fuel Process. Technol. **212**, 106626 (2021). https://doi.org/10.1016/j.fuproc.2020.106626
39. Y. Kharchenko, Z. Blikharskyy, V. Vira et al., Study of nanostructural changes in a Ni-containing cermet material during reduction and oxidation at 600 °C. Appl. Nanosci. **10**, 4535–4543 (2020). https://doi.org/10.1007/s13204-020-01391-1
40. S.S. Savka, D.I. Popovych, A.S. Serednytski, Molecular dynamics simulations of the formation processes of zinc oxide nanoclusters in oxygen environment. Nanophysics, nanomaterials, interface studies, and applications. Springer Proc. Phys. **195**, 145–156 (2017). https://doi.org/10.1007/978-3-319-56422-7_11
41. S. Jiang, X. Huang, Z. He et al., Phase transformation and lattice parameter changes of non-trivalent rare earth-doped YSZ as a function of temperature. J. Mater. Eng. Perform. **27**, 2263–2270 (2018). https://doi.org/10.1007/s11665-018-3159-3
42. M.Y. Smyrnova-Zamkova, V.P. Red'ko, O.K. Ruban et al., The properties of nanocrystalline powder of 90% Al_2O_3–10% ZrO_2 (wt.%) obtained via the hydrothermal synthesis/mechanical mixing. Nanosistemi. Nanomater. Nanotehnol. **15**(2), 309–317 (2017). https://doi.org/10.15407/nnn.15.02.0309
43. T. Sakuma, Y. Yoshizawa, The grain growth of zirconia during annealing in the cubic/tetragonal two-phase region. Mater. Sci. Forum **94–96**, 865–870 (1992). https://doi.org/10.4028/www.scientific.net/MSF.94-96.865
44. L. Li, O. Van Der Biest, P.L. Wang et al., Estimation of the phase diagram for the ZrO_2–Y_2O_3–CeO_2 system. J. Eur. Ceram. Soc. **21**(16), 2903–2910 (2001). https://doi.org/10.1016/S0955-2219(01)00218-7

45. J.A. Allemann, B. Michel, H.B. Märki et al., Grain growth of differently doped zirconia. J. Eur. Ceram. Soc. **15**(10), 951–958 (1995). https://doi.org/10.1016/0955-2219(95)00073-4
46. N.M. Ersoy, H.M. Aydoğdu, B.U. Değirmenci et al., The effects of sintering temperature and duration on the flexural strength and grain size of zirconia. Acta. Biomater. Odontol. Scand. **1**(2–4), 43–50 (2015). https://doi.org/10.3109/23337931.2015.1068126
47. B. Stawarczyk, A. Emslander, M. Roos et al., Zirconia ceramics, their contrast ratio and grain size depending on sintering parameters. Dent. Mater. J. **33**(5), 591–598 (2014). https://doi.org/10.4012/dmj.2014-056
48. P. Duràn, M. Villegas, J.F. Fernàndez et al., Theoretically dense and nanostructured ceramics by pressureless sintering of nanosized Y-TZP powders. Mater. Sci. Eng. **A232**, 168–176 (1997). https://doi.org/10.1016/S0921-5093(97)00099-3
49. M. Trunec, K. Castkova, P. Roupcova, Effect of phase structure on sintering behavior of zirconia nanopowders. J. Am. Ceram. Soc. **96**, 3720–3727 (2013). https://doi.org/10.1111/jace.12624
50. C.G. Soubelet, M.P. Albano, Differences in microstructure and mechanical properties between Y-TZP and Al_2O_3-doped Y-TZP/bioglass ceramics. Int. J. Appl. Ceram. Technol. **18**, 2237–2249 (2021). https://doi.org/10.1111/ijac.13864
51. L. An, A. Ito, T. Goto, Transparent yttria produced by spark plasma sintering at moderate temperature and pressure profiles. J. Eur. Ceram. Soc. **32**(5), 1035–1040 (2012). https://doi.org/10.1016/j.jeurceramsoc.2011.11.023
52. M. Ahsanzadeh-Vadeqani, R.S. Razavi, Spark plasma sintering of zirconia-doped yttria ceramic and evaluation of the microstructure and optical properties. Ceram. Int. **42**(16), 18931–18936 (2016). https://doi.org/10.1016/j.ceramint.2016.09.043
53. A. Gupta, S. Sharma, N. Mahato et al., Mechanical properties of spark plasma sintered ceria reinforced 8 mol% yttria-stabilized zirconia electrolyte. Nanomater. Energy **1**(NME5), 306–315 (2012). https://doi.org/10.1680/nme.12.00018
54. J. Grabis, D. Jankoviča, I. Šteins et al., Characteristics and sinterability of ceria stabilized zirconia nanoparticles prepared by chemical methods. Mater. Sci. **24**(3), 243–246 (2018). https://doi.org/10.5755/j01.ms.24.3.18288
55. M. Vojtko, V. Puchy, E. Múdra et al., Coarse-grain CeO_2 doped ZrO_2 ceramic prepared by spark plasma sintering. J. Eur. Ceram. Soc. **40**, 4844–4852 (2020). https://doi.org/10.1016/j.jeurceramsoc.2020.05.014
56. E.F. Garcia, C.F.G. Gonzalez, A. Fernandez et al., Processing and spark plasma sintering of zirconia/titanium cermets. Ceram. Int. **39**(6), 6931–6936 (2013). https://doi.org/10.1016/j.ceramint.2013.02.029
57. Á.G. López, C.L. Pernía, C.M. Ferreiro et al., Spark plasma sintered zirconia ceramic composites with graphene-based nanostructures. Ceramics **1**(1), 153–164 (2018). https://doi.org/10.3390/ceramics1010014
58. S. Kikuchi, K. Katahira, J. Komotori, Formation of titanium/zirconia based biomaterial fabricated by spark plasma sintering. J. Jpn. Inst. Met. Mater. **82**(9), 341–348 (2018). https://doi.org/10.2320/jinstmet.J2018017
59. A. Nastic, A. Merati, M. Bielawski et al., Instrumented and Vickers indentation for the characterization of stiffness, hardness and toughness of zirconia toughened Al_2O_3 and SiC armor. J. Mater. Sci. Technol. **31**(8), 773–783 (2015). https://doi.org/10.1016/j.jmst.2015.06.005
60. Y.G. Chabak, V.I. Fedun, K. Shimizu et al., Phase-structural composition of coating obtained by pulsed plasma treatment using eroded cathode of T1 high speed steel. Probl. At. Sci. Technol. **104**(4), 100–106 (2016)
61. O.M. Romaniv, I.V. Zalite, V.M. Simin'kovych et al., Effect of the concentration of zirconium dioxide on the fracture resistance of Al_2O_3–ZrO_2 ceramics. Mater. Sci. **31**(5), 588–594 (1996). https://doi.org/10.1007/BF00558793
62. O.M. Romaniv, B.D. Vasyliv, Some features of formation of the structural strength of ceramic materials. Mater. Sci. **34**(2), 149–161 (1998). https://doi.org/10.1007/BF02355530
63. V.V. Kulyk, Z.A. Duriagina, B.D. Vasyliv et al., Effects of yttria content and sintering temperature on the microstructure and tendency to brittle fracture of yttria-stabilized zirconia. Arch. Mater. Sci. Eng. **109**(2), 65–79 (2021). https://doi.org/10.5604/01.3001.0015.2625

64. V. Kulyk, Z. Duriagina, B. Vasyliv et al., The effect of sintering temperature on the phase composition, microstructure, and mechanical properties of yttria-stabilized zirconia. Materials **15**, 2707 (2022). https://doi.org/10.3390/ma15082707
65. Q.L. Li, Y.Y. Jiang, Y.R. Wei et al., The influence of yttria content on the microstructure, phase stability and mechanical properties of dental zirconia. Ceram. Int. **48**, 5361–5368 (2022). https://doi.org/10.1016/j.ceramint.2021.11.079
66. V. Kulyk, Z. Duriagina, A. Kostryzhev et al., The effect of yttria content on microstructure, strength, and fracture behavior of yttria-stabilized zirconia. Materials **15**, 5212 (2022). https://doi.org/10.3390/ma15155212
67. ASTM E 384-11, Standard test method for Knoop and Vickers hardness of materials. ASTM International (2011). https://doi.org/10.1520/E0384-11
68. ASTM C 1327-03, Standard test method for Vickers indentation hardness of advanced ceramics. ASTM International (2003). https://doi.org/10.1520/C1327-03
69. A.G. Evans, E.A. Charles, Fracture toughness determinations by indentation. J. Am. Ceram. Soc. **59**(7–8), 371–372 (1976). https://doi.org/10.1111/j.1151-2916.1976.tb10991.x
70. B.R. Lawn, E.R. Fuller, Equilibrium penny-like cracks in indentation fracture. J. Mater. Sci. **10**(12), 2016–2024 (1975). https://doi.org/10.1007/BF00557479
71. B.R. Lawn, M.V. Swain, Microfracture beneath point indentations in brittle solids. J. Mater. Sci. **10**(1), 113–122 (1975). https://doi.org/10.1007/BF00541038
72. B.R. Lawn, A.G. Evans, D.B. Marshall, Elastic/plastic indentation damage in ceramics: the median/radial crack system. J. Am. Ceram. Soc. **63**(9–10), 574–581 (1980). https://doi.org/10.1111/j.1151-2916.1980.tb10768.x
73. B.R. Lawn, *Fracture of Brittle Solids*, 2nd edn. Cambridge (1993). https://doi.org/10.1017/CBO9780511623127
74. R.F. Cook, G.M. Pharr, Direct observation and analysis of indentation cracking in glasses and ceramics. J. Am. Ceram. Soc. **73**(4), 787–817 (1990). https://doi.org/10.1111/j.1151-2916.1990.tb05119.x
75. K. Niihara, R. Morena, D.P.H. Hasselman, Evaluation of K_{Ic} of brittle solids by the indentation method with low crack-to-indent ratios. J. Mater. Sci. Lett. **1**(1), 13–16 (1982). https://doi.org/10.1007/BF00724706
76. K. Niihara, A fracture mechanics analysis of indentation-induced Palmqvist crack in ceramics. J. Mater. Sci. Lett. **2**(5), 221–223 (1983). https://doi.org/10.1007/BF00725625
77. K. Tanaka, Elastic/plastic indentation hardness and indentation fracture toughness: the inclusion core model. J. Mater. Sci. **22**(4), 1501–1508 (1987). https://doi.org/10.1007/BF01233154
78. J.E. Blendell, The origins of internal stresses in polycrystalline alumina and their effects on mechanical properties. Cambridge (1979). http://hdl.handle.net/1721.1/44234
79. J. Lankford, Indentation microfracture in the Palmqvist crack regime: implications for fracture toughness evaluation by the indentation method. J. Mater. Sci. Lett. **1**(11), 493–495 (1982). https://doi.org/10.1007/BF00721938
80. G.R. Anstis, P. Chantikul, B.R. Lawn et al., A critical evaluation of indentation techniques for measuring fracture toughness: I, direct crack measurement. J. Am. Ceram. Soc. **64**(9), 533–538 (1981). https://doi.org/10.1111/j.1151-2916.1981.tb10320.x
81. M.Y. Smyrnova-Zamkova, O.K. Ruban, O.I. Bykov et al., The influence of the ZrO_2 solid solution amount on the physicochemical properties of Al_2O_3–ZrO_2–Y_2O_3–CeO_2 powders. Powder Metall. Met. Ceram. **60**(3–4), 129–141 (2021). https://doi.org/10.1007/s11106-021-00222-4
82. B. Vasyliv, V. Kulyk, Z. Duriagina et al., Estimation of the effect of redox treatment on microstructure and tendency to brittle fracture of anode materials of YSZ-NiO(Ni) system. East Eur. J. Enterp. Technol. **108/6**(12), 67–77 (2020). https://doi.org/10.15587/1729-4061.2020.218291
83. J.W. Adams, R. Ruh, K.S. Mazdiyasni, Young's modulus, flexural strength, and fracture of yttria-stabilized zirconia versus temperature. J. Am. Ceram. Soc. **80**(4), 903–908 (1997). https://doi.org/10.1111/j.1151-2916.1997.tb02920.x

84. V.V. Kulyk, B.D. Vasyliv, Z.A. Duriagina et al., The effect of water vapor containing hydrogenous atmospheres on the microstructure and tendency to brittle fracture of anode materials of YSZ–NiO(Ni) system. Arch. Mater. Sci. Eng. **108**(2), 49–67 (2021). https://doi.org/10.5604/01.3001.0015.0254
85. J.M. Gere, S.P. Timoshenko, *Mechanics of Materials*, 4th edn. (PWS Publishing Company, Boston, MA, USA, 1997), p.912
86. B.D. Vasyliv, V.Y. Podhurska, O.P. Ostash et al., Effect of a hydrogen sulfide-containing atmosphere on the physical and mechanical properties of solid oxide fuel cell materials. Nanochem. Biotechnol. Nanomater. Appl. Springer Proc. Phys. **214**, 475–485 (2018). https://doi.org/10.1007/978-3-319-92567-7_30
87. ASTM E 399-20a, Standard test method for linear-elastic plane-strain fracture toughness of metallic materials. ASTM International (2020). https://doi.org/10.1520/E0399-20A
88. ASTM C 1421-18, Standard test methods for determination of fracture toughness of advanced ceramics at ambient temperature. ASTM International (2018). https://doi.org/10.1520/C1421-18
89. J. Kübier, Fracture toughness of ceramics using the SEVNB method: From a preliminary study to a standard test method, in *Fracture Resistance Testing of Monolithic and Composite Brittle Materials*. ed. by J. Salem et al., ASTM International, pp. 93–106 (2002). https://doi.org/10.1520/STP10473S
90. M. Radovic, E. Lara-Curzio, Mechanical properties of tape cast nickel-based anode materials for solid oxide fuel cells before and after reduction in hydrogen. Acta. Mater. **52**(20), 5747–5756 (2004). https://doi.org/10.1016/j.actamat.2004.08.023
91. T.B. Serbenyuk, T.O. Prikhna, V.B. Sverdun et al., Effect of the additive of Y_2O_3 on the structure formation and properties of composite materials based on AlN–SiC. J. Superhard. Mater. **40**(1), 8–15 (2018). https://doi.org/10.3103/S1063457618010021
92. M. Ettler, G. Blaβ, N.H. Menzler, Characterization of Ni–YSZ-cermets with respect to redox stability. Fuel Cells **7**(5), 349–355 (2007). https://doi.org/10.1002/fuce.200700007
93. B.D. Vasyliv, Initiation of a crack from the edge of a notch with oblique front in specimens of brittle materials. Mater. Sci. **38**(5), 724–728 (2002). https://doi.org/10.1023/A:1024222709514
94. M.Y. Smyrnova-Zamkova, O.K. Ruban, O.I. Bykov et al., Physico-chemical properties of fine-grained powder in Al_2O_3–ZrO_2–Y_2O_3–CeO_2 system produced by combined method. Comp. Theory Pract. **18**(4), 234–240 (2018). https://kompozyty.ptmk.net/pliczki/pliki/1290_2018t04_maria-y-smyrnova-zamkova-.pdf
95. O.P. Ostash, B.D. Vasyliv, V.Y. Podhurs'ka et al., Optimization of the properties of 10Sc1CeSZ-NiO composite by the redox treatment. Mater. Sci. **46**(5), 653–658 (2011). https://doi.org/10.1007/s11003-011-9337-1
96. Y. Wang, M.E. Walter, K. Sabolsky et al., Effects of powder sizes and reduction parameters on the strength of Ni–YSZ anodes. Solid State Ionics **177**, 1517–1527 (2006). https://doi.org/10.1016/j.ssi.2006.07.010
97. M. Andrzejczuk, O. Vasylyev, I. Brodnikovskyi et al., Microstructural changes in NiO–ScSZ composite following reduction processes in pure and diluted hydrogen. Mater. Charact. **87**, 159–165 (2014). https://doi.org/10.1016/j.matchar.2013.11.011
98. I. Danilenko, F. Glazunov, T. Konstantinova et al., Effect of Ni/NiO particles on structure and crack propagation in zirconia based composites. Adv. Mater. Lett. **5**(8), 465–471 (2014). https://doi.org/10.5185/amlett.2014.amwc1040II
99. G.A. Gogotsi, S.N. Dub, E.E. Lomonova et al., Vickers and Knoop indentation behaviour of cubic and partially stabilized zirconia crystals. J. Eur. Ceram. Soc. **15**(5), 405–413 (1995). https://doi.org/10.1016/0955-2219(95)91431-M
100. V.Y. Podhurs'ka, B.D. Vasyliv, O.P. Ostash et al., Structural transformations in the NiO-containing anode of ceramic fuel cells in the course of its reduction and oxidation. Mater. Sci. **49**(6), 805–811 (2014). https://doi.org/10.1007/s11003-014-9677-8
101. I. Danilenko, G. Lasko, I. Brykhanova et al., The peculiarities of structure formation and properties of zirconia-based nanocomposites with addition of Al_2O_3 and NiO. Nanoscale Res. Lett. **12**, 125 (2017). https://doi.org/10.1186/s11671-017-1901-7

102. B.D. Vasyliv, A procedure for the investigation of mechanical and physical properties of ceramics under the conditions of biaxial bending of a disk specimen according to the ring–ring scheme. Mater. Sci. **45**(4), 571–575 (2009). https://doi.org/10.1007/s11003-010-9215-2
103. B.D. Vasyliv, Improvement of the electric conductivity of the material of anode in a fuel cell by the cyclic redox thermal treatment. Mater. Sci. **46**(2), 260–264 (2010). https://doi.org/10.1007/s11003-010-9282-4
104. Z. Peng, J. Gong, H. Miao, On the description of indentation size effect in hardness testing for ceramics: analysis of the nanoindentation data. J. Eur. Ceram. Soc. **24**, 2193–2201 (2004). https://doi.org/10.1016/s0955-2219(03)00641-1
105. V.I. Zurnadzhy, V.G. Efremenko, I. Petryshynets et al., Mechanical properties of car-bide-free lower bainite in complex-alloyed constructional steel: effect of bainitizing treatment parameters. Kov. Mater. **58**(2), 129–140 (2020). https://doi.org/10.4149/km_2020_2_129
106. Y. Zhang, K. Shimizu, X. Yaer et al., Erosive wear performance of heat treated multi-component cast iron containing Cr, V, Mn and Ni eroded by alumina spheres at elevated temperatures. Wear **390–391**, 135–145 (2017). https://doi.org/10.1016/j.wear.2017.07.017

Structure and Functional Properties of Nanocomposites Based on Polyethylene Oxide and Silver Nanoparticles

E. A. Lysenkov and O. V. Strutskyi

Abstract The combination of polyethylene oxide (PEO) matrix and stabilized silver nanoparticles (AgNPs) in the new material is promising and requires further study. Therefore, we studied structure and functional properties of PEO-AgNPs materials using the methods of electronic microscopy, X-ray diffraction, dielectric relaxation and acoustic spectroscopy, and microbiologic study. It was established that the obtained nanoparticles are core–shell objects, inside the particle is a silver core covered with a hyperbranched ionic liquid shell. Stabilized silver nanoparticles significantly affect the functional properties of the nanocomposite material at relatively low concentrations of the nanofiller (1–2%). The use of stabilized AgNPs made it possible to achieve very low dielectric loss values in the materials. It is shown that the greatest effect on the dielectric properties is manifested at the content of 1% of the filler. The activation energy and relaxation time of PEO macromolecules, as well as the speed of ultrasound and the damping coefficient of the studied materials were extremely dependent on the content of the filler. This dependence is explained by aggregation processes of nanoparticles in the polymer matrix. It is shown that 1% of AgNPs in the system is optimal, and it is assumed that with this filler content, the functional characteristics will be the most acceptable for the use of such a material. The general electrical conductivity of the systems increases by two orders of magnitude when 2% filler is introduced. Nanocomposite material containing 2% of AgNPs shows significant activity against *Staphylococcus aureus* (inhibition zone is 22 mm). The novel approach for synthesis of stabilized silver nanoparticles open a new window of possibilities in the development of new antimicrobial materials, based on various polymer matrix: from thermoplastic to elastic.

E. A. Lysenkov (✉)
Petro Mohyla Black Sea National University, Mykolaiv, Ukraine
e-mail: ealysenkov@ukr.net

O. V. Strutskyi
Institute of Macromolecular Chemistry, NAS of Ukraine, Kyiv, Ukraine

O. Fesenko and L. Yatsenko (eds.), *Nanostructured Surfaces, Nanocomposites and Nanomaterials, and Their Applications*, Springer Proceedings in Physics 296,
https://doi.org/10.1007/978-3-031-42704-6_18

1 Introduction

In recent decades, polymer nanocomposite materials have become the object of intensive study because they demonstrate unique physicochemical properties that cannot be obtained using individual components. The growing interest in such materials is explained by the fact that they possess unique properties, such as high mechanical and thermal stability, accessibility, electrical conductivity, and improved optical and magnetic properties [1]. Therefore, polymer nanocomposites represent a wide area for their application [2–4]. The desired properties of nanocomposites are achieved by adding various nanosized fillers to the matrix.

Polymer nanocomposite materials created in this way combine the optimal properties of polymers and inorganic particles [5–7]. A special place among metal-polymer composites is occupied by materials with the inclusion of silver nanoparticles (AgNPs), which are characterized by improved mechanical properties and also exhibit high antimicrobial activity [8]. Due to the variety of their forms and properties, AgNPs are promising in the preparation of nanomaterials for use in applied high-tech fields, such as medicine, pharmacology, optics, photochemistry, biosensors, etc., and fundamental science in general [9]. Today, regardless of the method of obtaining AgNPs, there is a problem of stabilizing their surface to prevent aggregation, which significantly reduces their activity and prevents uniform distribution during the preparation of nanomaterials [9]. Among known approaches to solving this problem, the use of oligomers with a highly branched structure is promising [10, 11]. The use of branched derivatives of polyglycerols, polyethylene oxides, polyethyleneimines, polyamides, polyamidoamines, and polyester polyols of various generations, which are used as such, as well as in combination with traditional reductants in the preparation of AgNPs [10, 11], including in the form of copolymers, is described. Another promising approach to obtaining AgNPs is the use of ionic liquids as media that facilitate the formation of stable nanoparticles with controlled sizes and their narrow distribution [12].

Most of the functional characteristics of polymer nanocomposite materials significantly depend on the operating conditions (temperature, applied pressure, duration, action of external electromagnetic radiation, etc.). There are many publications that study the above-mentioned characteristics, but there are almost no works that would describe the mechanisms of influence of nanosized filler on the structure and internal processes.

The introduction of inorganic nanoparticles into the composition of polymers of various types, both thermoplastics and reactive plastics, leads to a significant impact on the structures and properties of the resulting nanocomposite material [13, 14]. Polymer-AgNPs materials are characterized by unique functional properties and have wide practical applications. References [15–17] describes the materials based on polyethylene oxide with silver nanoparticles and other fillers additionally doped with an inorganic salt which showed acceptable characteristics and were promising for the creation of solid polymer electrolytes for lithium current sources. Also, polymer nanocomposites containing silver nanoparticles are characterized by high dielectric permeability and low dielectric losses. Such materials are widely used in many fields, including high-density energy storage capacitors and high-speed integrated circuits, due to their flexibility and tailored dielectric properties [18]. To obtain materials with the required dielectric characteristics, silver nanoparticles with a modified surface (stabilized) are used. Thus, Dai et al. studied the effect of stabilized silver nanoparticles on the dielectric properties of a material based on polyvinylidene fluoride [19]. They found that after modification of Ag nanoparticles with TiO_2, the dielectric constant at 100 Hz of composites containing 50 vol.% of the filler was 61, and the dielectric loss decreased to 0.04, which is almost 96.4% lower than that of composites with unmodified pieces of silver.

The study of relaxation processes in a system based on polyvinyl alcohol filled with silver nanoparticles and graphene oxide made it possible to predict and explain the final properties of nanocomposite materials [20]. An increase in dielectric constant and electrical conductivity with increasing filler content in the system was recorded.

More et al. studied the relaxation properties of a composite material based on polyvinyl alcohol and zinc oxide particles [21]. The permittivity is shown to decrease with frequency but exhibits slightly different behavior as it varies with temperature, this may be due to the viscous nature of the polymer. It was established that the relaxation time values decrease with temperature.

Important information about the processes that occur in the polymer matrix can be obtained by studying the impact on the material using sound waves (acoustic relaxation) [22]. Zhang et al. [23] studied the propagation of ultrasound in composite systems based on silicon and colloidal aluminum particles. They established that the acoustic phase velocity and acoustic impedance of nanocomposite films increase with the content of aluminum oxide. This provided another option for adapting the acoustic resistance of the nanocomposite film to create coatings for high-quality materials used in high-frequency medical ultrasound transducers.

Recent advancements in nanotechnology-based medicines have opened new horizons for combating multidrug resistance in microorganisms. In particular, the use of silver nanoparticles as a potent antibacterial agent has received much attention [24]. AgNPs exhibit their antimicrobial potential through multifaceted mechanisms. AgNPs adhesion to microbial cells, penetration inside the cells, ROS and free radical generation, and modulation of microbial signal transduction pathways have been recognized as the most prominent modes of antimicrobial action.

Qasim et al. analyzed the antimicrobial activities of poly-*N*-isopropylacrylamide (pNIPAM)-based polymeric nanoparticles encapsulating AgNPs [25]. Polymeric nanoparticles showed significant bacteriostatic activities against Gram-negative (*Escherichia coli*) and Gram-positive (*Staphylococcus aureus*) bacteria depending on the nanoparticle size and amount of $AgNO_3$ used during fabrication.

To prevent microbial contamination in foods and medical tools, Cao et al. prepared a novel antibacterial material nanosilver loaded poly(styrene-co-acrylic) (nAg-PSA) composites by in situ reductions [26]. Results showed that the silver nanoparticles were successfully loaded onto PSA microspheres, highly dispersed on the surface of nAg-PSA nanoparticles. The nAg-PSA nanoparticles all showed significant antibacterial effects, among which nAg-PSA4 had maximum diameters of inhibition zones against the Gram-negative *E. coli* (1.18 mm) and the Gram-positive *S. aureus* (1.29 mm), respectively, which closely related to the size and density of silver nanoparticles covered on the surface.

Kausar prepared poly(acrylonitrile-co-methyl acrylate)/polyaniline (PANMA/PANI)-based nanocomposite with silver nanoparticle and cellulose-silver nanoparticle (Cell-AgNP) [27]. Antimicrobial activity of PANMA/PANI/Cell-AgNP nanocomposites with various nanoparticle content has been tested using exposure time against *S. aureus*. Both the exposure time and nanoparticle content in novel nanocomposites affected the antimicrobial activity.

Rujitanaroj et al. studied poly(ethylene oxide) (PEO) solutions containing various amounts of poly(ethylene glycol) (PEG) and various amount of silver nanoparticles [28]. The assessment for the antimicrobial activity of the PEO-5%/PEG-15% solutions containing 0.5–2.5% $AgNO_3$ against Gram-positive *S. aureus* and Gram-negative *Escherichia coli* (*E. coli*) showed that the antibacterial activity of these solutions was found to increase with increasing $AgNO_3$ content and the as-formed AgNPs were found to be more effective in the protection against *E. coli*.

Therefore, the study of the structure, relaxation characteristics, functional properties, and antimicrobial activity of nanocomposite materials based on polyethylene oxide and AgNPs is promising and will be informative for explaining the processes associated with the effect of nanoparticles on the polymer matrix. Therefore, the purpose of this work was to create a new material based on polyethylene oxide and silver nanoparticles, as well as to study its functional and antimicrobial characteristics.

2 The Experimental Part

Polymer nanocomposite materials based on polyethylene oxide and silver nanoparticles were used for the study.

Polyethylene oxide (PEO 1000), $HO[–CH_2–CH_2–O–]_nH$ ($n \approx 22$) with molecular weight $M_w = 1000$, manufactured by Aldrich.

In this work, for the first time, an anionic oligomeric ionic liquid (OIL) with a hyperbranched structure developed by us was used as a surface stabilizer in the synthesis of AgNPs. The proposed OIL is a product of exhaustive acylation of hyperbranched oligoester polyol of the 3rd generation (containing 32 primary aliphatic hydroxyl groups) with cyclic 2-sulfobenzoic anhydride and subsequent neutralization of the resulting sulfoacid compound with N-methylimidazole [29] (Fig. 1).

Synthesis of nanoparticles. Ag nanoparticles were synthesized by reducing Ag(I) ions in $AgNO_3$ with trisodium citrate in the presence of the OIL at 100 °C for an hour in an aqueous solution. The ratio of Ag(I) and ionic groups of the OIL were equal to 1:3 and the ratio of Ag(I) and trisodium citrate was 1:4. The water was evaporated

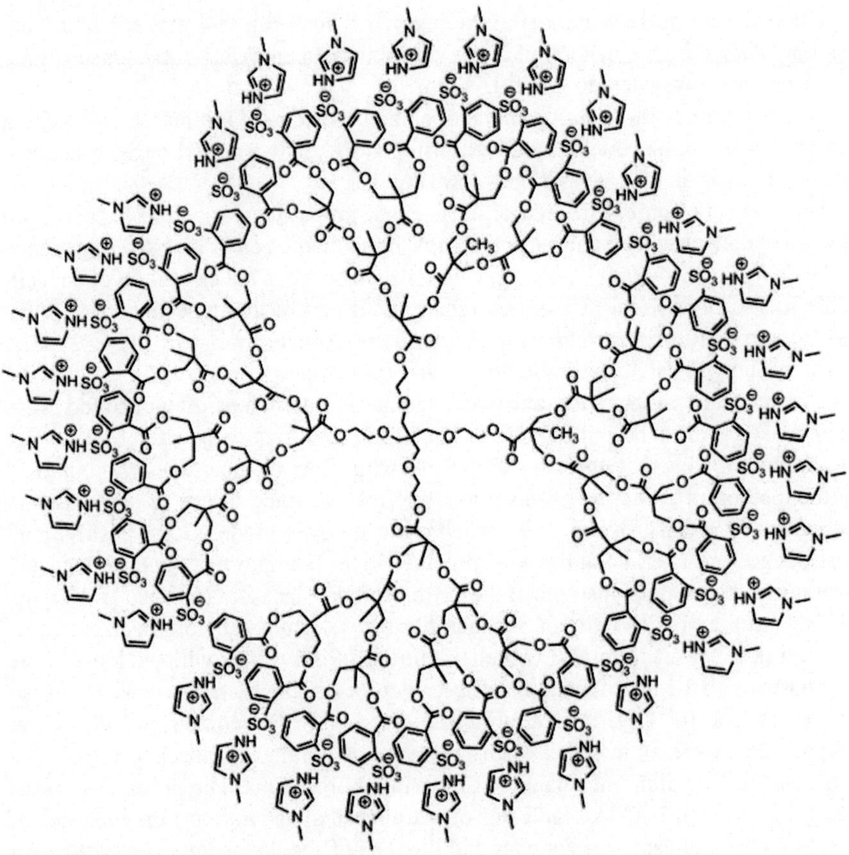

Fig. 1 Structure of oligomeric ionic liquid

and the residue was thoroughly washed with ethanol followed by vacuum drying (3–5 mm Hg). Yield 75%. The resulting AgNPs stabilized with OIL were obtained as a brown powder soluble in water. The latter significantly expands the possibilities of introducing AgNPs into various polymer systems, in particular water-soluble ones.

Preparation of nanocomposites. The nanocomposite materials were obtained by adding the appropriate weights of filler AgNPs to PEO in the liquid state (T = 80 °C) with subsequent 10 min sonication of the mixture using a UZD A650 ultrasonic disperser, in accordance with procedure essentially similar to the previously described [30, 31]. After sonication, the composites were cooled down to the solid state and kept at room temperature until further experiments. We investigated a series of samples with contents of fillers within 0.5–2 wt%.

Methods. The structural features of nanoparticles were obtained using a transmission electron microscope PEM-125K.

The structure of the materials in the region of small (up to 4 nm) spatial orders was established by the method of X-ray diffraction, using XRD-7000. Used Cu K_α radiation with a wavelength $\lambda = 0{,}154$ nm.

In order to study the OIL-Ag interaction, the IR spectra of initial OIL and AgNPs were obtained. IR spectra were recorded using a Bruker Tensor-37 Fourier transform IR spectrometer in the range of 600–4000 cm^{-1}.

The study of electrical, dielectric, and acoustic properties was performed using the method of dielectric relaxation spectroscopy, implemented on the basis of impedance meters E7-20 and Z-2000. The sample was placed between the electrodes of the cell, while measuring its real (Z') and imaginary (Z'') part of the impedance. Measurements were performed at room temperature in the frequency range of 10 Hz–2 MHz. The constant gap between the electrodes was 0.11 mm.

The speed of propagation and the attenuation coefficient of ultrasound were determined by the echo-pulse method. The ultrasound frequency was 5 MHz.

Antimicrobial activity of nanocomposites with silver nanoparticles was obtained by the method of reduction of metal ions against reference strains of opportunistic pathogens *S. aureus* and *E. coli*, which were used as model Gram-positive and Gram-negative bacteria. Studies were provided in the laboratory of microbiology and virology of the Medical Institute of the Petro Mohyla Black Sea National University.

The antimicrobial activity of nanocomposites was studied by diffusion into agar on solid nutrient medium MH (Mueller–Hinton). Petri dishes with MH agar were inoculated with 10 L of inoculum of test microorganisms *S. aureus* and *E. coli* at the rate of 2×10^5 CFU/ml. Prefabricated samples of nanocomposites with silver nanoparticles were cut into disks with a diameter of 10 mm and placed on the surface of the nutrient medium inoculated with test microorganisms. The plates were incubated for 24 h at 37 °C. An indicator of antimicrobial activity was the presence of a clear microorganism-free zone around the disc of the nanocomposite with silver nanoparticles. The larger area around the disk indicated the higher the inhibitory efficiency of the sample. The nanocomposite disk without silver nanoparticles served as a control. The experiment was repeated three times.

3 Results and Their Discussion

The final properties of the nanocomposite material strongly depend on the structure and distribution of the filler in the polymer matrix. To establish the influence of silver nanoparticles on the functional properties of the obtained materials, we studied their dielectric, relaxation, acoustic, electrical, and antimicrobial characteristics.

3.1 Structure of Silver Nanoparticles

Figure 2 presents FTIR spectra of the initial OIL and obtained nanoparticles. There is a significant bathochromic shift of the absorption bands of the valence vibrations of carbonyl groups (from 1720 cm^{-1} to 1566 cm^{-1}), a shift and change in the intensities of the absorption bands of the C–C valence vibrations (1360–1625 cm^{-1}) and C–H (3000–3100 cm^{-1}) bonds of the aromatic component and valence and combined vibrations of N^{+}–H (2300–3700 cm^{-1}) in the spectrum of the nanoparticles compared to that of initial OIL.

This indicates the adsorption of carbonyl and ionic (their cationic component) groups on the surface of the formed AgNPs [32] and the formation of host–guest complexes between OIL and residual silver ions. The formation of AgNPs and complexes between OIL and silver ions is also confirmed by X-ray analysis (Fig. 3).

The X-ray diffractograms of the obtained substances show reflections at 38.0°, 44.2°, 64.4°, and 77.3°, which relate to (111), (200), (220), and (311) reflections of crystal planes and indicate the formation of particles of silver with a face-centered cubic structure [33].

Also, on these diffractograms, reflexes appear at 7.2°, 14.2°, 18.2°, 22.2°, 29.7°, and 36.7°, which indicate the formation of complexes between OIL and silver ions, which form regular structures with long-range in order.

The formation of AgNPs is also confirmed by transmission electron microscopy (TEM) data (Fig. 4). On the TEM micrograph of the obtained AgNPs powder, nano-sized particles of the "core–shell" type are observed, where the AgNPs are the core,

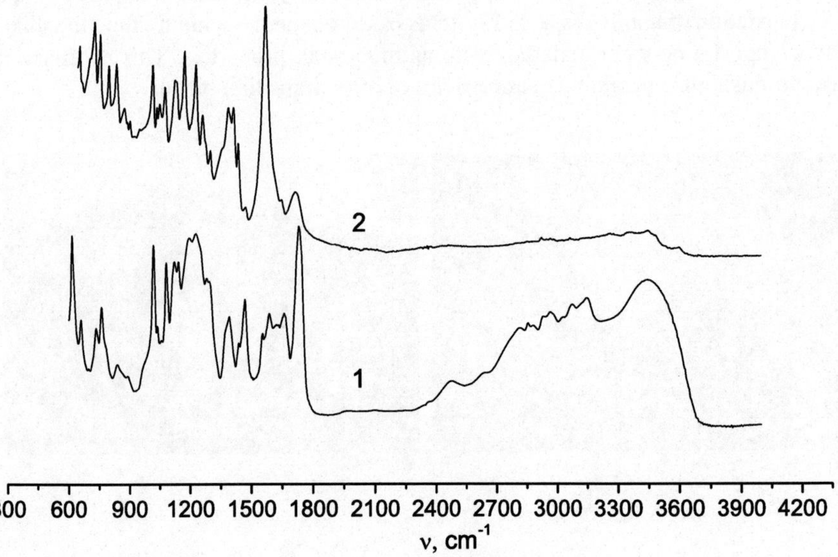

Fig. 2 FTIR spectra of the initial OIL (1) and AgNPs (2)

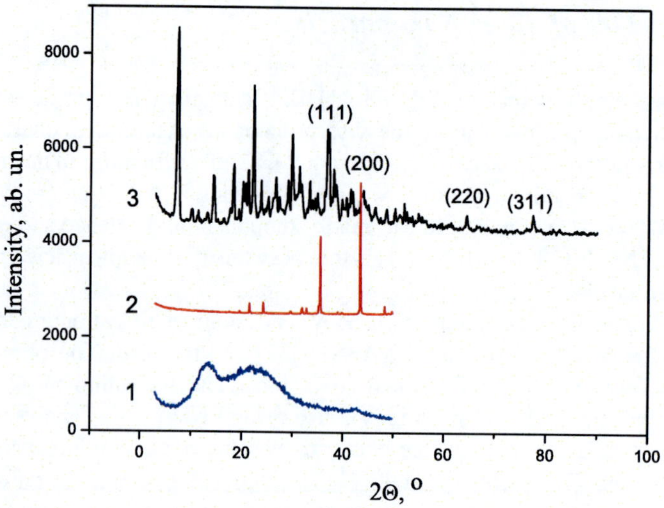

Fig. 3 X-ray diffraction patterns of OIL (1), $AgNO_3$ (2). and AgNPs (3)

and the OIL as a stabilizer forms the shell. At the same time, nanoparticles form structures in the form of swarms. Analysis of microphotographs showed that the obtained AgNPs are characterized by a narrow size distribution in the range of 5–10 nm.

The approach developed by us to the synthesis of AgNPs in the presence of an anionic OIL hyperdispersive structure as a surface stabilizer with certain functionalization of the latter opens up new opportunities in the synthesis of nanoparticles and the preparation of highly dispersed systems based on them, including functionalized nanocomposite polymer materials with antimicrobial properties. This approach is also promising for obtaining nanoparticles of other transition metals.

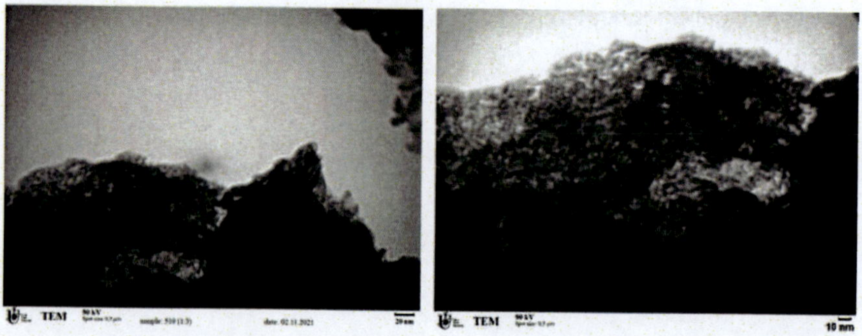

Fig. 4 TEM images of AgNPs powder

3.2 *Dielectric Properties of Polyethylene Oxide-Silver Nanoparticles Systems*

When studying the electrophysical properties of polymer composites, special attention is paid to issues related to charge transfer and accumulation processes. Taking into account the latest achievements in the development of experimental techniques, one of the methods of their evaluation is dielectric relaxation spectroscopy. This method measures the response of the studied system to a changing external electric field, which reflects its structural features and allows obtaining a number of dielectric parameters, such as complex [real (ε') and imaginary (ε')] dielectric permittivity, electrical conductivity (σ'), as well as the tangent of the dielectric loss angle (tg δ).

Figure 5 shows the dependence of dielectric constant on the content of silver nanoparticles for systems based on polyethylene oxide. It is seen, that the dielectric constant (ε') of composite materials exhibits non-linear behavior in a wide frequency range. In the frequency interval from 10^1 to 10^4 Hz, the dielectric constant decreases sharply, and in the interval 10^4–10^6 Hz it is almost independent of frequency. High values of dielectric constant at low frequencies and its sharp decrease are associated with the so-called electrode blocking effect at the boundary of the electrode–composite distribution.

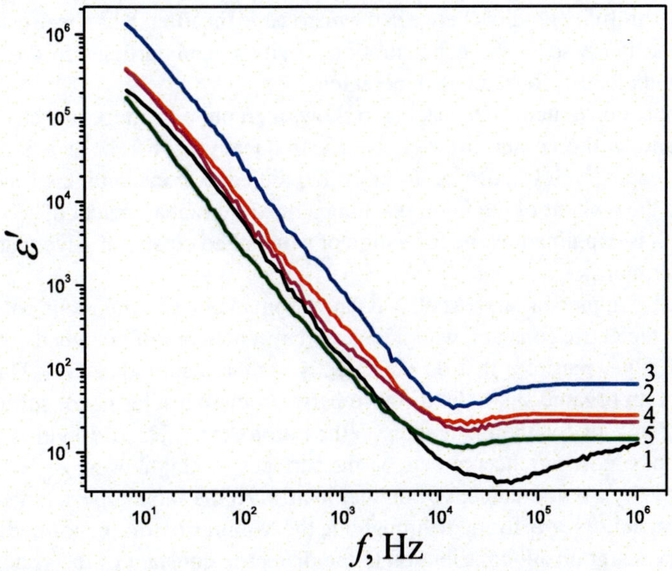

Fig. 5 Frequency dependences of dielectric constant for materials based on PEO, which contain AgNPs: 1–0%; 2–0.5%; 3–1%; 4–1.5%; 5–2%

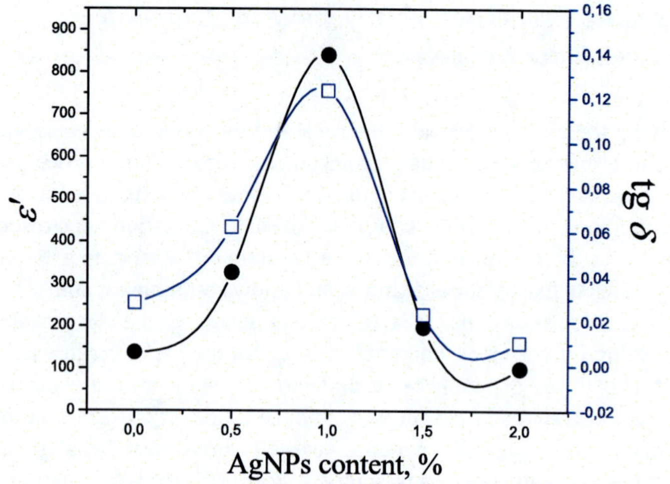

Fig. 6 Concentration dependence of dielectric permittivity and dielectric loss at a frequency of 1 kHz for materials based on PEO

The frequency invariance of the permittivity is likely related to the structure of the AgNPs, which have a silver core and a hyperbranched ionic liquid shell, creating a minimal permittivity gradient between the matrix and the filler. It has been established that with an increase in the concentration of silver nanoparticles in the polymer matrix, the dielectric constant also increases.

For a more convenient analysis, Fig. 6 shows a graph of the dependence of dielectric constant on the content of AgNPs at a frequency of 1 kHz. It is shown that the dependence graph is extreme. Initially, the dielectric constant increases with an increase in the content of the filler, reaching a maximum at a content of 1% AgNPs. This growth is explained by the formation of a branched cluster of silver particles in the polymer matrix.

Maxwell–Wagner-Sillars (MWS) polarization plays an important role in the increase of dielectric constant, which is caused by a network of branched conductive clusters of silver particles that do not directly contact each other [34]. Due to the presence of an organic shell, the distance between the particles is not sufficient for charge hopping or for their tunneling. When an external electric field is applied, charges opposite in sign accumulate on the surfaces of neighboring AgNPs, which are separated by a non-conductive polymer matrix. At the same time, a large number of small capacitors are formed throughout the volume of the nanofilled system, which make a significant contribution to the dielectric constant of the systems. The maximum surface of silver nanoparticles in the system is achieved with their content of 1%. After reaching this concentration, the dielectric constant begins to decrease with an increase in the filler content (Fig. 6). This effect is associated with intensive aggregation of silver nanoparticles in the PEO matrix. When the particles stick together, the surface of the particle–matrix interaction is significantly reduced, this leads to a decrease in the dielectric constant of the material.

Therefore, the relative frequency invariance of the dielectric constant and its high value allows the use of the obtained systems as materials with a high dielectric constant for capacitors and other electrical devices.

Dielectric losses (tg δ) of composites based on PEO and AgNPs are quite low in the frequency range from 10^2 to 10^5 Hz (Fig. 7). These values are lower than the dielectric loss for many two-phase composite systems [35].

For a nanocomposite material containing 2% of nanoparticles, tg δ was 0.01 at 1 kHz. The low value of tg δ for composites containing AgNPs is due to the presence of a stabilizer (macromolecules of a hyperbranched ionic liquid) on the surface of Ag nanoparticles, which forms an insulating shell around the particle. Since the presence of an insulating shell prevents direct contact between Ag nanoparticles, it significantly affects the value of tg δ. Larger values of tg δ in the frequency range up to 10^2 Hz for all studied systems are due to the effect of interphase polarization due to relaxation loss caused by surface polarization (MWS). The conclusion about the presence of interphase polarization is confirmed by the data for the dielectric constant (Fig. 5). A sharp increase in tg δ in the high-frequency range of 10^5–10^6 Hz indicates the presence of another type of relaxation, which is caused by dipole-orientational polarization [36].

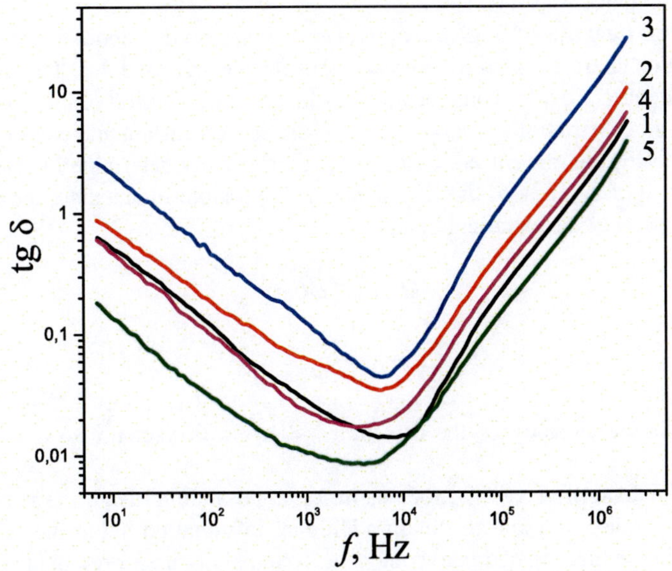

Fig. 7 Frequency dependences of the dielectric loss for materials based on PEO, which contain AgNPs: 1–0%; 2–0.5%; 3–1%; 4–1.5%; 5–2%

The behavior of the dependence of dielectric loss on the content of silver nanoparticles is similar to the dielectric constant (Fig. 6). At the same time, tg δ initially increases with increasing filler content, reaching a maximum at a concentration of 1% AgNPs, after which a decrease in dielectric loss is observed. This behavior is explained by the aggregation processes in the polymer composite material, namely by the change of the matrix-filler interaction surface. These data confirm the conclusion that with a 1% content of AgNPs in the system, the filler is distributed most evenly, which significantly affects the final dielectric properties of the material. Further processes of aggregation that occur with an increase in the concentration of the filler neutralize the effect of surface polarization effects, so the dielectric characteristics deteriorate.

3.3 Relaxation Processes in Polyethylene Oxide-Silver Nanoparticles Systems

Dielectric relaxation spectroscopy is one of the effective methods of studying the relaxation characteristics of various materials, in particular polymer nanocomposites. From the results of the frequency dependences of the imaginary part of the electric modulus and the tangent of the dielectric loss angle for polymer nanocomposite materials, it is possible to judge the mechanisms of movement of macromolecules. These mechanisms largely depend on the composition, structure, and morphology of polymer nanocomposites.

One of the quantities that gives comprehensive information about the behavior of the relaxation characteristics of the material is the electric modulus. The use of the electric modulus provides some advantages in the interpretation of bulk relaxation processes, since variations in large values of dielectric constant and conductivity at low frequencies are minimized. Using the experimentally determined values of the real and imaginary parts of the impedance, it is possible to calculate the real and imaginary parts of the electrical modulus:

$$M' = Z''\omega C_o, \tag{1}$$

$$M'' = Z'\omega C_o, \tag{2}$$

where C_o is the capacity of the capacitor “filled” with vacuum; ω is the cyclic frequency.

The formalism of electrical modules makes it possible to avoid the masking of conductivity relaxation and to obtain additional information about the relaxation processes in the investigated nanocomposite systems. Figure 8 presents the spectra of the imaginary part of the complex electrical modulus for nanocomposites based on PEO and AgNPs.

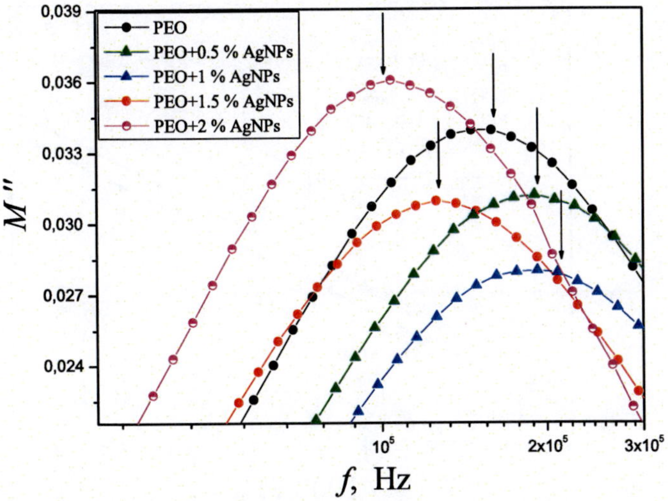

Fig. 8 Dependence of the imaginary part of the electric modulus on the frequency for nanocomposites based on PEO filled with AgNPs. Arrows show maxima that correspond to f_{max}

The analysis of the presented dependences shows that when the filler is introduced, in the area of concentrations up to 1% of AgNPs, there is a shift of the relaxation areas toward higher frequencies, which is associated with an increase in segmental mobility in composite systems. With a content of 1% of the filler, the maximum segmental mobility is observed. With an increase in the content of nanofiller (more than 1%), a shift of the relaxation regions toward lower frequencies is observed, which indicates a decrease in the segmental mobility of polyethylene oxide macromolecules.

Using the expression for the relaxation time, which corresponds to the maximum in Fig. 8 $\tau_{max} = 1/(2\pi f_{max})$, as well as the values f_{max} from the $M''(f)$ dependences, relaxation time spectra were calculated for nanocomposites with different filler contents. Figure 9 shows temperature dependences of relaxation times in semi-logarithmic coordinates.

As can be seen from Fig. 9, the dependences $\tau_{max}(1/T)$ are linear, which makes it possible to describe them using the Arrhenius equation for relaxation times [37]:

$$\tau_{max} = \tau_0 \exp\left(-\frac{E_a}{kT}\right), \quad (3)$$

where E_a is the activation energy; k is the Boltzmann constant, τ_0 is the relaxation time at $T \to \infty$.

The results of calculating the parameters of (3) from the experimental data are given in Table 1. The analysis of the obtained results shows that the activation energy strongly depends on the composition of the systems. With an increase in the content of the filler, a decrease in the relaxation time is first observed. This is evidence of an increase in molecular mobility. At the content of 1% of the nanofiller in the system,

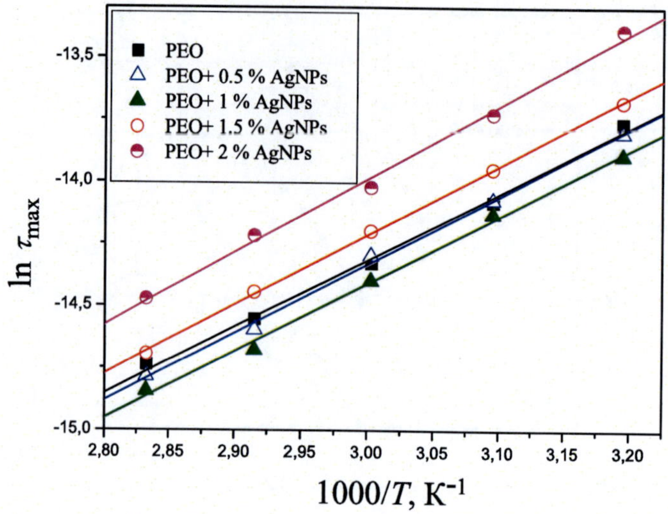

Fig. 9 Temperature dependence of τ_{max} for nanocomposites based on PEO and AgNPs

Table 1 Calculated relaxation characteristics of nanocomposites based on PEO and AgNPs

Sample	τ_0, × 10^{-10} s	E_a, × 10^{-19} J	E_a, eV
PEO	1.9	0.35	0.21
PEO + 0.5% AgNPs	1.5	0.38	0.24
PEO + 1% AgNPs	1.2	0.40	0.25
PEO + 1.5% AgNPs	1.6	0.37	0.23
PEO + 2% AgNPs	1.8	0.36	0.22

the minimum relaxation time and, accordingly, the maximum activation energy are observed. After reaching a critical concentration (1%) of AgNPs in the system, the relaxation time increases, and the activation energy decreases. Comparing the values of E_a and τ_{max} for unfilled PEO and nanocomposite systems based on it, we can conclude that with the introduction of the nanofiller the relaxation time of polyether macromolecules decreases and the activation energy increases. This fact is explained by the "loosening" of the system upon the introduction of AgNPs, which leads to greater mobility of macromolecules.

3.4 Acoustic Properties of Polyethylene Oxide-Silver Nanoparticles Systems

In order to study the influence of AgNPs on the acoustic characteristics of PEO-based nanocomposites, the peculiarities of the passage of ultrasound through the obtained samples were studied. The results of the research are shown in Fig. 10. It can be

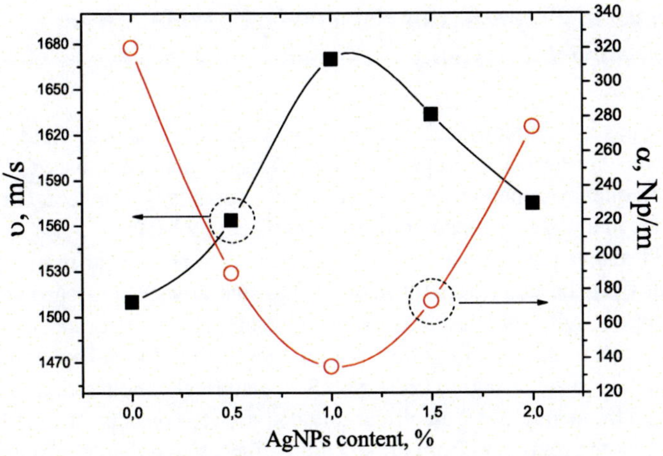

Fig. 10 Dependence of ultrasound propagation speed and attenuation coefficient on the content of AgNPs for PEO-based nanocomposites

seen that the velocity of ultrasound propagation (υ) increases with increasing filler content, exhibiting extreme behavior. At the same time, the speed of sound increases from 1510 m/s for unfilled PEO to almost 1690 m/s for the material containing 1% of AgNPs. A similar increase in the value of υ in the concentration range up to 1% was recorded by Klepko et al. [38] for a system based on polyethylene oxide and carbon nanotubes. After reaching the maximum, the value of υ decreases (Fig. 10).

Figure 10 shows the dependence of the ultrasound attenuation coefficient (α) on the content of silver nanoparticles. It is shown that this concentration dependence also exhibits extreme behavior. In the range of concentrations from 0 to 1%, there is a sharp decrease in the attenuation coefficient from 320 Np/m to 135 Np/m. Such decrease is associated with the formation of an "infinite" cluster of silver nanoparticles. During the formation of such a cluster, phonon transmission paths are formed. At the same time, the energy dissipation of ultrasonic waves is sharply reduced. After reaching a maximum, the damping coefficient shows an increasing tendency, which is obviously correlated with a decrease in the value of υ. These processes are explained by the aggregation of filler particles.

The distribution of filler particles significantly affects the relaxation and acoustic characteristics of the studied materials. With a relatively small amount of filler (less than 1%), AgNPs are distributed evenly, which leads to a significant effect of them on relaxation (E_a and τ_{max}) and acoustic characteristics (υ and α). The process of aggregation of AgNPs neutralizes the effect of their nanosize due to the formation of significantly larger particles. Therefore, the trend of changing the studied characteristics in the concentration range from 1 to 2% is the reverse of the previous one. Therefore, 1% of AgNPs in the system is optimal, so it can be predicted that with this content of the filler, the functional characteristics will be the most acceptable for the use of such a material.

3.5 Electrical Properties of Polyethylene Oxide-Silver Nanoparticles Systems

Since silver has a high electrical conductivity, it was necessary to study the effect of AgNPs on the electrical properties of PEO-based nanocomposites. Figure 11 shows the frequency dependence of electrical conductivity on the content of silver nanoparticles in a wide frequency range from 10 Hz to 2 MHz.

Figure 11 shows that the specified dependencies are non-linear in nature. These dependencies can be conditionally divided into two areas: the area of growth of electrical conductivity and the area of the plateau. The first region is related to the so-called electrode blocking effect and, in fact, is the Maxwell–Wagner–Sillars polarization. The plateau region corresponds to electrical conductivity, which does not depend on frequency and is electrical conductivity at a constant current. Figure 11 also shows that the transition from polarization effects to the direct current region is observed at different frequencies. At the same time, with an increase in the content of the filler, the frequency of such a change increases. This effect is associated with an increase in the area of the interfacial layer.

Analyzing the general level of direct current conductivity, it can be stated that this level increases with an increase in the AgNPs content. At the same time, the electrical conductivity reaches the level of 10^{-4} S/cm with a filler content equal to 2%. This effect is associated with an increase in the content of more electrically conductive particles in the system compared to the polymer matrix [39]. However, it should be noted that the obtained level of electrical conductivity is significantly lower than the

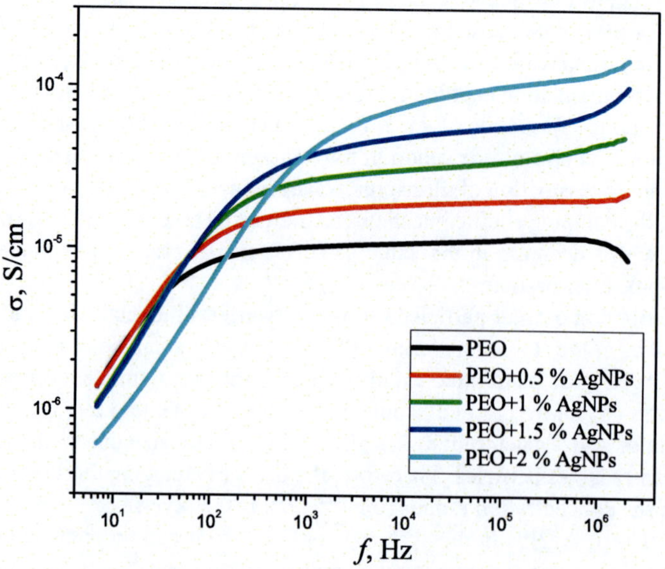

Fig. 11 Frequency dependence of electrical conductivity for PEO-AgNPs systems

electrical conductivity of pure silver. This is because the silver nanoparticles have been stabilized by an ionic liquid, which is essentially a hyperbranched polymer. The polymer creates a shell around the particle, which prevents the direct passage of charge carriers through the silver particles. Therefore, the influence of polarization effects increases for the studied nanocomposite systems, and the level of electrical conductivity is significantly lower than that of silver.

3.6 *Antimicrobial Properties of Polyethylene Oxide-Silver Nanoparticles Systems*

Wide practical application of the studied polymer systems in beauty industry is evidence of their loyalty to the environment and the human body in particular, which opens wide prospects for their use in biology, medicine, pharmacology, and more [40]. In view of this, the antimicrobial properties of these materials had to be studied. Nanocomposites based on PEO and silver nanoparticles show high antimicrobial activity against strains of *S. aureus* and *E. coli*. After 24 h of incubation at 37 °C, the presence of a clean zone around the contours of the films, which indicates inhibition of the growth of microorganisms (Fig. 12 and Table 2) was observed.

It was found that the antimicrobial activity of silver-containing nanocomposites depends on the content of nanoparticles. Thus, for a system containing 0.5% of nanocrystals, no inhibition of any microorganisms is observed, while for a filler content of 1% or more, the growth of all tested bacteria is inhibited. As expected, samples containing 2% nanosilver exhibit the highest antimicrobial activity.

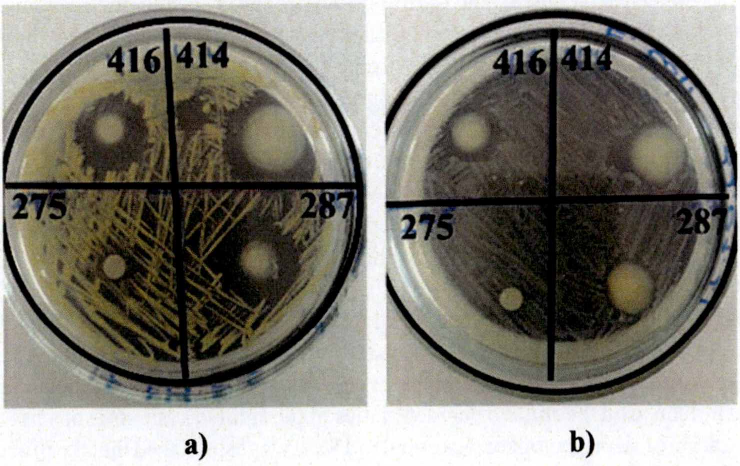

Fig. 12 Antimicrobial activity of silver-containing nanocomposites, based on PEO against strains of *S. aureus* (**a**) and *E. coli* (**b**)

Table 2 The values of the diameters of the zones of growth retardation of microorganisms (zones of inhibition) in the Mueller–Hinton medium

Sample	*Staphylococcus aureus* (mm)	*Escherichia coli* (mm)
275 (0.5% AgNPs)	–	–
287 (1.0% AgNPs)	22	13
416 (1.5% AgNPs)	20	16
414 (2.0% AgNPs)	20	16

For this sample, the characteristic zone of inhibition for *S. aureus* bacteria is 22 mm, which is very close to the effect of antibiotics. The obtained values of antimicrobial activity for *S. aureus* were significantly higher than the values obtained by Rujitanaroj et al. [28] for coatings based on polyethylene oxide and silver nanoparticles. In their work, the zone of inhibition at a filling of 2.5% silver nanoparticles was 16 mm. So, obtained materials have significant antimicrobic properties and high potential of applications.

4 Conclusions

In this work, a new approach to the synthesis of stabilized silver nanoparticles was proposed and the functional properties of polymer nanocomposite materials based on polyethylene oxide were investigated. The resulting nanoparticles are core–shell objects, inside the particle is a silver core covered with a hyperbranched ionic liquid shell. It was established that stabilized silver nanoparticles significantly affect the dielectric, dielectric, acoustic, and antimicrobic properties of the polymer matrix at relatively low concentrations of the nanofiller (~ 1–2%). At the same time, dielectric permittivity and dielectric loss exhibit extreme behavior with increasing filler content in the system. It was determined that the greatest effect on the dielectric properties of the studied systems is observed at a concentration of silver nanoparticles equal to 1%. It was established that two types of polarization occur in the investigated composite systems: interphase and dipole-orientational. Developed materials with improved dielectric properties can be used in the future as mediums for capacitors with high energy storage density, as well as coatings with the required dielectric properties.

The relaxation characteristics of the polymer matrix (activation energy and relaxation time) and acoustic (sound speed and attenuation coefficient) exhibit extreme behavior with increasing filler content in the system. It was determined that the greatest influence on the studied characteristics of the studied systems is observed at a concentration of silver nanoparticles equal to 1%. It was established that the processes of aggregation of the filler, which reduce the nanosize of silver particles, have a decisive influence on the relaxation and acoustic characteristics of nanocomposite systems based on PEO.

It is shown that the electrical conductivity of the studied systems increases with the increase in the content of nanoparticles in the system. With a content of 2 wt% filler electrical conductivity reaches the level of 10^{-4} S/cm. Nanocomposite containing 2 wt% of silver nanoparticles shows significant activity against *S. aureus* (inhibition zone is 22 mm). The developed materials with improved functional characteristics can be used in the future as antimicrobial coatings, antimicrobial films, and creams, as well as building products with the required functional characteristics and antimicrobial function.

References

1. M. Muhammed Shameem, S.M. Sasikanth, R. Annamalai, R. Ganapathi Raman, A brief review on polymer nanocomposites and its applications. Mater. Today Proc. **45**(2), 2536 (2021)
2. E.A. Lysenkov, N.G. Leonova, S.V. Zhiltsova, Effect of the silicon-containing phase on the hierarchy of the structure of epoxy-silica nanocomposites. Theor. Exper. Chem. **55**(4), 250 (2019)
3. E.A. Lysenkov, V.V. Klepko, Influence of anisometric fillers on electrical properties of polypropylene glycol-based nanocomposites. Ukr. J. Phys.. J. Phys. **56**(5), 484 (2011)
4. E.A. Lysenkov, Z.O. Haholkina, E.V. Lobko, M.H. Tkalich, V.V. Klepko, Influence of carbon nanotubes on the mechanical properties of cross-linked polyurethanes. Mater. Sci. **53**(1), 14 (2017)
5. J. Wan, B. Fan, S.H. Thang, Sonochemical preparation of polymer–metal nanocomposites with catalytic and plasmonic properties. Nanoscale Adv. **3**, 3306 (2021)
6. D. Giliopoulos, A. Zamboulis, D. Giannakoudakis, D. Bikiaris, K. Triantafyllidis, Polymer/metal organic framework (MOF) nanocomposites for biomedical applications. Molecules **25**(1), 185 (2020)
7. E.A. Lysenkov, V.V. Klepko, I.P. Lysenkova, Features of microstructure and percolation behavior of polypropylene glycol, filled by multiwalled carbon nanotubes. J. Nano Electron. Phys. **9**(5), 05021 (2017)
8. T. Bruna, F. Maldonado-Bravo, P. Jara, N. Caro, Silver nanoparticles and their antibacterial applications. Int. J. Mol. Sci. **22**, 7202 (2021)
9. S. Dawadi, S. Katuwal, A. Gupta, U. Lamichhane, R. Thapa, S. Jaisi, G. Lamichhane, D.P. Bhattarai, Parajuli N (2021) Current research on silver nanoparticles: synthesis, characterization, and applications. J. Nanomater.Nanomater. **6**, 6687290 (2021)
10. Y. Stetsyshyn, K. Awsiuk, V. Kusnezh, J. Raczkowska, B.R. Jany, A. Kostruba, K. Harhay, H. Ohar, Y. Shymborska, O. Lishchynskyi, Y. Kryvenchuk, F. Krok, A. Budkowski, Shape-controlled synthesis of silver nanoparticles in temperature-responsive grafted polymer brushes for optical applications. Appl. Surf. Sci. **463**, 1124 (2019)
11. P. Lodeiro, E.P. Achterberg, C. Rey-Castro, M.S. El-Shahawi, Effect of polymer coating composition on the aggregation rates of Ag nanoparticles in NaCl solutions and seawaters. Sci. Total. Environ. **631–632**, 1153 (2018)
12. E. Husanu, C. Chiappe, A. Bernardini, V. Cappello, M. Gemmi, Synthesis of colloidal Ag nanoparticles with citrate based ionic liquids as reducing and capping agents. Colloid Surf. Physicochem. Eng. Aspect **538**, 506 (2018)
13. E.A. Lysenkov, V.V. Klepko, I.P. Lysenkova, Features of structural organization of nanodiamonds in the polyethylene glycol matrix. J Nano Electron. Phys. **12**(4), 04006 (2020)
14. E.A. Lysenkov, I.P. Lysenkova, Influence of nanodiamonds on the structure and thermophysical properties of polyethylene glycol-based systems. Funct. Mater. **27**(4), 774 (2020)

15. K.N. Kumar, M. Kang, K. Sivaiah, M. Ravi, Y.C. Ratnakaram, Enhanced electrical properties of polyethylene oxide (PEO) + polyvinylpyrrolidone (PVP):Li^+ blended polymer electrolyte films with addition of Ag nanofiller. Ionics **22**, 815 (2015)
16. E.A. Lysenkov, V.V. Klepko, Y.V. Yakovlev, Specifics of percolation behavior in the polyether–carbon nanotube systems doped with LiClO4. Surf. Eng. Appl. Electrochem.Electrochem. **52**(2), 186 (2016)
17. E.A. Lysenkov, Y.V. Yakovlev, V.V. Klepko, The influence of the $LiClO_4$ salt on the percolation behavior of the systems based on the polypropylene glycol and carbon nanotubes. J. Phys. Stud. **17**(1), 1 (2013)
18. Q. Chen, Y. Shen, S. Zhang, Q.M. Zhang, Polymer-based dielectrics with high energy storage density. Ann. Rev. Mater. Res. **45**, 433 (2015)
19. J. Dai, S. Meng, C. Yang, W. Lu, X. Chen, Y. Yin, F. Liang, Effect of nano silver modification on the dielectric properties of Ag@TiO_2/PVDF composites. J. Wuhan Univ. Technol. Mater. Sci. Ed. **36**, 303 (2021)
20. G. Sahu, M. Das, M. Yadav, B.P. Sahoo, J. Tripathy, Dielectric relaxation behavior of silver nanoparticles and graphene oxide embedded poly(vinyl alcohol) nanocomposite film: an effect of ionic liquid and temperature. Polymers **12**, 374 (2020)
21. S. More, R. Dhokne, S. Moharil, Dielectric relaxation and electric modulus of polyvinyl alcohol–Zinc oxide composite films. Mater. Res. Express **4**, 055302 (2017)
22. V. Shilov, V. Sperkach, Y. Sperkach, A. Strybulevych, Acoustic relaxation of liquid poly(tetramethylene oxide) with hydroxyl and acyl terminal groups. Polym. J.. J. **34**(8), 565 (2002)
23. R. Zhang, W. Cao, Q. Zhou, J.H. Cha, K.K. Shung, Y. Huang, Acoustic properties of alumina colloidal/polymer nano-composite film on silicon. IEEE Trans. Ultrason. Ferroelectr. Freq. Control **54**(3), 467 (2007)
24. T.C. Dakal, A. Kumar, R.S. Majumdar, V. Yadav, Mechanistic basis of antimicrobial actions of silver nanoparticles. Front. Microbiol.Microbiol. **7**, 1831 (2016)
25. M. Qasim, N. Udomluck, J. Chang, H. Park, K. Kim, Antimicrobial activity of silver nanoparticles encapsulated in poly-N-isopropylacrylamide-based polymeric nanoparticles. Int. J. Nanomed.Nanomed. **13**, 235 (2018)
26. R. Cao, X. Zhai, X. Li, X. Zhao, Antibacterial properties of a novel nano-silver loaded poly(styrene-co-acrylic) composites. Polym. Polym. Compos. **29**(9_suppl), S1017 (2021)
27. A. Kausar, Polymer/silver nanoparticle nanocomposite as antimicrobial materials. Front. Sci. **7**(2), 31 (2017)
28. P. Rujitanaroj, N. Pimpha, P. Supaphol, Preparation of ultrafine poly (ethylene oxide) / poly (ethylene glycol) fibers containing silver nanoparticles as antibacterial coating, in *2nd IEEE International Conference on Nano/Micro Engineered and Molecular Systems*, p. 1065 (2007)
29. V.V. Shevchenko, A.V. Stryutsky, N.S. Klymenko, M.A. Gumenna, A.A. Fomenko, V.N. Bliznyuk, V.V. Trachevsky, V.V. Davydenko, V.V. Tsukruk, Protic and aprotic anionic oligomeric ionic liquids. Polymer **55**(16), 3349 (2014)
30. N.I. Lebovka, E.A. Lysenkov, A.I. Goncharuk, Yu.P. Gomza, V.V. Klepko, Yu.P. Boiko, Phase behaviour, microstructure, and percolation of poly (ethylene glycol) filled by multiwalled carbon nanotubes and organophilic montmorillonite. J. Compos. Mater. **45**(24), 2555 (2011)
31. E.A. Lysenkov, V.V. Klepko, I.P. Lysenkova, Influence of modification of carbon nanotubes on the microstructure and percolation behavior of the systems based on polyethylene glycol. J. Phys. Stud. **21**(4), 4701 (2017)
32. B. Kumar, K. Smita, L. Cumbal, A. Debut, Green synthesis of silver nanoparticles using Andean blackberry fruit extract. Saudi J. Biol. Sci. **24**(1), 45 (2017)
33. S.B. Aziz, Z.H.Z. Abidin, A.K. Arof, Effect of silver nanoparticles on the DC conductivity in chitosan-silver triflate polymer electrolyte. Phys. B Condens. Matter **405**(21), 4429 (2010)
34. M. Samet, A. Kallel, A. Serghei, Maxwell-Wagner-Sillars interfacial polarization in dielectric spectra of composite materials: Scaling laws and applications. J. Compos. Mater. **56**(20), 3197 (2022)

35. Z. Wang, M. Fang, H. Li, Y. Wen, C. Wang, Y. Pu, Enhanced dielectric properties in poly(vinylidene fluoride) composites by nanosized $Ba(Fe_{0.5}Nb_{0.5})O_3$ powders. Compos. Sci. Technol. **117**, 410 (2015)
36. G. Chen, X. Wang, J. Lin, W. Yang, H. Li, Y. Wen, Interfacial polarity modulation of $KTa_{0.5}Nb_{0.5}O_3$ nanoparticles and its effect on dielectric loss and breakdown strength of poly(vinylidene fluoride) nanocomposites with high permittivity. J. Phys. Chem. C **120**, 28423 (2016)
37. Y. Shangguan, F. Chen, E. Jia, Y. Lin, J. Hu, Q. Zheng, New insight into time-temperature correlation for polymer relaxations ranging from secondary relaxation to terminal flow: application of a universal and developed WLF equation. Polymers **9**, 567 (2017)
38. V.V. Klepko, B.B. Kolupaev, E.A. Lysenkov, M.O. Voloshyn, Viscoelastic properties of filled polyethylene glycol in the megahertz frequency band. Mater. Sci. **47**(1), 14 (2011)
39. E.A. Lysenkov, V.V. Klepko, Analysis of percolation behavior of electrical conductivity of the systems based on polyethers and carbon nanotubes. J. Nano Electron. Phys **8**(1), 01017 (2016)
40. A. Haider, I.K. Kang, Preparation of silver nanoparticles and their industrial and biomedical applications: a comprehensive review. Adv. Mater. Sci. Eng. **2015**, 165257 (2015)

Thermal Properties of Chitosan-Based Interpenetrating Polymer Network Hydrogels

O. Nadtoka, O. Vashchenko, and N. Kutsevol

Abstract Interpenetrating polymer networks (IPNs) were synthesized when acrylamide was subjected to radical polymerization and simultaneous cross-linking in the presence of chitosan. IPN samples were prepared as hydrogels using ammonium cerium (IV) nitrate as a redox initiator and *N*,*N*′-methylene-bis-acrylamide as a cross-linking agent. The structure of the obtained polymer networks was confirmed by FTIR spectra. The effect of chitosan content on the thermal properties of IPNs was investigated by the thermogravimetry and differential thermogravimetry analysis.

1 Introduction

Polyacrylamide (PAA) hydrogels are cross-linked hydrophilic polymer networks that are able to hold a significant amount of water and aqueous solutions in their pores. This makes them promising biomaterials for various biomedical applications, such as drug delivery systems, wound dressings, tissue engineering, as well as water purification, removal of heavy metals and dyes, etc. [1, 2]. Although polyacrylamide hydrogels are widely used in various fields, there is a demand for materials with improved characteristics and functionality. Since the approval of new polymers for introduction and use in biomedicine requires complex tests regarding their safety, combining known polymers with natural ones is a simple and effective idea for improving the properties of hydrogels according to the goals. Each network retains its individual properties caused by the structure of the polymer, so it is possible to observe synergistic improvements in the physical and chemical properties of combined systems [3].

O. Nadtoka (✉) · N. Kutsevol
Taras Shevchenko National University of Kyiv, 64, Volodymyrska str., Kyiv 01033, Ukraine
e-mail: oksananadtoka@ukr.net; nadtoka_oksana@knu.ua

O. Vashchenko
Institute for Scintillation Materials of NAS of Ukraine, Ave. Nauky, 60, Kharkiv 61172, Ukraine

O. Fesenko and L. Yatsenko (eds.), *Nanostructured Surfaces, Nanocomposites and Nanomaterials, and Their Applications*, Springer Proceedings in Physics 296,
https://doi.org/10.1007/978-3-031-42704-6_19

Polysaccharides as natural polymers are known for their high biodegradability, biocompatibility, and bioactivity, while they are cheap materials for developing new-generation hydrogels. Among the polysaccharides that have recently attracted increasing attention in creating hydrogel materials, chitosan (pure or in combination with natural/synthetic polymers) has been widely investigated for use in the biomedical field.

Chitosan is one of the most common biopolymers in nature. In cationic form, chitosan is obtained after deacetylation of chitin in an alkaline medium [4]. The chain of this biopolymer consists mainly of β-[1-4]-2-acetamido-2-deoxy-D-glucose units. The presence of the NH_2 group in the C-2 position of the repeating unit of D-glucosamine determines the protonation reaction and conversion of the polysaccharide into a polyelectrolyte in an acidic solution. This leads to the ability of chitosan to dissolve well in dilute acids.

One of the well-known methods of improving the mechanical, thermal, and other properties of hydrogels is the creation of materials based on an interpenetrating polymer networks (IPNs). The proposed methods of obtaining IPN based on chitosan involve the cross-linking of the acrylamide-chitosan mixture using *N,N′*-methylene-bis-acrylamide as a cross-linker [5]. The grafting reaction of acrylamide on carboxymethylchitosan using cerium ammonium nitrate as an initiator [6] indicates the possibility of obtaining a cross-linked polymer network based on grafted copolymers.

In this article, we describe the preparation of interpenetrating polymer networks based on chitosan and PAA obtained by radical polymerization of acrylamide in the presence of polymer chitosan and cross-linker. These IPN-hydrogels are promising biomaterials that can be used for the production of wound dressings. The obtained copolymers were characterized by FTIR spectroscopy to analyze their structure. DTGA study of the influence of the mass ratio of chitosan and polyacrylamide on the thermal properties of polymer networks is a tool for considering the effect of intermolecular interactions on the peculiarities of the morphological structure of hydrogels.

2 Experimental

2.1 Materials

Acrylamide (AA) obtained from Aldrich was twice re-crystallized from chloroform and dried under vacuum. Cerium (IV) ammonium nitrate (CAN), *N,N′*-methylene-bis-acrylamide (MBA), acetic acid, and concentrated nitric acid were purchased from Aldrich, without additional purification. Chitosan (Ch) of average molecular weight 250,000, viscosity 30–100 cP, and deacetylation degree > 90% was obtained from Glentham Life Sciences. Distilled water was used throughout the experiments and as polymerization medium during hydrogel synthesis.

2.2 Polymer Network Synthesis

To prepare the initial reaction mixture, 0.05 g (0.1 g or 0.2 g) of chitosan was dissolved in 25 ml of an aqueous solution of acetic acid (2%). Then, acrylamide 3.5 g and *N,N′*-methylene-bis-acrylamide 0.4% (wt/monomer) were added to the resulting viscous chitosan solution and stirred for 30 min to achieve uniform distribution of the reagents. The reaction mixture was heated to 40 °C and purged with argon bubbling within 25 min. Cerium (IV) ammonium nitrate 0.0164 g was dissolved in 1 ml of 0.1 M solution of nitric acid and was added instantly. After thorough stirring in an argon environment for 2 min, the reaction mixture was left for a day for gelation. The resulting hydrogels were immersed in distilled water at room temperature for 48 h to remove unreacted chemicals. Obtained IPN-hydrogels had different mass ratios of chitosan and polyacrylamide. They were designated as PAA-Ch-0.05; PAA-Ch-0.1; PAA-Ch-0.2, according to the amount of chitosan used during the reaction.

For comparison, a hydrogel based on cross-linked polyacrylamide (designated throughout as PAA) was obtained [7].

2.3 Characterization of Polymer Networks

FTIR spectra of the obtained IPNs were recorded in KBr using a BrukerVector-22 spectrometer in the wavenumber range of 400–4000 cm^{-1}. The samples were made 0.25 mm thick in the form of KBr tablets with polymer samples.

Thermal properties of the hydrogel samples were studied using thermoanalytical equipment Mettler TA 3000 (thermogravimetry and differential thermogravimetry analysis, DTGA). Atmosphere gas (1 bar, relative humidity 40%) was applied in all the thermoanalytical measurements.

The samples were examined after vacuum drying at 4×10^{-7} bar and 60 °C using concentrator Concentrator Plus. Drying procedure has been performed until stabilization of sample mass was achieved (during 90 min). Sample mass was controlled by means of Mettler XP26 microbalance. After drying, the samples were placed in a desiccator where they were further incubated for 24 h. Average values of weight loss were obtained for 3 samples of each type.

TGA and DTG curves of the dried samples (4–5 mg) were obtained in temperature range of 30–500 °C at scanning rate of 5 °C/min using 40 μl crucibles of aluminum oxide. TGA and DTG profiles were also obtained for fully hydrated (original) samples weighting 40–50 mg in temperature range of 30–105 °C at scanning rate of 1 °C/min using 150 μl crucibles of aluminum oxide. DTGA data were processed after baseline subtraction. Proper baselines were obtained in blank scans (using empty crucible) under corresponding experiment conditions. Weight loss of samples was calculated relatively to initial sample mass in each measurement.

3 Results and Discussion

3.1 IPNs Synthesis and Characterization

Interpenetrating polymer networks based on chitosan and PAA were produced by the radical polymerization method to obtain materials with the functionality of both polymers. Here, the polymer chitosan was dissolved in an aqueous solution of monomers, which were then polymerized to form a cross-linked polymer network called an interpenetrating polymer network IPN. The scheme of possible simultaneous reactions is described in Fig. 1. It indicates the formation of free radicals both on the acrylamide monomer and on the chitosan macromolecule. To the chitosan macroradical, the acrylamide molecule is attached and initiates the formation of the grafted chain. Cross-linking during the simultaneous processes of polymerization of acrylamide and grafting of polyacrylamide on chitosan can lead to interpenetrating networks of cross-chitosan-g-polyacrylamide and cross-polyacrylamide, which were designated as PAA-Ch-x.

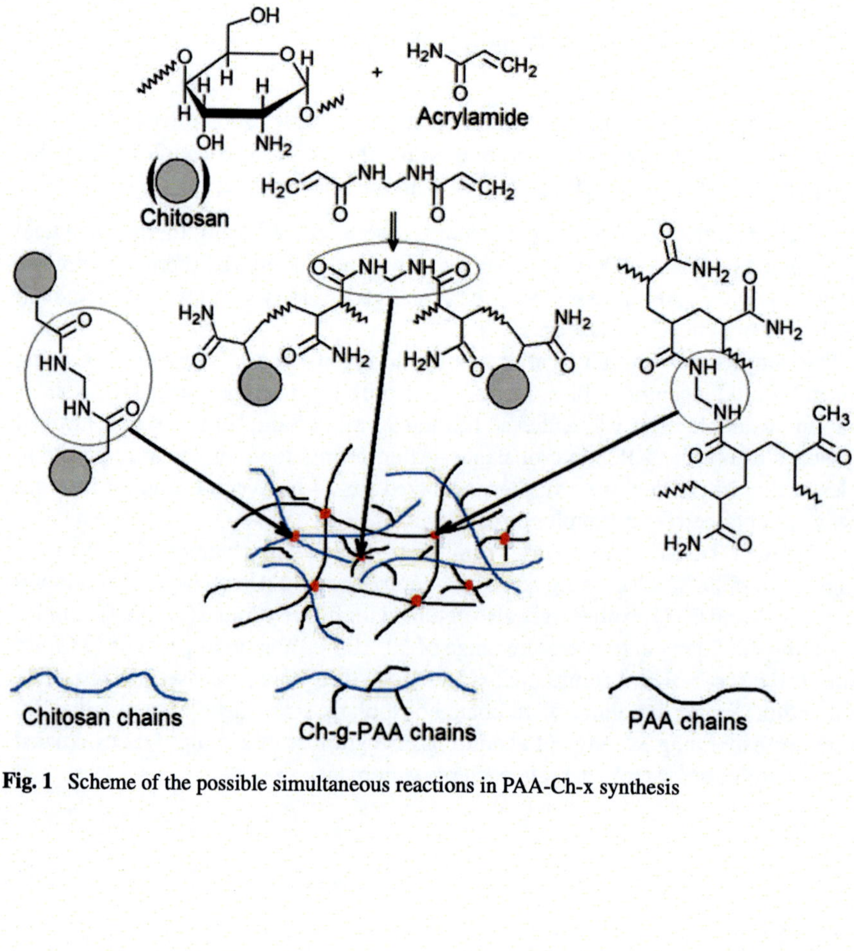

Fig. 1 Scheme of the possible simultaneous reactions in PAA-Ch-x synthesis

The structure of polymer networks based on chitosan and polyacrylamide was characterized by FTIR spectra (Fig. 2). As an example, the spectrum of sample PAA-Ch-0.1 is presented. For comparison, FTIR spectra of chitosan and PAA were obtained. Since the broad absorption band from 3600 to 2850 cm^{-1} corresponds to the valence vibrations associated with the hydrogen bonds of OH and NH_2 groups of chitosan, and the absorption maximum of water lies at 3400 cm^{-1}, all these bands largely overlap. Thus, this region of the FTIR spectra was not used for the comparative analysis of the studied cross-linked polymers. PAA samples show characteristic absorption peaks at 1665 cm^{-1} (ν(C=O), amide I) and 1616 cm^{-1} (δ(N–H), amide II) [7]. In the spectrum of the chitosan characteristic peaks at 1655 cm^{-1} and 1597 cm^{-1} can be assigned to carbonyl stretching and the stretching vibrations of the amino group (amide bands) of the aminoacetyl group of chitosan [8]. Absorption peak at 1382 cm^{-1} was assigned to the vibrations of C–H. In this region, asymmetric stretching mode of CH_2 group of PAA can be found through a medium sharp peak at 1450 cm^{-1}. In contrast, the peak at 1413 cm^{-1} corresponds to the inplane C–O–H bonding mode of PAA. Both of these respective for PAA peaks are shifted to a higher wave number of 1417 and 1454 cm^{-1} in the PAA-Ch-0.1 network. It may confirm the interaction between NH_2 of chitosan and oxygen of PAA.

In chitosan the broad peak at 1082 cm^{-1} is from C–O stretching vibrations. Furthermore, the peaks at 1159 cm^{-1} can be assigned to C–N stretching in the amino group of chitosan. These peaks are present in the spectrum of the PAA-Ch-0.1, but they are less intense. The rest of the characteristic peaks of polymers are CH_2 rocking mode from 794 to 792 cm^{-1} of the polymer chains.

3.2 Thermal Properties

For thermal studies, polymer samples were dried in a vacuum drying oven followed by incubation in a desiccator until a constant mass was reached.

TGA and DTG thermograms of cross-linked PAA (for comparison) and PAA-Ch-x are shown in Figs. 3, 4, 5 and 6 and tabulated in Table 1.

Each thermogram contains a region of initial weight loss and a set of peaks before decomposition onset. As seen from the figures, the first weight loss step begins at the temperature of 30 °C and corresponds to the loss of residual moisture. The following events up to about 300 °C seem rather sporadic by value and temperature, so they could be attributed to the release of residual bound water [9, 10]. Nevertheless, some regularity could be traced in these data. Namely, the sum of weight loss values in this temperature region reduces significantly, from 23% for PAA to 12% for PAA-Ch-0.2. So, DTGA data of the dried samples allow one to conclude that increased chitosan content leads to a reduced amount of bonded water in the samples. This can be explained by the fact that the hydrophilicity of chitosan is lower than that of PAA.

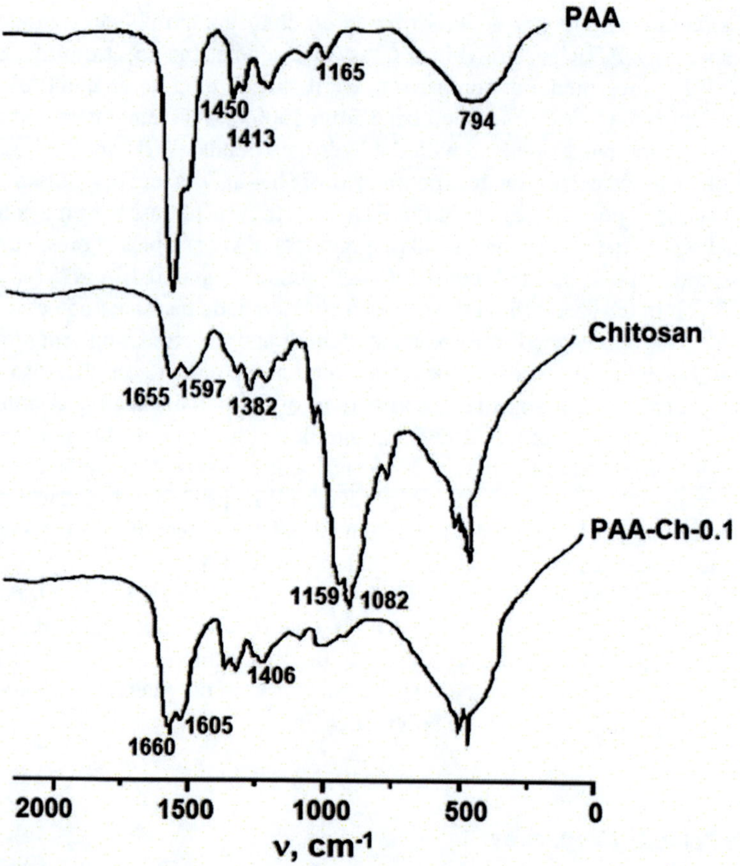

Fig. 2 FTIR spectra of PAA, chitosan and PAA-Ch-0.1

For the PAA sample, one can observe a prominent DTG peak at 393 °C which corresponds to the polyacrylamide decomposition [11]. Notably, the chitosan addition to the structure of polymer networks has shifted the main peak of PAA destruction from 393 to 408 °C and elevates the thermal stability of polyacrylamide by 10–15 °C. The fraction of this peak is gradually decreased with chitosan concentration (see Table 1), from 38% for PAA to 25% for PAA-Ch-0.2 with the largest chitosan content.

The stage of the thermal destruction of chitosan begins at about 300 °C, which contributed to the depolymerization and decomposition of the basic unit of polymer polysaccharides [12]. Correspondingly, a fraction of the DTG peak of chitosan decomposition increases from 0 to 12.5% with increasing chitosan concentration in samples. The increase of the peak of the chitosan thermal decomposition probably reflects its higher thermostability when combined with polyacrylamide.

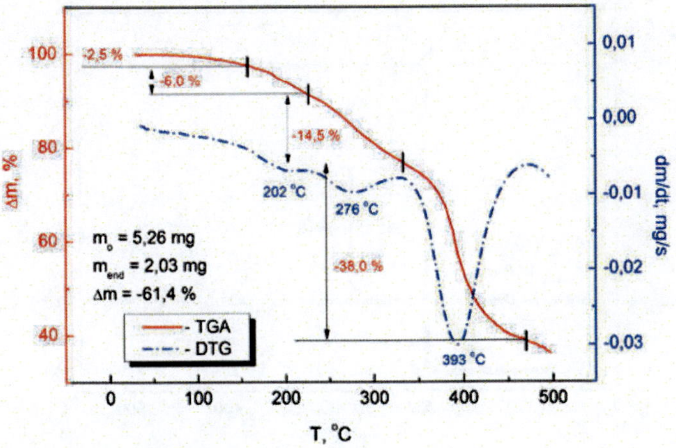

Fig. 3 TGA and DTG profiles of cross-linked PAA

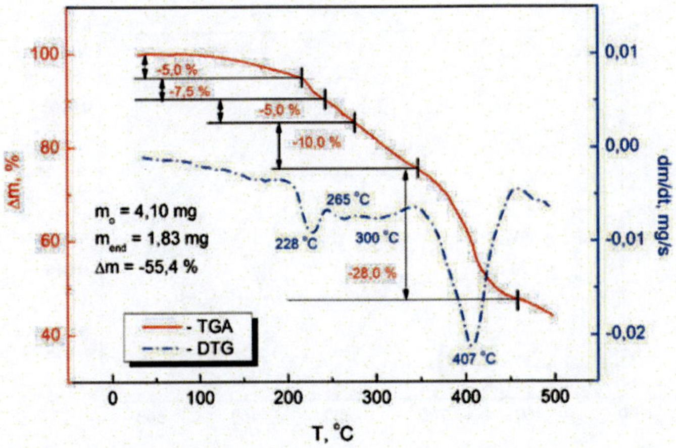

Fig. 4 TGA and DTG profiles of PAA-Ch-0.05

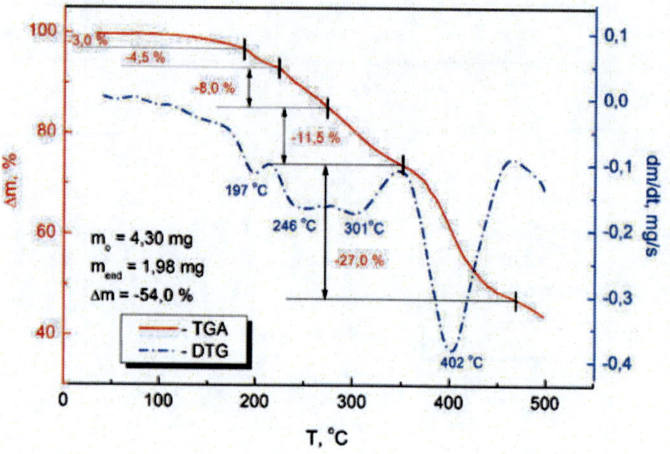

Fig. 5 TGA and DTG profiles of PAA-Ch-0.1

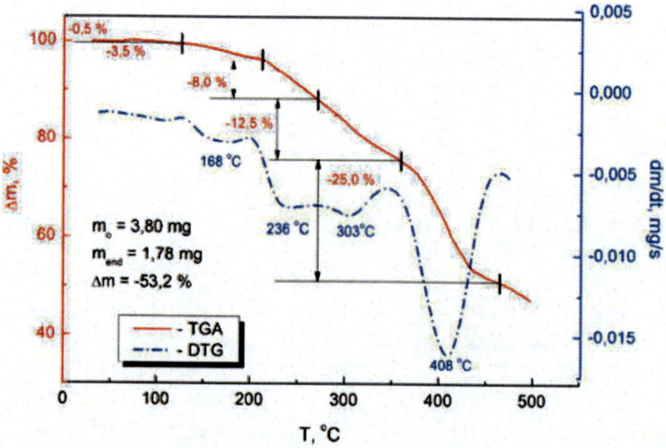

Fig. 6 TGA and DTG profiles of PAA-Ch-0.2

Table 1 Results of the TG analysis of polymer networks

Samples	Stage	T_{onset}, °C	T_{peak}, °C	T_{endset}, °C	ΔT, °C	Weight loss, %	Residual water, %
PAA	I	30	–	158	128	2.5	23
	II	158	202	226	68	6.0	
	III	226	276	333	107	14.5	
	IV	333	393	469	136	38.0	
PAA-Ch-0.05	I	30	–	217	187	5.0	17.5
	II	217	228	243	26	7.5	
	III	243	265	277	34	5.0	
	IV	277	300	347	70	10.0	
	V	347	407	457	110	28.0	
PAA-Ch-0.1	I	30	–	188	158	3.0	15.5
	II	188	197	224	36	4.5	
	III	224	246	280	56	8.0	
	IV	280	301	351	71	11.5	
	V	351	402	469	118	27.0	
PAA-Ch-0.2	I	30	–	127	97	0.5	12
	II	127	168	212	85	3.5	
	III	212	236	271	59	8.0	
	IV	271	303	359	88	12.5	
	V	359	408	464	105	25.0	

4 Conclusion

Interpenetrating polymer networks based on chitosan and PAA were produced by the radical polymerization method. During the simultaneous processes of polymerization of acrylamide and grafting of polyacrylamide on chitosan, a cross-linking reaction occurred. As a result, cross-linked chitosan-g-polyacrylamide and polyacrylamide networks were formed. FTIR spectra confirmed the structure of IPNs. DTGA studies have shown that with an increase in the content of chitosan in the polymer network, the thermal stability of the materials increases. This can be explained by the formation of more rigid structures as a result of the intermolecular interaction of polymer chains of PAA and chitosan. Also, the thermograms of IPN and PAA contain a region of initial weight loss before the onset of decomposition, which is associated with the release of residual bound water. Analysis of DTGA data shows the effect of chitosan on the amount of bound water in the samples. With an increase in the content of chitosan in IPN, the hydrophilicity of the networks decreases significantly. The obtained IPN-hydrogels are proposed as promising biomaterials for the production of wound dressings.

References

1. E.M. Ahmed, Hydrogel: preparation, characterization, and applications: a review. J. Adv. Res. **6**, 105–121 (2015)
2. Zh. Zhao, Y. Huang, Y. Wu, Sh. Li, H. Yin, J. Wang, α-ketoglutaric acid modified chitosan/ polyacrylamide semi-interpenetrating polymer network hydrogel for removal of heavy metal ions. Colloids Surf. A **628**, 127262 (2021)
3. L. Ignat, A. Stanciu, Advanced polymers: interpenetrating polymer networks, in *Handbook of Polymer Blends and Composites* (Rapra Technology, 2003), pp. 275–280
4. C.T.G.V.M.T. Pires, J.A.P. Vilela, C. Airoldi, The effect of chitin alkaline deacetylation at different condition on particle properties. Procedia Chem. **9**, 220–225 (2014)
5. V.R. Makarand, R.B. Ramesh, Polyacrylamide-chitosan hydrogels: in vitro biocompatibility and sustained antibiotic release studies. Drug Deliv. **7**(2), 69–75 (2000)
6. J.M. Joshi, V.K. Sinha, Ceric ammonium nitrate induced grafting of polyacrylamide onto carboxymethyl chitosan. Carbohyd. Polym. **67**(3), 427–435 (2007)
7. O. Nadtoka, P. Virych, S. Nadtoka, N. Kutsevol, Synthesis and performance of hybrid hydrogels loaded with methylene blue and its use for antimicrobial photodynamic inactivation. J. Chem. **2020**(Article ID 6679960) (2020)
8. R.K.S.V. Krishna, K.N.B. Vijaya, M.C.S. Subha, M. Sairam, T.M. Aminabhavi, Novel chitosan-based pH-sensitive interpenetrating network microgels for the controlled release of cefadroxil. Carbohydr. Polym. **66**, 333–344 (2006)
9. O. Nadtoka, N. Kutsevol, Thermal analysis of cross-linked hydrogels based on PVA and D-g-PAA obtained by various methods. Mol. Cryst. Liq. Cryst. **661**(1), 52–57 (2018)
10. O. Nadtoka, N. Kutsevol, A. Onanko, V. Neimash, Mechanical and thermal characteristics of irradiation cross-linked hydrogels. Springer Proc. Phys. **214**, 205–214 (2018)
11. N.A. Rahman, S.A. Hanifah, N.N. Mobarak, A. Ahmad, L.K. Shyuan, L.T. Khoon, Synthesis and characterizations of o-nitrochitosan based biopolymer electrolyte for electrochemical devices. PLoS One **14**(2), e0212066 (2019)
12. M. Kurek, C.H. Brachais, C.M. Nguimjeu, A. Bonnotte, A. Voilley, K. Galic, J.-P. Couvercelle, F. Debeaufort, Structure and thermal properties of a chitosan coated polyethylene bilayer film. Polym. Degrad. Stab. **97**, 1232–1240 (2012)

Microstructure, High-Temperature Strength, and Fracture Toughness of Ti–Si–X Composites Containing Refractory Phases

B. D. Vasyliv, V. V. Kulyk, Z. A. Duriagina, P. Ya. Lyutyy, T. M. Kovbasiuk, A. M. Trostianchyn, V. V. Vira, V. I. Vavrukh, and V. A. Vynar

Abstract State-of-the-art Ti–Si–X composites are being developed for applications in modern aircraft and rocket engines as well as power equipment owing to their comparatively low specific weight and high strength and fracture toughness in a temperature range of 20–650 °C. This work is aimed at improving mechanical characteristics of the composites, namely strength and fracture toughness and increasing their operating temperature range up to 700–800 °C. Ti–Si–X composites (X = Al, Zr, and others) were manufactured with electron arc smelting. The content of alloying elements varied in a wide range (0.8–2.5 wt% Al, 5–8 wt% Zr). Strength tests of specimen series were carried out under three-point bending in a temperature range of 20–1000 °C. Fracture toughness tests of single-edge notch beam specimens were performed in a temperature range of 20–900 °C. The chemical and phase compositions of the composites were determined, as well as their microstructure and failure micromechanisms in relation to mechanical behavior were analyzed. Based on the

B. D. Vasyliv (✉)
Department of Hydrogen Technologies and Alternative Energy Materials, Karpenko Physico-Mechanical Institute of the NAS of Ukraine, 5 Naukova str., Lviv 79060, Ukraine
e-mail: mechengin1111@gmail.com

V. V. Kulyk · Z. A. Duriagina · P. Ya. Lyutyy · T. M. Kovbasiuk · A. M. Trostianchyn · V. I. Vavrukh
Department of Materials Science and Engineering, Lviv Polytechnic National University, 12 S. Bandera str., Lviv 79013, Ukraine

Z. A. Duriagina
Department of Materials Engineering, The John Paul II Catholic University of Lublin, 14 Racławickie Al., 20-950 Lublin, Poland

V. V. Vira
Department of Strength of Materials and Structural Mechanics, Lviv Polytechnic National University, 12 S. Bandera str., Lviv 79013, Ukraine

V. A. Vynar
Department of Corrosion and Corrosion Protection, Karpenko Physico-Mechanical Institute of the NAS of Ukraine, 5 Naukova str., Lviv 79060, Ukraine

O. Fesenko and L. Yatsenko (eds.), *Nanostructured Surfaces, Nanocomposites and Nanomaterials, and Their Applications*, Springer Proceedings in Physics 296,
https://doi.org/10.1007/978-3-031-42704-6_20

dependences of strength and fracture toughness on testing temperature for the specimen series as well as the microstructure and failure micromechanism analyses, it was found that the Ti_5Si_3 and complex $(Ti, Zr)_5Si_3$ refractory phases play a crucial role in enhancing fracture toughness of all the specimen series in the high-temperature (700–800 °C) range.

1 Introduction

Novel high-strength materials like advanced ceramics and composites resistible to high-temperature oxidation (alumina and zirconia based ceramics, silicon carbide, silicon nitride, ceramic matrix composites, etc.) are recently being used to ensure required lifetime of products in hard operating conditions. They can operate under high temperature, high pressure, corrosion, radiation, etc. [1–12]. Ti-based composites combine high fracture toughness and strength in a temperature range 20–650 °C as well as low specific weight. Therefore, they have an advantage over conventional titanium alloys which can operate at a temperature up to 300–500 °C. Such composites can find a potential application in equipment for power generation (gas turbines, compressors, fuel cells, etc.) and components of aircraft engines [13–26]. However, for reliable operation of products made of Ti-based composites, their operating temperature should be increased to 700–850 °C. The materials should possess high fracture toughness, strength, and oxidation resistance under these conditions [14, 25–29]. This concerns both the new developed composites and known materials with modified microstructure [30–38]. The improvement of processing and treatment modes as well as chemical composition should be related to the microstructure, phase compositions, and mechanical behavior under high-temperature conditions of the developed materials [39–53].

There are lots of works on high-temperature behavior of MAX phases as ternary compounds comprising transition metals of the d-group, elements of the p-group, and carbon or nitrogen (hereinafter: M, A, and X, respectively) and described as $M_{n+1}AX_n$ ($n = 1, 2, 3 \ldots$) [54]. Ti_3SiC_2 is a well-known MAX phase for high-temperature applications [55–67]. In a number of works, it was shown that $Ti_3Si_{1-x}Al_xC_2$ solid solutions may be formed due to the partial replacement of the Si atoms in this MAX phase with Al [22, 67, 68] that allows the self-healing effect to be reached. Such an effect is due to the rapid diffusion of aluminum and its oxidation. As a result, aluminum oxide is formed, and oxidation resistance of the material is improved. Besides, in the case of Sn addition to Ti_2AlC MAX phase, SnO_2 can be already formed at a temperature of about 600 °C which provides the perfect crack filling [22, 60, 67, 69]. However, low mechanical strength of the SnO_2 phase itself is a reason of lower strength of the bulk composite [67, 69, 70].

High strength and fracture toughness of bulk MAX phases along specific directions are related to the lamellar crystal structure [54, 61, 67, 68, 71]. Such advantages make them applicable in harsh operating conditions.

To estimate the lifetime of products made of ceramics and composites, various test methods are used [63, 72–74]. In addition to those, other methods should be developed to examine material in terms of their resistance to microstructural changes in harsh conditions including environmental effects [75–78], as these changes may result in the degradation of materials [79–83]. The indentation test known as a simple mechanical method [74, 84, 85] is utilized for examination of the material hardness and crack growth resistance. During the last decades, this technique has been used to develop formulas for calculation of fracture toughness of ceramics and composites [80, 84, 86, 87]. Therefore, a combined estimation of material bearing capacity including strength and fracture toughness tests is preferable when developing ceramic and composite materials with required mechanical and functional properties.

The purpose of this work is to evaluate the influence of chemical composition on the formation of refractory phases in microstructure of Ti–Si–X composites and estimate their high-temperature strength, fracture toughness, and dominant fracture micromechanisms.

2 Materials and Methods

In this work, the composites of the systems Ti–Si–Al–Zr–Fe (composite 1) and Ti–Si–Al–Zr–Mo–Fe–Cr (composites 2 and 3) of various chemical compositions have been studied (Table 1).

The composites were arc melted from raw constituents in argon on a copper hearth with water cooling [50, 63]. The elements with the purities Ti > 99.6 at% and Si, Al, Zr, Mo, Fe, Cr > 99.9 at% were used. After melting, the annealing procedure was performed at 1210 °C for 10 min.

The sets of beam specimens approximately 5 × 7.5 × 48 mm^3 in size were machined from the ingots, grinded, and polished.

The specimens were tested in flexure in a temperature range of 20–1000 °C using a three-point bend scheme. The fracture stress (σ_f) was calculated based on the "load–flexure" experimental graphs [29, 33, 40, 77].

For characterization of the material resistance to crack growth (fracture toughness), the critical stress intensity factor (SIF) K_{Ic} is widely used [88–92]. Among the methods of estimating the fracture toughness of brittle materials, an indentation method implementing various formulas for the SIF calculation [93–102] is the simplest one. Values of the SIF calculated by these formulas seem to be invariant

Table 1 Marking of the investigated composites and their chemical compositions

Composite marking	System and chemical composition, wt%
1	Ti–2.1Si–1.6Al–4.9Zr–0.1Fe
2	Ti–5.4Si–0.8Al–6.5Zr–1.7Mo–0.5Fe–0.3Cr
3	Ti–2.3Si–2.5Al–8.1Zr–0.5Mo–0.3Fe–0.1Cr

for many ceramics, as they are consistent with those obtained by traditional fracture mechanics methods [84, 101, 103]. One of the conventional methods is a single-edge notch beam (SENB) test [104–106]. This test is one of the most popular for estimating fracture toughness of a lot of ceramics [100, 103].

The SENB test method was implemented in our work for testing specimens under three-point bending in a temperature range of 20–900 °C. The critical SIF of the studied composites was calculated using corresponding formulas [104–106].

The microstructure of specimens and their fracture surfaces after mechanical tests were investigated with a scanning electron microscope (SEM) Carl Zeiss EVO-40XVP. Back-scattered electron (BSE) and secondary electron (SE) imaging modes were used. An energy-dispersive X-ray (EDX) microanalysis using an INCA Energy 350 system was performed for determination of chemical compositions in local areas of the specimens.

An X-ray diffractometer (Aeris, Malvern Panalytical) with Cu Kα radiation operating at 15 mA and 40 kV was used for obtaining X-ray diffraction (XRD) patterns. The angular range was 20–90°, and a step was 0.0217°. Highscore software was used for the XRD analysis with a reference to the International Center for Diffraction Data (ICDD). For the refinements and calculation of phase fractions, the WinCSD software was used [107].

3 Results and Discussion

We studied microstructure and mechanical properties (flexural strength and fracture toughness) of the composites in a wide range of temperatures.

XRD Analysis

The XRD patterns (Fig. 1) were used for determination of phase fractions in the studied materials.

Ti–Si–Al–Zr–Fe composite (1). The XRD pattern of composite 1 (Ti–Si–Al–Zr–Fe system) contains two types of peaks corresponding to the α-Ti and Ti_5Si_3 phases (Fig. 1a). The fractions of these phases were found to be as follows (Table 2): α-Ti phase (78.7 wt%) and Ti_5Si_3 phase (21.3 wt%). The image of the Ti–Si–Al–Zr–Fe microstructure made at a low magnification exhibited a network of "ribs" of dark-gray color and small round-shaped inclusions of the same color more or less uniformly distributed in the matrix of light-gray color (Fig. 2a). At a higher magnification, it can be seen that a fragment of the network (a rib) consists of fine particles which were formed by growing in specific directions (Fig. 2b). According to EDX analysis (spectrum 1 in Fig. 2b and Table 3), this material contained 87.62 wt% Ti, 2.51 wt% Si, 2.08 wt% Al, and 7.79 wt% Zr. Thus, this material is a metal-matrix composite of Ti–Si–Al–Zr–Fe system possibly comprising the α-Ti matrix phase and complex $(Ti, Zr)_5Si_3$ refractory phase as a ceramic skeleton in this matrix [50, 62, 63].

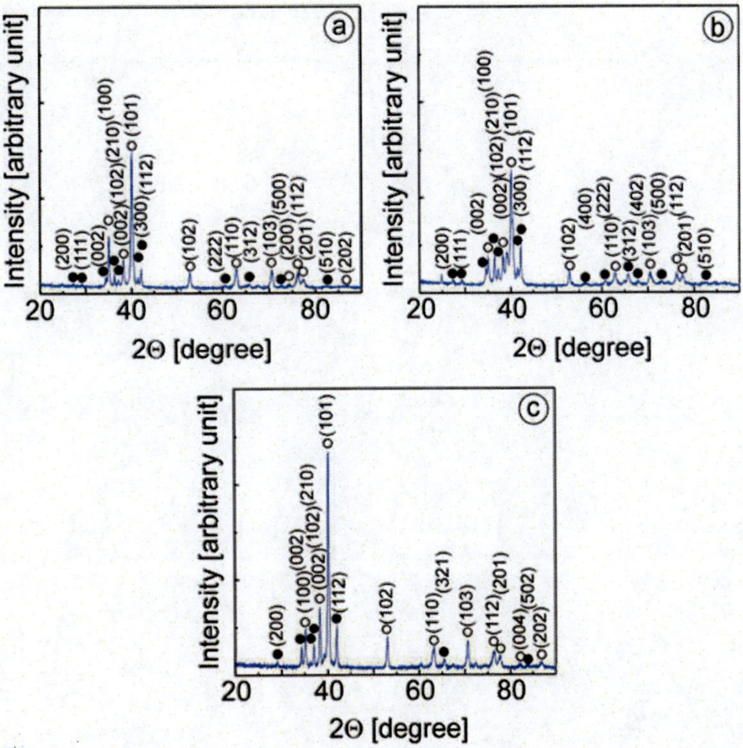

Fig. 1 XRD patterns of the studied composites **a** 1, **b** 2 and **c** 3 (Table 1) showing corresponding peaks for the α-Ti (light circles) and Ti_5Si_3 phases (dark circles) and corresponding Miller indices (in parentheses)

Table 2 Phase compositions of the investigated materials

Composite marking	Phase composition (wt%)
1	α-Ti (78.7); Ti_5Si_3 (21.3)
2	α-Ti (50.48); Ti_5Si_3 (49.52)
3	α-Ti (75.17); Ti_5Si_3 (24.83)

In the α-Ti matrix phase (92.23 wt% Ti, spectrum 2 in Fig. 2b and Table 3), silicon (0.98 wt%), aluminum (3.61 wt%), and zirconium (3.18 wt%) were revealed. The total amount of the α-Ti phase (Fig. 2a) was estimated as 72–75 vol%.

Round-shaped and needle-shaped particles of dark-gray color in the wide size range (2–50 μm, spectrum 3 in Fig. 2b) which form ribs represent the $(Ti, Zr)_5Si_3$ phase. The lamellae of α-Ti are adjacent to the areas occupied by this silicide phase. The $(Ti, Zr)_5Si_3$ phase occupies the total area of about 25–28 vol% (Fig. 2a).

Ti–Si–Al–Zr–Mo–Fe–Cr composite (2). We obtained the XRD pattern of composite 2 (Ti–Si–Al–Zr–Mo–Fe–Cr system) similar to that of composite 1. This

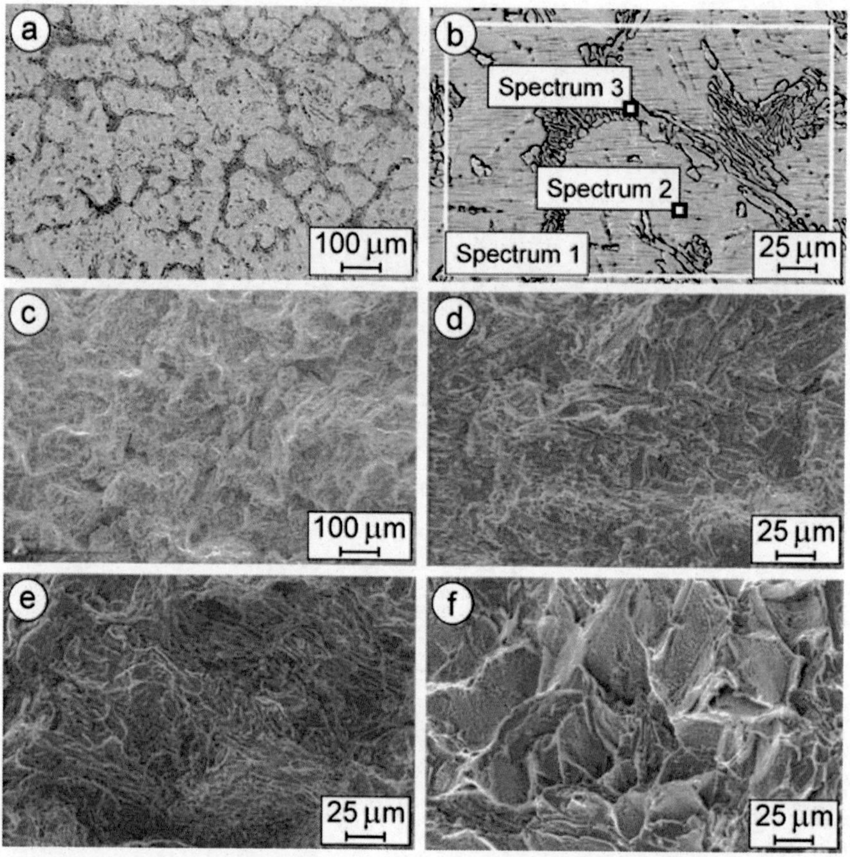

Fig. 2 SEM **a**, **b** microstructures (BSD images) with marked zones of **b** general (spectrum 1) and local (spectra 2 and 3) EDX analyses, and **c**–**f** fractography (SE images) of specimens of composite 1 after fracture toughness tests at **c**, **d** 20 °C, **e** 700 °C, and **f** 800 °C (Table 1)

Table 3 Data of the EDX spectra 1–3 marked in Fig. 2 for a specimen of composite 1

Chemical element and X-ray series	Spectra					
	1		2		3	
	wt%	at%	wt%	at%	wt%	at%
Al K	2.08	3.70	3.61	6.27	0.98	1.57
Si K	2.51	4.29	0.98	1.65	22.97	35.42
Ti K	87.62	87.90	92.23	90.44	62.64	56.64
Zr L	7.79	4.11	3.18	1.64	13.41	6.37

composite also contains the peaks of the α-Ti and Ti_5Si_3 phases (Fig. 1b). Its phase balance is as follows (Table 2): α-Ti phase (50.48 wt%) and Ti_5Si_3 phase (49.52 wt%). No resemblance to the microstructure of composite 1 was found for composite 2. At a low magnification, it exhibited a homogeneous microstructure (Fig. 3a).

The magnified microstructure image (Fig. 3b) presents more or less uniformly ordered particles of the $(Ti, Zr)_5Si_3$ phase in the α-Ti matrix. The chemical composition of this material estimated by EDX analysis (spectrum 1 in Fig. 3b and Table 4) is as follows: 79.48 wt% Ti, 9.94 wt% Si, 1.5 wt% Al, 7 wt% Zr, 1.54 wt% Mo, 0.48 wt% Fe, and 0.06 wt% Cr. Similarly to composite 1, this material presents the α-Ti matrix phase and $(Ti, Zr)_5Si_3$ phase, but in contrast to that composite, the $(Ti, Zr)_5Si_3$ phase is uniformly distributed here [50, 62, 63].

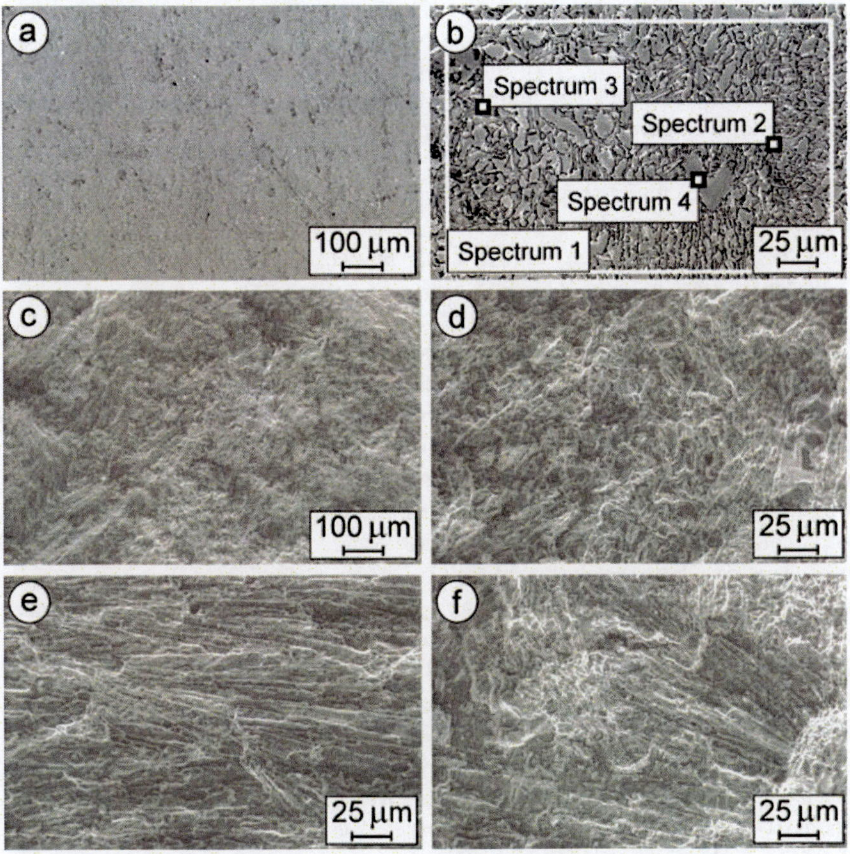

Fig. 3 SEM **a**, **b** microstructures (BSD images) with marked zones of **b** general (spectrum 1) and local (spectra 2, 3, and 4) EDX analyses, and **c–f** fractography (SE images) of specimens of composite 2 after fracture toughness tests at **c**, **d** 20 °C, **e** 700 °C, and **f** 800 °C (Table 1)

Table 4 Data of the EDX spectra 1–4 marked in Fig. 3 for a specimen of composite 2

Chemical element and X-ray series	Spectra							
	1		2		3		4	
	wt%	at%	wt%	at%	wt%	at%	wt%	at%
Al K	1.50	2.57	1.65	3.01	2.24	3.97	0.85	1.35
Si K	9.94	16.30	0.74	1.30	0.65	1.11	23.30	35.64
Ti K	79.48	76.42	86.50	88.87	91.85	91.80	63.89	57.30
Cr K	0.06	0.05	1.14	1.08	0.31	0.29	0.23	0.19
Fe K	0.48	0.39	1.47	1.29	0.82	0.70	–	–
Zr L	7.00	3.53	3.48	1.87	2.61	1.37	11.73	5.52
Mo L	1.54	0.74	5.02	2.58	1.52	0.76	–	–

We also revealed in the α-Ti phase (86.5 wt% Ti, spectrum 2 in Fig. 3b and Table 4) silicon (0.74 wt%), aluminum (1.65 wt%), zirconium (3.48 wt%), molybdenum (5.02 wt%), iron (1.47 wt%), and chromium (1.14 wt%). The total amount of the α-Ti phase (Fig. 3b) was found to be about 64–67 vol%.

Predominantly, round-shaped dark-gray particles of the $(Ti, Zr)_5Si_3$ phase are 4–30 μm in size (spectrum 4 in Fig. 3b). In contrast to composite 1, they neither adjoin each other nor form any network. In total, the $(Ti, Zr)_5Si_3$ phase occupies about 33–36 vol% (Fig. 3b).

Ti–Si–Al–Zr–Mo–Fe–Cr composite (3). In the XRD pattern of composite 3 (Ti–Si–Al–Zr–Mo–Fe–Cr system), we identified peaks of the α-Ti and Ti_5Si_3 phases (Fig. 1c) similarly to those of composite 2. The corresponding phase fractions were estimated (Table 2): α-Ti phase (75.17 wt%) and Ti_5Si_3 phase (24.83 wt%). We detected some resemblance in microstructure of composites 1 and 3. The microstructure of composite 3 at a low magnification exhibited a network of clusters, or ribs, of light-gray color and small round particles uniformly distributed in the matrix of darker color (Fig. 4a). In the magnified image of microstructure (Fig. 4b), grains of the α-Ti matrix adjoin the clusters of the $(Ti, Zr)_5Si_3$ phase. These clusters comprise thicker particles than those of composite 1. EDX analysis (spectrum 1 in Fig. 4b and Table 5) showed the following chemical composition of this material: 80.1 wt% Ti, 5.32 wt% Si, 5.09 wt% Al, 8.22 wt% Zr, 0.7 wt% Mo, 0.41 wt% Fe, and 0.16 wt% Cr. Thus, this Ti–Si–Al–Zr–Mo–Fe–Cr composite consists of the α-Ti matrix phase and a ceramic skeleton formed by the $(Ti, Zr)_5Si_3$ phase [50, 62, 63].

In the α-Ti matrix phase (87.16 wt% Ti, spectrum 2 in Fig. 4b and Table 5), we detected silicon (0.94 wt%), aluminum (6.34 wt%), zirconium (5.01 wt%), molybdenum (0.41 wt%), and chromium (0.14 wt%). The total amount of the α-Ti phase (Fig. 4a) was determined as 68–72 vol%.

Light-gray particles of the $(Ti, Zr)_5Si_3$ phase are 8–50 μm in size (spectrum 3 in Fig. 4b). The total amount of these particles is about 28–32 vol% (Fig. 4a).

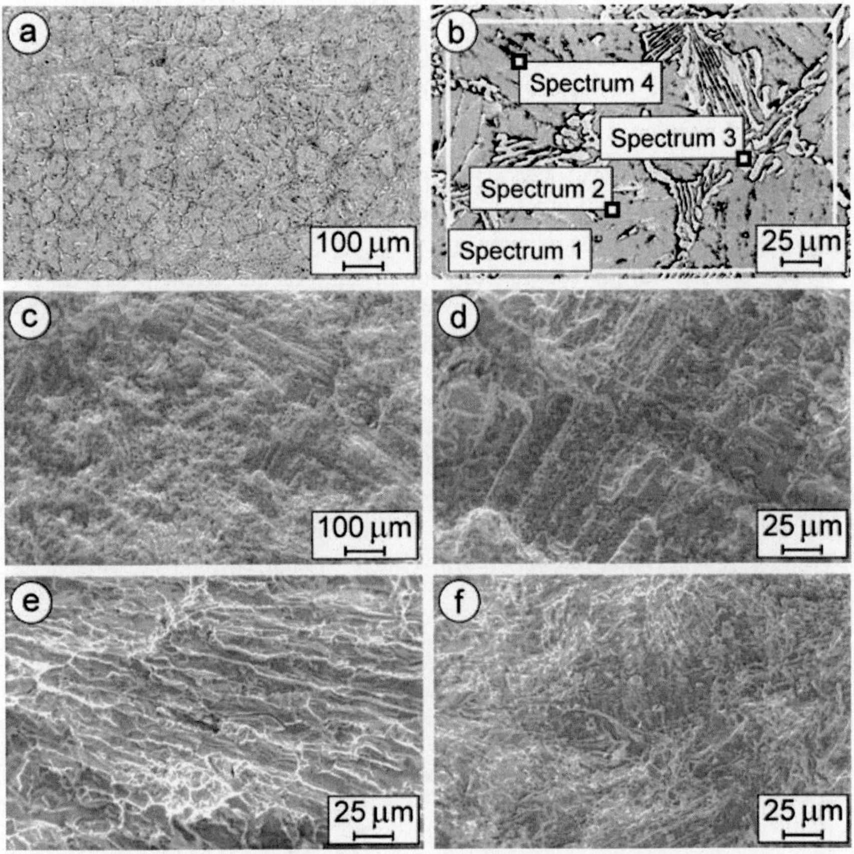

Fig. 4 SEM **a, b** microstructures (BSD images) with marked zones of **b** general (spectrum 1) and local (spectra 2, 3, and 4) EDX analyses, and **c–f** fractography (SE images) of specimens of composite 3 after fracture toughness tests at **c, d** 20 °C, **e** 700 °C, and **f** 800 °C (Table 1)

Table 5 Data of the EDX spectra 1–4 marked in Fig. 4 for a specimen of composite 3

Chemical element and X-ray series	Spectra							
	1		2		3		4	
	wt%	at%	wt%	at%	wt%	at%	wt%	at%
Al K	5.09	8.73	6.34	10.93	1.85	3.06	4.39	7.80
Si K	5.32	8.77	0.94	1.55	20.91	33.35	0.31	0.52
Ti K	80.10	77.49	87.16	84.64	57.69	53.94	86.43	86.52
Cr K	0.16	0.15	0.14	0.13	–	–	0.37	0.35
Fe K	0.41	0.34	–	–	0.12	0.11	1.17	1.00
Zr L	8.22	4.18	5.01	2.55	19.43	9.54	5.77	3.03
Mo L	0.70	0.34	0.41	0.20	–	–	1.56	0.78

Besides, we detected tiny particles enriched with zirconium, molybdenum, chromium, and iron on the α-Ti grain boundaries (spectrum 4 in Fig. 4b and Table 5) which were not recognized by XRD analysis.

Mechanical Behavior of the Composites

Obtained temperature dependences of both flexural strength and fracture toughness of the studied composites showed distinct changes in their mechanical behavior caused by transformations in microstructure (Fig. 5).

As can be seen, all the composites showed high flexural strength in a testing temperature range of 20–600 °C with its gradual increase while increasing the temperature up to 600 °C (Fig. 5a). The lowest values of strength were obtained at 20 °C. A similar trend in this temperature range was revealed for fracture toughness of the composites (Fig. 5b). However, distinct differences in mechanical behaviors were revealed with a further increase in a testing temperature up to 800 °C.

Ti–2.1Si–1.6Al–4.9Zr–0.1Fe composite showed, in the temperature range of 20–600 °C, fracture along the boundaries of α-Ti lamellae packets and brittle fracture of the ribs of the $(Ti, Zr)_5Si_3$ phase (Fig. 2c, d). Such fracture micromechanism provides high strength (above 1000 MPa) and fracture toughness (above 18 MPa m$^{1/2}$)

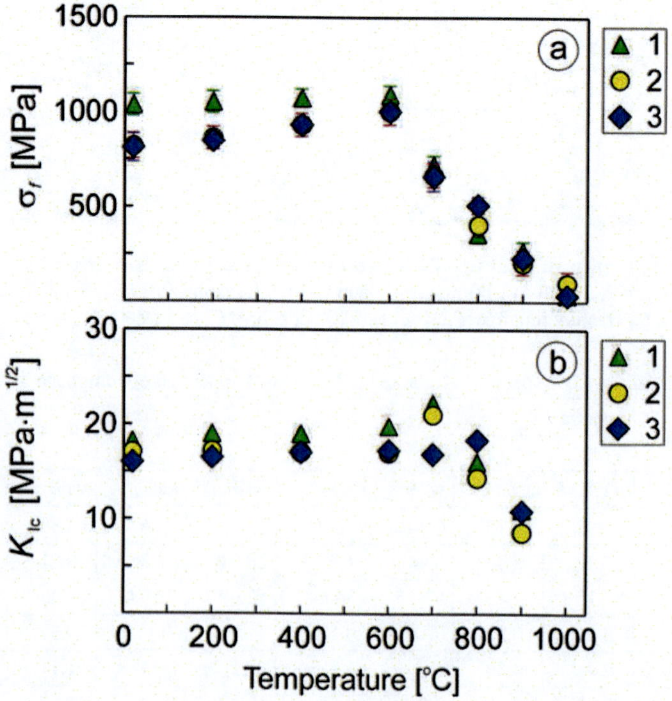

Fig. 5 Temperature dependences of **a** flexural strength and **b** fracture toughness of the studied composites (Table 1)

of this composite (Fig. 5a). With a further increase in a testing temperature up to 700 °C, a remarkable increase in its fracture toughness (up to 22 MPa $m^{1/2}$) occurred due to plastic behavior of the (Ti, $Zr)_5Si_3$ refractory phase (Fig. 2e), in contrast to drastically lowered strength (Fig. 5a). A further increase in a testing temperature up to 800 °C caused a significant lowering of both fracture toughness and strength of the composite due to high-temperature plasticization of α-Ti grains and implementation of high-temperature fracture micromechanism (Fig. 2f).

Mechanical behavior of Ti–5.4Si–0.8Al–6.5Zr–1.7Mo–0.5Fe–0.3Cr composite in the temperature range of 20–600 °C is similar to that of Ti–2.1Si–1.6Al–4.9Zr–0.1Fe composite. However, the composite with higher percentage of Si and Zr showed steeper increase in strength with a temperature increase up to 600 °C (Fig. 5a) due to fracture along the boundaries of α-Ti lamellae packets and plastic elongation of the (Ti, $Zr)_5Si_3$ refractory phase particles (Fig. 3c, d). Such fracture became more pronounced while increasing temperature up to 700 °C (Fig. 3e). As a result, this composite showed almost the same level of fracture toughness (about 21 MPa $m^{1/2}$) at this temperature. With an increase in a testing temperature up to 800 °C, strength and fracture toughness of the composite lowered significantly due to the implementation of high-temperature fracture micromechanism (Fig. 3f) similarly to that found in Ti–2.1Si–1.6Al–4.9Zr–0.1Fe composite.

Ti–2.3Si–2.5Al–8.1Zr–0.5Mo–0.3Fe–0.1Cr composite showed fracture character similar to that for Ti–5.4Si–0.8Al–6.5Zr–1.7Mo–0.5Fe–0.3Cr composite in the temperature range of 20–600 °C (Fig. 4c, d). However, more pronounced signs of the orientation of the (Ti, $Zr)_5Si_3$ refractory phase particles can be observed on fracture surface (Fig. 4d). In contrast to two other composites, fracture character of this material remained brittle while increasing temperature up to 700 °C (Fig. 4e), and its fracture toughness did not increase (Fig. 5b). A distinct microstructural change in this composite occurred with a further increase in a testing temperature up to 800 °C. As an evidence of this change, ductile fracture of a specimen of this composite tested at 800 °C can be mentioned (Fig. 4f) along with an increase in its fracture toughness up to 19 MPa $m^{1/2}$ (Fig. 5b). Strength of this composite at 800 °C (about 500 MPa) is higher than that of two others. It should be noted that this composite also showed an advantage in fracture toughness over two other composites at a temperature of 900 °C (Fig. 5b).

Therefore, we can conclude that Ti–2.3Si–2.5Al–8.1Zr–0.5Mo–0.3Fe–0.1Cr composite exhibited improved fracture toughness in a high-temperature (800–900 °C) range.

4 Conclusions

1. The phase balance and microstructure peculiarities of the studied Ti–Si–X composites have been substantiated.

2. All the studied composites exhibited comparatively high strength (above 750 MPa) and fracture toughness (above 15 MPa $m^{1/2}$) in a temperature range of 20–600 °C.
3. As compared to Ti–2.1Si–1.6Al–4.9Zr–0.1Fe and Ti–5.4Si–0.8Al–6.5Zr–1.7Mo–0.5Fe–0.3Cr composites, Ti–2.3Si–2.5Al–8.1Zr–0.5Mo–0.3Fe–0.1Cr composite exhibited both the increased strength (about 500 MPa) and fracture toughness (about 19 MPa $m^{1/2}$) at a temperature of 800 °C.

Acknowledgements The authors are thankful to the staff of the Scientific Equipment Collective Use Center "Laboratory of Advanced Technologies, Creation and Physicochemical Analysis of a New Substances and Functional Materials" at Lviv Polytechnic National University (https://lpnu.ua/ckkno) for their kind help in performing X-ray diffraction studies.

References

1. L.A. Dobrzański, L.B. Dobrzański, A.D. Dobrzańska-Danikiewicz, Additive and hybrid technologies for products manufacturing using powders of metals, their alloys and ceramics. Arch. Mater. Sci. Eng. **102**(2), 59–85 (2020). https://doi.org/10.5604/01.3001.0014.1525
2. V. Podhurska, B. Vasyliv, O. Ostash et al., Influence of treatment temperature on microstructure and properties of YSZ–NiO anode materials. Nanoscale Res. Lett. **11**, 93 (2016). https://doi.org/10.1186/s11671-016-1306-z
3. D. Li, Y. Dong, Z. Zhang et al., An as-cast Ti-V-Cr-Al light-weight medium entropy alloy with outstanding tensile properties. J. Alloy. Compd. **877**, 160199 (2021). https://doi.org/10.1016/j.jallcom.2021.160199
4. T.A. Prikhna, O.P. Ostash, A.S. Kuprin et al., A new MAX phases-based electroconductive coating for high-temperature oxidizing environment. Compos. Struct. **277**, 114649 (2021). https://doi.org/10.1016/j.compstruct.2021.114649
5. Z. Liu, J. Yang, Y. Qian et al., In-situ reaction synthesis and mechanical properties of quaternary MAX phase $(Cr_{2/3}Ti_{1/3})_3AlC_2$. Ceram. Int. **46**(14), 22854–22860 (2020). https://doi.org/10.1016/j.ceramint.2020.06.055
6. B.D. Vasyliv, V.Y. Podhurs'ka, O.P. Ostash et al., Influence of reducing and oxidizing media on the physicomechanical properties of ScCeSZ–NiO and YSZ–NiO ceramics. Mater. Sci. **49**(2), 135–144 (2013). https://doi.org/10.1007/s11003-013-9593-3
7. H.A. Shabri, M.H.D. Othman, M.A. Mohamed et al., Recent progress in metal-ceramic anode of solid oxide fuel cell for direct hydrocarbon fuel utilization: a review. Fuel Process. Technol. **212**, 106626 (2021). https://doi.org/10.1016/j.fuproc.2020.106626
8. T.B. Serbenyuk, T.O. Prikhna, V.B. Sverdun et al., Effect of the additive of Y_2O_3 on the structure formation and properties of composite materials based on AlN–SiC. J. Superhard Mater. **40**(1), 8–15 (2018). https://doi.org/10.3103/S1063457618010021
9. T.S. Cherepova, H.P. Dmytrieva, O.I. Dukhota et al., Properties of nickel powder alloys hardened with titanium carbide. Mater. Sci. **52**(2), 173–179 (2016). https://doi.org/10.1007/s11003-016-9940-2
10. V. Kulyk, Z. Duriagina, A. Kostryzhev et al., The effect of yttria content on microstructure, strength, and fracture behavior of yttria-stabilized zirconia. Materials **15**, 5212 (2022). https://doi.org/10.3390/ma15155212
11. M.H. Bocanegra-Bernal, S. Díaz de la Torre, Phase transitions in zirconium dioxide and related materials for high performance engineering ceramics. J. Mater. Sci. **37**, 4947–4971 (2002). https://doi.org/10.1023/A:1021099308957

12. L.Y. Ropyak, M.V. Makoviichuk, I.P. Shatskyi et al., Stressed state of laminated interference-absorption filter under local loading. Funct. Mater. **27**(3), 638–642 (2020). https://doi.org/10.15407/fm27.03.638
13. Y. Komatsu, A. Sciazko, N. Shikazono, Isostatic pressing of screen printed nickel-gadolinium doped ceria anodes on electrolyte-supported solid oxide fuel cells. J. Power. Sources **485**, 229317 (2021). https://doi.org/10.1016/j.jpowsour.2020.229317
14. W.M. Budzianowski, J. Milewski, Solid-oxide fuel cells in power generation applications: a review. Recent Patents Eng. **5**(3), 165–189 (2011). https://doi.org/10.2174/187221211797636926
15. B.D. Vasyliv, A procedure for the investigation of mechanical and physical properties of ceramics under the conditions of biaxial bending of a disk specimen according to the ring–ring scheme. Mater. Sci. **45**(4), 571–575 (2009). https://doi.org/10.1007/s11003-010-9215-2
16. S.N. Perevislov, T.V. Sokolova, V.L. Stolyarova, The Ti_3SiC_2 max phases as promising materials for high temperature applications: formation under various synthesis conditions. Mater. Chem. Phys. **267**, 124625 (2021). https://doi.org/10.1016/j.matchemphys.2021.124625
17. B.D. Vasyliv, Improvement of the electric conductivity of the material of anode in a fuel cell by the cyclic redox thermal treatment. Mater. Sci. **46**(2), 260–264 (2010). https://doi.org/10.1007/s11003-010-9282-4
18. E. Tabares, S.C. Cifuentes, A. Jiménez-Morales et al., Injection moulding of porous MAX phase Ti_3SiC_2 without using space-holder. Powder Technol. **380**, 96–105 (2021). https://doi.org/10.1016/j.powtec.2020.11.022
19. E.B. Kashkarov, N.S. Pushilina, M.S. Syrtanov et al., Preceramic paper-derived SiC_f/$Ti_3Al(Si)C_2$ and SiC_f/Ti_3SiC_2 MAX-phase based laminates fabricated using spark plasma sintering. Scripta Mater. **194**, 113696 (2021). https://doi.org/10.1016/j.scriptamat.2020.113696
20. Y. Tan, Y. Xia, Z. Teng et al., Synthesis and enhanced mechanical properties of compositionally complex MAX phases. J. Eur. Ceram. Soc. **41**(8), 4658–4665 (2021). https://doi.org/10.1016/j.jeurceramsoc.2021.03.027
21. L. Silvestroni, C. Melandri, J. Gonzalez-Julian, Exploring processing, reactivity and performance of novel MAX phase/ultra-high temperature ceramic composites: the case study of Ti_3SiC_2. J. Eur. Ceram. Soc. **41**(12), 6064–6069 (2021). https://doi.org/10.1016/j.jeurceramsoc.2021.05.029
22. W. Yu, X. Pi, W. Chen et al., Effects of A-site atoms in Ti_2AlC and Ti_3SiC_2 MAX phases reinforced Mg composites: interfacial structure and mechanical properties. Mater. Sci. Eng. A **826**, 141961 (2021). https://doi.org/10.1016/j.msea.2021.141961
23. V. Podhurska, B. Vasyliv, Influence of NiO reduction on microstructure and properties of porous Ni–ZrO_2 substrates, in *Proceedings of the 3rd International Conference on Oxide Materials for Electronic Engineering (OMEE-2012)*, Lviv, Ukraine (2012), pp. 293–294. https://doi.org/10.1109/OMEE.2012.6464761
24. I. Danilenko, G. Lasko, I. Brykhanova et al., The peculiarities of structure formation and properties of zirconia-based nanocomposites with addition of Al_2O_3 and NiO. Nanoscale Res. Lett. **12**, 125 (2017). https://doi.org/10.1186/s11671-017-1901-7
25. Y. Kharchenko, Z. Blikharskyy, V. Vira et al., Nanostructural changes in a Ni/NiO cermet during high-temperature reduction and reoxidation, in *Nanomaterials and Nanocomposites, Nanostructure Surfaces, and Their Applications*. Springer Proceedings in Physics, vol. 246 (2021), pp. 219–229. https://doi.org/10.1007/978-3-030-51905-6_17
26. D. Yu, Y. Tan, Oxidation behaviors of compositionally complex MAX phases in air. Ceram. Int. **47**(21), 30188–30193 (2021). https://doi.org/10.1016/j.ceramint.2021.07.198
27. V.V. Kulyk, B.D. Vasyliv, Z.A. Duriagina et al., The effect of water vapor containing hydrogenous atmospheres on the microstructure and tendency to brittle fracture of anode materials of YSZ–NiO(Ni) system. Arch. Mater. Sci. Eng. **108**(2), 49–67 (2021). https://doi.org/10.5604/01.3001.0015.0254
28. L.A. Dobrzański, L.B. Dobrzański, A.D. Dobrzańska-Danikiewicz, Manufacturing technologies thick-layer coatings on various substrates and manufacturing gradient materials using

powders of metals, their alloys and ceramics. J. Achievements Mater. Manuf. Eng. **99**(1), 14–41 (2020). https://doi.org/10.5604/01.3001.0014.1598
29. B.D. Vasyliv, V.Y. Podhurska, O.P. Ostash et al., Effect of a hydrogen sulfide-containing atmosphere on the physical and mechanical properties of solid oxide fuel cell materials, in *Nanochemistry, Biotechnology, Nanomaterials, and Their Applications*. Springer Proceedings in Physics, vol. 214 (2018), pp. 475–485. https://doi.org/10.1007/978-3-319-92567-7_30
30. S.S. Savka, D.I. Popovych, A.S. Serednytski, Molecular dynamics simulations of the formation processes of zinc oxide nanoclusters in oxygen environment, in *Nanophysics, Nanomaterials, Interface Studies, and Applications*. Springer Proceedings in Physics, vol. 195 (2017), pp. 145–156. https://doi.org/10.1007/978-3-319-56422-7_11
31. J. Milewski, J. Lewandowski, Solid oxide fuel cell fuelled by biogases. Arch. Thermodyn. **30**(4), 3–12 (2009). https://www.imp.gda.pl/fileadmin/doc/imp_publishing/wimp/archives%20of%20thermodynamics/C_09_4.pdf
32. Y. Zhang, K. Shimizu, X. Yaer et al., Erosive wear performance of heat treated multi-component cast iron containing Cr, V, Mn and Ni eroded by alumina spheres at elevated temperatures. Wear **390–391**, 135–145 (2017). https://doi.org/10.1016/j.wear.2017.07.017
33. M. Andrzejczuk, O. Vasylyev, I. Brodnikovskyi et al., Microstructural changes in NiO–ScSZ composite following reduction processes in pure and diluted hydrogen. Mater. Charact. **87**, 159–165 (2014). https://doi.org/10.1016/j.matchar.2013.11.011
34. J. Milewski, J. Lewandowski, A. Miller, Reducing CO_2 emissions from a coal fired power plant by using a molten carbonate fuel cell, in *Proceedings of the ASME Turbo Expo*, vol. 2, pp. 389–395 (2008). https://doi.org/10.1115/GT2008-50100
35. G. Witz, V. Shklover, W. Steurer et al., Phase evolution in yttria-stabilized zirconia thermal barrier coatings studied by Rietveld refinement of X-ray powder diffraction patterns. J. Am. Ceram. Soc. **90**(9), 2935–2940 (2007). https://doi.org/10.1111/j.1551-2916.2007.01785.x
36. D.R. Clarke, C.G. Levi, Material design for the next generation thermal barrier coatings. Annu. Rev. Mater. Res. **33**, 383–417 (2003). https://doi.org/10.1146/annurev.matsci.33.011403.113718
37. L.A. Dobrzański, L.B. Dobrzański, A.D. Dobrzańska-Danikiewicz, Overview of conventional technologies using the powders of metals, their alloys and ceramics in Industry 4.0 stage. J. Achievements Mater. Manuf. Eng. **98**(2), 56–85 (2020). https://doi.org/10.5604/01.3001.0014.1481
38. J. Milewski, J. Kupecki, A. Szczęśniak et al., Hydrogen production in solid oxide electrolyzers coupled with nuclear reactors. Int. J. Hydrog. Energy **46**(72), 35765–35776 (2021). https://doi.org/10.1016/j.ijhydene.2020.11.217
39. M.Y. Smyrnova-Zamkova, V.P. Red'ko, O.K. Ruban et al., The properties of nanocrystalline powder of 90% Al_2O_3–10% ZrO_2 (wt.%) obtained via the hydrothermal synthesis/mechanical mixing. Nanosistemi Nanomater. Nanotehnol. **15**(2), 309–317 (2017). https://doi.org/10.15407/nnn.15.02.0309
40. B. Vasyliv, J. Milewski, V. Podhurska et al., Study of the degradation of a fine-grained YSZ–NiO anode material during reduction in hydrogen and reoxidation in air. Appl. Nanosci. **12**, 965–975 (2022). https://doi.org/10.1007/s13204-021-01768-w
41. Y.G. Chabak, V.I. Fedun, K. Shimizu et al., Phase-structural composition of coating obtained by pulsed plasma treatment using eroded cathode of T1 high speed steel. Probl. At. Sci. Technol. **104**(4), 100–106 (2016)
42. M.Y. Smyrnova-Zamkova, O.K. Ruban, O.I. Bykov et al., Physico-chemical properties of fine-grained powder in Al_2O_3–ZrO_2–Y_2O_3–CeO_2 system produced by combined method. Comp. Theor. Pract. **18**(4), 234–240 (2018). https://kompozyty.ptmk.net/pliczki/pliki/1290_2018t04_maria-y-smyrnova-zamkova-.pdf
43. O.V. Sukhova, Influence of mechanisms of structure formation of interfaces in composites on their properties. Metallofiz. Noveishie Tekhnol. **31**(7), 1001–1012 (2009)
44. B.D. Vasyliv, A.D. Ivasyshyn, O.P. Ostash et al., Kinetics of corrosion-fatigue cracks in Ti–Si cermet composite. Mater. Sci. **38**(2), 220–224 (2002). https://doi.org/10.1023/A:1020990103898

45. O.V. Dudnik, I.O. Marek, O.K. Ruban et al., Effect of heat treatment on the structure and phase composition of the nanosized powder based on a ZrO_2 solid solution. Powder Metall. Met. Ceram. **59**(1–2), 1–8 (2020). https://doi.org/10.1007/s11106-020-00132-x
46. M. Kujawa, R. Suwak, L.A. Dobrzański et al., Thermal characterization of halloysite materials for porous ceramic preforms. Arch. Mater. Sci. Eng. **107**(1), 5–15 (2021). https://doi.org/10.5604/01.3001.0014.8189
47. O.P. Ostash, A.D. Ivasyshyn, B.D. Vasyliv et al., Influence of the structure and asymmetry of loading cycles on the cyclic crack resistance of Ti–Si composites. Mater. Sci. **38**(1), 55–61 (2002). https://doi.org/10.1023/A:1020120714703
48. I.O. Marek, O.K. Ruban, V.P. Redko et al., Physicochemical properties of hydrothermal nanocrystalline ZrO_2–Y_2O_3–CeO_2 powders. Powder Metall. Met. Ceram. **58**(3–4), 125–132 (2019). https://doi.org/10.1007/s11106-019-00055-2
49. X.W. Zhou, Y.F. Shen, H.M. Jin, Effect of deposition mechanism and microstructure of nano-ceria oxide addition on Ni-P coating by pulse electrodeposition. Adv. Mater. Res. **326**, 151–156 (2011). https://doi.org/10.4028/www.scientific.net/AMR.326.151
50. V.I. Mazur, Y.N. Taran, S.V. Kapustnikova et al., Titanium matrix composites. US Patent, No. 5366570, 22.11.1994 (1994)
51. A.V. Shevchenko, V.V. Lashneva, A.K. Ruban et al., Synthesis and study of high-purity nanocrystalline powder of a solid solution of CeO_2 and Y_2O_3 in zirconium dioxide. Powder Metall. Met. Ceram. **54**(9–10), 548–553 (2016). https://doi.org/10.1007/s11106-016-9748-5
52. B. Vasyliv, A. Ivasyshyn, O. Ostash et al., High-temperature fatigue crack growth resistance of thermo-mechanically and heat treated cast Ti–Si–Al–Zr composites, in *Metallic Materials with High Structural Efficiency* (Kluwer Academic Publishers, The Netherlands, 2004), pp. 235–240. https://doi.org/10.1007/1-4020-2112-7_24
53. A.S. Doroshkevich, E.B. Asgerov, A.V. Shylo et al., Direct conversion of the water adsorption energy to electricity on the surface of zirconia nanoparticles. Appl. Nanosci. **9**, 1603–1609 (2019). https://doi.org/10.1007/s13204-019-00979-6
54. M.W. Barsoum, The $M_{n+1}AX_n$ phases: a new class of solids. Prog. Solid State Chem. **28**(1–4), 201–281 (2000). https://doi.org/10.1016/S0079-6786(00)00006-6
55. F.W. Crossman, A.S. Yue, Unidirectionally solidified Ti-TiB and Ti-Ti_5Si_3 eutectic composites. Met. Trans. **2**, 1545–1555 (1971)
56. W.T. Chiu, T. Ishigaki, N. Nohira et al., Effect of Cr additions on the phase constituent, mechanical properties, and shape memory effect of near-eutectoid Ti-4Au towards the biomaterial applications. J. Alloy. Compd. **867**, 159037 (2021). https://doi.org/10.1016/j.jallcom.2021.159037
57. H. Zhao, L. Hu, C. Li et al., Influence of metallic Cr addition on the phase structure and mechanical properties of plasma-sprayed Ti-Si-C coatings. Ceram. Int. **47**(12), 17570–17579 (2021). https://doi.org/10.1016/j.ceramint.2021.03.075
58. X. Zhao, M. Sokol, M.W. Barsoum et al., Effect of grain orientation on the compressive response of highly oriented MAX phase Ti_3SiC_2. Mater. Sci. Eng. A **809**, 140869 (2021). https://doi.org/10.1016/j.msea.2021.140869
59. O.P. Ostash, B.D. Vasyliv, V.Y. Podhurs'ka et al., Optimization of the properties of 10Sc1CeSZ-NiO composite by the redox treatment. Mater. Sci. **46**(5), 653–658 (2011). https://doi.org/10.1007/s11003-011-9337-1
60. W.T. Chiu, K. Wakabayashi, A. Umise et al., Enhancement of the shape memory effect by the introductions of Cr and Sn into the β-Ti alloy towards the biomedical applications. J. Alloy. Compd. **875**, 160088 (2021). https://doi.org/10.1016/j.jallcom.2021.160088
61. Y. Kharchenko, Z. Blikharskyy, V. Vira et al., Study of nanostructural changes in a Ni-containing cermet material during reduction and oxidation at 600 °C. Appl. Nanosci. **10**, 4535–4543 (2020). https://doi.org/10.1007/s13204-020-01391-1
62. G. Frommeyer, R. Rosenkranz, C. Ludecke, Microstructure and properties of the refractory intermetallic Ti_5Si_3 compound and the unidirectionally solidified eutectic Ti-Ti_5Si_3 alloy. Metallkunde **81**, 307–313 (1990). https://doi.org/10.1515/ijmr-1990-810501

63. O.P. Ostash, A.D. Ivasyshyn, B.D. Vasyliv et al., High-temperature and cyclic corrosion crack resistance of alloys of the Ti–Si–Al–Zr system. Mater. Sci. **42**(3), 330–343 (2006). https://doi.org/10.1007/s11003-006-0087-4
64. W.T. Chiu, K. Wakabayashi, A. Umise et al., Enhancement of mechanical properties and shape memory effect of Ti-Cr-based alloys via Au and Cu modifications. J. Mech. Behav. Biomed. Mater. **123**, 104707 (2021). https://doi.org/10.1016/j.jmbbm.2021.104707
65. A. Shylo, A. Doroshkevich, A. Lyubchyk et al., Electrophysical properties of hydrated porous dispersed system based on zirconia nanopowders. Appl. Nanosci. **10**, 4395–4402 (2020). https://doi.org/10.1007/s13204-020-01471-2
66. A. Kehal, N. Saoula, S.E.H. Abaidia et al., Effect of Ar/N_2 flow ratio on the microstructure and mechanical properties of Ti-Cr-N coatings deposited by DC magnetron sputtering on AISI D2 tool steels. Surf. Coat. Technol. **421**, 127444 (2021). https://doi.org/10.1016/j.surfcoat.2021.127444
67. Z. Zhang, X. Duan, D. Jia et al., On the formation mechanisms and properties of MAX phases: a review. J. Eur. Ceram. Soc. **41**(7), 3851–3878 (2021). https://doi.org/10.1016/j.jeurceramsoc.2021.02.002
68. R. Liu, M. Tane, H. Kimizuka et al., Elastic isotropy originating from heterogeneous interlayer elastic deformation in a Ti_3SiC_2 MAX phase with a nanolayered crystal structure. J. Eur. Ceram. Soc. **41**(4), 2278–2289 (2021). https://doi.org/10.1016/j.jeurceramsoc.2020.11.026
69. M.S. Park, W.T. Chiu, N. Nohira et al., Effects of Cr and Sn additives on the martensitic transformation and deformation behavior of Ti-Cr-Sn biomedical shape memory alloys. Mater. Sci. Eng. A **822**, 141668 (2021). https://doi.org/10.1016/j.msea.2021.141668
70. K.H. Omran, M.S.A. El-sadek, M. Mostafa et al., Influence of PbO phase content on structural and optical properties of PZT nanopowders. Appl. Nanosci. **10**, 2315–2327 (2020). https://doi.org/10.1007/s13204-020-01390-2
71. Z. Wang, G. Ma, L. Liu et al., High-performance Cr_2AlC MAX phase coatings: oxidation mechanisms in the 900–1100°C temperature range. Corros. Sci. **167**, 108492 (2020). https://doi.org/10.1016/j.corsci.2020.108492
72. R.N. Gupta, Study of pulse electrodeposition parameters for nano YSZ-Ni coatings and its effect on tribological and corrosion characteristics. Appl. Nanosci. **11**, 173–185 (2021). https://doi.org/10.1007/s13204-020-01567-9
73. A.I. Yurkova, S.O. Nakonechnyi, V.V. Cherniavsky et al., Nanostructured AlCoFeCrVNi and AlCoFeCrVTi high-entropy alloys resulted from mechanical alloying and sintering. Appl. Nanosci. **12**, 849–860 (2022). https://doi.org/10.1007/s13204-021-01856-x
74. K.W. Jeong, J.-S. Han, G.-U. Yang et al., Influence of preaging temperature on the indentation strength of 3Y–TZP aged in ambient atmosphere. Materials **14**, 2767 (2021). https://doi.org/10.3390/ma14112767
75. S. Buchaniec, A. Sciazko, M. Mozdzierz et al., A novel approach to the optimization of a solid oxide fuel cell anode using evolutionary algorithms. IEEE Access **7**, 34361–34372 (2019). https://doi.org/10.1109/ACCESS.2019.2904327
76. V.I. Zurnadzhy, V.G. Efremenko, I. Petryshynets et al., Mechanical properties of carbide-free lower bainite in complex-alloyed constructional steel: effect of bainitizing treatment parameters. Kov. Mater. **58**(2), 129–140 (2020). https://doi.org/10.4149/km_2020_2_129
77. O.M. Romaniv, B.D. Vasyliv, Some features of formation of the structural strength of ceramic materials. Mater. Sci. **34**(2), 149–161 (1998). https://doi.org/10.1007/BF02355530
78. A. Sciazko, T. Shimura, Y. Komatsu et al., Ni-GDC and Ni-YSZ electrodes operated in solid oxide electrolysis and fuel cell modes. J. Therm. Sci. Technol. **16**(1), JTST0013 (2021). https://doi.org/10.1299/jtst.2021jtst0013
79. A.S. Basaleh, R.M. Mohamed, Photodegradation of thiophene over ZrO_2–SiO_2 nanoparticles: impact of copper decoration on their photocatalytic activity. Appl. Nanosci. **9**, 2051–2058 (2019). https://doi.org/10.1007/s13204-019-00992-9
80. S. Tao, J. Yang, M. Zhai et al., Thermal stability of YSZ thick thermal barrier coatings deposited by suspension and atmospheric plasma spraying. Crystals **10**(11), 984 (2020). https://doi.org/10.3390/cryst10110984

81. V.Y. Podhurs'ka, B.D. Vasyliv, O.P. Ostash et al., Structural transformations in the NiO-containing anode of ceramic fuel cells in the course of its reduction and oxidation. Mater. Sci. **49**(6), 805–811 (2014). https://doi.org/10.1007/s11003-014-9677-8
82. O.V. Perlova, Y.S. Dzyazko, A.V. Palchik et al., Composites based on zirconium dioxide and zirconium hydrophosphate containing graphene-like additions for removal of U(VI) compounds from water. Appl. Nanosci. **10**, 4591–4602 (2020). https://doi.org/10.1007/s13204-020-01313-1
83. A.B. Hernandez, D. Cortés-Arriagada, H.C. García et al., Quantum molecular study on doping effect in titanium and vanadium clusters: their application to remove some chemical species. Appl. Nanosci. **10**, 37–49 (2020). https://doi.org/10.1007/s13204-019-01072-8
84. O.M. Romaniv, I.V. Zalite, V.M. Simin'kovych et al., Effect of the concentration of zirconium dioxide on the fracture resistance of Al_2O_3–ZrO_2 ceramics. Mater. Sci. **31**(5), 588–594 (1996). https://doi.org/10.1007/BF00558793
85. P. Khajavi, P.V. Hendriksen, J. Chevalier et al., Improving the fracture toughness of stabilized zirconia-based solid oxide cells fuel electrode supports: effects of type and concentration of stabilizer(s). J. Eur. Ceram. Soc. **40**(15), 5670–5682 (2020). https://doi.org/10.1016/j.jeurceramsoc.2020.05.042
86. B.D. Vasyliv, Initiation of a crack from the edge of a notch with oblique front in specimens of brittle materials. Mater. Sci. **38**(5), 724–728 (2002). https://doi.org/10.1023/A:1024222709514
87. A.D. Ivasyshyn, B.D. Vasyliv, Effect of the size and form of specimens on the diagram of growth rates of fatigue cracks. Mater. Sci. **37**(6), 1002–1004 (2001). https://doi.org/10.1023/A:1015669913601
88. R.F. Cook, G.M. Pharr, Direct observation and analysis of indentation cracking in glasses and ceramics. J. Am. Ceram. Soc. **73**(4), 787–817 (1990). https://doi.org/10.1111/j.1151-2916.1990.tb05119.x
89. V. Kulyk, Z. Duriagina, B. Vasyliv et al., The effect of sintering temperature on the phase composition, microstructure, and mechanical properties of yttria-stabilized zirconia. Materials **15**, 2707 (2022). https://doi.org/10.3390/ma15082707
90. A. Nastic, A. Merati, M. Bielawski et al., Instrumented and Vickers indentation for the characterization of stiffness, hardness and toughness of zirconia toughened Al_2O_3 and SiC armor. J. Mater. Sci. Technol. **31**(8), 773–783 (2015). https://doi.org/10.1016/j.jmst.2015.06.005
91. ASTM E 384-11, Standard test method for Knoop and Vickers hardness of materials (ASTM International, 2011). https://doi.org/10.1520/E0384-11
92. ASTM C 1327-03, Standard test method for Vickers indentation hardness of advanced ceramics (ASTM International, 2003). https://doi.org/10.1520/C1327-03
93. B.R. Lawn, M.V. Swain, Microfracture beneath point indentations in brittle solids. J. Mater. Sci. **10**(1), 113–122 (1975). https://doi.org/10.1007/BF00541038
94. G.R. Anstis, P. Chantikul, B.R. Lawn et al., A critical evaluation of indentation techniques for measuring fracture toughness: I, Direct crack measurement. J. Am. Ceram. Soc. **64**(9), 533–538 (1981). https://doi.org/10.1111/j.1151-2916.1981.tb10320.x
95. A.G. Evans, E.A. Charles, Fracture toughness determinations by indentation. J. Am. Ceram. Soc. **59**(7–8), 371–372 (1976). https://doi.org/10.1111/j.1151-2916.1976.tb10991.x
96. M.Y. Smyrnova-Zamkova, O.K. Ruban, O.I. Bykov et al., The influence of the ZrO_2 solid solution amount on the physicochemical properties of Al_2O_3–ZrO_2–Y_2O_3–CeO_2 powders. Powder Metall. Met. Ceram. **60**(3–4), 129–141 (2021). https://doi.org/10.1007/s11106-021-00222-4
97. B.R. Lawn, A.G. Evans, D.B. Marshall, Elastic/plastic indentation damage in ceramics: the median/radial crack system. J. Am. Ceram. Soc. **63**(9–10), 574–581 (1980). https://doi.org/10.1111/j.1151-2916.1980.tb10768.x
98. O.N. Grigoriev, V.B. Vinokurov, T.V. Mosina et al., Kinetics of shrinkage, structurization, and the mechanical characteristics of zirconium boride sintered in the presence of activating additives. Powder Metall. Met. Ceram. **55**(11–12), 676–688 (2017). https://doi.org/10.1007/s11106-017-9855-y

99. G.A. Gogotsi, S.N. Dub, E.E. Lomonova et al., Vickers and Knoop indentation behaviour of cubic and partially stabilized zirconia crystals. J. Eur. Ceram. Soc. **15**(5), 405–413 (1995). https://doi.org/10.1016/0955-2219(95)91431-M
100. I. Danilenko, F. Glazunov, T. Konstantinova et al., Effect of Ni/NiO particles on structure and crack propagation in zirconia based composites. Adv. Mater. Lett. **5**(8), 465–471 (2014). https://doi.org/10.5185/amlett.2014.amwc1040II
101. B. Vasyliv, V. Kulyk, Z. Duriagina et al., Estimation of the effect of redox treatment on microstructure and tendency to brittle fracture of anode materials of YSZ-NiO(Ni) system. East Eur. J. Enterp. Technol. **108/6**(12), 67–77 (2020). https://doi.org/10.15587/1729-4061.2020.218291
102. J. Lankford, Indentation microfracture in the Palmqvist crack regime: implications for fracture toughness evaluation by the indentation method. J. Mater. Sci. Lett. **1**(11), 493–495 (1982). https://doi.org/10.1007/BF00721938
103. V.V. Kulyk, Z.A. Duriagina, B.D. Vasyliv et al., Effects of yttria content and sintering temperature on the microstructure and tendency to brittle fracture of yttria-stabilized zirconia. Arch. Mater. Sci. Eng. **109**(2), 65–79 (2021). https://doi.org/10.5604/01.3001.0015.2625
104. ASTM E 399-20a, Standard test method for linear-elastic plane-strain fracture toughness of metallic materials (ASTM International, 2020). https://doi.org/10.1520/E0399-20A
105. ASTM C 1421-18, Standard test methods for determination of fracture toughness of advanced ceramics at ambient temperature (ASTM International, 2018). https://doi.org/10.1520/C1421-18
106. J. Kübier, Fracture toughness of ceramics using the SEVNB method: from a preliminary study to a standard test method, in *Fracture Resistance Testing of Monolithic and Composite Brittle Materials*, ed. by J. Salem et al. (ASTM International, 2002), pp. 93–106. https://doi.org/10.1520/STP10473S
107. L. Akselrud, Y. Grin, WinCSD: software package for crystallographic calculations (Version 4). J. Appl. Crystallogr. **47**, 803–805 (2014). https://doi.org/10.1107/S1600576714001058

Study of the Role of Alloying Elements Cr, Ni, and Al in the Improvement of High-Temperature (700–800 °C) Fracture Toughness of Ti-Based Composites

B. D. Vasyliv, V. V. Kulyk, Z. A. Duriagina, P. Ya. Lyutyy, T. M. Kovbasiuk, V. I. Vavrukh, A. M. Trostianchyn, and V. V. Vira

Abstract Advanced Ti-based composites are being used for manufacturing components of state-of-the-art power equipment as well as aircraft and rocket engines owing to their high strength-to-weight ratio and comparatively high fracture toughness in a temperature range of 20–650 °C. These materials have an advantage over conventional titanium alloys, an operating temperature of which is limited by 350–550 °C. This work is aimed at studying the role of alloying elements Cr, Ni, and Al in the improvement of high-temperature (700–800 °C) fracture toughness of Ti-based composites. Ti–Al–X composites (X = Cr and/or Ni and others) were manufactured with electron arc smelting. The contents of alloying elements marked with X varied in a range of 1.7–41 wt%. Fracture toughness tests of single-edge notch beam specimens were carried out in a temperature range of 20–800 °C. The microstructure and phase composition in relation to fracture toughness data and failure micromechanisms of the composites were analyzed. The role of alloying elements Cr, Ni, and Al in the improvement of high-temperature fracture toughness of Ti-based composites

B. D. Vasyliv (✉)
Department of Hydrogen Technologies and Alternative Energy Materials, Karpenko Physico-Mechanical Institute of the NAS of Ukraine, 5 Naukova str., Lviv 79060, Ukraine
e-mail: mechengin1111@gmail.com

V. V. Kulyk · Z. A. Duriagina · P. Ya. Lyutyy · T. M. Kovbasiuk · V. I. Vavrukh · A. M. Trostianchyn
Department of Materials Science and Engineering, Lviv Polytechnic National University, 12 S. Bandera str., Lviv 79013, Ukraine

Z. A. Duriagina
Department of Materials Engineering, The John Paul II Catholic University of Lublin, 14 Racławickie Al., 20-950 Lublin, Poland

V. V. Vira
Department of Strength of Materials and Structural Mechanics, Lviv Polytechnic National University, 12 S. Bandera str., Lviv 79013, Ukraine

O. Fesenko and L. Yatsenko (eds.), *Nanostructured Surfaces, Nanocomposites and Nanomaterials, and Their Applications*, Springer Proceedings in Physics 296,
https://doi.org/10.1007/978-3-031-42704-6_21

was substantiated. It was found that both the Ti–Al–Cr–X and Ti–Al–Ni–X composites of the optimal phase composition are promising in terms of fracture toughness for high-temperature (700–800 °C) applications.

1 Introduction

For operation in harsh conditions (high stress, high-temperature gaseous atmospheres, corrosion, etc.), novel composites and advanced ceramics possessing high-strength and bearing capacity are being used [1–13]. Owing to high strength and crack growth resistance at temperatures up to 550–600 °C and a low specific weight titanium composites are promising for applications in various high-temperature components of aircraft and power generation equipment [14–28]. However, Ti-based composites should have high oxidation resistance, strength, and crack growth resistance at increased operating temperatures (650–800 °C) to meet corresponding requirements [16, 29–31]. Optimization of microstructure is a key issue in this direction [32–36], and the knowledge on the phase composition and microstructure related mechanical properties serves as a driving force in designing such materials [37–58].

It is known that Cr and Ti-based composites contain high-temperature phases like aluminides, silicides, or MAX phases [59–66]. The authors of a series of works have shown that MAX phases can be used in many branches of the industry (mechanical engineering, aerospace, power generation, etc.) both as the bulk materials [67, 68] and high-temperature microstructural constituents [69–72]. MAX phases are ternary compounds described by the formula $M_{n+1}AX_n$ ($n = 1, 2, 3 \ldots$). Here, a transition metal of the d-group (Sc, Ti, V, Cr, Zr, Nb, Mo, Hf, Ta) is marked as M, an element of the p-group (Si, Ge, Al, Ga, S, P, Sn, As, Cd, I, Tl, Pb) is marked as A, and carbon or nitrogen is marked as X [73].

Various fabrication techniques, in particular, direct hot pressing (HP) and spark plasma sintering (SPS) allow providing the hexagonal crystal system in bulk MAX phases [24, 65, 69, 73–75]. The materials obtained under external conditions (e.g., a pressure or a magnetic field) consist of preferentially oriented plate-like grains [69, 73, 76]. When the fabrication conditions are not appropriate for the formation of highly textured bulk MAX phases, the mentioned HP and SPS techniques still allow forming MAX phase areas along the boundaries of matrix grains in composites based on Cr and Ti [23, 37, 51, 69].

Degradation of materials as a result of destructive microstructural changes [77–81] is a cause of implementation of various test methods for predicting a service life of various composites and ceramics taking into account the influence of different environments [68, 80–82].

Crack growth resistance of brittle materials is an important property that allows estimating the lifetime of products made of these materials. Various methods for determination of crack growth resistance parameters [83, 84] and numerous formulas for calculation of fracture toughness of structural materials have been developed

[83, 85–88]. Therefore, a fracture toughness test serves as a useful tool for estimating material bearing capacity in the stage of designing the material with required functional and mechanical properties.

This study aims at estimating the effect of chemical composition on resulting phase balance in Ti–Al–X composites, their microstructure, and high-temperature fracture toughness.

2 Materials and Methods

We have studied Ti-based composites of the following systems: Ti–Al–Si–Mo–Zr–Cr–V–C (composite 1), Ti–Al–Ni (composite 2), and Ti–Al–Cr–C (composite 3). Their chemical compositions are given in Table 1.

The ingots were produced by arc melting in an argon shielding atmosphere on a water cooled copper hearth [54, 68]. The raw constituents and their purities were as follows: Ti > 99.6 at% and Al, Si, Mo, Zr, Cr, V, Ni, C > 99.9 at%. To prepare reinforced composites, corresponding MAX phases were added to the melt. The ingots were annealed after melting at 1190 °C for 10 min.

Beam specimens $4 \times 5 \times 32\ mm^3$ in size machined from the ingots by an electric-discharge method were then grinded and polished.

It is well known that crack growth resistance is an important property of materials [9, 79, 80]. According to various standards and plenty of scientific works [48, 61, 86], the critical stress intensity factor (SIF) K_{Ic} as a measure of fracture toughness is widely used [89–92]. For many brittle materials like ceramics and ceramic-matrix composites, an indentation method as one of the simplest tests [93–101] provides close results to those obtained by other methods of fracture mechanics [83, 98, 102]. Among the last ones, a single-edge notch beam (SENB) method is applied for testing many ceramic materials [100, 102–105].

We performed the high-temperature (from 20 °C up to 800 °C) SENB testing of composite specimens using a three-point bend scheme and calculated their critical SIF using appropriate formulas given in [103–105].

A scanning electron microscope (SEM) Carl Zeiss EVO-40XVP in a back-scattered electron (BSE) imaging mode for the microstructural analysis and in secondary electron (SE) mode for studying fracture surface morphology was used in combination with an INCA Energy 350 system for analyzing chemical compositions of materials in local points due to their energy-dispersive X-ray (EDX) microanalysis.

Table 1 Marking of the studied composites and their chemical compositions

Composite marking	System and chemical composition, wt%
1	Ti–1.6Al–1.7Si–3.1Mo–4.3Zr–6.5Cr–8V–1.2C
2	Ti–1.8Al–40Ni
3	Ti–1.5Al–41Cr–1.8C

X-ray diffraction (XRD) patterns were obtained on Malvern Panalytical X-ray diffractometer operated in the following mode: Cu Kα radiation, 40 kV, 15 mA, the angular range of 20–90°, a step of 0.0217°. For the XRD analysis, a software with a reference to the International Center for Diffraction Data (ICDD) was used. The WinCSD software was utilized for the refinements and determination of phase fractions [106].

3 Results and Discussion

Phase composition and microstructure related characteristics of crack growth resistance of the composites in a wide temperature range have been studied.

Phase Balance in Ti–Al–X Composites

Phase fractions in the studied materials were estimated using the XRD patterns (Fig. 1).

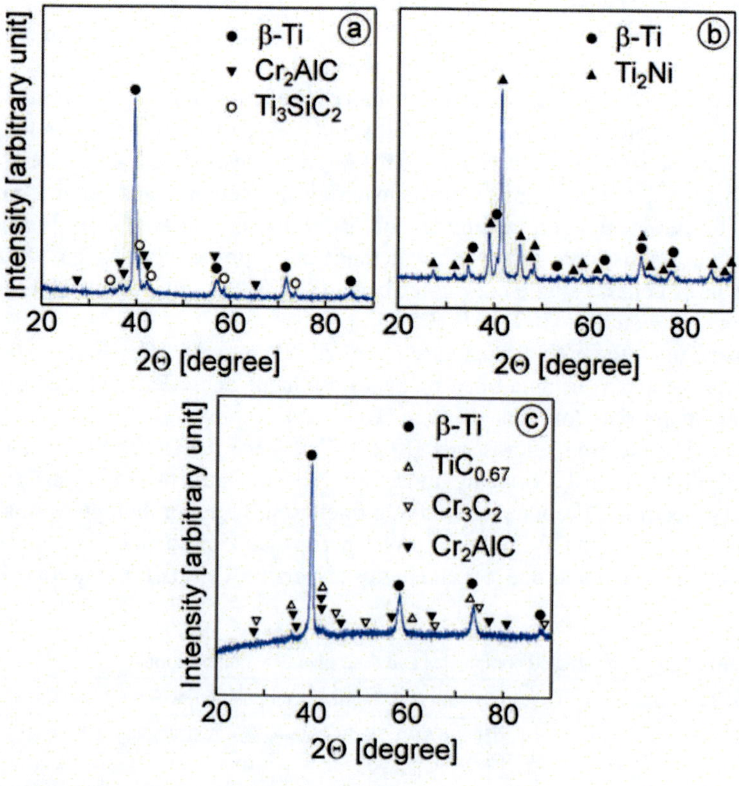

Fig. 1 XRD patterns of the investigated composites **a** 1, **b** 2, and **c** 3 (Table 1)

Ti–Al–Si–Mo–Zr–Cr–V–C. In composite 1 (Table 1), according to the XRD pattern (Fig. 1a), three main phases dominate, namely β-Ti, Cr_2AlC MAX phase (or/and $(Cr_{2/3}Ti_{1/3})_3AlC_2$ MAX phase), and Ti_3SiC_2 MAX phase [1, 24, 76]). The fractions of these phases are as follows (Table 2): β-Ti (91.65 wt%), Cr_2AlC (4.12 wt%), and Ti_3SiC_2 (4.23 wt%). Like many composites of the Ti–Al–Si–Zr system [49, 51, 67, 68], in which α-Ti is the matrix phase and $(Ti, Zr)_5Si_3$ phase serves as a ceramic skeleton, the microstructure image of composite 1 made at a low magnification exhibited a similar skeleton of dark-gray color and the matrix of light-gray color (Fig. 2a).

The high magnification image shows fragments of this skeleton, which are adjacent to the packets of matrix phase grains (Fig. 2b). EDX analysis showed (spectrum 1 in Fig. 2b and Table 3) that composite 1 contains 71.78 wt% Ti, 3.12 wt% Al, 3.34 wt% Si, 7.85 wt% V, 5.84 wt% Cr, 5.03 wt% Zr, and 3.04 wt% Mo. Thus, in this material, packets of β-Ti grains present a matrix phase, whereas both the Cr_2AlC and Ti_3SiC_2 MAX phases form a skeleton.

The β-Ti matrix phase (spectrum 2 in Fig. 2b and Table 3) contains titanium (74.09 wt%), aluminum (3.40 wt%), silicon (1.84 wt%), vanadium (7.89 wt%), chromium (6.19 wt%), zirconium (3.44 wt%), and molybdenum (3.15 wt%). The total amount of the β-Ti phase (Fig. 2a) is 71–74 vol%.

Dark-gray fragments of the skeleton 25–60 μm in length (spectrum 3 in Fig. 2b) represent both the Cr_2AlC and Ti_3SiC_2 MAX phases adjacent to the matrix phase grains. A comparatively high percentage of silicon detected by EDX analysis (59.52 wt% Ti, 1.58 wt% Al, 16.08 wt% Si, 3.72 wt% V, 1.9 wt% Cr, 15.79 wt% Zr, and 1.41 wt% Mo) evidences that, along with the Cr_2AlC MAX phase, the Ti_3SiC_2 MAX phase is present in a ceramic skeleton. Both these phases occupy the total area of 26–29 vol% (Fig. 2a). It should be also noted that no carbon was detected by EDX analysis.

Ti–Al–Ni. The composite 2 (Table 1) showed the peaks of the β-Ti and Ti_2Ni phases in the XRD pattern (Fig. 1b). Its phase composition was determined as follows (Table 2): β-Ti phase (4.28 wt%) and Ti_2Ni phase (95.72 wt%) [107]. The microstructure of this composite differs from that of composite 1. A homogeneous microstructure with the domination of sharp-angle light-gray grains of the Ti_2Ni intermetallic phase 10–20 μm in size can be observed at a low magnification (Fig. 3a).

Comparatively small areas occupied by the β-Ti/Ti_2Ni eutectics are clearly seen in the magnified microstructure image (Fig. 3b). The eutectics area consists of fine round-shaped precipitations of β-Ti of dark-gray color and ultrafine Ti_2Ni phase of

Table 2 Phase compositions of the investigated materials

Composite marking	Phase composition (wt%)
1	β-Ti (91.65); Cr_2AlC (4.12); Ti_3SiC_2 (4.23)
2	β-Ti (4.28); Ti_2Ni (95.72)
3	β-Ti(Cr) (78); Cr_3C_2 (4); Cr_2AlC (6); $TiC_{0.67}$ (12)

Comment Phase weight percentage is given in parentheses

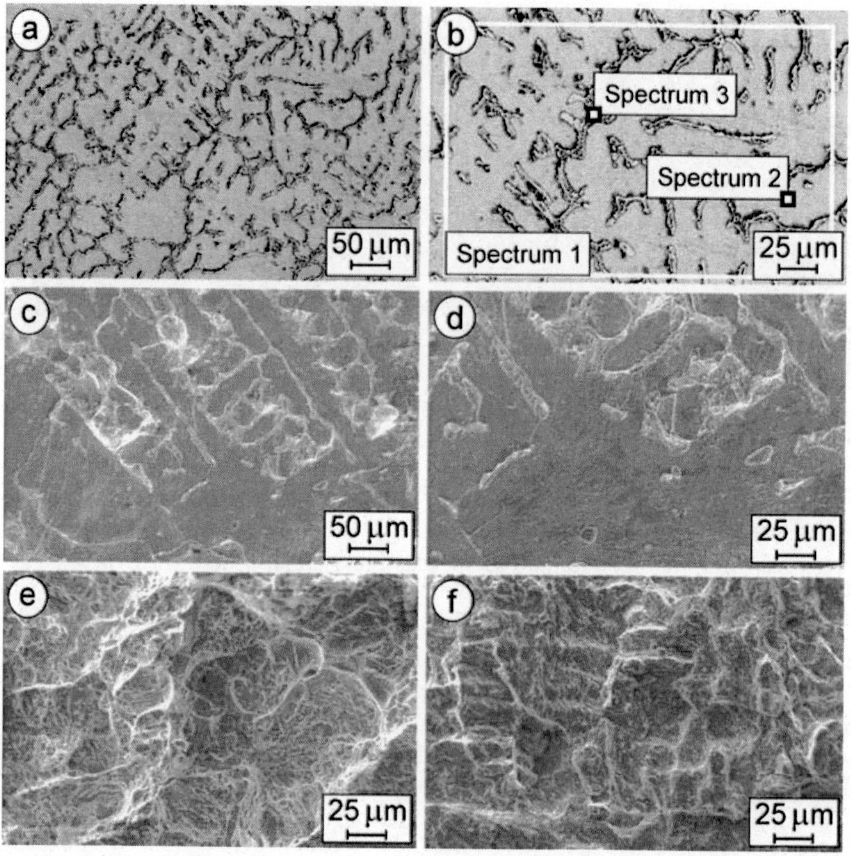

Fig. 2 SEM **a**, **b** microstructures (BSD images) with marked zones of **b** general (spectrum 1) and local (spectra 2 and 3) EDX analyses, and **c**–**f** fractography (SE images) of specimens of composite 1 after fracture toughness tests at **c**, **d** 20 °C, **e** 600 °C, and **f** 700 °C (Table 1)

Table 3 Data of the EDX spectra 1–3 marked in Fig. 2 for a specimen of composite 1

Chemical element and X-ray series	Spectra					
	1		2		3	
	wt%	at%	wt%	at%	wt%	at%
Al K	3.12	5.54	3.40	6.05	1.58	2.70
Si K	3.34	5.70	1.84	3.15	16.08	26.36
Ti K	71.78	71.83	74.09	74.27	59.52	57.26
V K	7.85	7.39	7.89	7.43	3.72	3.36
Cr K	5.84	5.38	6.19	5.72	1.90	1.68
Zr L	5.03	2.64	3.44	1.81	15.79	7.97
Mo L	3.04	1.52	3.15	1.57	1.41	0.67

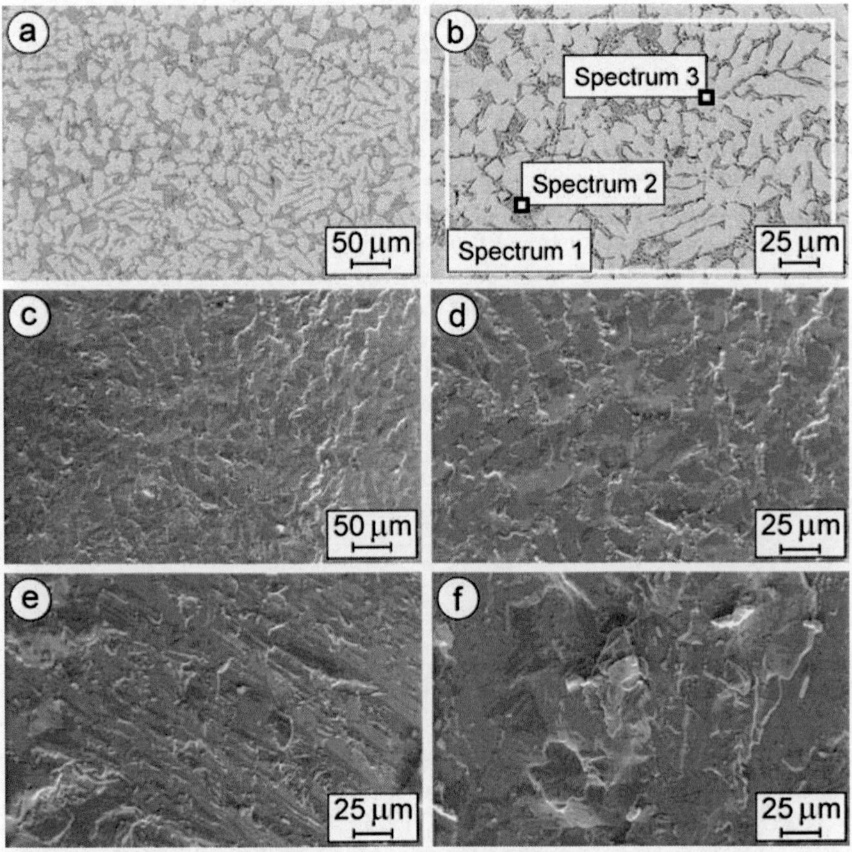

Fig. 3 SEM **a**, **b** microstructures (BSD images) with marked zones of **b** general (spectrum 1) and local (spectra 2, 3, and 4) EDX analyses, and **c–f** fractography (SE images) of specimens of composite 2 after fracture toughness tests at **c**, **d** 20 °C, **e** 700 °C, and **f** 800 °C (Table 1)

light-gray color [107]. Using EDX analysis (spectrum 1 in Fig. 3b and Table 4), it was found that composite 2 comprises 60.18 wt% Ti, 3.76 wt% Al, and 36.06 wt% Ni. The Ti_2Ni phase dominates here.

Table 4 Data of the EDX spectra 1–3 marked in Fig. 3 for a specimen of composite 2

Chemical element and X-ray series	Spectra					
	1		2		3	
	wt%	at%	wt%	at%	wt%	at%
Al K	3.76	6.94	4.38	7.99	3.38	6.28
Ti K	60.18	62.50	62.42	64.17	58.12	60.84
Ni K	36.06	30.56	33.20	27.84	38.50	32.88

In the β-Ti/Ti_2Ni eutectics, we revealed titanium (62.42 wt%), nickel (33.2 wt%), and aluminum (4.38 wt%) according to spectrum 2 in Fig. 3b and Table 4. The total amount of the eutectics (Fig. 4a) was determined as 14–18 vol%.

According to EDX analysis, Ti_2Ni intermetallic phase grains of light-gray color (spectrum 3 in Fig. 4b and Table 5) contain 58.12 wt% Ti, 3.38 wt% Al, and 38.5 wt% Ni. The total amount of this phase is 82–86 vol% (Fig. 4a).

Ti–Al–Cr–C. The composite 3 (Table 1) contains the β-Ti(Cr), Cr_2AlC, Cr_3C_2, and $TiC_{0.67}$ phases revealed in the XRD pattern (Fig. 1c). Its phase balance is approximately as follows (wt%): 78% β-Ti(Cr), 12% $TiC_{0.67}$, 6% Cr_2AlC, and 4% Cr_3C_2. We can observe distinct regions (or grains) of the Ti–Al–Cr–C composite in a microstructure image at a low magnification (Fig. 4a). In the image at a high magnification (Fig. 4b), uniformly distributed small particles (about 1–2.5 μm in size) can be observed inside regions of a matrix phase with the fringe-like boundaries. Small

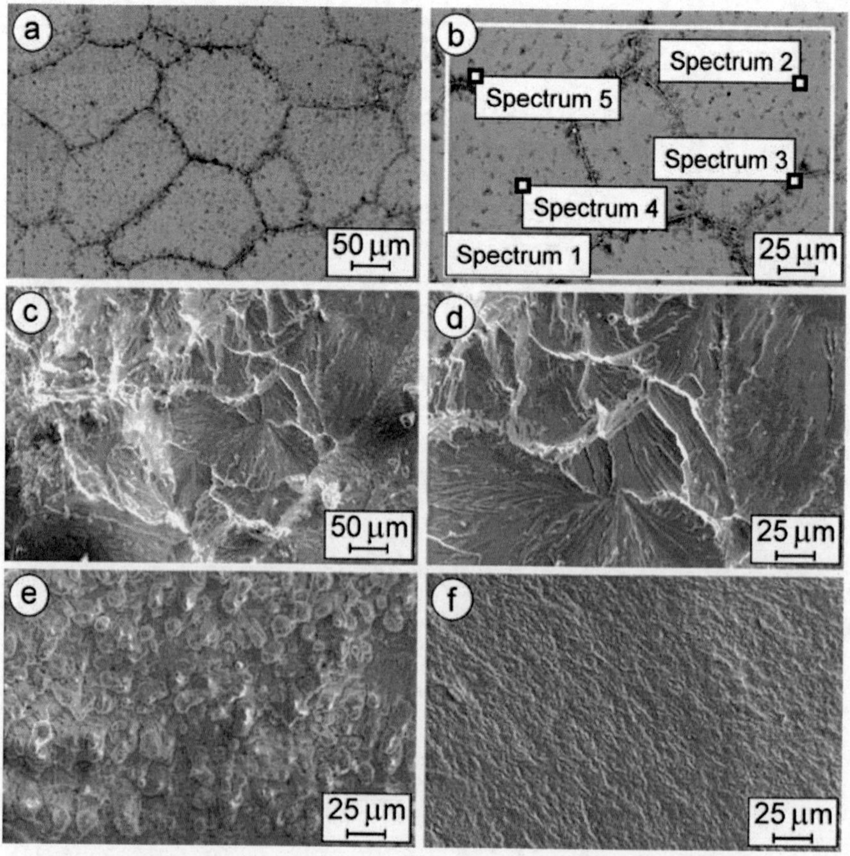

Fig. 4 SEM **a**, **b** microstructures (BSD images) with marked zones of **b** general (spectrum 1) and local (spectra 2, 3, and 4) EDX analyses, and **c**–**f** fractography (SE images) of specimens of composite 3 after fracture toughness tests at **c**, **d** 20 °C, **e** 700 °C, and **f** 800 °C (Table 1)

Table 5 Data of the EDX spectra 1–5 marked in Fig. 4 for a specimen of composite 3

Chemical element and X-ray series	Spectra									
	1		2		3		4		5	
	wt%	at%	wt%	at%	wt%	at%	wt%	at%	wt%	at%
C K	5.12	17.78	–	–	14.30	35.77	11.99	35.30	6.29	19.95
O K	–	–	–	–	8.08	15.17	–	–	3.86	9.20
Al K	3.14	4.85	3.50	6.23	3.31	3.69	2.98	3.91	2.74	3.87
Ti K	54.40	47.40	57.30	57.52	49.17	30.84	50.98	37.63	50.29	40.00
Cr K	37.34	29.97	39.20	36.25	25.14	14.53	34.05	23.16	36.82	26.98

needle-shaped particles of different colors constitute these boundaries. EDX analysis (spectrum 1 in Fig. 4b and Table 5) exhibited the following chemical composition of this composite: 54.4 wt% Ti, 3.14 wt% Al, 37.34 wt% Cr, and 5.12 wt% C. Thus, this metal-matrix composite consists of the β-Ti(Cr) matrix phase, Cr_2AlC or/and $(Cr_{2/3}Ti_{1/3})_3AlC_2$ MAX phase [1, 24, 76], Cr_3C_2 phase, chromium oxide phase, and $TiC_{0.67}$ phase [49, 51, 67]. However, peaks of the Al_2O_3 phase were not identified in the XRD pattern of composite 3 as its percentage was too small (Fig. 1c).

In the β-Ti(Cr) phase, we detected titanium (57.3 wt%), chromium (39.2 wt%), and aluminum (3.5 wt%) according to spectrum 2 in Fig. 4b and Table 5. The total amount of this phase (Fig. 4a) was determined as 66–72 vol%.

Dark-gray particles at the grain boundaries (spectrum 3 in Fig. 4b) contain 49.17 wt% Ti, 3.31 wt% Al, 25.14 wt% Cr, 8.08 wt% O, and 14.3 wt% C. We suggest that this composite contains chromium oxide and titanium carbide at the grain boundaries [76], as the maximum amounts of oxygen and carbon were detected there. The total amount of these particles is about 4–6 vol% (Fig. 4b).

According to EDX analysis, small dark-gray particles distributed in the matrix (spectrum 4 in Fig. 4b and Table 5) consist of 50.98 wt% Ti, 2.98 wt% Al, 34.05 wt% Cr, and 11.99 wt% C. Probably, they correspond to the Cr_3C_2 or $TiC_{0.67}$ phase identified by XRD analysis. Their total amount is 10–12 vol%.

Needle-shaped particles 5–15 μm in length (spectrum 5 in Fig. 4b) situated along the grain boundaries and forming fringe-like areas contain 50.29 wt% Ti, 2.74 wt% Al, 36.82 wt% Cr, 3.86 wt% O, and 6.29 wt% C. Probably, such an area is a mixture of the Cr_2AlC or/and $(Cr_{2/3}Ti_{1/3})_3AlC_2$ MAX phase [1, 24, 76], Al_2O_3, and $TiC_{0.67}$ phases. We suggest that the MAX phase forms textured microareas [1, 76] in combination with the Al_2O_3 phase along β-Ti(Cr) grain boundaries. However, as mentioned above, the Al_2O_3 phase was not detected by XRD analysis. In total, these phases occupy the area of 14–16 vol% (Fig. 4a).

Fracture Toughness of Ti–Al–X Composites

We have studied mechanical behavior of Ti–Al–X composites in a wide temperature range based on fracture toughness data (Fig. 5).

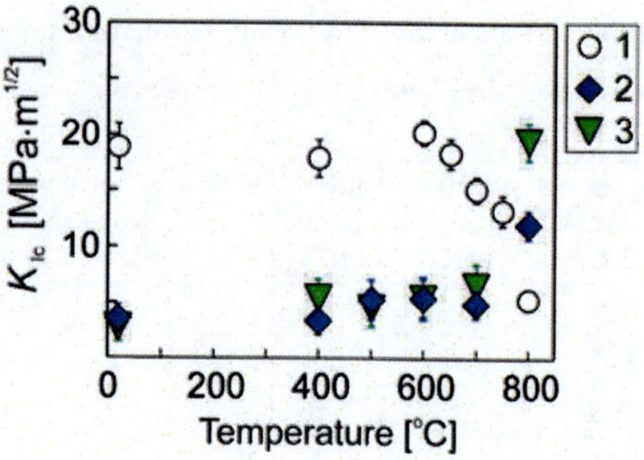

Fig. 5 Temperature dependences of fracture toughness of Ti–Al–X composites (Table 1)

In a testing temperature range of 20–400 °C, fracture toughness of the composite 1 was found to be about 20 MPa $m^{1/2}$ not changing significantly with a gradual increase in the temperature up to 400 °C (Fig. 5). A fracture pattern of a specimen of this composite tested at 20 °C displays a ceramic skeleton fractured during crack growth along appropriate cleavage planes of the β-Ti matrix phase (Fig. 2c, d).

Although no change in fracture toughness was revealed while increasing a testing temperature up to 600 °C (Fig. 5), the fracture micromechanism at 600 °C changed significantly. We can observe for these conditions signs of material plastification as a result of implementation of quasi-ductile fracture micromechanism (Fig. 2e). This micromechanism is related to the comparatively high fracture toughness.

With a further increase of a testing temperature up to 700 °C, fracture toughness changed drastically with an approximately 25% lowering (Fig. 5). Such behavior of the material is obviously associated with its significant plastification at this temperature and implementation of high-temperature fracture micromechanism. A fracture pattern of the specimen tested at 700 °C shows fragments of a ceramic skeleton disintegrated due to the significant plastic deformation of adjacent grains of the β-Ti matrix phase (Fig. 2f).

In contrast to this material, composite 2 (Ti–Al–Ni system) has four times lower fracture toughness in the temperature range of 20–600 °C (Fig. 5). This composite showed a general trend toward a slight increase in fracture toughness values with a gradual increase in temperature in this range. However, the values of fracture toughness of composite 2 do not exceed 6 MPa $m^{1/2}$ even at 600 °C. At 20 °C, a specimen of this composite during the SENB test exhibited a clear cleaved surface corresponding to brittle fracture (Fig. 3c, d).

No change in values of fracture toughness of composite 2 was revealed with a temperature increase from 600 °C up to 700 °C (Fig. 5). The fracture micromechanism at 700 °C remained almost the same not showing any sign of the material plastification (Fig. 3e).

Surprisingly, with an increase of the testing temperature from 700 °C up to 800 °C, we obtained for this composite an approximately 2.5-fold increase in fracture toughness (up to 13 MPa $m^{1/2}$, Fig. 5). Obviously, this is associated with changes in the fracture micromechanism caused by phase transformations in the composite [49, 67, 68, 79]. We can see distinct relief of crack path with variations of cleavage planes on the fracture surface of a specimen of this material tested at 800 °C (Fig. 3f). Such behavior of composite 2 is promising in terms of its high-temperature reliability.

Temperature dependence of fracture toughness of composite 3 (Ti–Al–Cr–C system) exhibited a gradual increase in the temperature range of 20–600 °C similarly to that of composite 2 (Ti–Al–Ni). However, the critical SIF values for the Cr-containing composite 3 are slightly higher in this range (Fig. 5). In a specimen of this composite tested at 20 °C, we observed transgranular cleavage fracture of the β-Ti(Cr) matrix phase and separate areas corresponding to the grain boundary regions with the titanium/chromium carbide and chromium oxide phase particles (Fig. 4c, d). Tiny carbide particles can also be seen on the cleaved surface.

Compared to composite 2, steeper increase in fracture toughness of composite 3 with a temperature increase from 600 °C up to 700 °C was observed (Fig. 5). In a specimen of composite 3 tested at 700 °C, fracture surface morphology is different from that of this composite tested at 20 °C, as it presents multiple titanium/chromium carbide particles embodied into the β-Ti(Cr) matrix plastificated at 700 °C (Fig. 4e). This evidences a ductile fracture micromechanism implemented in the composite at this temperature. Some increase in its fracture toughness (Fig. 5) is related to a relaxation of stress ahead of the advancing crack due to the plasticization of the β-Ti(Cr) matrix phase.

An approximately fourfold increase in fracture toughness of this composite (up to 20 MPa $m^{1/2}$) at 800 °C (Fig. 5) was obtained, which is associated with high-temperature interdiffusion of some elements, in particular aluminum, chromium, and oxygen [24, 69, 75, 76] between the matrix phase, Cr_2AlC MAX phase, and chromium oxide phase. As a result, phase transformations occur. Such microstructural changes are accompanied by corresponding changes in the fracture micromechanism [67, 68, 79]. In a specimen of composite 3 tested at 800 °C, we can observe signs (striations) of quasi-ductile high-temperature fracture (Fig. 4f) similar to the fatigue fracture surface of high-strength alloys [5].

Therefore, based on fracture toughness tests results for Ti–Al–X composites in a wide temperature range it can be stated that both Ti–Al–Ni–X and Ti–Al–Cr–X composites are promising for high-temperature (700–800 °C) applications.

4 Conclusions

Based on the fracture toughness test and microstructure analysis of Ti–Al–X composites in a temperature range of 20–800 °C, the following conclusions were drawn:

1. Phase composition has a crucial effect on the microstructure of the studied composites. Fracture toughness parameters can be successfully used for the characterization of high-temperature (700–800 °C) mechanical stability of the Ti-based composites.
2. The morphology of microstructural components of the composites was investigated in detail and dominant fracture micromechanisms in the investigated temperature range were found. The best fracture toughness of Ti–Al–Cr–C composite was revealed at 800 °C.
3. Both the Ti–Al–Cr–X and Ti–Al–Ni–X composites of the optimal phase compositions are promising in terms of fracture toughness for high-temperature (700–800 °C) applications.

Acknowledgements The authors are thankful to the staff of the Scientific Equipment Collective Use Center "Laboratory of Advanced Technologies, Creation and Physicochemical Analysis of a New Substances and Functional Materials" at Lviv Polytechnic National University (https://lpnu.ua/ckkno) for their kind help in performing X-ray diffraction studies.

References

1. Z. Liu, J. Yang, Y. Qian et al., In-situ reaction synthesis and mechanical properties of quaternary MAX phase $(Cr_{2/3}Ti_{1/3})_3AlC_2$. Ceram. Int. **46**(14), 22854–22860 (2020). https://doi.org/10.1016/j.ceramint.2020.06.055
2. L.A. Dobrzański, L.B. Dobrzański, A.D. Dobrzańska-Danikiewicz, Additive and hybrid technologies for products manufacturing using powders of metals, their alloys and ceramics. Arch. Mater. Sci. Eng. **102**(2), 59–85 (2020). https://doi.org/10.5604/01.3001.0014.1525
3. V. Podhurska, B. Vasyliv, O. Ostash et al., Influence of treatment temperature on microstructure and properties of YSZ–NiO anode materials. Nanoscale Res. Lett. **11**, 93 (2016). https://doi.org/10.1186/s11671-016-1306-z
4. K. Nonaka, M. Teramae, G. Pezzotti, Evaluation of the effect of high-speed sintering and specimen thickness on the properties of 5 mol% yttria-stabilized dental zirconia sintered bodies. Materials **15**, 5685 (2022). https://doi.org/10.3390/ma15165685
5. D. Li, Y. Dong, Z. Zhang et al., An as-cast Ti-V-Cr-Al light-weight medium entropy alloy with outstanding tensile properties. J. Alloy. Compd. **877**, 160199 (2021). https://doi.org/10.1016/j.jallcom.2021.160199
6. T.A. Prikhna, O.P. Ostash, A.S. Kuprin et al., A new MAX phases-based electroconductive coating for high-temperature oxidizing environment. Compos. Struct. **277**, 114649 (2021). https://doi.org/10.1016/j.compstruct.2021.114649
7. B.D. Vasyliv, V.Y. Podhurs'ka, O.P. Ostash et al., Influence of reducing and oxidizing media on the physicomechanical properties of ScCeSZ–NiO and YSZ–NiO ceramics. Mater. Sci. **49**(2), 135–144 (2013). https://doi.org/10.1007/s11003-013-9593-3
8. H.A. Shabri, M.H.D. Othman, M.A. Mohamed et al., Recent progress in metal-ceramic anode of solid oxide fuel cell for direct hydrocarbon fuel utilization: a review. Fuel Process. Technol. **212**, 106626 (2021). https://doi.org/10.1016/j.fuproc.2020.106626
9. V. Kulyk, Z. Duriagina, A. Kostryzhev et al., The effect of yttria content on microstructure, strength, and fracture behavior of yttria-stabilized zirconia. Materials **15**, 5212 (2022). https://doi.org/10.3390/ma15155212

10. T.S. Cherepova, H.P. Dmytrieva, O.I. Dukhota et al., Properties of nickel powder alloys hardened with titanium carbide. Mater. Sci. **52**(2), 173–179 (2016). https://doi.org/10.1007/s11003-016-9940-2
11. T.B. Serbenyuk, T.O. Prikhna, V.B. Sverdun et al., Effect of the additive of Y_2O_3 on the structure formation and properties of composite materials based on AlN–SiC. J. Superhard Mater. **40**(1), 8–15 (2018). https://doi.org/10.3103/S1063457618010021
12. L.Y. Ropyak, M.V. Makoviichuk, I.P. Shatskyi et al., Stressed state of laminated interference-absorption filter under local loading. Funct. Mater. **27**(3), 638–642 (2020). https://doi.org/10.15407/fm27.03.638
13. M.H. Bocanegra-Bernal, S. Díaz de la Torre, Phase transitions in zirconium dioxide and related materials for high performance engineering ceramics. J. Mater. Sci. **37**, 4947–4971 (2002). https://doi.org/10.1023/A:1021099308957
14. S.N. Perevislov, T.V. Sokolova, V.L. Stolyarova, The Ti_3SiC_2 max phases as promising materials for high temperature applications: formation under various synthesis conditions. Mater. Chem. Phys. **267**, 124625 (2021). https://doi.org/10.1016/j.matchemphys.2021.124625
15. B.D. Vasyliv, Improvement of the electric conductivity of the material of anode in a fuel cell by the cyclic redox thermal treatment. Mater. Sci. **46**(2), 260–264 (2010). https://doi.org/10.1007/s11003-010-9282-4
16. W.M. Budzianowski, J. Milewski, Solid-oxide fuel cells in power generation applications: a review. Recent Patents Eng. **5**(3), 165–189 (2011). https://doi.org/10.2174/187221211797636926
17. B.D. Vasyliv, A procedure for the investigation of mechanical and physical properties of ceramics under the conditions of biaxial bending of a disk specimen according to the ring–ring scheme. Mater. Sci. **45**(4), 571–575 (2009). https://doi.org/10.1007/s11003-010-9215-2
18. Y. Komatsu, A. Sciazko, N. Shikazono, Isostatic pressing of screen printed nickel-gadolinium doped ceria anodes on electrolyte-supported solid oxide fuel cells. J. Power Sources **485**, 229317 (2021). https://doi.org/10.1016/j.jpowsour.2020.229317
19. Y. Tan, Y. Xia, Z. Teng et al., Synthesis and enhanced mechanical properties of compositionally complex MAX phases. J. Eur. Ceram. Soc. **41**(8), 4658–4665 (2021). https://doi.org/10.1016/j.jeurceramsoc.2021.03.027
20. V. Podhurska, B. Vasyliv, Influence of NiO reduction on microstructure and properties of porous Ni–ZrO_2 substrates, in *Proceedings of the 3rd International Conference on Oxide Materials for Electronic Engineering (OMEE-2012)*, Lviv, Ukraine (2012), pp. 293–294. https://doi.org/10.1109/OMEE.2012.6464761
21. L. Silvestroni, C. Melandri, J. Gonzalez-Julian, Exploring processing, reactivity and performance of novel MAX phase/ultra-high temperature ceramic composites: the case study of Ti_3SiC_2. J. Eur. Ceram. Soc. **41**(12), 6064–6069 (2021). https://doi.org/10.1016/j.jeurceramsoc.2021.05.029
22. E. Tabares, S.C. Cifuentes, A. Jiménez-Morales et al., Injection moulding of porous MAX phase Ti_3SiC_2 without using space-holder. Powder Technol. **380**, 96–105 (2021). https://doi.org/10.1016/j.powtec.2020.11.022
23. E.B. Kashkarov, N.S. Pushilina, M.S. Syrtanov et al., Preceramic paper-derived SiC_f/$Ti_3Al(Si)C_2$ and SiC_f/Ti_3SiC_2 MAX-phase based laminates fabricated using spark plasma sintering. Scripta Mater. **194**, 113696 (2021). https://doi.org/10.1016/j.scriptamat.2020.113696
24. W. Yu, X. Pi, W. Chen et al., Effects of A-site atoms in Ti_2AlC and Ti_3SiC_2 MAX phases reinforced Mg composites: interfacial structure and mechanical properties. Mater. Sci. Eng. A **826**, 141961 (2021). https://doi.org/10.1016/j.msea.2021.141961
25. I. Danilenko, G. Lasko, I. Brykhanova et al., The peculiarities of structure formation and properties of zirconia-based nanocomposites with addition of Al_2O_3 and NiO. Nanoscale Res. Lett. **12**, 125 (2017). https://doi.org/10.1186/s11671-017-1901-7
26. X. Wang, J. Jie, S. Shichao Liu et al., Growth mechanism of primary Ti_5Si_3 phases in special brasses and their effect on wear resistance. J. Mater. Sci. Technol. **61**, 138–146 (2021). https://doi.org/10.1016/j.jmst.2020.05.063

27. Y. Kharchenko, Z. Blikharskyy, V. Vira et al., Nanostructural changes in a Ni/NiO cermet during high-temperature reduction and reoxidation, in *Nanomaterials and Nanocomposites, Nanostructure Surfaces, and Their Applications. Springer Proceedings in Physics,* **246**, 219–229 (2021). https://doi.org/10.1007/978-3-030-51905-6_17
28. D. Yu, Y. Tan, Oxidation behaviors of compositionally complex MAX phases in air. Ceram. Int. **47**(21), 30188–30193 (2021). https://doi.org/10.1016/j.ceramint.2021.07.198
29. V.V. Kulyk, B.D. Vasyliv, Z.A. Duriagina et al., The effect of water vapor containing hydrogenous atmospheres on the microstructure and tendency to brittle fracture of anode materials of YSZ–NiO(Ni) system. Arch. Mater. Sci. Eng. **108**(2), 49–67 (2021). https://doi.org/10.5604/01.3001.0015.0254
30. L.A. Dobrzański, L.B. Dobrzański, A.D. Dobrzańska-Danikiewicz, Manufacturing technologies thick-layer coatings on various substrates and manufacturing gradient materials using powders of metals, their alloys and ceramics. J. Achievements Mater. Manuf. Eng. **99**(1), 14–41 (2020). https://doi.org/10.5604/01.3001.0014.1598
31. B.D. Vasyliv, V.Y. Podhurska, O.P. Ostash et al., Effect of a hydrogen sulfide-containing atmosphere on the physical and mechanical properties of solid oxide fuel cell materials, in *Nanochemistry, Biotechnology, Nanomaterials, and Their Applications.* Springer Proceedings in Physics, vol. 214 (2018), pp. 475–485. https://doi.org/10.1007/978-3-319-92567-7_30
32. J. Milewski, J. Lewandowski, A. Miller, Reducing CO_2 emissions from a coal fired power plant by using a molten carbonate fuel cell, in *Proceedings of the ASME Turbo Expo*, vol. 2 (2008), pp. 389–395. https://doi.org/10.1115/GT2008-50100
33. Y. Zhang, K. Shimizu, X. Yaer et al., Erosive wear performance of heat treated multicomponent cast iron containing Cr, V, Mn and Ni eroded by alumina spheres at elevated temperatures. Wear **390–391**, 135–145 (2017). https://doi.org/10.1016/j.wear.2017.07.017
34. M. Andrzejczuk, O. Vasylyev, I. Brodnikovskyi et al., Microstructural changes in NiO–ScSZ composite following reduction processes in pure and diluted hydrogen. Mater. Charact. **87**, 159–165 (2014). https://doi.org/10.1016/j.matchar.2013.11.011
35. J. Milewski, J. Lewandowski, Solid oxide fuel cell fuelled by biogases. Arch. Thermodyn. **30**(4), 3–12 (2009). https://www.imp.gda.pl/fileadmin/doc/imp_publishing/wimp/archives%20of%20thermodynamics/C_09_4.pdf
36. G. Witz, V. Shklover, W. Steurer et al., Phase evolution in yttria-stabilized zirconia thermal barrier coatings studied by Rietveld refinement of X-ray powder diffraction patterns. J. Am. Ceram. Soc. **90**(9), 2935–2940 (2007). https://doi.org/10.1111/j.1551-2916.2007.01785.x
37. Z. Tang, J.J. Williams, A.J. Thom et al., High temperature oxidation behavior of Ti_5Si_3-based intermetallics. Intermetallics **16**(9), 1118–1124 (2008). https://doi.org/10.1016/j.intermet.2008.06.013
38. S.S. Savka, D.I. Popovych, A.S. Serednytski, Molecular dynamics simulations of the formation processes of zinc oxide nanoclusters in oxygen environment, in *Nanophysics, Nanomaterials, Interface Studies, and Applications.* Springer Proceedings in Physics, vol. 195, pp. 145–156 (2017). https://doi.org/10.1007/978-3-319-56422-7_11
39. M.Y. Smyrnova-Zamkova, V.P. Red'ko, O.K. Ruban et al., The properties of nanocrystalline powder of 90% Al_2O_3-10% ZrO_2 (wt.%) obtained via the hydrothermal synthesis/mechanical mixing. Nanosistemi Nanomater. Nanotehnol. **15**(2), 309–317 (2017). https://doi.org/10.15407/nnn.15.02.0309
40. L.A. Dobrzański, L.B. Dobrzański, A.D. Dobrzańska-Danikiewicz, Overview of conventional technologies using the powders of metals, their alloys and ceramics in Industry 4.0 stage. J. Achievements Mater. Manuf. Eng. **98**(2), 56–85 (2020). https://doi.org/10.5604/01.3001.0014.1481
41. J. Milewski, J. Kupecki, A. Szczęśniak et al., Hydrogen production in solid oxide electrolyzers coupled with nuclear reactors. Int. J. Hydrog. Energy **46**(72), 35765–35776 (2021). https://doi.org/10.1016/j.ijhydene.2020.11.217
42. W. Huang, H. Qiu, Y. Zhang et al., Microstructure and phase transformation behavior of Al_2O_3–ZrO_2 under microwave sintering. Ceram. Int. (2022). https://doi.org/10.1016/j.ceramint.2022.09.376

43. B. Vasyliv, J. Milewski, V. Podhurska et al., Study of the degradation of a fine-grained YSZ–NiO anode material during reduction in hydrogen and reoxidation in air. Appl. Nanosci. **12**, 965–975 (2022). https://doi.org/10.1007/s13204-021-01768-w
44. Y.G. Chabak, V.I. Fedun, K. Shimizu et al., Phase-structural composition of coating obtained by pulsed plasma treatment using eroded cathode of T1 high speed steel. Probl. At. Sci. Technol. **104**(4), 100–106 (2016)
45. O.V. Sukhova, Influence of mechanisms of structure formation of interfaces in composites on their properties. Metallofiz. Noveishie Tekhnol. **31**(7), 1001–1012 (2009)
46. M.Y. Smyrnova-Zamkova, O.K. Ruban, O.I. Bykov et al., Physico-chemical properties of fine-grained powder in Al_2O_3–ZrO_2–Y_2O_3–CeO_2 system produced by combined method. Comp. Theor. Pract. **18**(4), 234–240 (2018). https://kompozyty.ptmk.net/pliczki/pliki/1290_2018t04_maria-y-smyrnova-zamkova-.pdf
47. W. Huang, Y. Zhang, J. Lu et al., Effect of sintering time on the microstructure and stability of Al_2O_3–ZrO_2 composite powders under microwave-assisted sintering. Ceram. Int. (2022). https://doi.org/10.1016/j.ceramint.2022.11.054
48. B.D. Vasyliv, A.D. Ivasyshyn, O.P. Ostash et al., Kinetics of corrosion-fatigue cracks in Ti–Si cermet composite. Mater. Sci. **38**(2), 220–224 (2002). https://doi.org/10.1023/A:1020990103898
49. M. Kujawa, R. Suwak, L.A. Dobrzański et al., Thermal characterization of halloysite materials for porous ceramic preforms. Arch. Mater. Sci. Eng. **107**(1), 5–15 (2021). https://doi.org/10.5604/01.3001.0014.8189
50. O.V. Dudnik, I.O. Marek, O.K. Ruban et al., Effect of heat treatment on the structure and phase composition of the nanosized powder based on a ZrO_2 solid solution. Powder Metall. Met. Ceram. **59**(1–2), 1–8 (2020). https://doi.org/10.1007/s11106-020-00132-x
51. O.P. Ostash, A.D. Ivasyshyn, B.D. Vasyliv et al., Influence of the structure and asymmetry of loading cycles on the cyclic crack resistance of Ti–Si composites. Mater. Sci. **38**(1), 55–61 (2002). https://doi.org/10.1023/A:1020120714703
52. E. Haurat, J.P. Crocombette, T. Schuler et al., Hydrogen diffusion coefficient in monoclinic zirconia in presence of oxygen vacancies. Int. J. Hydrog. Energy **47**(78), 33517–33529 (2022). https://doi.org/10.1016/j.ijhydene.2022.07.216
53. I.O. Marek, O.K. Ruban, V.P. Redko et al., Physicochemical properties of hydrothermal nanocrystalline ZrO_2–Y_2O_3–CeO_2 powders. Powder Metall. Met. Ceram. **58**(3–4), 125–132 (2019). https://doi.org/10.1007/s11106-019-00055-2
54. V.I. Mazur, Y.N. Taran, S.V. Kapustnikova et al., Titanium matrix composites. US Patent, No. 5366570, 22.11.1994 (1994)
55. F. Mayinger, R. Buser, M. Laier et al., Impact of the material and sintering protocol, layer thickness, and thermomechanical aging on the two-body wear and fracture load of 4Y-TZP crowns. Clin. Oral Invest. **26**, 6617–6628 (2022). https://doi.org/10.1007/s00784-022-04616-5
56. B. Vasyliv, A. Ivasyshyn, O. Ostash et al., High-temperature fatigue crack growth resistance of thermo-mechanically and heat treated cast Ti–Si–Al–Zr composites, in *Metallic Materials with High Structural Efficiency* (Kluwer Academic Publishers, The Netherlands, 2004), pp. 235–240. https://doi.org/10.1007/1-4020-2112-7_24
57. A.S. Doroshkevich, E.B. Asgerov, A.V. Shylo et al., Direct conversion of the water adsorption energy to electricity on the surface of zirconia nanoparticles. Appl. Nanosci. **9**, 1603–1609 (2019). https://doi.org/10.1007/s13204-019-00979-6
58. A.V. Shevchenko, V.V. Lashneva, A.K. Ruban et al., Synthesis and study of high-purity nanocrystalline powder of a solid solution of CeO_2 and Y_2O_3 in zirconium dioxide. Powder Metall. Met. Ceram. **54**(9–10), 548–553 (2016). https://doi.org/10.1007/s11106-016-9748-5
59. M. Fiore, F. Beneduce, C.R.F. Azevedo, Simplification of the thermodynamic description of the Ti-Si system. Tecnol. Metal Mater. Min. **13**(1), 91–97 (2016). https://doi.org/10.4322/2176-1523.1045
60. F.W. Crossman, A.S. Yue, Unidirectionally solidified Ti-TiB and Ti-Ti_5Si_3 eutectic composites. Met. Trans. **2**, 1545–1555 (1971)

61. K. Kasraee, M. Yousefpour, S.A. Tayebifard, Mechanical properties and microstructure of Ti_5Si_3 based composites prepared by combination of MASHS and SPS in Ti-Si-Ni and Ti-Si-Ni-C systems. Mater. Chem. Phys. **222**, 286–293 (2019). https://doi.org/10.1016/j.matchemphys.2018.10.024
62. H. Zhao, L. Hu, C. Li et al., Influence of metallic Cr addition on the phase structure and mechanical properties of plasma-sprayed Ti-Si-C coatings. Ceram. Int. **47**(12), 17570–17579 (2021). https://doi.org/10.1016/j.ceramint.2021.03.075
63. X. Zhao, M. Sokol, M.W. Barsoum et al., Effect of grain orientation on the compressive response of highly oriented MAX phase Ti_3SiC_2. Mater. Sci. Eng. A **809**, 140869 (2021). https://doi.org/10.1016/j.msea.2021.140869
64. Y. Kharchenko, Z. Blikharskyy, V. Vira et al., Study of nanostructural changes in a Ni-containing cermet material during reduction and oxidation at 600 °C. Appl. Nanosci. **10**, 4535–4543 (2020). https://doi.org/10.1007/s13204-020-01391-1
65. W.T. Chiu, K. Wakabayashi, A. Umise et al., Enhancement of the shape memory effect by the introductions of Cr and Sn into the β-Ti alloy towards the biomedical applications. J. Alloy Compd. **875**, 160088 (2021). https://doi.org/10.1016/j.jallcom.2021.160088
66. W.T. Chiu, T. Ishigaki, N. Nohira et al., Effect of Cr additions on the phase constituent, mechanical properties, and shape memory effect of near-eutectoid Ti-4Au towards the biomaterial applications. J. Alloy Compd. **867**, 159037 (2021). https://doi.org/10.1016/j.jallcom.2021.159037
67. G. Frommeyer, R. Rosenkranz, C. Ludecke, Microstructure and properties of the refractory intermetallic Ti_5Si_3 compound and the unidirectionally solidified eutectic Ti-Ti_5Si_3 alloy. Metallkunde **81**, 307–313 (1990). https://doi.org/10.1515/ijmr-1990-810501
68. O.P. Ostash, A.D. Ivasyshyn, B.D. Vasyliv et al., High-temperature and cyclic corrosion crack resistance of alloys of the Ti–Si–Al–Zr system. Mater. Sci. **42**(3), 330–343 (2006). https://doi.org/10.1007/s11003-006-0087-4
69. Z. Zhang, X. Duan, D. Jia et al., On the formation mechanisms and properties of MAX phases: a review. J. Eur. Ceram. Soc. **41**(7), 3851–3878 (2021). https://doi.org/10.1016/j.jeurceramsoc.2021.02.002
70. M. Fiore, F. Beneduce Neto, C.R.F. Azevedo, Assessment of the Ti-rich corner of the Ti-Si phase diagram using two sublattices to describe the Ti_5Si_3 phase. REM Int. Eng. J. **70**(2), 201–207 (2017). https://doi.org/10.1590/0370-44672016700073
71. O.P. Ostash, B.D. Vasyliv, V.Y. Podhurs'ka et al., Optimization of the properties of 10Sc1CeSZ-NiO composite by the redox treatment. Mater. Sci. **46**(5), 653–658 (2011). https://doi.org/10.1007/s11003-011-9337-1
72. A. Shylo, A. Doroshkevich, A. Lyubchyk et al., Electrophysical properties of hydrated porous dispersed system based on zirconia nanopowders. Appl. Nanosci. **10**, 4395–4402 (2020). https://doi.org/10.1007/s13204-020-01471-2
73. M.W. Barsoum, The $M_{n+1}AX_n$ phases: a new class of solids. Prog. Solid State Chem. **28**(1–4), 201–281 (2000). https://doi.org/10.1016/S0079-6786(00)00006-6
74. M.S. Park, W.T. Chiu, N. Nohira et al., Effects of Cr and Sn additives on the martensitic transformation and deformation behavior of Ti-Cr-Sn biomedical shape memory alloys. Mater. Sci. Eng. A **822**, 141668 (2021). https://doi.org/10.1016/j.msea.2021.141668
75. R. Liu, M. Tane, H. Kimizuka et al., Elastic isotropy originating from heterogeneous interlayer elastic deformation in a Ti_3SiC_2 MAX phase with a nanolayered crystal structure. J. Eur. Ceram. Soc. **41**(4), 2278–2289 (2021). https://doi.org/10.1016/j.jeurceramsoc.2020.11.026
76. Z. Wang, G. Ma, L. Liu et al., High-performance Cr_2AlC MAX phase coatings: oxidation mechanisms in the 900–1100°C temperature range. Corros. Sci. **167**, 108492 (2020). https://doi.org/10.1016/j.corsci.2020.108492
77. A.I. Yurkova, S.O. Nakonechnyi, V.V. Cherniavsky et al., Nanostructured AlCoFeCrVNi and AlCoFeCrVTi high-entropy alloys resulted from mechanical alloying and sintering. Appl. Nanosci. **12**, 849–860 (2022). https://doi.org/10.1007/s13204-021-01856-x
78. V.I. Zurnadzhy, V.G. Efremenko, I. Petryshynets et al., Mechanical properties of carbide-free lower bainite in complex-alloyed constructional steel: effect of bainitizing treatment parameters. Kov Mater. **58**(2), 129–140 (2020). https://doi.org/10.4149/km_2020_2_129

79. O.M. Romaniv, B.D. Vasyliv, Some features of formation of the structural strength of ceramic materials. Mater. Sci. **34**(2), 149–161 (1998). https://doi.org/10.1007/BF02355530
80. K.W. Jeong, J.-S. Han, G.-U. Yang et al., Influence of preaging temperature on the indentation strength of 3Y–TZP aged in ambient atmosphere. Materials **14**, 2767 (2021). https://doi.org/10.3390/ma14112767
81. V.Y. Podhurs'ka, B.D. Vasyliv, O.P. Ostash et al., Structural transformations in the NiO-containing anode of ceramic fuel cells in the course of its reduction and oxidation. Mater. Sci. **49**(6), 805–811 (2014). https://doi.org/10.1007/s11003-014-9677-8
82. A. Sciazko, T. Shimura, Y. Komatsu et al., Ni-GDC and Ni-YSZ electrodes operated in solid oxide electrolysis and fuel cell modes. J. Therm. Sci. Technol. **16**(1), JTST0013 (2021). https://doi.org/10.1299/jtst.2021jtst0013
83. O.M. Romaniv, I.V. Zalite, V.M. Simin'kovych et al., Effect of the concentration of zirconium dioxide on the fracture resistance of Al_2O_3–ZrO_2 ceramics. Mater. Sci. **31**(5), 588–594 (1996). https://doi.org/10.1007/BF00558793
84. P. Khajavi, P.V. Hendriksen, J. Chevalier et al., Improving the fracture toughness of stabilized zirconia-based solid oxide cells fuel electrode supports: effects of type and concentration of stabilizer(s). J. Eur. Ceram. Soc. **40**(15), 5670–5682 (2020). https://doi.org/10.1016/j.jeurceramsoc.2020.05.042
85. R.F. Cook, G.M. Pharr, Direct observation and analysis of indentation cracking in glasses and ceramics. J. Am. Ceram. Soc. **73**(4), 787–817 (1990). https://doi.org/10.1111/j.1151-2916.1990.tb05119.x
86. B.D. Vasyliv, Initiation of a crack from the edge of a notch with oblique front in specimens of brittle materials. Mater. Sci. **38**(5), 724–728 (2002). https://doi.org/10.1023/A:1024222709514
87. M.Y. Smyrnova-Zamkova, O.K. Ruban, O.I. Bykov et al., The influence of the ZrO_2 solid solution amount on the physicochemical properties of Al_2O_3–ZrO_2–Y_2O_3–CeO_2 powders. Powder Metall. Met. Ceram. **60**(3–4), 129–141 (2021). https://doi.org/10.1007/s11106-021-00222-4
88. A.D. Ivasyshyn, B.D. Vasyliv, Effect of the size and form of specimens on the diagram of growth rates of fatigue cracks. Mater. Sci. **37**(6), 1002–1004 (2001). https://doi.org/10.1023/A:1015669913601
89. A. Nastic, A. Merati, M. Bielawski et al., Instrumented and Vickers indentation for the characterization of stiffness, hardness and toughness of zirconia toughened Al_2O_3 and SiC armor. J. Mater. Sci. Technol. **31**(8), 773–783 (2015). https://doi.org/10.1016/j.jmst.2015.06.005
90. V. Kulyk, Z. Duriagina, B. Vasyliv et al., The effect of sintering temperature on the phase composition, microstructure, and mechanical properties of yttria-stabilized zirconia. Materials **15**, 2707 (2022). https://doi.org/10.3390/ma15082707
91. ASTM E 384-11, Standard test method for Knoop and Vickers hardness of materials (ASTM International, 2011). https://doi.org/10.1520/E0384-11
92. ASTM C 1327-03, Standard test method for Vickers indentation hardness of advanced ceramics (ASTM International, 2003). https://doi.org/10.1520/C1327-03
93. G.R. Anstis, P. Chantikul, B.R. Lawn et al., A critical evaluation of indentation techniques for measuring fracture toughness: I, Direct crack measurement. J. Am. Ceram. Soc. **64**(9), 533–538 (1981). https://doi.org/10.1111/j.1151-2916.1981.tb10320.x
94. B.R. Lawn, M.V. Swain, Microfracture beneath point indentations in brittle solids. J. Mater. Sci. **10**(1), 113–122 (1975). https://doi.org/10.1007/BF00541038
95. A.G. Evans, E.A. Charles, Fracture toughness determinations by indentation. J. Am. Ceram. Soc. **59**(7–8), 371–372 (1976). https://doi.org/10.1111/j.1151-2916.1976.tb10991.x
96. B.R. Lawn, A.G. Evans, D.B. Marshall, Elastic/plastic indentation damage in ceramics: the median/radial crack system. J. Am. Ceram. Soc. **63**(9–10), 574–581 (1980). https://doi.org/10.1111/j.1151-2916.1980.tb10768.x
97. G.A. Gogotsi, S.N. Dub, E.E. Lomonova et al., Vickers and Knoop indentation behaviour of cubic and partially stabilized zirconia crystals. J. Eur. Ceram. Soc. **15**(5), 405–413 (1995). https://doi.org/10.1016/0955-2219(95)91431-M

98. B. Vasyliv, V. Kulyk, Z. Duriagina et al., Estimation of the effect of redox treatment on microstructure and tendency to brittle fracture of anode materials of YSZ-NiO(Ni) system. East Eur. J. Enterp. Technol. **108/6**(12), 67–77 (2020). https://doi.org/10.15587/1729-4061.2020.218291
99. J. Lankford, Indentation microfracture in the Palmqvist crack regime: implications for fracture toughness evaluation by the indentation method. J. Mater. Sci. Lett. **1**(11), 493–495 (1982). https://doi.org/10.1007/BF00721938
100. I. Danilenko, F. Glazunov, T. Konstantinova et al., Effect of Ni/NiO particles on structure and crack propagation in zirconia based composites. Adv. Mater. Lett. **5**(8), 465–471 (2014). https://doi.org/10.5185/amlett.2014.amwc1040II
101. O.N. Grigoriev, V.B. Vinokurov, T.V. Mosina et al., Kinetics of shrinkage, structurization, and the mechanical characteristics of zirconium boride sintered in the presence of activating additives. Powder Metall. Met. Ceram. **55**(11–12), 676–688 (2017). https://doi.org/10.1007/s11106-017-9855-y
102. V.V. Kulyk, Z.A. Duriagina, B.D. Vasyliv et al., Effects of yttria content and sintering temperature on the microstructure and tendency to brittle fracture of yttria-stabilized zirconia. Arch. Mater. Sci. Eng. **109**(2), 65–79 (2021). https://doi.org/10.5604/01.3001.0015.2625
103. ASTM E 399-20a, Standard test method for linear-elastic plane-strain fracture toughness of metallic materials (ASTM International, 2020). https://doi.org/10.1520/E0399-20A
104. ASTM C 1421-18, Standard test methods for determination of fracture toughness of advanced ceramics at ambient temperature (ASTM International, 2018). https://doi.org/10.1520/C1421-18
105. J. Kübier, Fracture toughness of ceramics using the SEVNB method: from a preliminary study to a standard test method, in *Fracture Resistance Testing of Monolithic and Composite Brittle Materials*, ed. by J. Salem et al. (ASTM International, 2002), pp. 93–106. https://doi.org/10.1520/STP10473S
106. L. Akselrud, Y. Grin, WinCSD: software package for crystallographic calculations (Version 4). J. Appl. Crystallogr. **47**, 803–805 (2014). https://doi.org/10.1107/S1600576714001058
107. L. Sheng, Y. Yang, T. Xi, Effect of heat treatment on morphology evolution of Ti_2Ni phase in Ti-Ni-Al-Zr alloy. IOP Conf. Ser. Mater. Sci. Eng. **322**, 022040 (2018). https://doi.org/10.1088/1757-899X/322/2/022040

Magnetic Field-Controlled Light Transmission in Ferronematic Liquid Crystals

A. M. Korostil and M. M. Krupa

1 Introduction

Ferronematic liquid crystals (FNLCs) are characterized by combination of the orientation ordering of the anisotropic nematic molecules with the ferromagnetic ordering of impurity ferromagnetic nanoparticles anchored by interphase interaction with the nematic molecular crystal (NLC) [1, 2]. The strong effect of the external magnetic field on the orientation molecular ordering is related to its strong coupling with the nanoparticle magnetization. Such the reorientation results in a distortion of the average molecular direction represented by the nematic director $\boldsymbol{n}$ [3].

The orientation molecular structure is characterized by the orientation distribution function $f(\boldsymbol{x}, \boldsymbol{u}, t)$ describing the density of anisotropic molecules in the position $\boldsymbol{x}$ pointed along the unite vector $\boldsymbol{u}$ at the time t. This function determines the macroscopic characteristics of NLCs, in particular the nematic director $\boldsymbol{n}$. In the framework of microscopic approach [4, 5] accounting for the intermolecular interaction, this distribution function enters in a definition of the intermolecular potential, U, and related to it part of the free energy,

$$\frac{1}{2}\int_{\mathbb{R}^3}\int_{S^2} U_\varepsilon(\boldsymbol{x}, \boldsymbol{u}, t)\mathrm{d}\boldsymbol{u}\,\mathrm{d}\boldsymbol{x} = \int_{\mathbb{R}^3}\int_{S^2} B_\varepsilon(\boldsymbol{x}, \boldsymbol{x}', \boldsymbol{u}, \boldsymbol{u}') f(\boldsymbol{x}', \boldsymbol{u}', t)\mathrm{d}\boldsymbol{u}'\mathrm{d}\boldsymbol{x}'\mathrm{d}\boldsymbol{m}\mathrm{d}\boldsymbol{x}, \quad (1)$$

where $B_\varepsilon(\boldsymbol{x}, \boldsymbol{x}', \boldsymbol{u}, \boldsymbol{u}') = (1/\varepsilon^{3/2})g\big((x - x)/\sqrt{\varepsilon}\big)\alpha|\boldsymbol{u} \times \boldsymbol{u}|^2$ is the interaction kernel with the small ε parameter representing the typical interaction distance. The model approximation of the weak deviation of the NLC from the equilibrium state together

A. M. Korostil (✉) · M. M. Krupa
Institute of Magnetism of NAN of Ukraine and MES of Ukraine, Kyiv, Ukraine
e-mail: amand@imag.kiev.ua

O. Fesenko and L. Yatsenko (eds.), *Nanostructured Surfaces, Nanocomposites and Nanomaterials, and Their Applications*, Springer Proceedings in Physics 296,
https://doi.org/10.1007/978-3-031-42704-6_22

with a local Taylor expansion is reduced (1) to the representation in terms of the tensor operator, $\boldsymbol{Q} = S(\boldsymbol{n}\otimes\boldsymbol{n}-\boldsymbol{I}/3)$, of the nematic ordering and its derivatives. The quantity S is the strength of the nematic ordering describing the molecule deviation from the average orientation $\boldsymbol{n}$.

The orientation ordering states of the NLCs are formed by the dispersion forces of attraction and Van der Waals forces of repulsion providing their sensitivity with respect to external electric field. Their magnetic field control and sensitivity of the orientation molecular ordering are realized in FNLCs via ferromagnetic nanoparticles coupled with nematic molecules by interphase anchoring forces. This magnetic sensitivity is related to the properties of the spontaneous magnetization, $\boldsymbol{M}$, of the ferromagnetic nanoparticles [1–3].

The magnetic field-induced change of the magnetization results in a distortion of the orientation molecular ordering and the nematic director, $\boldsymbol{n}$, coinciding with the optical axis determining the optical properties of the liquid crystal. The passage of the light beam through the NLC, possessing by the optical anisotropy, accompanies by the changes of the refractive indices along the optical axis (nematic director), $n_{||}$, and perpendicular to it, $n_{\perp}$. This results in the change of the light polarization, phase delay and is manifested as the magneto-optic effect of the polarization twisting and the large birefringence of linear polarized light in the FNLC [4]. The transmission of the linearly polarized light through the FNLC sandwiched by two crossed polarized is characterized by the dependence of the transmitted light intensity on the applied magnetic field, that can be important applications in displays, photonics, and controllable light modulators [6].

The paper is organized as follows. In Sect. 2, the macroscopic model of the FNLC is considered as the coupled system of the magnetization, $\boldsymbol{M}$, and the nematic director, $\boldsymbol{n}$, under the external magnetic field. In Sect. 3, features of the light transmission through the FNLC cell and its dependence on the applied magnetic field are studied. Section 4 is devoted to a experimental investigation of the magneto-optic effect in the ferromagnetic nematic liquid crystals.

2 Macroscopic Model

The dynamic properties of the FNLCs are described in the framework of the model cell in the form of the plane-parallel capillary filled by the NLC with the suspense of ferromagnetic nanoparticles. In the given Cartesian coordinate system, the axis z is perpendicular to the cell surfaces and another two axes, x and y, lie in the cell plane. Interaction between magnetic moments is enough to onset of the spontaneous magnetization, $\boldsymbol{M}$. Alignment of the nematic director, $\boldsymbol{n}$, relative to the axis x is given by the interphase interaction between the nematic molecules and the plane-parallel cell surfaces. Collinearity of the magnetic moment of each of the ferromagnetic nanoparticles to the nematic director, $\boldsymbol{n}$, is given by its magnetic anisotropy. The external magnetic field, $\boldsymbol{H}$, applied perpendicular to the cell surface, causes a distortion of the nematic director and the optical axis parallel to it. This is manifested in the light passing through the FNLC cell.

The static behavior of the magnetization, $\boldsymbol{M}$, and the nematic director field, $\boldsymbol{n}$, are described by the equations of the variation problem for the functional of the free energy density. The free energy density can be represented in the form, $f = f_m + f_d + f_s$ [2, 7], where the first term describes the magnetic contribution

$$f_m = -\mu_0 \boldsymbol{M} \cdot \boldsymbol{H} - \frac{1}{2} A_1 (\boldsymbol{M} \cdot \boldsymbol{Q} \cdot \boldsymbol{M}) + \frac{1}{2} A_2 (|\boldsymbol{M}| - M_0)^2, \tag{2}$$

where μ_0 is the magnetic constant, $\boldsymbol{H} = H\hat{\boldsymbol{e}}_z$ is the applied magnetic field, $A_{1,2} > 0$ is a constant. The first term in (2) represents the coupling of the magnetization and external magnetic field. Since $H >> M_0$, the local magnetic field is equal to $\boldsymbol{H}$ independent on the $\boldsymbol{M}(\boldsymbol{r})$ configuration. The second term in (2) describes the static coupling between the director and the magnetization (originated from magnetic nanoparticles). The third term describes the energy connected with the deviation of the modulus from M_0.

The second term, f_d, describes distortion energy which can be represented via the gradients of the quadrupole tensors, $\partial Q_{ij}/\partial x_k$ [4, 5] as

$$2 f_d = L_1 \frac{\partial Q_{ij}}{\partial x_k} \frac{\partial Q_{ij}}{\partial x_k} + L_2 \frac{\partial Q_{ij}}{\partial x_j} \frac{\partial Q_{ik}}{\partial x_k} + L_3 Q_{ij} \frac{\partial Q_{kl}}{\partial x_i} \frac{\partial Q_{ik}}{\partial x_j}, \tag{3}$$

where L_i $(i = 1, 2, 3)$ is the elastic constant. In the term of the nematic vector $\boldsymbol{n}$,

$$f_d = \frac{1}{2} K_1 (\nabla \cdot \boldsymbol{n})^2 + \frac{1}{2} K_2 [\boldsymbol{n} \cdot (\nabla \times \boldsymbol{n})]^2 + \frac{1}{2} K_3 [\boldsymbol{n} \times (\nabla \times \boldsymbol{n})]. \tag{4}$$

Here, elastic constants K_i $(i = 1, 2, 3)$ for splay $(i = 1)$, twist $(i = 2)$, and band $(i = 3)$ are related to elastic constants L_i $(i = 1, 2, 3)$ by the relations, $L_1 = (K_3 + 2K_2 - K_1)/9S^2$, $L_{2(3)} = 4(K_{1(3)} - K_2)/9S^2$. In the single-constant approximation for the given model $K_i = K$ $(i = 1, 2, 3)$, the free energy density takes the form, $f_d = \frac{1}{2} K (\partial_z \theta)^2$.

The term, f_s, is the a finite surface energy associated with the anchoring of the director at the plates, $f_s = -W(\boldsymbol{n}_s Q \cdot \boldsymbol{n}_s)/2$, where W is the anchoring strength and $\boldsymbol{n}_s = \hat{\mathbf{e}}_z \sin\varphi_s + \hat{\mathbf{e}}_x \sin\varphi_s$ is the preferred direction specified by the pre-tilted angle φ_s. For the total free energy, $F = \int \mathrm{d}V (f_m + f_d) + \int \mathrm{d}S\, f_s$, the equilibrium condition requires $\delta F = 0$.

3 Magneto-controlled Light Transmission

In the framework of the given model, in equilibrium the magnetic field-distorted nematic director, $\boldsymbol{n}$, and magnetization, $\boldsymbol{M}$, are characterized by their polar angles $(\theta,\ \varphi)$ and $(\psi,\ \varphi)$, respectively, where θ and ψ are deviation angles relative to z-axis and φ deviation angle relative to the plane zx of the Cartesian coordinate system. Generally, these angles are functions of the z-coordinate along the light ray. The average magnetization z-component, $M_z = \int dz M \cos \psi(z)/d$, where d is thickness of the FNLC cell.

The electric component of the monochromatic linearly polarized light described as, $\boldsymbol{E} = E_0 \boldsymbol{j} \exp i(\boldsymbol{k}_i \boldsymbol{r} - \omega t)$, where E_0 is the electric field amplitude, $\boldsymbol{j}$ is the polarization vector $\boldsymbol{j}$, the wave vector $\boldsymbol{k}_i = k_0 \boldsymbol{n}_z$ ($k_0 = 2\pi/\lambda$, where λ is the light wavelength), and $\boldsymbol{n}_z$ is the z-component of the refractive vector. The effect of the nematic crystal cell on the light is manifested in change in its polarization, $\boldsymbol{j}$, as a result of a change in the refractive vector components [3, 8].

After the polarizer, the linear polarized light with the polarization vector $\boldsymbol{j} = \left(1/\sqrt{2}\right)(1,\, 1)^{\mathrm{T}}$ enters the liquid–crystal cell that is accompanied by changing the polarization along the axis, z. That is related to the z-dependent variation of the tilt angle of the optical axis coinciding with the nematic director $\boldsymbol{n}$.

Optical properties of the NLC cell for the monochromatic electromagnetic field are described by the Maxwell equations [8],

$$i\omega \boldsymbol{H} = c \operatorname{rot} \boldsymbol{E}, \quad i\omega \boldsymbol{D} = -c \operatorname{rot} \boldsymbol{H}, \tag{5}$$

where $\boldsymbol{H}$ and $\boldsymbol{E}$ are the magnetic and electric fields, $\boldsymbol{D}$ is the electric induction related to the electric field $\boldsymbol{E}$ via dielectric permittivity tensor, $\boldsymbol{\varepsilon}$ by the relation $\boldsymbol{D} = \boldsymbol{\varepsilon}\, \mathbf{E}$. Elimination the magnetic field with the substitution of the explicit expression for the electric field reduce (5) to the equation,

$$\boldsymbol{D} = \varepsilon \boldsymbol{E} = n^2 \boldsymbol{E} - \boldsymbol{n}(\boldsymbol{n}\boldsymbol{E}). \tag{6}$$

The solvability condition of (6) gives the so-called Fresnel equation [8].

$$\det\left|n^2 \delta_{ik} - n_i n_k - \varepsilon_{ik}\right| = 0, \tag{7}$$

where the indices $i,\ k$ take values $1, 2, 3$ corresponding to coordinates $x,\ y,\ z$. For the nematic molecular crystals belonging to uniaxial crystals, the one dielectric tensor principal value, $\varepsilon_e = \varepsilon_{33}$, corresponds to the principal axis parallel to the FNMC optical axis. The second dielectric tensor principal value, $\varepsilon_o = \varepsilon_{11} = \varepsilon_{22}$, corresponds to the principal axes perpendicular to the optical axes. In this case, (7) is reduced to the equation,

$$\left(n^2 - \varepsilon_o\right)\left(\varepsilon_e n_z^2 + \varepsilon_o\left(n_x^2 + n_y^2\right) - \varepsilon_o\varepsilon_e\right) = 0 \tag{8}$$

splitting into the two equations,

$$n^2 - \varepsilon_o = 0, \qquad \frac{n_z^2}{\varepsilon_{e0}} + \frac{n_x^2 + n_y^2}{\varepsilon_o} = 1. \tag{9}$$

The first and second equations in (9) correspond to the ordinary and extraordinary rays, respectively. Corresponding solutions for the ordinary and the effective extraordinary refractive indexes, n_o and n_e, respectively, have the form,

$$n_o = \sqrt{\varepsilon_o} \text{ and } n_e = \left(n_{e0}^{-2} \sin\theta(z) + n_0^{-2} \cos\theta(z)\right)^{-1/2}, \tag{10}$$

where $n_{e0} = \sqrt{\varepsilon_e}$ is the extraordinary refractive index, and $\theta(z)$ is the tilt angle of the nematic director (or the optic axis) with respect to the axis x perpendicular to the axis z of the light direction. With increasing the tilt angle of the nematic director with respect to the axis z, the refractive index for the ordinary ray does not change, but for the effective extraordinary ray, it decreases, i.e., $n_e \to n_o$. The difference in refractive indices for extraordinary and ordinary rays is caused by the anisotropic shape and molecular polarization.

Changes in the refractive indices of the light ray in the liquid–crystal cell lead to a change of its polarization state. That describes as converting the polarization vector $\boldsymbol{j}$ of the input ray into the polarization vector $\boldsymbol{j}'$ of the output ray under action of the transition matrix $\boldsymbol{R}$, or the Jones matrix, on each an element of the light ray path, Δ. At the zero tilt of the nematic director relatively the plane xz ($\varphi = 0$), the Jones matrix is described as [3, 8]

$$\boldsymbol{R}_t[z] = \left\| \delta_{ij} e^{i s_i k_0 (n_e(z) - n_0)\Delta/2} \right\|, \quad s_i = (-1)^{(i-1)}, \tag{11}$$

where $i, j = 1, 2$, and Δ is the enough small neighborhood of the point z, i.e., thin slice of the liquid–crystal cell. The angular deviation of the nematic director is described with help of the rotation matrix,

$$\boldsymbol{R}_r[\varphi] = \left\| \delta_{ij} \cos\varphi - \delta_{|i-j|,1} (-1)^i \sin\varphi \right\|, \quad i, j = 1, 2 \tag{12}$$

that determines the transfer matrix at the point z as, $\boldsymbol{R}(z) = \boldsymbol{R}_r[-\varphi]\boldsymbol{R}_t[z]\boldsymbol{R}_r[\varphi]$. The complete transfer matrix for the light transmission is the result of multiplication of the Jones matrices of all elements (slices) in the order in which the light passes them. For the whole liquid–crystal cell, $\boldsymbol{R} = \overleftarrow{\prod}_{z\in[0,d]} \boldsymbol{R}(z)$. Successive passage of the light ray through an analyzer at an angle β which describes by the transfer matrix

$$\boldsymbol{P}_\beta = \left\| \delta_{ij}\left(\delta_{i,1}\cos^2\beta + \delta_{i,2}\sin^2\beta\right) + \delta_{|i-j|,1} \sin 2\beta/2 \right\| \tag{13}$$

results in the transfer matrix converting the initial polarization vector of the light ray into the output one, $\boldsymbol{j}' = (\boldsymbol{P} \cdot \boldsymbol{R})\ \boldsymbol{j}$. In the considered case of the constant angular deviations, φ, for the analyzer angle, $\beta = -\pi/4$, the normalized intensity of output light ray is described as,

$$\frac{I'}{I} = \boldsymbol{j}^{*T} \boldsymbol{j}' = \cos^2 2\varphi \sin^2 \phi, \quad \phi = k_0 \int_0^d \mathrm{d}z\, (n_e(z) - n_0), \tag{14}$$

where d is the thickness of the liquid–crystal cell and ϕ implies the phase difference. The intensity of the output ray dominates at the zero deviation of the nematic director from the xz plane ($\beta = 0$). Its dependence on the applied external magnetic field in (14) is determined via the distortion angle $\theta(z)$ entering in determination of the refractive index $n_e(z)$ in (9). The intensity (13) is non-monotonic function of the external magnetic field and dependence on the frequency ω of electromagnetic field. The normalized phase difference defined as $r(H) = 1 - \frac{\phi(H)}{\phi_o}$, where ϕ_0 is the normalized phase difference at zero magnetic field, is characterized by the strong dependence on the magnetic field near its zero value.

The phase difference depends on the magnetic field H via the distortion angle $\theta(z)$ of the nematic director which is the solution of the variation problem for the free energy functional of the free energy density of the system. The solution $\theta(z)$ of the corresponding equation at low field is characterized by the linear field dependence [3],

$$\theta(z) = a(z)H + b, \tag{15}$$

where

$$a(z) = \frac{\mu_0 M_0}{2}\left(\frac{z(z-d)}{K} - \frac{d}{2W}\right), \quad b = \frac{\pi}{2} - \varphi_s. \tag{16}$$

Here, K is the elastic constant of the single-constant approximation. The angular deviation $\psi(z)$ of the magnetization $\boldsymbol{M}$ is characterized by the small shift relatively of the angle $\theta(z)$ which has the linear dependence on the magnetic field H. This shift is enough small at the strong anchoring between magnetic nanoparticles and anisotropic molecules. At the limit of the infinite anchoring (14) is reduced to the equation,

$$r(H) = b_2 H^2 + b_1 H \tag{17}$$

(where b_1 and b_2 are functions of such parameters as $r_0 = n_{e0}(n_{e0} + n_o)/(2n_{e0}^2)$, K, d, and φ_s), which describes the normalized deference phase as the quadratic function of the applied magnetic field.

4 Measurement of Magneto-optic Effects

Magneto-optic properties of pure and impure doped by ferromagnetic nanoparticles' NLCs have studied. These NLCs were confined by two plane-parallel glass plates spaced 250–300 μ. For their planar orientation on the surface of transparent glass (or fused quartz), substrates were applied a layer of polyamide varnish or polyvinyl alcohol with a 1 μm thickness. The directional molecular orientation of the NLCs was created by the orientation surface rubbing of the glass plates.

Ferromagnetic Fe_3O_4 nanoparticles were added into the NLC solution. The concentration of such nanoparticles was in the range 2–7% by the solution weight. The ferromagnetic nanoparticles were distributed by vibration evenly in the NLC solution. The quality of the NLCs and the uniformity of the nanoparticle distribution were controlled with help of a polarizing microscope and an atomic force microscope with a magnetic measuring head.

The sizes of the NLCs were 18 × 18 and 3 × 3 mm. Larger samples were used for spectral studies. Smaller samples were used to study the sensitivity and dynamic characteristics of the magneto-sensitive layer of the FNLC. Electrodes were placed on two plane-parallel surfaces of the small samples.

The dependence of the rotation angle of the NLCs on the magnetic field was investigated (Fig. 1). During the measurements, the transmission plane of the polarizer was at an angle, $\alpha = \pi/2$, with respect to the direction of the nematic director, and the analyzer (Senarmon prism) was set so that the transmission plane of one channel was parallel and the transmission plane of the second channel was perpendicular to the polarization transmission plane.The magnetic field acted in the plane of the sample and its direction was at an angle $\alpha = \pi/2$ with respect to the direction of the nematic director. The magnitude of the signal for the first and second photodetectors is determined in the absence of a magnetic field by the general formula,

$$I_i = A_i I_0 \left[\delta_{i,1} + (-1)^i \frac{1}{\sqrt{2}} \cdot \sin^2(\phi/2) \right], \quad i = 1, 2. \tag{18}$$

Here, I_1 and I_2 are the signal values from the two photodetectors in the cases of parallel and perpendicular orientations of the Senarmon prism transmission plane relative to the transmission plane of the linear polarizer, respectively, and ϕ is the phase difference between extraordinary and ordinary polarized lights passed through the FNLC in the absence of an external magnetic field. The system (18) describes this phase difference.

At given measurement conditions, the value of photodetector signals corresponding to the magneto-induced rotation of the nematic director at an angle, θ, is described by the system,

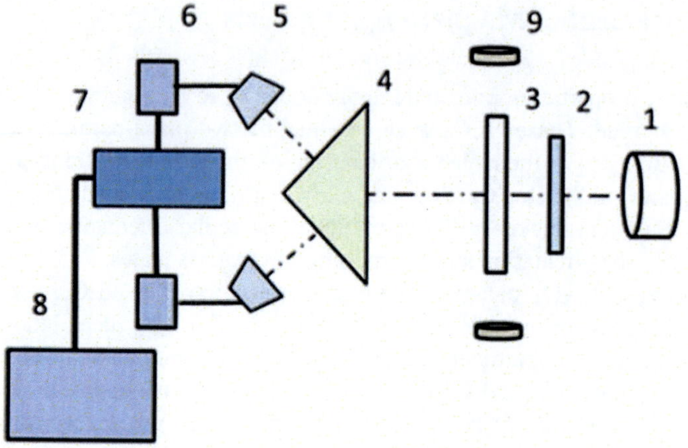

Fig. 1 Scheme of measurements of characteristics of the magnetically sensitive layer with ferromagnetic nanoparticles: 1—semiconductor laser, 2—polarizer, 3—liquid–crystal sample, 4—polarization prism Senarmon, 5—two photodetectors, 6—two amplifiers, 7—two-channel analog–digital converter, 8—personal computer, 9—Helmholtz coils

$$\frac{I_1(H) - I_1(0)}{I_2(0)} = 1 - \cos 2\theta \cdot (\cos\theta + \sin\theta),$$
$$I_1(H)/I_2(0) = \cos 2\theta \cdot (\cos\theta + \sin\theta), \tag{19}$$

where the first and second lines correspond to the cases in perpendicular and parallel directions of the magnetic field relative to the transmission plane of the linear polarizer.

Using the measurement scheme (Fig. 1) with the help of Eqs. (18) and (19), it was obtained the characteristic magnetic field dependences of the rotation angle of the nematic director of the FNLC, $\theta = \theta(H)$, (Fig. 2). As can be seen from the presented graphics, the dependence $\theta = \theta(H)$ is characterized by the strong nonlinear dependence on the applied magnetic field.

These measurements were performed on the FNLC sample of size 18 × 18 mm. The reference magnetic field had a fixed frequency of 1 Hz and variable amplitude H. The magnetic field was generated by a sequence of electric pulses of the same polarity, with the duration of 5 ms, and the pulse interval at 10 ms. The mentioned experimental results indicate a strong magnetic sensitivity of the NLCs doped by ferromagnetic nanoparticles which is orders of magnitude more than of pure NLCs. So, the FNLCs can be considered as perspective elements for data visualization systems.

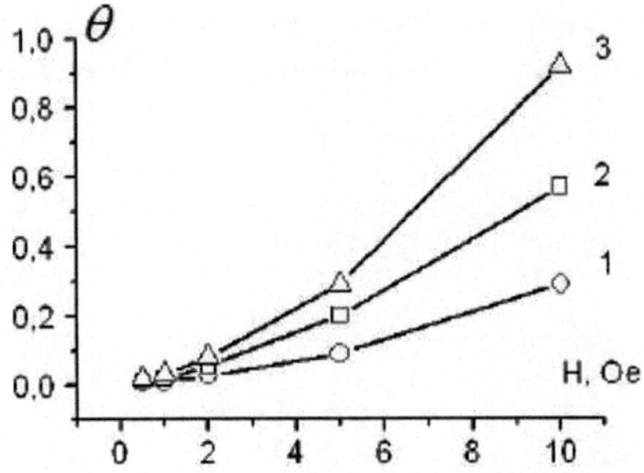

Fig. 2 Dependence of the angle θ of the nematic director rotation on the applied magnetic field H at the different values of the reference magnetic field, H_r. 1 – H_r=1 Oe, 2 – H_r = 10 Oe, 3 – H_r = 20 Oe

Acknowledgements This work was supported by the European project H2020-"MSCA-RISE-2017-778308-SPINMULTIFILM".

References

1. A. Mertelj, D. Lisjak, Ferromagnetic nematic liquid crystals. Liquid Cryst. Rev. **5**, 1 (2017)
2. A. Mertelj, N. Osterman, D. Lisjak, M. Copič, Magneto-optic and converse magneto-electric effects in a ferromagnetic liquid crystal. Soft Matter **10**, 45 (2014)
3. T. Potisk, A. Mertelj, N. Sebastian, N. Osterman, D. Lisjak, H.R. Brand, H. Pleiner, D. Svenšek, Magneto-optic dynamics in a ferromagnetic nematic liquid crystal. Phys. Rev. E **97**, 012701 (2018)
4. J. Han, Y. Luo, W. Wang, P. Zhang, Z. Zhang, Archive for rational mechanics and analysis **215**, 741 (2014)
5. P.G. de Gennes, M. Prost, *The Physics of Liquid Crystals* (Clarendon Press, Oxford, 1995)
6. K. Yang, S.T. Wu, *Fundamentals of Liquid Crystal Devices* (Wiley, West Sussex, 2006)
7. N. Podolyak, O. Buchnev, G.D. Alessandro, M. Kachmarerek, Y. Reznikov, T.J. Sluckin, Macroscopic optical effects in low concentration ferronematics. Soft Matter **7**, 4742 (2011)
8. A.A. Zvezdin, V.A. Kotov, *Modern Magnetooptics and Magnetooptical Materials* (CRC Press, Boca Raton, 2020)

On Spintronic Torque Effect in Multilayer Nanostructures

A. M. Korostil

Abstract In the framework of the tight-binding one-band model and a modified theory of non-equilibrium Green functions, the features of the spin torque effect are successively studied in the AF tunnel nanostructures. Based on the Kadanoff–Baym equations for the one-particle Green's functions, the spin-dependent electron transport through the tunnel barrier layer with rectangular electron potential is considered. The main role of closest spins next to interface AF/B in formation of the spin-dependent tunnel current is described. The spin torque and TMR effect are described as the results of the exchange interaction between magnetic moments of the itinerant electrons and localized magnetic moments. In the model of randomly distributed one-site energies, the robustness of the AF junction against disorder is considered.

1 Introduction

Spintronics of multilayer magnetic nanostructures is based on the electric control and detection of localized magnetic states via the exchange interaction of the electric-induced polarized spin current and localized magnetic moments. Regardless of the type of the magnetic exchange interaction [ferromagnetic (FM) or antiferromagnetic (AF)], the impact on the magnetic states occurs via the spin torques exerting on the localized magnetic moments. In the FM and AF cases, these spin torques can result in the FM and AF Néel order dynamics and switching [1-5]. In the case of FM nanostructures, it can be exhibited in the form of the known giant magnetoresistance effect which was applied for new kinds of field-sensing and magnetic memory devices [6–8].

The above-mentioned spintronic effects in the multilayer magnetic nanostructures are based on the electric field-controlled interaction between the spin current and localized magnetic states. Spatial scales of the effects are determined by the

A. M. Korostil (✉)
Institute of Magnetism of NAN of Ukraine and MES of Ukraine, Kyiv, Ukraine
e-mail: amand@imag.kiev.ua

O. Fesenko and L. Yatsenko (eds.), *Nanostructured Surfaces, Nanocomposites and Nanomaterials, and Their Applications*, Springer Proceedings in Physics 296,
https://doi.org/10.1007/978-3-031-42704-6_23

conditions of spin coherence and the magnetic state robust with respect to magnetic field perturbations and defect perturbations. The time scales are determined by the exchange interaction between localized magnetic moments of the system. The spin current-induced changes of magnetic states, manifesting via magnetic dynamics and switching, are realized through the spin torque of the localized magnetic moments at threshold spin current densities sufficient to overcome the magnetic coercitive energy [1, 9].

The application of the spintronic effect for magnetic information processing involves the limit reduction of the spatial sizes and the limit increasing of a magnetic dynamic frequency at the magnetic state and also robustness with respect to external field perturbations at low energy consumption. Due to the intrinsic magnetic structure, the mentioned conditions are most consistent with the multilayer AF nanostructures [2, 3]. Magnetic state control in magnetic nanostructures upon ultrafast time scales and ultra-small space scales at low power consumption is the top problem of the physics of magnetism.

Solving this problem involves modeling non-equilibrium spin dynamics and developing the physical principles of manipulation of localized magnetic states in the multilayer magnetic nanostructures. This constitutes the background for enhancement of functionalities of magnetic nanostructures, specifically, for utilizing in information technologies.

Characteristic properties of the magnetic dynamics and electric control of the magnetic states in the AF nanostructures are described in the framework of the microscopic spin model with the exchange interaction between the staggered spin current density and the local magnetic moments [2, 3]. A self-consistence analysis of this model is based on the approach of non-equilibrium real-time Green functions [10–12]. It is shown the role of the staggered spin torque in the electric control of the dynamics of AF order parameter and magnetic order switching.

2 Model Three-Layer AF Nanostructure

Electric current-controlled magnetic states in magnetic multilayer nanostructures both ferro- and AF types are based on the spin torque effect of the exchange interaction between the spin current density, $\boldsymbol{Q}$, and the localized magnetic moments. The latter in the AF case are related to the sublattices with opposite magnetizations, $\boldsymbol{M}$.

The characteristic magnetic properties of the AF nanostructures are exhibited in the model AF tunnel junction consisting of two metal AF layers separated by the nonmagnetic insulating spacer, B. The scheme of such the model tunnel junctions is represented as in Fig. 1 [13, 14].

In the typical AF tunnel junction, the semi-infinite leads are modeled as a chain compensated localized spins with translation invariance in the plane perpendicular to the electron flow. The sublattice magnetization $\boldsymbol{M}_\lambda(\lambda = 1,\ 2$ denotes number of the magnetic sublattice) in the reference left lead is directed at an angle θ to the axis z. Before the passage of electric current through the AF junction, the sublattice

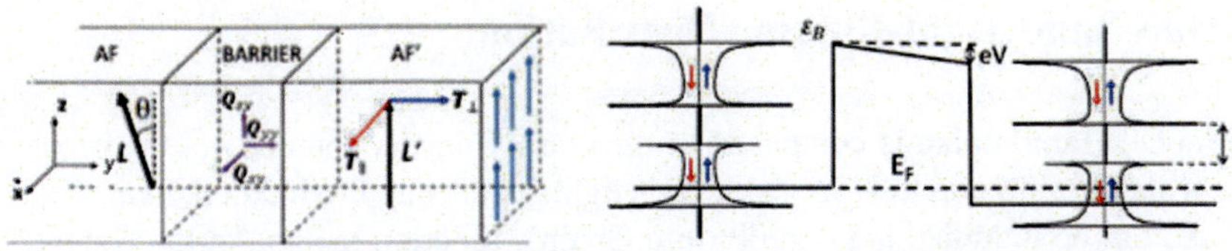

Fig. 1 (Left) Scheme of the AF tunnel junction of the form AF/Barrier/AF. $\boldsymbol{Q}_{ij}$ and $\boldsymbol{T}_\perp, \boldsymbol{T}_{||}$ are the spin current density and the torque components, respectively. $\boldsymbol{L}$ and $\boldsymbol{L}'$ are the AF order parameters, θ is the tilt angle. (Right) The density of states and the potential profile, ε_B, is the in-site energy in the barrier, Δ is the staggered spin splitting, E_F and V are the Fermi level and applied voltage

magnetization $\boldsymbol{M}'_\lambda$ in the free right lead is collinear with the axis, z. The electron flow induced by the applied voltage difference V undergoes the spin polarization characterizing by the staggered spin current density caused by the *s–d* exchange interaction between the itinerant spins and localized magnetic moments with alternating orientation. Herewith, the spin splitting Δ alternates in sign from one-site to the next one according to the alternating orientation of the localized magnetic moments. The spin-dependent on-site energy in the leads is described by the sum, $\varepsilon_i = \varepsilon_{0i} + \Delta_i$, where i denotes the position of the site, first and the second terms denote the orbital energy and the spin splitting energy, Δ_i, respectively.

The one-site energy in the barrier with a height of $\alpha\, t$ at zero bias has the form, $\varepsilon_i = \varepsilon_0 + \alpha t$. The applied potential difference results in the one-site energy of the form, $\varepsilon_i = \varepsilon_0 + \alpha\, t = eV[1/2 - x_i(N_b - 1)]$, where N_b is the number of on-sites in the barrier and $x_i = 0$ implies the first one-site in the barrier.

Under the applied voltage difference the exchange *s–d* interaction between the spin current density and localized one-site magnetizations causes the effect of on-site spin torques delivered in the right lead. This on-site torque, $\boldsymbol{T}$, consists of the in-plane, $\boldsymbol{T}_{||}$, and out-of-plane $\boldsymbol{T}_\perp$, components induced by the spin density current, Q. Generally, these torque components are defined via the AF order parameters of the reference layer (the left AF lead), $\boldsymbol{p}$, and the free layer, $\boldsymbol{n}$ (the right AF lead) [2, 3] as

$$\boldsymbol{T}_{||} = T_{IP}\, \boldsymbol{n} \times \boldsymbol{n} \times \boldsymbol{p}, \quad \boldsymbol{T}_\perp = T_{\text{op}}\, \boldsymbol{n} \times \boldsymbol{p}, \tag{1}$$

where the coefficients $\boldsymbol{T}_{||}$ and $T_\perp$ are amplitudes of the in-plane and out-of-plane torque components dependence of the characteristics of the system including the electron band spectrum, the profile of the tunnel barrier, and the applied voltage difference.

Calculation of the mentioned torque amplitudes is based on the microscopic system description in the framework of the theory of non-equilibrium Green functions and more specifically, on real-time propagation of the Kadanoff-Baym equations [11]. This approach corresponds to the accepted tight-binding model of the AF tunnel junction [12].

3 One-Band Tight-Binding Distribution

The model Hamiltonian is composed of the two AF layers coupled via the barrier layer which atomic structures modeled by in tight-binding approximation. Its analysis is based on the modified non-equilibrium Green's function method in the Keldysh formalism [10, 11].

The model Hamiltonian is represented by the sum

$$H = H_L + H_R + H_C + H_T, \tag{2}$$

with subscripts corresponding to the AF left L and right R layers, nonmagnetic barrier C layer, and the interlayer coupling, T. In the tight-binding approximation and the second quantization representation of creation, $\boldsymbol{c}_{i\uparrow}^{\dagger}$, and annihilation, $\boldsymbol{c}_i$, operators (i and σ are the site and spin), the isolated and interaction contribution are described as

$$H_\Omega = \sum_{i\in\Omega} \varepsilon_i \boldsymbol{c}_i^{\dagger} \boldsymbol{c}_i + \sum_{i\in\Omega} \Delta_i c_i^{\dagger} \boldsymbol{m}_i \cdot \sigma\, \boldsymbol{c}_{i'} + \sum_{i,i'\in\Omega} t_{ii'} \boldsymbol{c}_i^{\dagger} \boldsymbol{c}_{i'}, \tag{3}$$

$$H_{\text{int}} = t_{a\alpha} \boldsymbol{c}_a^{\dagger} \boldsymbol{c}_\alpha + t_{b\alpha'} \boldsymbol{c}_b^{\dagger} \boldsymbol{c}_{\alpha'} + H \cdot c. \tag{4}$$

where $\boldsymbol{c}_i^{\dagger} = \left(\boldsymbol{c}_{i\uparrow}^{\dagger}, \boldsymbol{c}_{i\downarrow}^{\dagger}\right)$, $L, R, C \in \Omega$, ε_i is the on-site energy, $t_{ii'}$ is the hopping parameter, Δ_i is the exchange energy between the staggered local magnetic moment $\boldsymbol{m}_i$ on site i and the itinerant electron spin σ, notations $\uparrow, \downarrow$ refer spin projections along the quantization axis, subscripts, a and b refer to the end on-sites in the barrier layer at interfaces, α and α' denote the end on-sites of L and R of AF layers near interfaces.

Observables are described via the one-particle non-equilibrium Green's functions defined as [12]

$$G(t, t') = \theta(t, t') \left\| G_{ii'}^{>}(t, t') \right\| + \theta(t', t) \left\| G_{ii'}^{<}(t, t') \right\|, \tag{5}$$

$$G_{ii'}^{>}(t, t') = -i\left\langle c_{i\sigma}(t) c_{i'\sigma'}^{\dagger}(t') \right\rangle, \quad G_{ii'}^{>}(t, t') = i\left\langle c_{i'\sigma'}^{\dagger}(t') c_{i\sigma}(t) \right\rangle, \tag{5a}$$

where $\langle \cdots \rangle$ implies quantum-statistical average with the model Hamiltonian (2). These Green's functions describe transport properties of the staggered spin-polarized electron current flowing along the AF tunnel junction (Fig. 2). Passage of the spin current through the barrier layer is described by the Green's functions with off-diagonal coordinate indices in the sense of their belonging to not coincidental parts, L, R, C of the system.

Here, $h_{\alpha\beta}$ ($\alpha, \beta = L, R$) is the matrix elements of the Hamiltonian (2), and $\sum_{CC}^{MB}$ is the matrix element of the self-energy operator which (as it be shown below) enter the equation of motion for the one-particle Green functions.

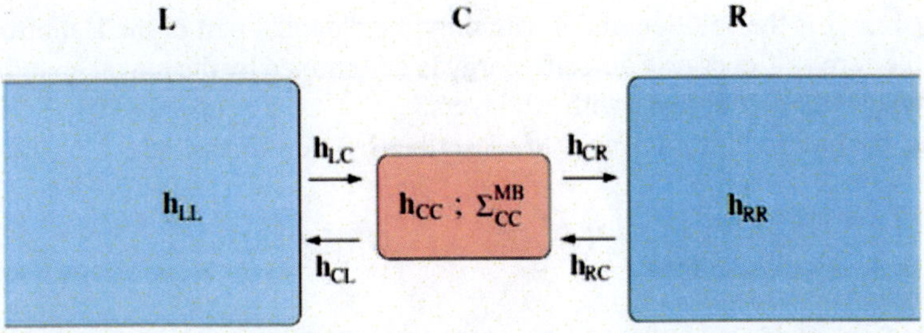

Fig. 2 Topology of a quantum transport system of the MTJ with the central region (*C*) coupled to the left (*L*) and right (*R*) regions via the tunneling Hamiltonians $h_{C\alpha}$ and $h_{\alpha C}$ ($\alpha = L, R$)

4 The Spin Density and Spin Torque

The one-electron Green's function with the Hamiltonian (2) obeys the embedded Kadanoff-Baym equations [11, 12] describing the spin-dependent transmission of electrons through the AF junction with interface scattering and a formation of the spin torque on the local AF order.

The equation of motion for the one-electron Green function (3) in the site basis representation is

$$\{i\partial_z - \boldsymbol{h}(z)\}G(z, z') = 1\delta(z, z') + \int d\overline{z}\, \Sigma(\overline{z}, z')\boldsymbol{G}(\overline{z}, z'). \tag{6}$$

Here, $\boldsymbol{h} = \|h_{mn}\|$, $\boldsymbol{\Sigma} = \left\|\Sigma_{mn}\delta_{m,n=C}\right\|$, $m, n = (L, C, R)$, $\|\cdot\cdot\cdot\|$ denotes the matrix. The matrix, Σ, implies the block matrix of the self-energy describing the electron interface scattering; subscripts refer to the matrix elements in different subspaces. The quantity, $\boldsymbol{h}$, is the isolated part of the Hamiltonian without its coupling with the barrier. The variable, z, implies the generalized time variable on the Keldysh contour composed of forward and backward real-time branches [10–12].

The matrix (3) is reduced to the system equations for elements of the block matrix Green functions, $\boldsymbol{G}_{\mathrm{mn}}$. The one-particle Green function projected on the central region C, the (6), takes the form,

$$\{i\partial_z - \boldsymbol{h}_{\mathrm{CC}}(z)\}\boldsymbol{G}_{\mathrm{CC}}(z, z') = \mathbf{1}\delta(z, z') + \int d\overline{z}\, \boldsymbol{G}_{\mathrm{CC}}(z, \overline{z})\left[\boldsymbol{\Sigma}_{\mathrm{CC}} + \boldsymbol{\Sigma}_{\mathrm{em,\,CC}}\right](\overline{z}, z') \tag{7}$$

with the introduced embedding self-energy

$$\boldsymbol{\Sigma}_{\mathrm{em,\,CC}}(z, z') = \sum_{\alpha=L,R} \boldsymbol{h}_{C\alpha}\boldsymbol{g}_{\alpha\alpha}(z, z')\boldsymbol{h}_{\alpha C}, \tag{8}$$

where $\boldsymbol{g}(z, z')$ is the solution of (7) with only the diagonal part of the Hamiltonian, $\boldsymbol{h}$, and the zero self-energy. This self-energy is determined by the tunnel coupling of the nonmagnetic barrier layer with AF layers.

Then, the charge current density, J, describes by the expression,

$$J = \frac{t}{8\pi^3\hbar} \int \mathrm{Tr}_\sigma \left[G^{<\sigma\sigma'}_{\lambda',\lambda'+\nu} - G^{<\sigma\sigma'}_{\lambda'+\nu,\lambda'} \right] \mathrm{d}E\, \mathrm{d}\boldsymbol{k}_{||}, \tag{9}$$

where $\boldsymbol{e}$ is the electron charge and $\boldsymbol{k}_{||}$ is the in-plane wave vector component. The local torque, $T_{\lambda'}$, in the left AF lead of the AF junction expresses via the spin current density, Q, as

$$\boldsymbol{T}_{\lambda'} = Q_{\lambda'+1,\lambda'} - Q_{\lambda',\lambda'+1} = \frac{t}{16\pi^3\hbar} \int \mathrm{d}E\mathrm{d}\boldsymbol{k}_{||} \mathrm{Tr}_\sigma \sum_{\nu=,1,-1} \left[G^{<\sigma\sigma'}_{\lambda',\lambda'+\nu} - G^{<\sigma\sigma'}_{\lambda'+\nu,\lambda'} \right] \cdot \sigma \mathrm{d}E\mathrm{d}\boldsymbol{k}_{||}. \tag{10}$$

The local spin density is determined via the non-equilibrium Green function as

$$\boldsymbol{S}_{\lambda'} = \frac{-i\hbar}{16\pi^3} \int \mathrm{d}E\mathrm{d}\boldsymbol{k}_{||} \mathrm{Tr}_\sigma \left[G^{<\sigma\sigma}_{\lambda',\lambda} \cdot \boldsymbol{\sigma} \right]. \tag{11}$$

The (9) and (10) determine the direct dependence of the spin torque on the spin density. Herewith, in-plane ($T_{||}$) and out-of-plane $T_{\perp}$ spin torque component direct dependence on the out-of-plane $S_{\perp}$ and in-plane $S_{||}$ components of spin density, respectively. In the one-dimensional model, this is expressed by the relations, $T_{||\lambda'} = (-1)^{\lambda'}(2\Delta_{\lambda'}/\hbar)S_{\perp\lambda'}$ and $T_{\perp\lambda'} = (-1)^{\lambda'}(2\Delta_{\lambda'}/\hbar)S_{||\lambda'}$. The in-plane spin torque component has the staggered distribution making the main contribution to the spin torque effect. The spin torque and spin density characterize by the spatial localization right next to the interfaces in the region with uncompensated magnetic moments. In this region, polarized density states (DOS) of the itinerant electrons are provided by the closest spins next to the barrier's interface.

The propagation of these states through the magnetic tunnel junctions depend on the electron structures of the AF leads, the tunnel barrier and their compatibility at interfaces [15]. The corresponding tunnel current due to (9) depends on the angle, θ, between the magnetic moments of the left and right leads. It expresses as the tunnel magnetoresistance effect defined as the ration, TMR $= (J(\theta = 0) - J(\theta = \pi))/J(\theta = \pi)$, with the monotonically increasing and symmetrical dependence on the voltage, V, [13].

The interfacial character of the effects of the spin density, spin torque, and TMR together with AF exchange interaction determine their robust against disorder. The influence of disorder is modeled by introducing random variations in the on-site atomic energies of width, γ, within AF leads [14]. In this case, the disorder configurationally average results in the dependence of the (7) with its Green's on the disorder strength, γ. For the spin density, spin torques, and TMR effect, the dependence on the disorder strength is weakly expressed.

5 Conclusions

In the framework of the tight-binding model based on the modified theory of non-equilibrium Green functions features of the spin torque effect is successively studied in the tunnel AF nanostructures. On the basis of the Kadanoff-Baym equations and its one-particle Green functions, the spin-dependent electron transport through the tunnel magnetic junction is derived. The essential role of closest spins next to the barrier's spin in the spin-dependent tunneling is considered. The interfacial character of the spin density, spin torque, and TMR effect in the tunnel junction is shown. In the model of the randomly distributed on-site energies taking into account AF exchange interaction, the robustness of the spin torque and TMR effect against disorder is considered.

Acknowledgements This work was supported by the European project H2020-"MSCA-RISE-2017-778308-SPINMULTIFILM."

References

1. D.C. Ralph, M.D. Stiles, Spin Transfer tongues. J. Mag. Mag. Mater. **320**, 1190 (2008)
2. B. Baltz, A. Manchon, M. Tsoi, T. Munyama, T. Ono, Y. Tserkovnyak, Antiferromagnetic spintronics. Rev. Mod. Phys. Mod. Phys. **90**, 015003 (2018)
3. A. Manchon, J. Zelezny, I.M. Miron, T. Jungwirz, J. Sinova, A. Thiaville, K. Garello, P. Gambardella, Current-induced spin-orbit torques in ferromagnetic and antiferromagnetic systems Rev. Mod. Phys. **91**, 035005 (2019)
4. S. Yuass, D.D. Djayaprawira, Giant tunnel magnitoresistance in magnetic tunnel junctions with a crystalline MgO(001) barrier. J. Phys. D Appl. Phys. **40**, R337 (2007)
5. I. Žutic, J. Fauber, S.D. Sarma, Spintronics: fundamentals and applications. Rev. Mod. Phys. **76**, 323 (2004)
6. A.H. MacDonald, M. Tsoi, Antiferromagnetic metal spintronics. R. Soc. A **369**, 3098 (2014)
7. S.A. Wolf, D.D. Awshalom, R.A. Buhrman, I.M. Daughton, S. Molna, M.I. Roukes, A.Y. Chtchelkanowa, D.M. Treger, Spintronics: a spin-based electronics vision for the feature. Science **294**, 1488 (2001)
8. C. Chappter, A. Fert, F.N. Van Dau, The emergence of spin electronics in data storage. Nat. Mater. **6**, 813 (2007)
9. J.C. Slonczewski, Current-driven excitation of magnetic multilayers. Magn. Magn. Mater. **159**, L1 (1996)
10. A. Stan, R.V. Leeuwen, N.E. Dahlen, R.V. Leeuwen, The propagation of the Kadanoff-Baym equations for inhomogeneous systems. Chem. Phys. **130**, 114105 (2009)
11. L.P. Kadanoff, G. Baym, *Quantum Statistical Mechanics* (Benjamin, New York, 1996)
12. P. Myöhänen, A. Stan, G. Stefanucci, R. Wan Leeuwenn, Kadanoff-Baym approach to quantum transport through interacting nanoscale systems: from the transient to the steady-state regime. Phys. Rev. B **80**, 115107 (2009)
13. P. Merodrio, A. Kalitsov, H. Bea, V. Baltz, M. Chshiev, Spin-modulated torque waves in ferrimagnetic tunnel junctions. Phys. Rev. B **90**, 224422 (2014)
14. H.B.M. Saidoui, X. Waintal, A. Manchon, Robust spin transfer torque in antiferromagnetic tunnel junctions. Phys. Rev. B **95**, 134424 (2017)
15. E.Y. Tsymbal, O.N. Mryasov, P.R. LeClair, Spin_Dependent tunneling in magnetic tunnel junctions. J. Phys. Cond. Matt. **15**, R109 (2003)

Internal Structural Water-Depended Effects in Nanoporous Humidity-Sensitive Ceramics Sintered at Different Temperatures

H. Klym

Abstract The effects of internal structural water on nanoporous humidity-sensitive $MgAl_2O_4$ ceramics were investigated by a combination of methods including X-ray diffraction, scanning electron microscopy, Hg porosimetry, and positron annihilation lifetime spectroscopy. The results revealed the presence of two or three phases in the ceramics, which had a well-developed structure of grains, grain boundaries, and pores with a trimodal pore size distribution. The capture of positrons on defects near the grain boundaries with additional phases was found to be more intense in the presence of water. When ceramics sintered at 1100 °C were immersed in water, the trapping of positrons and ortho-positronium occurred more intensely, and drying of the ceramics showed that a portion of the water remained based on the annihilation spectrum parameters of the third component. Adsorption processes in the ceramics were found to cause fragmentation of nanopores, while desorption caused their agglomeration. The presence of water in ceramics sintered at 1200 and 1300 °C led to fragmentation of the free volume of nanopores and additional decomposition of ortho-positronium atoms in the adsorbed water.

1 Introduction

The $MgAl_2O_4$ spinel ceramics are widely used in sensor electronics due to their easy production, flexibility, and stable operation [1–6]. However, the optimization of technological conditions is necessary to obtain a material with desired properties, which requires achieving an optimal structure of the internal components (grains, grain boundaries, pores) that would maximize the adsorption and desorption of water while being resistant to external influences [7–9].

H. Klym (✉)
Lviv Polytechnic National University, S. Bandera Str., 12, Lviv 79013, Ukraine
e-mail: halyna.i.klym@lpnu.ua; halyna.klym@lnu.edu.ua

Ivan Franko National University of Lviv, Drahomanova Str., 50, Lviv 79005, Ukraine

O. Fesenko and L. Yatsenko (eds.), *Nanostructured Surfaces, Nanocomposites and Nanomaterials, and Their Applications*, Springer Proceedings in Physics 296,
https://doi.org/10.1007/978-3-031-42704-6_24

To address this issue, numerous studies have been conducted to investigate the structural properties of ceramics. Most of these studies have focused on examining the microstructure, phase composition, and open porosity of the material [6, 10–16]. However, due to their limited utility, these traditional methods may not provide complete information about the internal free volume and nanopores [2, 4, 17]. As a result, alternative methods are employed to analyze the components of the free volume. One such method is positron annihilation lifetime spectroscopy (PALS), which is primarily used to study vacancies, clusters, and other defects in solids [18–20]. Additionally, the PALS technique can estimate the size of nanopores and investigate the processes occurring within them under the influence of external factors, including adsorbed water.

Previous research on $MgAl_2O_4$ nanostructured ceramics has examined the processes occurring at grain boundaries and nanopores using the PALS method [21–23]. PALS spectra for humidity-sensitive ceramics were typically divided into three or four components, depending on the characteristics of the material and the experiment.

A model of positron capture in functional materials was also proposed, in which the first component describes the features of spinels, the second component is associated with extended defects and processes at grain boundaries, and the third and fourth components relate to the decay of ortho-positronium (o-Ps) atoms in nanopores and adsorbed water [21–26].

Research has also been conducted on the effect of sintering temperature on the free volume (defects, nanopores) of ceramics and the influence of different amounts of water on the transformation of internal voids [27, 28]. This study aims to investigate the water-dependent effects on the internal structure of nanoporous humidity-sensitive $MgAl_2O_4$ ceramics sintered at different temperatures (1100, 1200, and 1300 °C) for 2 h under the influence of water, as well as the properties of the ceramics microstructure and phase composition.

2 Sintering and Investigation of Structural and Free-Volume Transformations in $MgAl_2O_4$ Ceramics Sintered at 1100 °C

2.1 Preparation of Experimental Setup for Phase Composition of $MgAl_2O_4$ Ceramics Sintered at 1100 °C

The synthesis of $MgAl_2O_4$ ceramics was carried out according to traditional ceramic technology using oxide method [29]. The raw materials MgO with a specific surface area of 10.7 m^2/g and Al_2O_3 with a specific surface area of 12.4 m^2/g were taken in a molar ratio of 1:1, mixed and ground in a ball mill in acetone, at a drum rotation speed of 138–140 rpm for 96 h. The resulting mixture was dried in evaporating cups for 2 h. Press powder was prepared in several stages. First, the powder was sifted through a kapron sieve, and then a plasticizer (5% solution of polyvinyl alcohol in distilled water) was introduced in the ratio "100 g of powder":"70 ml of plasticizer". The resulting mixture was dried at a temperature of 120 °C for 2 h. After that, the powder was again sifted alternately through a coarse and fine nylon sieve (the average size of the sieve cells is about 76 μm). Ceramic blanks were pressed under a pressure of 166.4 MPa into tablets with a diameter of 15 mm and a thickness of 1 mm. Sintering of ceramics was carried out at an isothermal sintering temperature of 1100 °C for 2 h [30].

Investigation of the phase composition using the X-ray diffraction (XRD) method (HZG-4a diffractometer with CuKα radiation) showed that $MgAl_2O_4$ ceramics contain the main spinel phase and two additional phases: MgO and Al_2O_3. A detailed crystallographic analysis of the obtained phases is given in Table 1 [30].

The study of the microstructure of ceramics (visual observation of grains, grain boundaries, and pores) was carried out using the SEM method on a LEO 982 microscope with the Oxford ISIS system on a fresh fracture of the studied samples.

The size distribution of pores in $MgAl_2O_4$ ceramics sintered at 1100 °C for 2 h was studied by mercury porometry using a POROSIMETR 4000 Hg porometer. Pre-weighed samples placed in a dilatometer were filled with Hg using a PASCAL 140, which operates in the low-pressure region (up to 400 kPa) and fills only macro- and mesopores with a size of at least 50 nm. The sample-filled mercury dilatometer is then weighed and placed in the porosimeter, which operates over a wide pressure range from atmospheric to 4000 bar. Pore size and particle distribution in the range from 2 to 900 nm for the studied ceramic samples were obtained using the MILES 200 program.

2.2 *Structural and Free Volume in $MgAl_2O_4$ Ceramics Sintered at 1100 °C*

To gather information on the microstructure of MgAl2O4 ceramics sintered at 1100 °C, a SEM analysis was performed. The microphotographs of the fracture surfaces (Fig. 1) revealed a branched grain structure with varying sizes and shapes of grain boundary pores. The elemental composition of both grains and grain boundaries was also examined using the EDX method. The results demonstrated that the ceramics exhibit an insufficiently formed grain structure and porosity. The average grain size was found to be less than 200 nm. Moreover, the distribution of additional MgO and Al_2O_3 phases was uneven throughout the ceramics, primarily concentrated near the grain boundaries adjacent to pores that were hundreds of nanometers in size.

Table 1 Crystallographic data for $MgAl_2O_4$ ceramics sintered at 1100 °C for 2 h

Spinel phase	$MgAl_2O_4$
Space group	Fd $\bar{3}$ m (setting 2)
Lattice parameter *a*, Å	8.0705(4)
Cell volume (Å^3)	525.66(4)
Radiation and wavelengths (Å)	Cu Kα 1.54056 1.54439
Angular range $2\theta_{min}$–$2\theta_{max}$ (degrees)	15.00–145.00 (step 0.05°2θ, 20 s/point)
Number of measured reflections	40
Number of fitted parameters	32
Mode of refinement	Full-profile Rietveld method
Profile parameters, U, V, W	0.66(8), –0.09(7), 0.17(2)
Reliability factors (in %):	
$R_B = \Sigma\|I_h(obs)–I_h(calc)\|/\Sigma\|I_h(obs)\|$	6.50
$R_F = \Sigma\|\|F_h(obs)\|–\|F_h(calc)\|\|/\Sigma\|F_h(obs)\|$	7.32
$R_p = \Sigma\|y_i(obs)–y_i(calc)\|/\Sigma y_i(obs)$	5.36
$R_{wp} = [\Sigma w_i\|y_i(obs)–y_i(calc)\|^2/\Sigma\ w_iy_i(obs)^2]^{1/2}$	7.01
$\chi^2 = \{R_{wp}/R_{exp}\}^2$	4.11
Fraction (wt. %)	***80.61(1.11)***
Additional phase	MgO
Space group	Fm $\bar{3}$ m
Lattice parameter *a*, Å	4.2108(1)
Cell volume (Å^3)	74.660(3)
Profile parameters, U, V, W	0.010(5), –0.005(9), 0.017(4)
R_{Bragg}, R_F (in %)	3.06 2.58
Fraction (wt. %)	***11.25(0.26)***
Addition phase	Al_2O_3
Space group	R $\bar{3}$ c
Lattice parameter *a*, *c*, Å	4.7585(4) 12.992(2)
Cell volume (Å^3)	254.77(4)
Profile parameters, U, V, W	0.02(1), –0.01(2), 0.024(8)
R_{Bragg}, R_F (in %)	17.5 15.3*
Fraction (wt. %)	***8.13(0.40)***

Ceramics typically feature open and closed pores. For efficient water sorption processes, a heterogeneous porous structure that includes macro- and mesopores with sizes ranging from tens to hundreds of nanometers, which act as water channels, as well as nanopores with sizes up to several nanometers, where capillary processes occur, is necessary. Figure 2 illustrates the distribution of open pores and particle sizes in the examined $MgAl_2O_4$ ceramics.

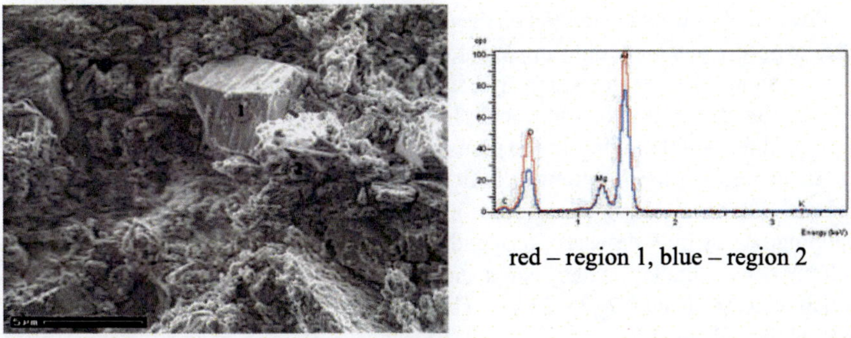

Fig. 1 Structure and reflexes from grain and grain boundary for $MgAl_2O_4$ ceramics sintered at 1100 °C for 2 h

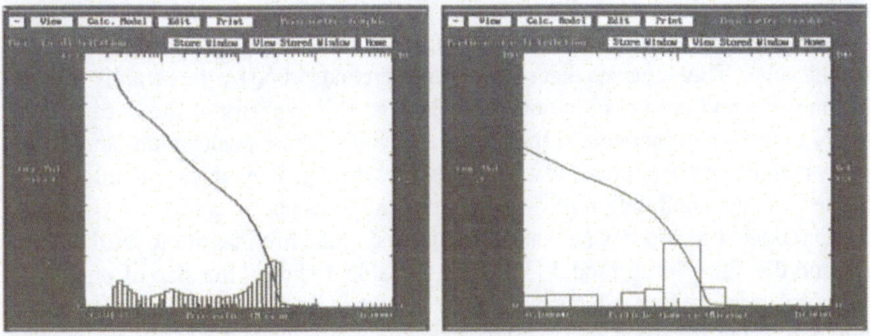

Fig. 2 Distribution of pore sizes and particles for the $MgAl_2O_4$ ceramics sintered at a temperature of 1100 °C for a duration of 2 h

The $MgAl_2O_4$ ceramics sintered at 1100 °C for 2 h exhibit a trimodal pore size distribution consisting of nano-sized pores (maximum radius of 2.8 nm), mesopores (radius of 14 nm), and macropores (size of 300 nm) [30]. The particle diameter is centered around 1 μm. However, the most noteworthy feature is the capillary condensation process that occurs in the nanopores. To investigate the effects of water on the free volumes and nanopores of $MgAl_2O_4$ ceramics, PALS was employed.

Previous studies [21, 23, 31] have shown that $MgAl_2O_4$ ceramics have a multichannel model of positron annihilation, with positron capture by volume defects and decay of o-Ps atoms being the primary channels. The short-term component of the positron lifetime reflects the microstructural features of spinel ceramics, while the middle component corresponds to volume defects localized near intergranular pores. The third component is related to o-Ps annihilation in water-filled nanopores.

PALS studies were conducted on the original $MgAl_2O_4$ ceramic samples, samples after exposure to water, and samples after drying for a month in natural conditions. Table 2 shows that ceramics drying results in irreversible structural changes due to the possible dissolution of additional phases at the grain boundaries adjacent to the pores and residual humidity in the pores ($I_3 = 9\%$).

An increase in the I_2 intensity indicates a higher positron capture rate by bulk defects. After immersing the samples in water, there is an increase in the positron trapping rate by bulk defects. Consequently, the parameters of positron trapping by volume defects undergo changes (as shown in Table 3), particularly the rate of κ_d, which increases from 0.7 to 1.2 ns^{-1}. This, in turn, alters the positron capture center through bulk defects.

As previously mentioned, the third component of the PALS spectrum is responsible for the annihilation of o-Ps in ceramic nanopores. The lifetime of the third component τ_3 decreases in ceramics after exposure to water to 1.88 ns, as shown in Table 2. This decrease indicates the annihilation of o-Ps in water. The lifetime increases again after the ceramic dries, but its value does not return to the original ceramic value. This suggests that some of the water adsorbed by the ceramic blocked its nanopores and did not come out after drying. The intensity of the third component I_3 indicates an increase in the number of voids where positronium annihilates in water-immersed ceramics and a decrease after drying. The increase in intensity to 15% reflects the annihilation of o-Ps in both nanopores and water.

The radius of nanopores R_3 was calculated using the lifetime of the third component and the Tao-Eldrup model [32, 33]. As Table 3 shows, the size of nanopores where positronium annihilates first decreases and then increases. These results suggest that the presence of adsorbed water in ceramics leads to the fragmentation of nanopores, and drying leads to the agglomeration of nanopores. However,

Table 2 Fitting parameters calculated by LT program the $MgAl_2O_4$ ceramics sintered at a temperature of 1100 °C for 2 h

Sample history	τ_1, ns	I_1, %	τ_2, ns	I_2, %	τ_3, ns	I_3, %
Initial	0.24	68	0.50	30	2.59	2
After immersion in water	0.24	56	0.50	29	1.88	15
After drying	0.18	56	0.42	35	2.02	9

Table 3 Trapping parameters and radius of nanopores for the $MgAl_2O_4$ ceramics sintered at 1100 °C for 2 h

Sample history	τ_{av}, ns	τ_b, ns	κ_d, ns^{-1}	$\tau_2 - \tau_b$, ns	τ_2/τ_b	R_3, nm
Initial	0.32	0.28	0.6	0.21	1.7	0.338
After immersion in water	0.33	0.29	0.7	0.21	1.7	0.276
After drying	0.27	0.23	1.2	0.19	1.8	0.289

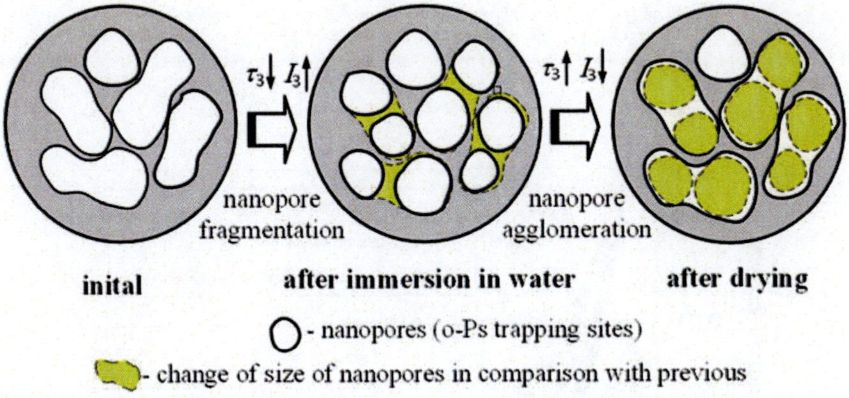

Fig. 3 Illustration of the changes in the volume of nanopores in $MgAl_2O_4$ ceramics sintered at 1100 °C for 2 h after immersion in water

some water remains in the nanopores. Figure 3 shows a schematic representation of the processes of changing the volume of nanopores in ceramics after immersion in water and drying [30].

After drying the ceramics for a month in natural conditions, an increase in the background is observed in the peak-normalized positron lifetime spectra of the initial $MgAl_2O_4$ ceramics sintered at 1100 °C, as shown in Fig. 4.

Obviously, when water is released from the meso- and macropores of ceramics, its accompanying free volume increases contributing to the annihilation of free positrons.

3 Sintering and Study of Nanostructural and Free-Volume Evolution of $MgAl_2O_4$ Ceramics Sintered at 1200 °C for 2 h

3.1 Preparation of Phase Composition of Experimental Setup for $MgAl_2O_4$ Ceramics Sintered at 1200 °C

The ceramics used in this study were prepared using the traditional oxide method. MgO and Al_2O_3 powders were mixed in a ball mill using acetone for 96 h. The resulting mixture was dried and formed in several stages, with the addition of a plasticizer and drying at 120 °C for 2 h. Ceramic samples were formed under a pressure of 166.4 MPa and sintered at a maximum temperature of 1200 °C for 2 h.

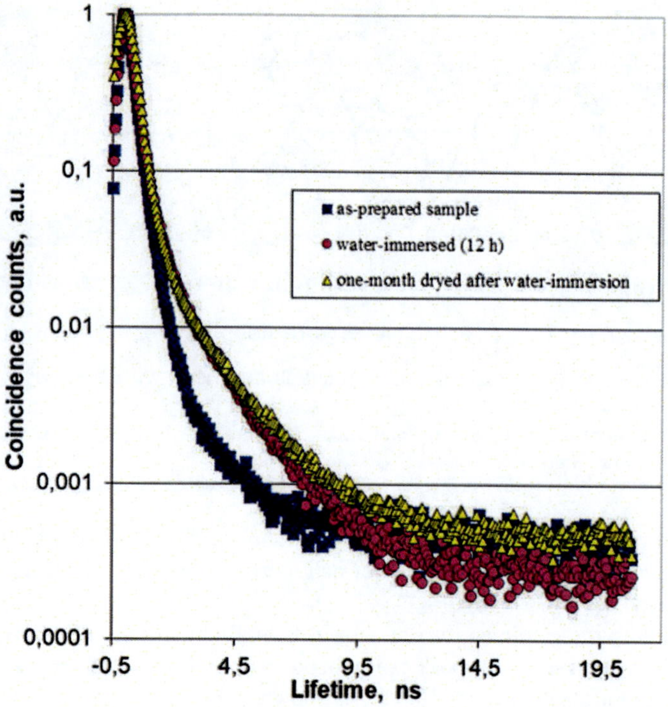

Fig. 4 Peak-normalized spectra for $MgAl_2O_4$ ceramics initially sintered at 1100 °C, after soaked in water for 12 h and after drying for one month in natural conditions

X-ray diffraction analysis (using CuKα radiation and a HZG-4a diffractometer) revealed that the sintered $MgAl_2O_4$ ceramics contain three phases: the main spinel phase (88.12%), as well as smaller amounts of MgO (6.06%) and Al_2O_3 (5.82%). A detailed presentation of the XRD results can be found in Fig. 5 and Table 4.

Freshly fractured samples of $MgAl_2O_4$ ceramics sintered at 1200 °C were analyzed for microstructure of grains, grain boundaries, and pores using an Oxford ISIS system equipped LEO 982 microscope.

Size distribution of open pores in the ceramics was studied using Hg-porometry (POROSIMETR 4000), and ceramic samples were analyzed for pore and particle sizes ranging from 2 to 900 nm using the MILES 200 program [34].

The PALS technique was utilized to investigate the nanostructural characteristics of $MgAl_2O_4$ ceramics sintered at 1200 °C, specifically, the free volume and nanopore sizes, as well as the impact of water absorption. The experimental procedure is outlined in [21, 35, 36]. Both the pristine ceramics and those soaked in water for 12 h were examined. The obtained PALS spectra were dissected into three components using the LT software package [37].

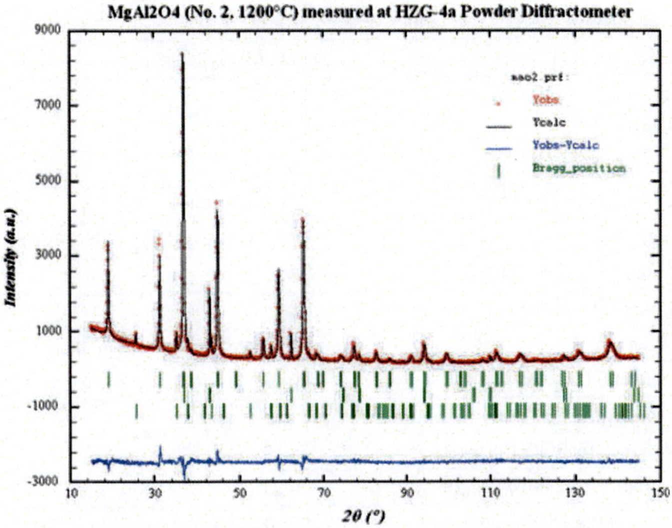

Fig. 5 X-ray diffraction patterns of $MgAl_2O_4$ ceramics sintered at 1200 °C: experimental (rings), theoretical (solid line), and differential (bottom) results (upper row is $MgAl_2O_4$, middle row is MgO, lower row is Al_2O_3)

3.2 *Nanostructural and Free-Volume Transformation in $MgAl_2O_4$ Ceramics Sintered at 1200 °C*

As mentioned earlier, the microstructure of $MgAl_2O_4$ ceramics sintered at 1200 °C was examined using SEM. The results (shown in Fig. 6) indicate that the ceramic has a complex structure consisting of grains with various shapes and sizes in contact with pores. The porous structure is also intricate, and pores of different shapes and sizes can be observed. Furthermore, EDX analysis was conducted to determine the elemental composition of the grains and grain boundaries.

The grain structure of the ceramic fired at a relatively low temperature of 1200 °C, as well as the porosity, is not yet fully developed (Fig. 6). The average grain size is between 100 and 200 nm. Additional phases are not evenly dispersed throughout the ceramic's volume, and they are primarily concentrated along the grain boundaries near pores of varying sizes.

It is established that an effective process of water adsorption and desorption necessitates a heterogeneous structure of open pores, including macro- and mesopores ranging from 10 to 100 nm in size, which serve as channels for water conduction, as well as nanopores with sizes up to several nm, where capillary condensation processes occur. Figure 7 illustrates the distribution of open pores based on the size of the samples and particles of the $MgAl_2O_4$ ceramics that were originally sintered at 1200 °C.

Table 4 XRD data for $MgAl_2O_4$ ceramics sintered at 1200 °C

Spinel phase	$MgAl_2O_4$
Space group	$Fd\bar{3}m$
Lattice parameter *a*, Å	8.0796(3)
Cell volume (Å^3)	527.44(3)
Radiation and wavelengths (Å)	Cu Kα 1.54056 1.54439
Angular range $2\theta_{min}$–$2\theta_{max}$ (degrees)	15.00–145.00 (step 0.05°2θ, 10 s/point)
Number of measured reflections	39
Number of fitted parameters	40
Mode of refinement	Full-profile Rietveld method
Profile parameters, U, V, W	0.18(2), – 0.03(2), 0.068(5)
Reliability factors (in %)	
$R_B = \Sigma\lvert I_h(obs)–I_h(calc)\rvert/\Sigma\lvert I_h(obs)\rvert$	5.49
$R_F = \Sigma\lvert\lvert F_h(obs)\rvert–\lvert F_h(calc)\rvert\rvert/\Sigma\lvert F_h(obs)\rvert$	6.27
$R_p = \Sigma\lvert y_i(obs)–y_i(calc)\rvert/\Sigma y_i(obs)$	5.54
$R_{wp} = [\Sigma w_i\lvert y_i(obs)–y_i(calc)\rvert^2/\Sigma\ w_i y_i(obs)^2]^{1/2}$	7.00
$\chi^2 = \{R_{wp}/R_{exp}\}^2$	2.26
Fraction (wt. %)	88.12(0.88)
Addition phase	MgO
Space group	$Fm\bar{3}m$
Lattice parameter *a* (Å)	4.2112(2)
Cell volume (Å^3)	74.684(5)
Profile parameters U, V, W	– 0.002(9), 0.03(2), 0.004(6)
R_{Bragg}, R_F, in %	6.73 5.62
Fraction (wt. %)	5.82(0.19)
Addition phase	Al_2O_3
Space group	$R\bar{3}c$
Lattice parameter *a*, *c* (Å)	4.7585(5) 12.991(2)
Cell volume (Å^3)	254.75(6)
Profile parameters U, V, W	0.01(1), 0.01(3), 0.02(1)
R_{Bragg}, R_F, in %	24.1 19.0
Fraction (wt. %)	6.06(0.37)

Figure 7 illustrates that the $MgAl_2O_4$ ceramics sintered at 1200 °C display a pore size distribution with three modes, including 3.1 nm nanopores, 28 nm mesopores, and 320 nm macropores. The ceramics contain particles with a maximum diameter of 1.2 μm.

Capillary condensation processes are known to occur in nanopores, but these processes cannot be studied with conventional methods. To study the size of nanopores and analyze the processes taking place within them, an alternative PALS

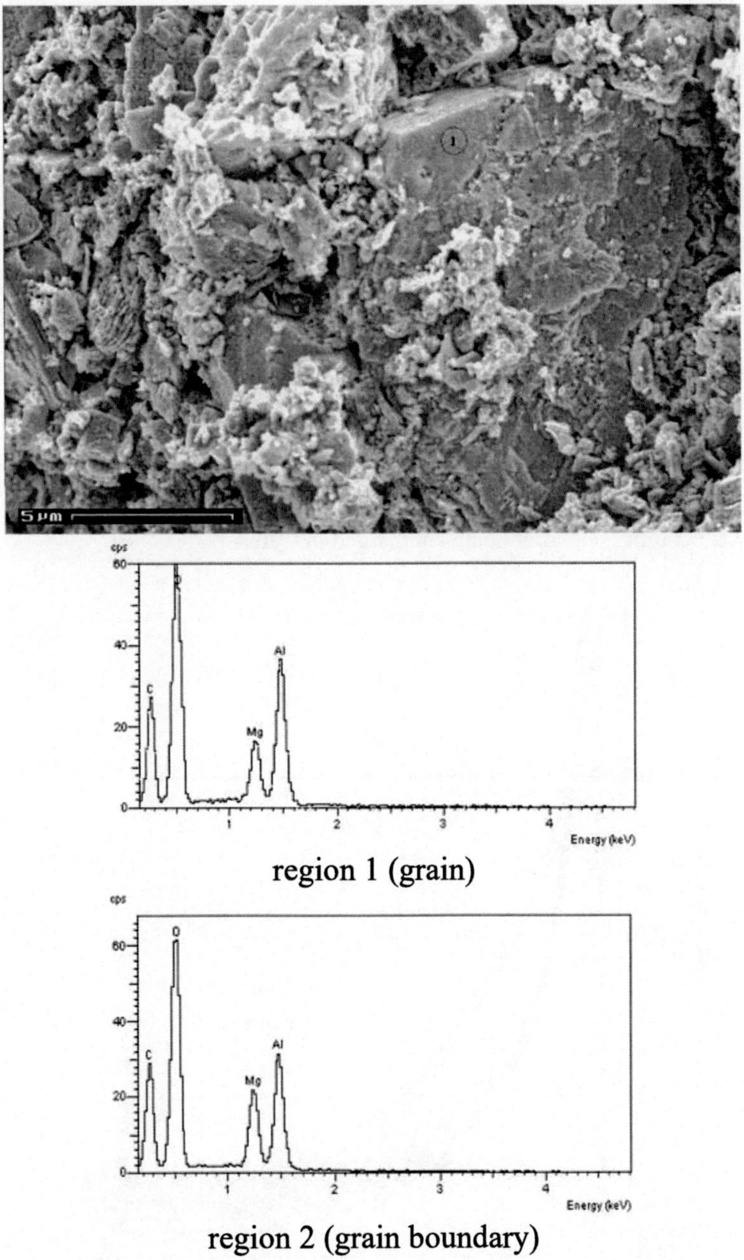

Fig. 6 Microstructural analysis and elemental composition of grains and grain boundaries in $MgAl_2O_4$ ceramics sintered at 1200 °C

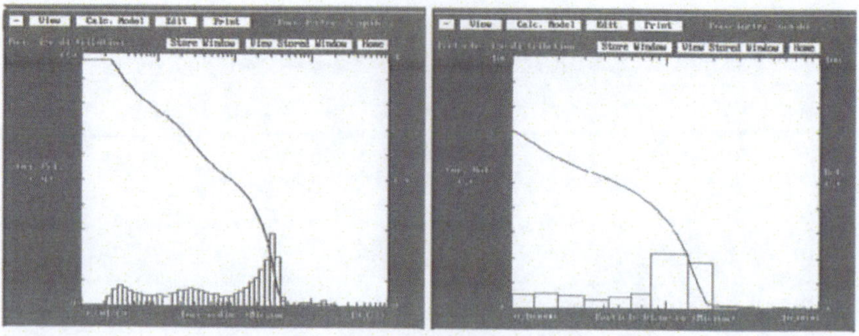

Fig. 7 Characterization of pore and particle size distribution in $MgAl_2O_4$ ceramics sintered at 1200 °C

method was employed. This method also allows for the assessment of free-volume defects in functional materials. Figure 8 presents the PALS spectra for the original $MgAl_2O_4$ ceramics and after being submerged in water.

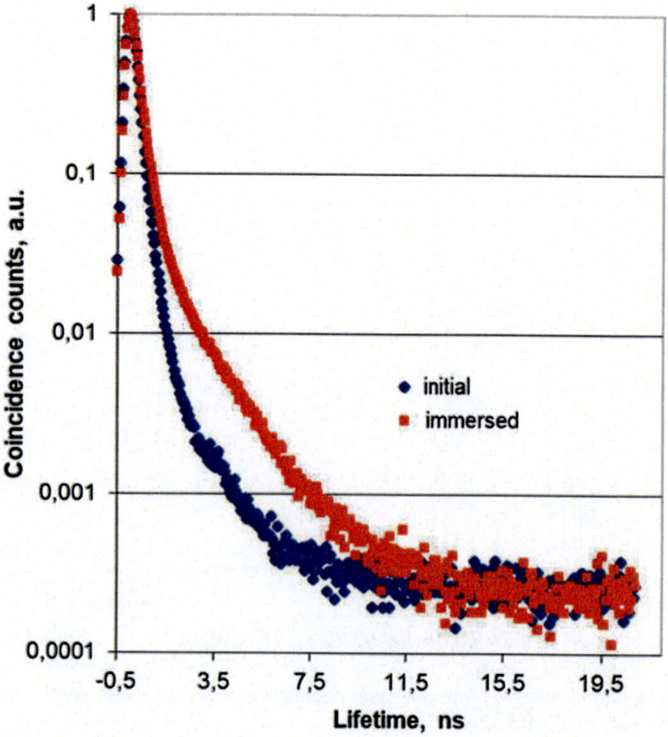

Fig. 8 PALS spectra analysis for $MgAl_2O_4$ ceramics sintered at 1200 °C: comparing initial and water-immersed samples

As stated earlier, the PALS spectra of the ceramics were analyzed for the initial and water-immersed samples using the three-component scattering method, which is common for these materials. As documented in [22, 38], the annihilation processes in $MgAl_2O_4$ spinel ceramics follow a multi-channel model, which comprises channels for positron trapping by defects and the decay of ortho-positronium atoms in nanopores. In this model, the first component accounts for the spinel microstructure with tetrahedral and octahedral vacancies, while the second component analyzes the processes in bulk defects at the grain boundaries, and the third describes the processes in nanopores, including those that occur when the ceramic is saturated with water.

Table 5 shows the intensities and lifetimes (fitting parameters) of the three components for the original and water-immersed ceramics.

Table 5 presents the fitting parameters for positron trapping modes and the radii of nanopores for both the original and water-immersed $MgAl_2O_4$ ceramics sintered at 1200 °C. The results show that the lifetime and intensity of the first and second components (τ_1 and τ_2) remain unchanged before and after wetting the ceramics. In contrast, the lifetime of the third component τ_3 significantly decreases to 1.88 ns, and the intensity I3 of this component increases from 2 to 13%, indicating the fragmentation of free space in nanopores due to the presence of water. Moreover, the lifetime value of 1.88 ns suggests the decay of ortho-positronium atoms o-Ps in water.

The positron capture parameters, such as the average positron lifetime τ_{av}, τ_b, κ_d, $\tau_2 - \tau_b$, and the ratios τ_2/τ_b, do not show any significant changes, except for the increase in the positron capture rate κ_d, which indicates more intense processes of their annihilation in the wetted ceramic at the grain boundaries, where additional phases precipitate.

The nanopore radii were calculated using the Tao-Eldrup model [32, 33] with the lifetime of the third component τ_3, and the results show that the size of the nanopores in the wetted ceramic is smaller. The increase in the I_3 intensity and the decrease in the size of the nanopores suggest that the presence of adsorbed water in ceramics leads to the fragmentation of nanopores. Figure 9 illustrates the schematic change in the volume of nanopores in ceramics after wetting.

Table 5 Nanopore radii and fitting parameters of positron trapping modes for $MgAl_2O_4$ ceramics sintered at 1200 °C

Sample prehistory	τ_1, ns	I_1, %	τ_2, ns	I_2, %	τ_3, ns	I_3, %
Initial	0.23	70	0.47	28	2.39	2
Water-immersed	0.23	59	0.48	28	1.88	13
Sample prehistory	$\tau_{av.}$, ns	τ_b, ns	κ_d, ns^{-1}	$\tau_2 - \tau_b$, ns	τ_2/τ_b	R_3, nm
Initial	0.30	0.27	0.6	0.2	1.7	0.322
Water-immersed	0.32	0.28	0.7	0.2	1.7	0.267

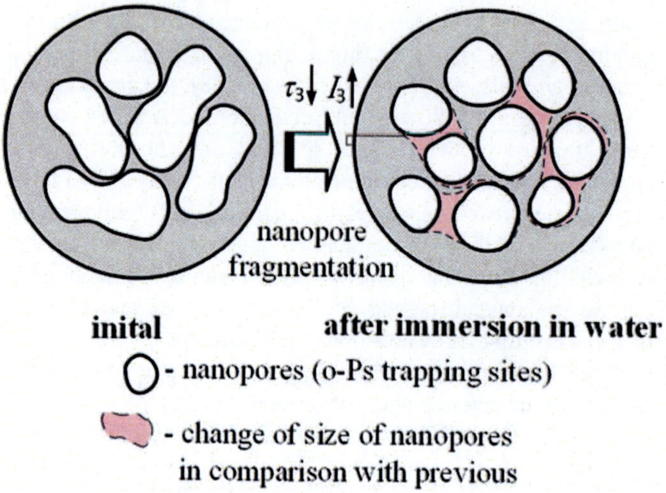

Fig. 9 Illustration of changes in nanopore volume of $MgAl_2O_4$ ceramics after immersion in water

The PALS technique is a non-invasive approach used to investigate the free volume in various materials [39–41]. The results obtained from the PALS measurements were further verified by repeating the ceramic pore size distribution analysis using Hg porosimetry. As shown in Fig. 10, some slight discrepancies may arise due to potential blockages within the internal structure of ceramics, which is commonly observed in materials sintered at 1200 °C with an underdeveloped structure.

Fig. 10 Pore size distribution analysis of $MgAl_2O_4$ ceramics sintered at 1200 °C using PALS method

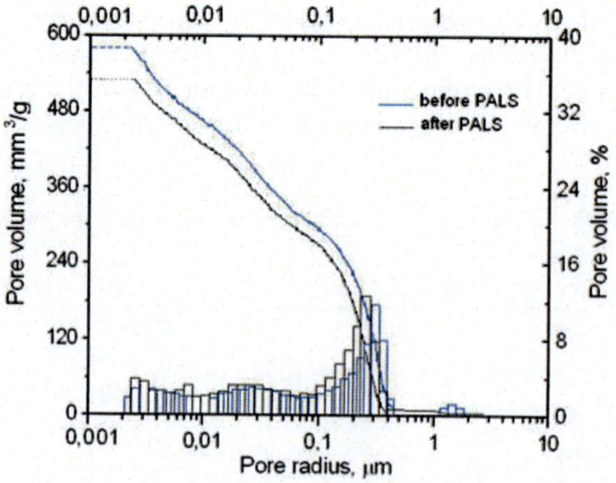

4 Sintering and Free-Volume Investigation of $MgAl_2O_4$ Ceramics Sintered at 1300 °C for 2 h

4.1 Preparation, Phase Composition of Experimental for $MgAl_2O_4$ Ceramics Sintered at 1300 °C

The $MgAl_2O_4$ ceramic samples were synthesized through the conventional oxide method using MgO (with a specific surface area of 10.7 m^2/g) and Al_2O_3 (with a specific surface area of 12.4 m^2/g) powders [42]. The powders were mixed in a planetary ball mill for 96 h in acetone at a 1:1 ratio. Subsequently, the mixture was dried for 2 h and passed through a coarse and fine sieve. The $MgAl_2O_4$ ceramic samples were then formed under a pressure of 166.4 MPa and sintered at 1300 °C for 2 h.

Phase composition and crystallographic data were analyzed by X-ray diffraction (XRD) using CuKα radiation (HZG-4a diffractometer). Microstructure analysis of the main components of the fresh ceramic samples was conducted using an LEO 982 microscope (Oxford ISIS system) with EDX analysis of grains and grain boundaries. Hg porosimetry (Hg porometer POROSIMETR 4000) was used to investigate the pore size distribution in the $MgAl_2O_4$ ceramics. The ceramic samples were loaded into a dilatometer and filled with Hg, and pore and particle sizes ranging from 2 to 900 nm were analyzed using MILES 200.

The PALS method, a non-destructive method for investigating free volume in materials, was used to study extended free-volume defects and nanopores in the $MgAl_2O_4$ ceramics, as well as their evolution under the influence of adsorbed water. An ORTEC spectrometer was used for the PALS measurements, as detailed in [22, 23]. The PALS spectra obtained were analyzed by the LT program, which decomposed the spectra into three components. The original ceramic samples were analyzed using the PALS method, as well as after soaking the samples in water for 12 h. The Hg porosimetry studies after the PALS measurements confirmed the obtained results. Minor deviations in the obtained data could be attributed to clogging of the internal structure of the ceramics obtained at 1200 °C, since this temperature results in the formation of a material with an underdeveloped structure.

4.2 Nanostructure Free Volumes of $MgAl_2O_4$ Ceramics Sintered at 1300 °C

Based on the XRD analysis, it was found that the $MgAl_2O_4$ ceramics consist of two phases, namely the spinel phase as the main phase with a fraction of 96.52%, and an additional phase of MgO with a fraction of 3.48% [42]. The XRD data, including detailed information, are presented in Fig. 11, Tables 6 and 7.

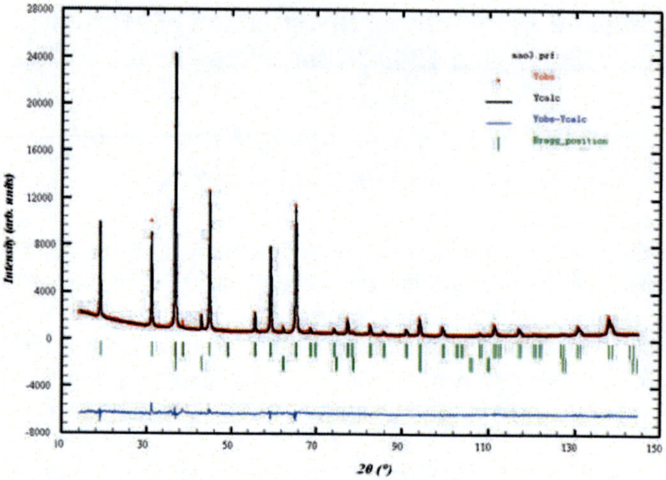

Fig. 11 X-ray diffraction patterns of $MgAl_2O_4$ ceramics sintered at 1300 °C: experimental (rings), theoretical (solid line), and differential (bottom) results (upper row is $MgAl_2O_4$, middle row is MgO, lower row is Al_2O_3)

Table 6 XRD data for $MgAl_2O_4$ ceramics sintered at 1300 °C

Spinel phase	$MgAl_2O_4$
Space group	$Fd\,\overline{3}\,m$ (setting 2)
Lattice parameter *a*, Å	8.0822(2)
Cell volume (Å^3)	527.95 (2)
Radiation and wavelengths (Å)	Cu Kα 1.54056 1.54439
Angular range $2\theta_{min}$–$2\theta_{max}$ (degrees)	14.00–145.00 (step 0.05°2θ)
Number of measured reflections	39
Number of fitted parameters	24
Reliability factors (in %)	
$R_B = \Sigma\lvert I_h(obs) – I_h(calc)\rvert/\Sigma\lvert I_h(obs)\rvert$	4.76
$R_F = \Sigma\lvert\lvert F_h(obs)\rvert – \lvert F_h(calc)\rvert\rvert/\Sigma\lvert F_h(obs)\rvert$	7.42
$R_p = \Sigma\lvert y_i(obs) – y_i(calc)\rvert/\Sigma y_i(obs)$	4.43
$R_{wp} = [\Sigma w_i\lvert y_i(obs) – y_i(calc)\rvert^2/\Sigma\ w_i y_i(obs)^2]^{1/2}$	5.79
$\chi^2 = \{R_{wp}/R_{exp}\}^2$	3.34
Fraction (%)	96.52(0.88)
Addition phase	MgO
Space group	$Fm\,\overline{3}\,m$
Lattice parameter *a* (Å)	4.2114 (3)
R_{Bragg}, R_F, in %	16.1 11.0
Fraction (%)	3.48(13)

Table 7 Atomic parameters for $MgAl_2O_4$ ceramics sintered at 1300 °C

Atom	Site	*x*	*y*	*z*	$B_{iso} \times 10^2$, nm^2	*G*
Mg	8*a*	1/8	1/8	1/8	1.24(4)	1
Al	16*d*	1/2	1/2	1/2	1.41(3)	1
O	32*e*	0.2605(1)	0.2605(1)	0.2605(1)	1.48(5)	1

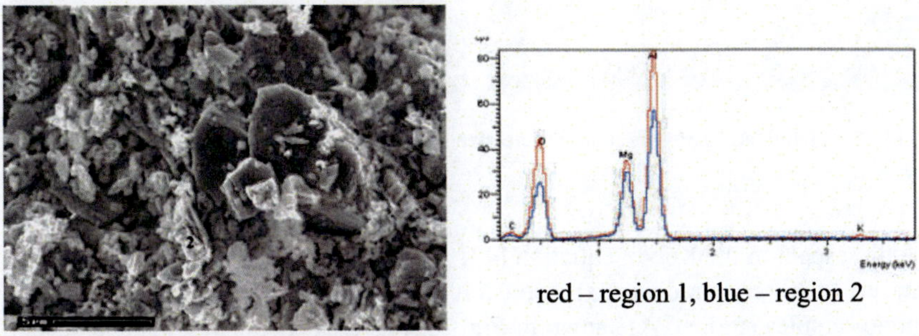

Fig. 12 Microstructural analysis and elemental composition of grains and grain boundaries in $MgAl_2O_4$ ceramics sintered at 1300 °C

Figure 12 illustrates the microstructure characteristics of the investigated $MgAl_2O_4$ ceramics. The ceramic structure displays a high degree of perfection, with an increase in the contact area between the grains leading to an increase in specific surface area. As a result, the grains amalgamate into agglomerates, and the content of open pores increases. These pores first acquire a spherical and subsequently a cylindrical shape, positioned along the grain boundaries. The average grain size also increases to 300–500 nm. Additionally, small pores in the ceramics solidify, resulting in the formation of a closed porosity that does not take part in water adsorption. Furthermore, the composition of grains and grain boundaries was analyzed using the EDX method (Fig. 12).

The $MgAl_2O_4$ ceramics contain an additional MgO phase that is distributed unevenly throughout its volume. It is concentrated mainly near pores of varying sizes and shapes located along the grain boundaries.

The porosity of the ceramics is characterized by both open and closed pores. To facilitate water sorption, an effective flow requires a heterogeneous porous structure that contains macro- and mesopores ranging in size from tens to hundreds of nanometers, which allow water-conducting channels to be used for capillary processes to take place. It has been found that sintering temperature affects the formation of internal open pores (pore size distribution).

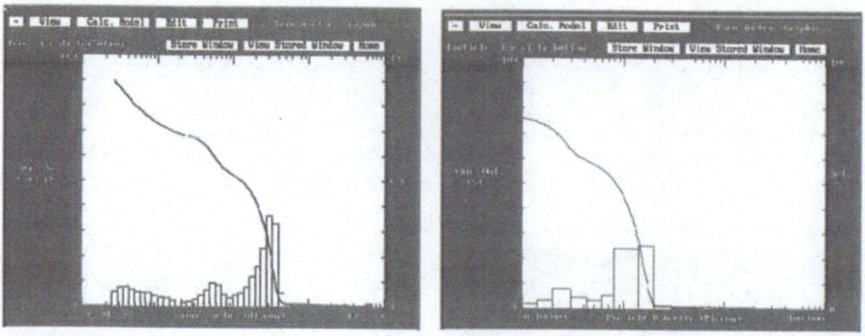

Fig. 13 Characterization of pore and particle size distribution in $MgAl_2O_4$ ceramics sintered at 1300 °C

With respect to the Kelvin equation [43], cylindrical pores ranging from approximately 1 to 20 nm in radius are responsible for capillary water condensation in the ceramic samples studied. As shown in Fig. 13, this area covers the first and second peaks. All samples of $MgAl_2O_4$ ceramics contain meso- and macropores with a radius of more than 20 nm (the second and third peaks), which are necessary for effective water conduction throughout the sample [44–47].

In summary, the pore size distribution in the $MgAl_2O_4$ ceramics studied is trimodal, with nanopores exhibiting a peak of 3.5 nm, mesopores with a radius of 43 nm, and macropores with a size of 320 nm (Fig. 13). The maximum particle diameter is approximately 1.2 μm.

Although capillary condensation processes occur in ceramic nanopores, traditional porosimetry methods cannot study the entire spectrum. To fully analyze the sizes of nanopores and the processes occurring in them, the PALS method was used as an alternative. This method also enables the evaluation of free-volume defects in functional materials. Figure 14 shows the PALS spectra of $MgAl_2O_4$ ceramics sintered at 1300 °C before and after immersion in water.

The LT program was used to decompose the PALS spectra into three components for both cases, which is a common method for analyzing spinel porous ceramics. Previous studies have indicated that the annihilation processes in $MgAl_2O_4$ ceramics can be explained by a multi-channel model, which involves channels for trapping positrons through defects and the decay of positronium atoms in nanopores. The first component of this model provides insights into the spinel microstructure, including information on octahedral and tetrahedral vacancies. The second component relates to the capture of positrons by extended defects at grain boundaries, while the third component reflects the decay of positron atoms in nanopores and analyzes the processes within them, such as those influenced by the presence of adsorbed water. Table 8 presents the parameters for positron capture and the modes of positron trapping.

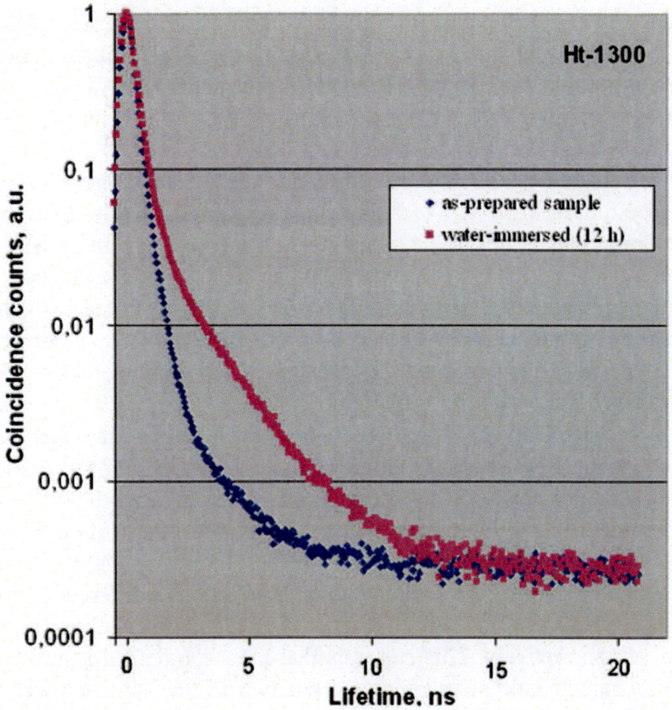

Fig. 14 PALS spectra analysis for $MgAl_2O_4$ ceramics sintered at 1300 °C: comparing initial and water-immersed samples

Table 8 Investigating fitting parameters, positron trapping modes, and nanopore radii of $MgAl_2O_4$ ceramics sintered at 1300 °C

Sample prehistory	τ_1, ns	I_1, %	τ_2, ns	I_2, %	τ_3, ns	I_3, %
Before water immersion	0.22	72	0.44	26	2.19	2
After water immersion	0.22	54	0.46	32	1.88	15
Sample prehistory	$\tau_{av.}$, ns	τ_b, ns	κ_d, ns^{-1}	$\tau_2 - \tau_b$, ns	τ_2/τ_b	R_3, nm
Before water immersion	0.27	0.25	0.6	0.19	1.7	0.305
After water immersion	0.31	0.27	0.9	0.19	1.7	0.267

Table 8 displays that the lifetime of the first component τ_I does not change when the ceramic samples are exposed to water, but its intensity I1 decreases. The lifetime of the second component τ_2 and intensity I_2 both slightly increase, which suggests that there are more positron capture processes at the grain boundaries in immersion ceramics where the additional MgO phase is concentrated. Meanwhile, the lifetime of the third component τ_3 decreases to 1.88 ns, and its intensity I_3 increases significantly from 2 to 15%.

According to previous studies, the lifetimes are linked to the size of defects or nanopores, while the intensities indicate their quantity [22]. The changes in the third component suggest that the free volume in nanopores undergoes fragmentation after immersion in water. The value of $\tau_3 = 1.88$ nm indicates additional processes of o-Ps decay in water.

It is evident that changes in the lifetimes and intensities of these components of the PALS spectrum of $MgAl_2O_4$ ceramics due to water adsorption in its nanopores would result in modifications to the parameters of positron capture by bulk defects. Table 8 presents the positron capture parameters, such as average positron lifetime τ_{av}, lifetime τ_b linked to positron annihilation in a defect-free area of the material, rate of positron capture into defects κ_d, $\tau_2 - \tau_b$ (average size of defects where positrons are trapped), and ratios τ_2/τ_b (which reflects the nature of bulk defects) for ceramic samples before and after immersion in water.

The data presented in Table 8 indicate that after immersion in water, there is a slight increase in the average positron lifetime τ_{av} and τ_b for $MgAl_2O_4$ ceramics. The increase in the rate of positron capture in defects κ_d from 0.6 to 0.9 ns^{-1} suggests more intense positron annihilation processes in ceramics, particularly at the grain boundaries where additional MgO phase is present. It appears that the presence of adsorbed water does not alter the geometry of bulk defects. The difference $\tau_2 - \tau_b$ and the ratio τ_2/τ_b remain unchanged for both the original and water-immersed samples, at 0.19 and 1.7, respectively. This suggests that the same defect positron trapping centers, with characteristic sizes ranging from two to four atomic vacancies, are predominant in the studied $MgAl_2O_4$ ceramics, regardless of the presence of adsorbed water. The nanopore radii were calculated using the Tao-Eldrup model [32, 33] and the lifetime of the third component τ_3. As shown in Table 8, nanopore size is larger for ceramics that have been immersed in water. Therefore, the presence of adsorbed water in ceramics leads to the fragmentation of internal free volumes of nanopores, as evidenced by the increase in I_3 intensity. This finding was confirmed by repeat studies of the pore size distribution in $MgAl_2O_4$ ceramics using Hg porosimetry after PALS measurements (see Fig. 15).

5 Conclusions

It was studied $MgAl_2O_4$ ceramics with different sintering temperatures (1100, 1200, and 1300 °C for 2 h) for their microstructure and sensitivity to humidity using both conventional microstructural methods and alternative positron annihilation lifetime spectroscopy. Our findings show that the ceramics have a well-developed structure with grains of varying sizes, intergrain boundaries with additional phases, and pores of different shapes and sizes. We describe the porous structure of ceramics using a trimodal distribution of open pores by size.

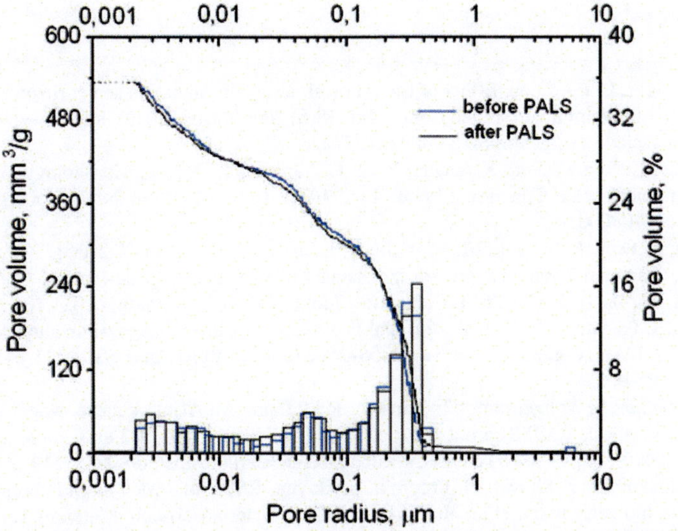

Fig. 15 Distribution of pore sizes in examined $MgAl_2O_4$ ceramics before and after PALS analysis

We investigated the changes in free volume and evolution of nanopores (both open and closed) in $MgAl_2O_4$ ceramics due to the adsorption and desorption of water using the PALS method. Our results indicate that the presence of water in ceramics leads to more effective positron capture by defects at grain boundaries, as reflected in the second component of the PALS spectrum and positron capture parameters.

The third component of the PALS spectrum describes processes in nanopores. An increase in the intensity of the third component is observed with an increase in water in ceramics, and the characteristic lifetime value of 1.88 ns indicates the annihilation of o-Ps in water. We observed that some nanopores remain blocked by water even after drying the ceramic samples, and the parameters do not return to their initial values. We established that water adsorption causes fragmentation of nanopores, while desorption leads to the agglomeration of nanopores in the ceramics. The lifetime value of the third component of 1.88 ns in water-immersed ceramics also indicates the decay of o-Ps in water.

Acknowledgements H.K. would like to express gratitude to the Ministry of Education and Science of Ukraine for providing support. The author also thanks Prof. O. Shpotyuk for valuable discussions and Dr. I. Hadzaman for assistance with the preparation of the samples.

References

1. M. Zhang, J. Liu, J. Zhou, Effect of Eu_2O_3 on sintering densification and corrosion resistance of magnesium aluminate spinel. Ceram. Int. **48**(6), 7557–7563 (2022). https://www.sciencedirect.com/science/article/pii/S0272884221037226
2. X. Zhao, X. Ren, C. Sun, X. Zhang, Y. Si, C. Yan, J. Xu, D. Xue, Morphology evolution at nano-to micro-scale. Functional Mater. Lett. **1**(03), 167–172 (2008). https://doi.org/10.1142/S1793604708000393
3. Q. Wu, G. Feng, F. Jiang, L. Miao, W. Jiang, J. Liang, J. Liu, J. Tan, X. Zhang, X. Lao, Preparation, characterisation, and growth mechanism of mesoporous petal-like $MgAl_2O_4$ spinel. Ceram. Int. **48**(3), 3351–3361 (2022). https://doi.org/10.1016/j.ceramint.2021.10.110
4. J.G. Li, T. Ikegami, J.H. Lee, T. Mori, Fabrication of translucent magnesium aluminum spinel ceramics. J. Am. Ceram. Soc. **83**(11), 2866–2868 (2000). https://doi.org/10.1111/j.1151-2916.2000.tb01648.x
5. A. Laobuthee, S. Wongkasemjit, E. Traversa, R.M. Laine, $MgAl_2O_4$ spinel powders from oxide one pot synthesis (OOPS) process for ceramic humidity sensors. J. Eur. Ceram. Soc. **20**(2), 91–97 (2000). https://www.sciencedirect.com/science/article/abs/pii/S0955221999001533
6. Y. Mordekovitz, Y. Shoval, N. Froumin, S. Hayun, Effect of structure and composition of non-stoichiometry magnesium aluminate spinel on water adsorption. Materials **13**(14), 3195 (2020). https://www.mdpi.com/1996-1944/13/14/3195
7. S. Sinhamahapatra, P. Das, K. Dana, H.S. Tripathi, Magnesium aluminate spinel: structure, properties, synthesis and applications. Trans. Indian Ceram. Soc. **81**(3), 97–120 (2022). https://doi.org/10.1080/0371750X.2022.2137701
8. Z. Shi, Q. Zhao, B. Guo, T. Ji, H. Wang, A review on processing polycrystalline magnesium aluminate spinel ($MgAl_2O_4$): sintering techniques, material properties and machinability. Mater. Des. **193**, 108858 (2020). https://doi.org/10.1016/j.matdes.2020.108858
9. M. Mouyane, B. Jaber, B. Bendjemil, J. Bernard, D. Houivet, J.G. Noudem, Sintering behavior of magnesium aluminate spinel $MgAl_2O_4$ synthesized by different methods. Int. J. Appl. Ceram. Technol. **16**(3), 1138–1149 (2019). https://doi.org/10.1111/ijac.13172
10. V. Yatsuk, T. Bubela, Y. Pokhodylo, Y. Yatsuk, R. Kochan, Improvement of data acquisition systems for the measurement of physical-chemical environmental properties, in *9th IEEE International Conference on Intelligent Data Acquisition and Advanced Computing Systems: Technology and Applications (IDAACS)*, vol. 1 (2017), pp. 41–46. https://ieeexplore.ieee.org/abstract/document/8095046
11. B. Stadnyk, S. Yatsyshyn, O. Seheda, Research in nanothermometry. Part 6. Metrology of Raman Thermometer with universal calibration artifacts. Sens. Transducers **142**(7), 1–9 (2012). https://www.sensorsportal.com/HTML/DIGEST/P_1002.htm
12. I. Karbovnyk, I. Bolesta, I. Rovetskii, S. Velgosh, H. Klym, Studies of CdI_2-Bi_3 microstructures with optical methods, atomic force microscopy and positron annihilation spectroscopy. Mater. Sci. Poland **32**(3), 391–395 (2014). https://doi.org/10.2478/s13536-014-0215-z
13. V.P. Savchyn, A.I. Popov, O.I. Aksimentyeva, H. Klym, Y.Y. Horbenko, V. Serga, A. Moskina, I. Karbovnyk, Cathodoluminescence characterization of polystyrene-$BaZrO_3$ hybrid composites. Low Temp. Phys. **42**(7), 597–600 (2016). https://doi.org/10.1063/1.4959020
14. Y. Bobalo, Z. Kolodiy, B. Stadnyk, S. Yatsyshyn, Development of noise measurements. Part 3. Passive method of electronic elements quality characterization. Sens. Transducers **152**(5), 164 (2013). https://www.proquest.com/openview/d49488d2c0030221018bf09605dd10b2/1?pq-origsite=gscholar&cbl=52938
15. T. Bubela, M. Mykyychuk, A. Hunkalo, O. Boyko, O. Basalkevych, A study of uncertainty of expert measurement results in the quality management system. Eastern-Eur. J. Enterprise Technol. **3**(3), 4–11 (2016)
16. I. Karbovnyk, I. Borshchyshyn, Y. Vakhula, I. Lutsyuk, H. Klym, I. Bolesta, Impedance characterization of Cr^{3+}, Y^{3+} and Zr^{4+} activated forsterite nanoceramics synthesized by sol–gel method. Ceram. Int. **42**(7), 8501–8504 (2016). https://doi.org/10.1016/j.ceramint.2016.02.075

17. Z. Wang, C.L. Chang, X. Zhao, W. Qian, X. Zhang, Z. Xie, B.-H. Hwang, C. Hu, J. Shen, R. Hui, $MgAl_2O_4$-based humidity-sensing material for potential application in PEM fuel cells. J. Power Sources **190**(2), 351–355 (2009). https://www.sciencedirect.com/science/article/abs/pii/S037877530801762X
18. R. Krause-Rehberg, H.S. Leipner, *Positron Annihilation in Semiconductors. Defect Studies* (Springer, Heidelberg/New York, 1999), p. 378
19. Y.C. Jean, P.E. Mallon, D.M. Schrader, *Principles and Application of Positron and Positronium Chemistry* (Word Scientific, Singapore, 2003)
20. O. Shpotyuk, J. Filipecki, Free Volume in Vitreous Chalcogenide Semiconductors: Possibilities of Positron Annihilation Lifetime Study: Wyd-wo WSP w Czestochowie, Czestochowa (2003)
21. H. Klym, I. Hadzaman, R. Vila, A.I. Popov, Extended positron–positronium trapping defects in the $MgAl_2O_4$ spinel ceramics. Physica Status Solidi (B) **259**(8), 2100473 (2022). https://doi.org/10.1002/pssb.202100473
22. H. Klym, I. Karbovnyk, S. Piskunov, A.I. Popov, Positron annihilation lifetime spectroscopy insight on free volume conversion of nanostructured $MgAl_2O_4$ ceramics. Nanomaterials **11**(12), 3373 (2021). https://doi.org/10.3390/nano11123373
23. H. Klym, A. Ingram, O. Shpotyuk, I. Hadzaman, Water-sorption processes in nanostructured ceramics for sensor electronics studied with positron annihilation instruments, in *28th International Conference on Microelectronics Proceedings* (2012), pp. 155–158. https://ieeexplore.ieee.org/abstract/document/6222821
24. H. Klym, I. Karbovnyk, A. Luchechko, Y. Kostiv, V. Pankratova, A.I. Popov, Evolution of free volumes in polycrystalline $BaGa_2O_4$ ceramics doped with Eu^{3+} ions. Crystals **11**(12), 1515 (2021). https://doi.org/10.3390/cryst11121515
25. H. Klym, I. Hadzaman, Y. Kostiv, S. Yatsyshyn, B. Stadnyk, Free-volume defects/nanopores conversion of temperature-sensitive $Cu_{0.1}Ni_{0.8}Co_{0.2}Mn_{1.9}O_4$ ceramics caused by addition phase and monolithization process. Appl. Nanosci. **12**(4), 1347–1354 (2022). https://doi.org/10.1007/s13204-021-01963-9
26. H. Klym, I. Hadzaman, V. Gryga, Combined study of internal nanovoids in $Cu_{0.1}Ni_{0.1}Co_{1.6}Mn_{1.2}O_4$-based thick-film layers formed near grain boundaries. Appl. Nanosci. **12**(4), 1257–1262 (2022). https://doi.org/10.1007/s13204-021-01801-y
27. J. Shah, R.K. Kotnala, Humidity sensing exclusively by physisorption of water vapors on magnesium ferrite. Sens. Actuators B: Chem. **171**, 832–837 (2012). https://www.sciencedirect.com/science/article/abs/pii/S0925400512005497
28. J.T.W. Yeow, J.P.M. She, Carbon nanotube-enhanced capillary condensation for a capacitive humidity sensor. Nanotechnology **17**(21), 5441 (2006). https://doi.org/10.1088/0957-4484/17/21/026
29. I. Hadzaman, H. Klym, O. Shpotuyk, M. Brunner, Temperature sensitive spinel-type ceramics in thick-film multilayer performance for environment sensors. Acta Physica Polonica-Ser. A General Phys. **117**(1), 234 (2010). http://przyrbwn.icm.edu.pl/APP/PDF/117/a117z148.pdf
30. H. Klym, I. Karbovnyk, Y. Kostiv, Effect changing of nanopore volumes in $MgAl_2O_3$ ceramics under the influence of water, in *IEEE 3rd Ukraine Conference on Electrical and Computer Engineering (UKRCON)* (2021), pp. 500–503. https://ieeexplore.ieee.org/abstract/document/9575283
31. H.I. Klym, A.I. Ivanusa, Yu.M. Kostiv, D.O. Chalyy, T.I. Tkachuk, R.B. Dunets, I.I. Vasylchyshyn, Methodology and algorithm of multicomponent analysis of positron annihilation spectra for nanostructured functional materials. J. Nano-Electron. Phys. **9**(3), 03037-1-6 (2017). https://doi.org/10.21272/jnep.9(3).03037
32. S.J. Tao, Positronium annihilation in molecular substance. J. Chem. Phys. **56**(11), 5499–5510 (1972). https://doi.org/10.1063/1.1677067
33. M. Eldrup, D. Lightbody, J.N. Sherwood, The temperature dependence of positron lifetimes in solid pivalic acid. Chem. Phys. **63**, 51–58 (1981). https://doi.org/10.1016/0301-0104(81)80307-2

34. H. Klym, I. Hadzaman, Micro-, nanostructural properties of the $MgAl_2O_3$ ceramics studied by combined methods, in *IEEE 16th International Conference on Advanced Trends in Radioelectronics, Telecommunications and Computer Engineering (TCSET)* (2022), pp. 833–836. https://ieeexplore.ieee.org/abstract/document/9766853
35. H. Klym, A. Ingram, O. Shpotyuk, I. Karbovnyk, Influence of CsCl addition on the nanostructured voids and optical properties of $80GeS_2$-$20Ga_2S_3$ glasses. Opt. Mater. **59**, 39–42 (2016). https://doi.org/10.1016/j.optmat.2016.03.004
36. H. Klym, A. Ingram, O. Shpotyuk, Free-volume nanostructural transformation in crystallized GeS_2–Ga_2S_3–CsCl glasses. Materialwiss. Werkstofftech. **47**(2–3), 198–202 (2016). https://doi.org/10.1002/mawe.201600476
37. J. Kansy, Microcomputer program for analysis of positron annihilation lifetime spectra. Nucl. Instrum. Methods Phys. Res. Sect. A **374**(2), 235–244 (1996)
38. P.M.G. Nambissan, C. Upadhyay, H.C. Verma, Positron lifetime spectroscopic studies of nanocrystalline $ZnFe_2O_4$. J. Appl. Phys. **93**, 6320 (2003). https://doi.org/10.1063/1.1569973
39. K. Bennewitz, M. Haaks, T. Staab, S. Eisenberg, T. Lampe, K. Maier, Positron annihilation spectroscopy—a non-destructive method for lifetime prediction in the field of dynamical material testing. Int. J. Mater. Res. **93**(8), 778–783 (2002). https://doi.org/10.3139/ijmr-2002-0131
40. Y. Uematsu, T. Kakiuchi, H. Nishizawa, A. Nakai, A. Ohtani, N. Uesugi, K. Hattori, Non-destructive fatigue damage detection of carbon fiber-reinforced thermoplastics using positron annihilation method. Fatigue Fract. Eng. Mater. Struct. **45**(10), 2784–2793 (2022). https://doi.org/10.1111/ffe.13775
41. J. Čížek, Characterization of lattice defects in metallic materials by positron annihilation spectroscopy: a review. J. Mater. Sci. Technol. **34**(4), 577–598 (2018). https://doi.org/10.1016/j.jmst.2017.11.050
42. H. Klym, I. Karbovnyk, I. Vasylchyshyn, Physical & chemical water-sorption processes in the $MgAl_2O_4$ ceramics, in *IEEE 12th International Conference Nanomaterials: Applications & Properties (NAP)*. NSS17-1 (2022). https://ieeexplore.ieee.org/abstract/document/9934270
43. A.C. Mitropoulos, The kelvin equation. J. Colloid Interface Sci. **317**(2), 643–648 (2008). https://doi.org/10.1016/j.jcis.2007.10.001
44. I. Leifer, R.K. Patro, The bubble mechanism for methane transport from the shallow sea bed to the surface: a review and sensitivity study. Continental Shelf Res. **22**(16), 2409–2428 (2002). https://www.sciencedirect.com/science/article/abs/pii/S0278434302000651
45. S. Ljunggren, J.C. Eriksson, The lifetime of a colloid-sized gas bubble in water and the cause of the hydrophobic attraction. Colloids Surf. A: Physicochem. Eng. Aspects **129**, 151–155 (1997). https://www.sciencedirect.com/science/article/abs/pii/S0927775797000332
46. A. Grosman, C. Ortega, Nature of capillary condensation and evaporation processes in ordered porous materials. Langmuir **21**, 10515–10521 (2005). https://doi.org/10.1021/la051030o
47. G. Dlubek, Yu. Yang, R. Krause-Rehberg, W. Beichel, S. Bulut, N. Pogodina, I. Krossing, Ch. Friedrich, Free volume in imidazolium triflimide ($[C_3MIM][NTf_2]$) ionic liquid from positron lifetime: amorphous, crystalline, and liquid states. J. Chem. Phys. **133**, 124502 (2010). https://doi.org/10.1063/1.3487522

Influence of the Number of Cycles of Severe Plastic Deformation on Nanostructure Formation in Low-alloyed Steels

Olha Maksymiv, Volodymyr Kyryliv, Borys Chaikovs'kyi, Volodymyr Gurey, Olha Zvirko, Yuriy Kulyk, and Ivan Yaroshovych

Abstract Increasing the durability of machine components during operation is achieved by obtaining strengthened surface layers with nanocrystalline structure (NCS). They are generated during surface hardening by mechanical-pulse treatment (MPT) due to the severe plastic deformation (SPD) caused by high-speed friction between the tool and the treated surface. NCS in the surface layers of metal is formed during the interaction of the tool's working surface with the treated surface by a certain number of cycles, which are determined by the ratio of the tool's width and the feed rate. Experimental studies on the effect of the cycle number of SPD during surface hardening on the structural-phase state of the metal were performed on samples made of steel 080A67. Grain refinement occurred mostly during the first three cycles of SPD, and the grain size was slightly influenced by a further increasing in a number of cycles. A martensitic–austenitic structure of the treated surface layer metal was obtained. The highest microhardness and cementite content in the surface layer with NCS were achieved after three cycles of SPD.

O. Maksymiv (✉) · V. Kyryliv · O. Zvirko
Karpenko Physico-Mechanical Institute of the NAS of Ukraine, 5, Naukova St., Lviv 79060, Ukraine
e-mail: okyryliv@yahoo.com

B. Chaikovs'kyi · I. Yaroshovych
Stepan Gzhytskyi National University of Veterinary Medicine and Biotechnologies of Lviv, 50, Pekarska St., Lviv 79010, Ukraine

V. Gurey
Lviv Politechnic National University, Lviv 12, Bandera St., 79013, Ukraine

Y. Kulyk
Ivan Franko National University of Lviv, 1, Universytetska St., Lviv 79000, Ukraine

O. Fesenko and L. Yatsenko (eds.), *Nanostructured Surfaces, Nanocomposites and Nanomaterials, and Their Applications*, Springer Proceedings in Physics 296,
https://doi.org/10.1007/978-3-031-42704-6_25

1 Introduction

The problem of purposeful change of the metal surface state in order to provide it with the necessary physical and mechanical properties to improve the efficiency of machine components is the subject of study by scientists and engineers for a long time [1–4]. Some progress in solving this problem has been achieved through the surface NCS formation. The most popular method of forming the surface layer with NCS is SPD [5–8]. Those methods include MPT, which belongs to surface attrition treatment. The source of SPD during MPT is high-speed friction [6, 7]. The hardened (strengthened) surface layer with NCS is formed during MPT changing structural-phase state [6], as well as the chemical composition of the surface layer metal due to doping from special technological media [9]. As a result, the MPT of low-alloyed medium carbon steels increases their surface microhardness and wear resistance [10, 11], fatigue, corrosion [12–14], and contact fatigue resistance [13, 14], reduces hydrogen permeability, since NCS serves as a barrier to hydrogen penetration into the matrix material [11].

The kinematic of MPT is identical to the grinding processes. It is performed on lathes and grinding machines for cylindrical and flat surfaces using special hardening (strengthening) tools. The hardening process takes place in the friction contact (FC) zone of the tool and the component due to the combined action of high pressure and temperature. Important technological parameters are: the working width of the tool b and the value of the longitudinal feed S for the treatment of the cylindrical surfaces and the feed on the double stroke of the table S for the treatment of the flat surfaces (Fig. 1a). Therefore, each point of the treated component is subjected to SPD several times, and a certain number of the treatment cycles n determined by Formula (1):

$$n = \frac{b}{S}. \tag{1}$$

The number of cycles can vary from 3 to 10 in different treatment regimes, and it indicates the number of cycles of SPD. To increase the productivity of MPT, higher feed S should be applied, resulting in decreasing the number of cycles n. In order to improve the physical and mechanical properties of the surface layer, namely its depth and microhardness, it is necessary to work with the minimum width of the tool b. The width b of the tool's working part can be changed, which will also affect the number of cycles. As the tool's working width b decreases, the specific pressure P and the temperature T in the FC zone increase (Fig. 1b), which are decisive for the formation of surface NCS.

For MPT regime optimization, it is necessary to investigate the influence of the number of cycles n on the dispersion of the structure, the formation of the phase state, and the physical and mechanical properties of the surface nanostructure.

The aim of the work is to investigate the influence of the number of cycles of SPD on the structural-phase state and properties of the surface layer with NCS during MPT of the steel 080A67.

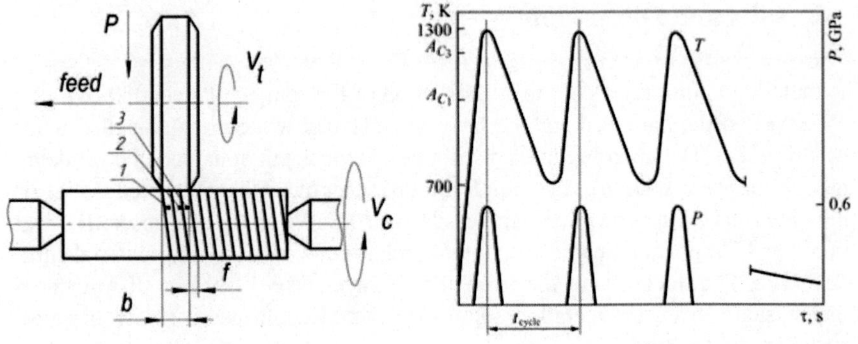

Fig. 1 Scheme of MPT of cylindrical surface (**a**) and cyclic change of temperature T and pressure P in the FC zone during MPT (**b**): V_t—velocity of the tool; V_C—velocity of the component; S—longitudinal feed of the tool; b—width of the working part of the tool; P, T—pressure and temperature in the FC zone; *1, 2, 3*—the number of cycles in the FC zone

2 Materials and Experimental Methods

The flat samples of 3 × 25 × 30 mm made of steel 080A67 (0.65C-1Mn) in the as-received state were subjected to MPT and then investigated. This steel is widely used, in particular, in the agricultural sector for the manufacture of discs for seeding machine furrow-openers. MPT was performed on the modernized grinding machine (model SPC-20A (Jotes, Poland)), by using the tool made of steel 41Cr4 (0.4C–1Cr), 250 mm in diameter, and a width of the working part b—4 mm. The treatment regimes of MPT were as follows: the linear rotation velocity of the hardening tool V_t—60 m/s, velocity of the component V_C—0.017 m/s (corresponds to the working speed of the table), cutting depth—0.35 mm, which corresponds to a pressure P in the FC zone of 0.7 GPa. Since the width of the zone of contact on the tested samples corresponds to the feed rate, the number of cycles, which was varied from 1 to 6, was obtained by the corresponding number of working passes of the machine table. Mineral oil with the addition of low molecular weight polyethylene was supplied as a technological medium during MPT [10].

The phase composition of the steel surface layers after MPT was studied by using the diffractometer DRON-3 with a CuKα X-ray source (voltage of 30 kV and intensity of 20 mA), spacing of 0.05°, and exposition of 4 s. The diffractograms were post-processed using the software CSD [15]. The diffraction patterns were analyzed using the JCPDS-ASTM files [16]. The grains sizes on the surface were determined by the method of X-ray phase diffraction analysis according to the half-width of the peaks. The structure was investigated using an optical microscope Neophot-21. The microhardness of the samples after MPT was measured by using the microhardness tester PMT-3 at the load of 100 g.

3 Results and Discussion

The metallographic studies of the cross-section of flat sample of steel 080A67 after MPT revealed the hardened surface layer as unetched white area (so-called white layer) (Fig. 2a). The microhardness distribution in the depth of hardened layer determined on the cross-section of the sample after 6 cycles of MPT is presented in Fig. 2b. It indicated that the surface microhardness H_μ of 10.5 GPa for the steel after MPT was more than 3.5 higher compared with the microhardness of the matrix material equal to 2.8 GPa. The thickness of the strengthened layer was ~ 130 μm. It was determined visually from the metallographic studies and confirmed by microhardness distribution in the depth.

According to X-ray studies, the hardened NCS surface layer of the steel 080A67 formed by MPT had mainly martensitic–austenitic–cementite structure. The oxides of Fe_3O_4 and FeO were also formed on the surface. However, the ratio of the phase composition changed with an increase in the number of cycles (Fig. 3). Starting from the third cycle, the lines of oxides Fe_3O_4 and FeO as well as the lines of cementite Fe_3C, become clearer on X-ray diffraction patterns (Fig. 3).

The relative microdeformation curve of the martensite lattice has two maxima: after the second and fifth cycles (Fig. 4). This can be explained by the high content of carbon in the steel and its additional diffusion from the technological medium during the first cycle of MPT, which contributes to the formation of additional martensite distorted by SPD. In the following cycles, carbon is "pulled out" by movable dislocations from the austenite lattice [17] with the formation of the maximum amount of cementite after the third cycle of MPT (Fig. 5), which is confirmed by the maximum surface microhardness H_μ of 10.8 GPa (Table 1).

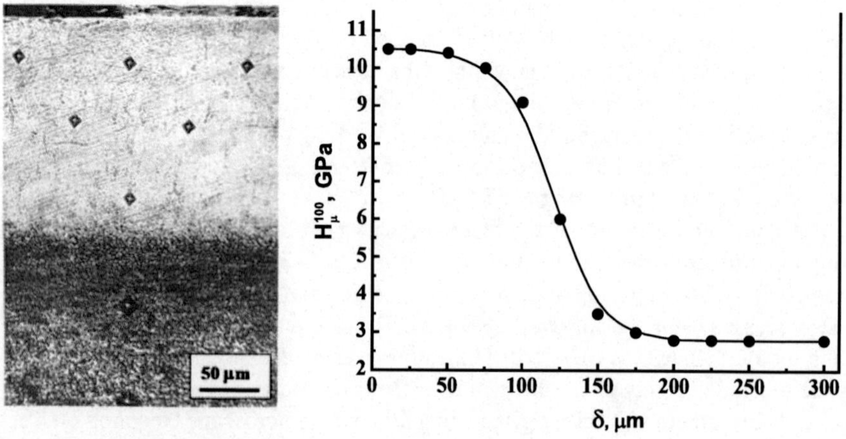

a b

Fig. 2 Microstructure (**a**) and microhardness distribution H_μ in the depth of the hardened layer from the surface (**b**) of steel 080A67 after six cycles of MPT

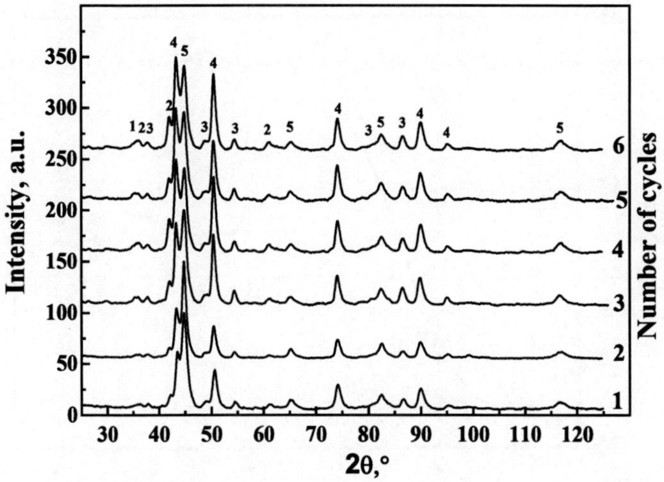

Fig. 3 Fragments of diffractograms of surface layer on steel 080A67 samples with different number of cycles of MPT: 1—Fe_3O_4, 2—FeO, 3—Fe_3C, 4—γ(Fe), 5—α(Fe)

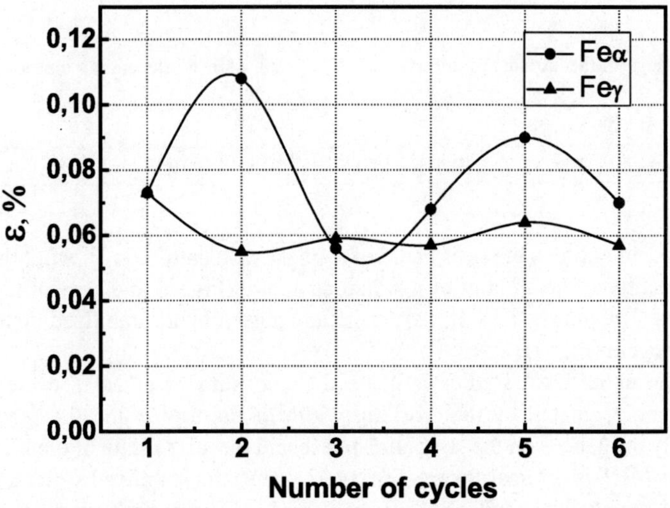

Fig. 4 Dependence of relative microdeformation of the lattice ε on the number of MPT cycles

It should be noted that the number of movable dislocations decreases with each next cycle of SPD, and starting from the fourth cycle, carbon could be "pulled out" by movable dislocations from the cementite lattice. This process is possible as far as the binding energy of carbon atoms in the cementite lattice 0.4–0.42 eV is much lower than the binding energy of dislocation with interstitial impurity (0.8 eV) [18]. The kinetics of "pulling-out" is determined by SPD and the process of transfer of carbon

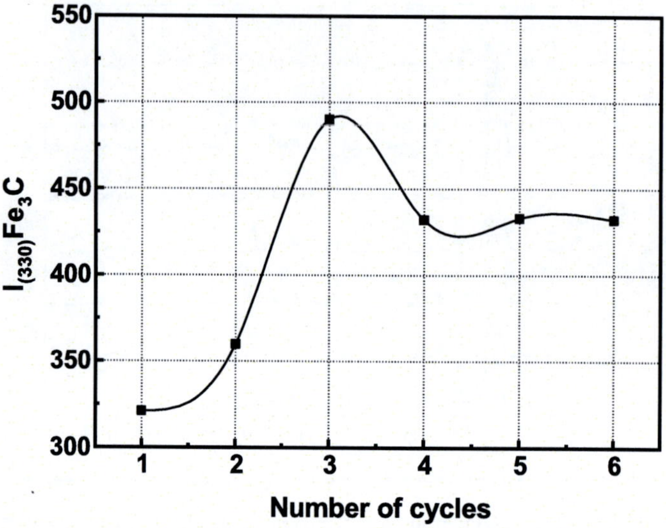

Fig. 5 Dependence of the integral intensity of the maximum (330) $I_{(330)}$ of cementite on the number of MPT cycles

Table 1 Change in the surface microhardness H_μ of steel 080A67 depending on the number of MPT cycles

The number of MPT cycles	1	2	3	4	5	6
Microhardness H_μ, GPa	10.4	10.5	10.8	10.5	10.5	10.5

atoms from cementite to the stress field of α-phase dislocations [18]. After the fourth cycle of treatment, the amount of residual austenite (Fig. 6) and cementite (Fig. 5) is stabilized. The grains sizes of martensite and austenite are stabilized mainly after the second cycle (Fig. 7).

Thus, the carried out studies indicate that the number of MPT cycles during the treatment, regulated by the working width of the tool b and the feed rate S, significantly influences on the structural-phase and the stress state of the NCS layer. Four cycles MPT is optimal for steel 080A67 due to the stabilized structural-phase state of the surface layer with NCS (Figs. 5 and 6). For optimization the physical and mechanical properties, in particular to increase the surface microhardness and depth of hardening, which are important for contact fatigue of drill bits, abrasive wear resistance of discs for seeding machine furrow-openers and also for machine components required finishing operations, it is advisable to use a reduced tool width b and increased feed rate S. At the same time, the unit pressure P and the temperature T in the FC zone will increase, which leads to a decrease in the grain size of the NCS surface layer. However, the optimal number of MPT cycles n will differ for different steels, and it will depend on its structural-phase state, as well as the required

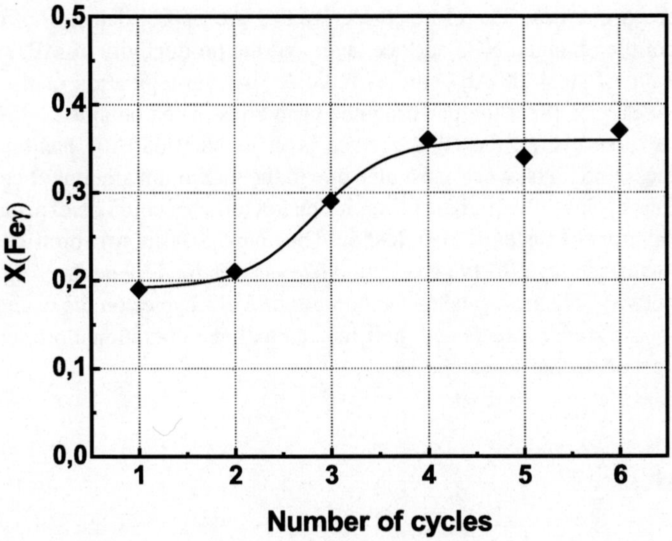

Fig. 6 Dependence of the volume fraction of the austenitic phase X(Fe_{γ}) on the number of MPT cycles

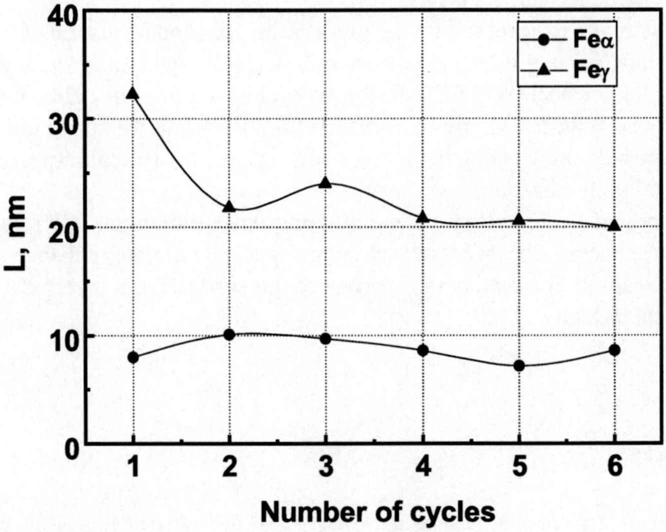

Fig. 7 Dependence of grains sizes L on the number of MPT cycles

performance properties. Therefore, these studies are appropriate to optimize the properties of the obtained NCS surface layer and the productivity of MPT.

The results of steel 080A67 studies indicate that the regulation of the number of MPT cycles will affect the performance properties of the obtained NCS surface layer. After three cycles, the NCS surface layer of steel 080A67 has the highest microhardness and, thus, wear resistance due to the maximum amount of cementite; and after four cycles—the highest plasticity characteristics could be expected to the maximum amount of austenite [19]. It should be noted, that the structural-phase state of the NCS surface layer of the steel 080A67 was obtained by using MPT with its high productivity. The influence of the number of MPT cycles on the change of the structural-phase state of steels and their mechanical and operational properties will be the subject of further research.

4 Conclusions

1. During the formation of surface NCS using MPT, the treated surface is processed by a certain number of SPD cycles, which are determined by the tool's width ratio and the feed rate. The number of SPD cycles affects the parameters of the formed hardened surface layers.
2. Dispersion of structure, stabilizing of the grain size, and formation of structural-phase state, namely martensitic–austenitic structure, take place during the first two cycles on steel 080A67. After three cycles, the lines of oxides Fe_3O_4 and FeO and cementite Fe_3C are more clearly identified, and the maximum hardness of the surface layer is obtained. After four cycles, the structural-phase state of the metal is stabilized.
3. The conducted research provides an opportunity to optimize MPT regimes in order to generate the necessary structural and phase state, enhance physical, mechanical, and operational properties of the surface NCS layer, and increase treatment capacity.

References

1. L. Quintino, Overview of coating technologies, in *Surface Modification by Solid State Processing*, ed. by R. Miranda (Elsevier, Cambridge, 2014), pp. 1–24. https://doi.org/10.1533/9780857094698.1
2. W. Kuang, Q. Miao, W. Ding et al., A short review on the influence of mechanical machining on tribological and wear behavior of components. Int. J. Adv. Manuf. Technol. **120**, 1401–1413 (2022). https://doi.org/10.1007/s00170-022-08895-w
3. T. Grosdidier, M. Novelli, Recent developments in the application of surface mechanical attrition treatments for improved gradient structures: processing parameters and surface reactivity. Mater. Trans. **60**, 1344–1355 (2019). https://doi.org/10.2320/matertrans.MF201929

4. M.M. Student, V.M. Posuvailo, H.H. Veselivs'ka et al., Corrosion resistance of plasma-electrolytic layers on alloys and coatings of the Al-Cu-Mg system for various models of heat treatment. Mater. Sci. **56**, 820–829 (2018). https://doi.org/10.1007/s11003-018-0137-8
5. K. Edalati, A. Bachmaier, V. Beloshenko et al., Nanomaterials by severe plastic deformation: review of historical developments and recent advances. Mater. Res. Lett. **10**, 163–256 (2022). https://doi.org/10.1080/21663831.2022.2029779
6. H. Nykyforchyn, V. Kyryliv, O. Maksymiv, O. Zvirko, Mechanical fabrication methods of nanostructured surfaces, in *Handbook of Modern Coating Technologies. Fabrication Methods and Functional Properties*, ed. by M. Aliofkhazraei, N. Ali, M. Chipara et al. (Elsevier, Amsterdam, 2021), pp. 25–67. https://doi.org/10.1016/C2012-0-06047-4
7. T.O. Olugbade, J. Lu, Literature review on the mechanical properties of materials after surface mechanical attrition treatment (SMAT). Nano Mater. Sci. **2**, 3–31 (2020). https://doi.org/10.1016/j.nanoms.2020.04.002
8. Y. Kyryliv, V. Kyryliv, B. Tsizh et al., Resistance of surface nanostructures and ultrafine grain structures on steel 40Kh steel to wear and cavitation-erosive destruction. Appl. Nanosci. **14**(4), 1085–1090 (2022). https://doi.org/10.1023/A:1021742204654
9. V.I. Kyryliv, Y.M. Koval', Surface alloying of steels from special process media. Mater. Sci. **37**, 816–819 (2001). https://doi.org/10.1023/A:1015060829045
10. H. Nykyforchyn, V. Kyryliv, O. Maksymiv et al., Wear resistance of the surface nanocrystalline structure under an action of diethylene glycol medium. Appl. Nanosci. **9**, 1085–1090 (2019). https://doi.org/10.1007/s13204-018-0690-3
11. H.M. Nykyforchyn, E. Lunarska, V.I. Kyryliv, O.V. Maksymiv, Hydrogen permeability of the surface nanocrystalline structures of carbon steel. Mater. Sci. **50**, 67–73 (2015). https://doi.org/10.1007/s11003-015-9774-3
12. V. Kyryliv, B. Chaikovs'kyi, O. Maksymiv, B. Mykytchak, Fatigue and corrosion fatigue of the roll steels with surface nanostructure. J. Nano Res. **51**, 92–97 (2018). https://doi.org/10.4028/www.scientific.net/JNanoR.51.92
13. V.I. Kyryliv, B.P. Chaikovs'kyi, O.V. Maksymiv et al., Serviceability of 60KH2M roll steel with surface nanostructure. Mater. Sci. **52**(6), 848–853 (2017). https://doi.org/10.1007/s11003-017-0030-x
14. V. Gurey, I. Hurey, Influence of surface hardened nanocrystalline layers on the resistance of contact fatigue destruction, in *Proceedings of the 3rd International Conference on Design, Simulation, Manufacturing: The Innovation Exchange*. Lecture Notes in Mechanical Engineering. Advances in Design, Simulation and Manufacturing III. DSMIE-2020 (June 9–12, 2020, Kharkiv, Ukraine) (2020), pp. 483–491. https://doi.org/10.1007.978-3-030-50794-7_47
15. W. Kraus, G. Nolze, Powder cell—a program for the representation and manipulation of crystal structures and calculation the resulting X-ray powder patterns. J. Appl. Cryst. **29**, 301–303 (1996). https://doi.org/10.1107/S0021889895014920
16. Powder Diffraction File Search Manual: Alphabetical Listing and Search Section of Frequently Encountered Phases. Inorganic 1974. JCPDS, Philadelphia, 839 p
17. V.G. Gavrilyuk, Decomposition of cementite in pearlitic steel due to plastic deformation. Mat. Sci. Eng. A **4**(3), 531–551 (2003). https://doi.org/10.1016/S0921-5093(02)00358-1
18. E.S. Gorkunov, S.V. Grachev, S.V. Smirnow et al., Effect of large deformations during drawing on the physicomechanical properties of patented steel wire. Phys. Met. Metallogr. **98**(5), 521–532 (2004). https://doi.org/10.1016/j.actamat.2015.06.025
19. B. Schuh, F. Mendez-Martin, B. Völker et al., Mechanical properties, microstructure and thermal stability of a nanocrystalline CoCrFeMnNi high-entropy alloy after severe plastic deformation. Acta Mater. **96**(1), 258–268 (2015). https://doi.org/10.1016/j.actamat.2015.06.025

Dielectric Anomaly and Space Charge Polarization in Single Crystals of $Na_{0.5}Bi_{0.5}TiO_3$ and $Na_{0.5}Bi_{0.5}TiO_3$–$BaTiO_3$

V. M. Sidak and M. P. Trubitsyn

Abstract The presence of nanometer size polar regions within nonpolar crystal matrix determines electrical properties of sodium bismuth titanate $Na_{0.5}Bi_{0.5}TiO_3$ and the solid solutions based on it. In the paper, slow dielectric relaxation was studied in the single crystals of $Na_{0.5}Bi_{0.5}TiO_3$ and $(1-x)$ $Na_{0.5}Bi_{0.5}TiO_3$–$x$$BaTiO_3$ ($x =$ 0.13) which were heat-treated in vacuum (1070 K, 2 h). The intense relaxation peak of permittivity was appeared in the temperature range 700–730 K for AC field frequencies below ~ 10 kHz. The permittivity peak vanished after subsequent heating up to ~ 800 K in air and could be restored by the next treatment in vacuum. Temperature and frequency behavior of ε was described by using Cole–Cole model. Kinetic equation was used to account thermally stimulated decay of the non-stable polarization. Great value of the dielectric anomaly in maximum (~ 10^4–10^5 at $f = 1$ kHz) more likely evidenced for space charge polarization mechanism related to the presence of polar nanoregions distributed in nonpolar structural matrix of $Na_{0.5}Bi_{0.5}TiO_3$ and the solid solution based on it. Oxygen vacancies V_O and electrons localized on titanium ions Ti^{3+} are considered as possible charged defects accumulating near the structural inhomogeneities.

1 Introduction

Relaxor ferroelectrics are widely used as functional materials in devices of micro- and nanoelectronics due to the extremely high response to external fields. Lead-containing relaxor such as lead zirconate-titanate Pb(Zr–Ti)O_3 (PZT) with mixing components in B-site of ABO_3 perovskite structure today is a basic material used

V. M. Sidak (✉)
Department of Biomedical Physics and Informatics, Dnipro State Medical University, Vernadsky str. 9, Dnipro 49044, Ukraine
e-mail: vasylsidak@gmail.com

M. P. Trubitsyn
Institute for Energy Efficient Technologies and Materials Sciences, Oles Honchar Dnipro National University, Gagarina ave. 72, Dnipro 49010, Ukraine

O. Fesenko and L. Yatsenko (eds.), *Nanostructured Surfaces, Nanocomposites and Nanomaterials, and Their Applications*, Springer Proceedings in Physics 296,
https://doi.org/10.1007/978-3-031-42704-6_26

for industrial production of electromechanical transducers, multilayer capacitors, etc. The promising materials often can be obtained by mixing the elements in A- or B-site in the lattice of perovskites which under the same external conditions belong to different structural phases. The most efficient compositions are usually found near morphotropic phase boundary (MPB) in the range of which changes of the ratio between mixed elements ($Zr_{0.52}$–$Ti_{0.48}$ for PZT) cause change of the structure. Neighboring to MPB structural phases are characterized by close values of free energy that makes the lattice unstable and extremely sensitive with respect to external impacts. Consequently, in the range of MPB relaxors show high generalized susceptibility including dielectric permittivity, piezoelectric coefficients, etc.

The relaxors demonstrate common features which are quite different from the properties observed around usual ferroelectric phase transitions. As it was pointed out for the first time in [1] in a broad temperature interval, relaxors show intense permittivity peak with maximal value at T_m and with specific frequency dispersion. With increase of AC field frequency, the value of ε in maximum decreases, while T_m shifts to high-temperature side. At the same time, broad $\varepsilon(T)$ maximum at T_m does not mark the transition from para- to ferroelectric phase and is not accompanied by appearance of long-range polar ordering as it is observed below points of usual ferroelectric phase transitions. In relaxors, polar ordering appears only in *polar nanometer-scaled regions* (PNR) which are distributed in nonpolar structural matrix of high-temperature phase and are practically independent from each other [2, 3]. PNR appear at the "Burns temperature" T_D, which is believed to be much higher than T_m. The polar phase ordering exists only in nanometer scale within each PNR, and since they do not interact, appearance of PNR does not induce long-range polar order observed at conventional ferroelectric phase transition. The directions of the dipole moments of PNR are distributed chaotically through the directions allowed by symmetry. Consequently, on passing through T_m from above macroscopically relaxors keep the symmetry of high-temperature phase and in the absence of external fields remain nonpolar. Nevertheless, PNR show high sensitivity and can be oriented by an external electric field. Dynamic response of PNR on AC field causes high dielectric permittivity in the broad interval around T_m. One should add that though the averaged value of polarization of randomly oriented PNR is equal to zero $\langle P \rangle = 0$, mean square of total PNR's polarization is nonzero $\langle P^2 \rangle \neq 0$. Thus, Burns temperature T_D can be detected by studying temperature behavior of the physical parameters, which are proportional to $\langle P^2 \rangle$, such as thermal expansion coefficients, refractive index n, and others [4–6]. For example, for one of the most systematically studied lead-based relaxor $PbMg_{1/3}Nb_{2/3}O_3$ (PMN), Burns temperature T_D was found about 350 K above the temperature of broad $\varepsilon(T)$ maximum $T_m \approx 270$ K ($f = 10^4$ Hz, [7]). Subsequently, starting of PNR formation at T_D was confirmed by appearance of diffuse peak in neutron scattering experiment [8, 9]. Slowing down of PNR re-orientational dynamics was observed by dielectric spectroscopy in [3]. It was shown that PNR's typical size in PMN grows from a few of nm at T_D to dozens of nm on further cooling.

2 Lead-Free Relaxor Sodium Bismuth Titanate $Na_{0.5}Bi_{0.5}TiO_3$

Up to now, lead-based relaxors are used widely and consequently have been studied in more details [2, 3, 5, 10, 11]. Lead-free relaxors with perovskites ABO_3 structure were studied less intensively. However, current ecological requirements stimulate efforts aimed to design alternative friendly for environment materials, and during the last decade, interest to lead-free relaxors has grown significantly [4, 12]. Currently, among the most promising candidates for applications in piezotechnics there are alkali metal niobates (K, Na)NbO_3 (KNN), the solid solutions based on barium titanate $BaTiO_3$ (BT) and alkaline bismuth titanates, such as $Na_{0.5}Bi_{0.5}TiO_3$ (NBT) and $K_{0.5}Bi_{0.5}TiO_3$ (KBT). One of the most popular objects of study is lead-free relaxor sodium bismuth titanate NBT [13–21]. NBT and especially solid solutions $Na_{0.5}Bi_{0.5}TiO_3$–$BaTiO_3$ (NBT–BT) are expected to substitute PZT in the industrial production of piezoelectric materials [12].

On cooling, NBT undergoes the following sequence of phase transitions [15]. Above ~ 813 K, the structure belongs to cubic symmetry (space group $Pm\bar{3}m$) (Fig. 1). On further cooling, the transition to the tetragonal phase (group *P4bm*) occurs, and the coexistence of the cubic and the tetragonal phases is observed in the interval of about 40 K (773–813 K). The tetragonal phase is stable in the range 673–773 K. At lower temperatures, the regions with the rhombohedral symmetry (group *R3c*) appear, grow in volume, and coexist with the tetragonal structure in a broad interval of about 145 K (528–673 K). The absence of the intermediate orthorhombic phase typical to many perovskites allows to classify the transition from the tetragonal to the rhombohedral phase as the diffuse one. On further temperature decreasing below T_d (which was estimated as ~ 528 in [15] and ~ 473 K in [16]), the hump of $\varepsilon(T)$ dependence with pronounced frequency dispersion was associated with the transition to rhombohedral ferroelectric phase (*R3c*) which remains stable up to 5 K [15]. Nevertheless, the data of various experimental techniques evidenced that the structural changes between T_d and T_m could be more complex as discussed in [15]. A double hysteresis loop of polarization observed in this range evidenced for appearance of antiferroelectric type ordering (see [4] and the refs therein). As it was argued in [16], in the range and above T_m the structure of NBT is nonpolar. Detailed study of transmission electron microscopy [16, 17] allowed to propose the model in which the rhombohedral to tetragonal phase transition occurred in two steps through intermediate modulated phase which is formed from approximately 500 K and existed up to T_m. The modulated phase consists of the orthorhombic sheets (group *Pnma*) distributed within the rhombohedral phase matrix which in the range of T_m immediately transforms into the tetragonal one. The orthorhombic to tetragonal phase transition was determined as antiferroelectric-to-paraelectric structural transformation. The symmetry group of long-range ordering within the tetragonal phase was refined as nonpolar (*P4/mbm*).

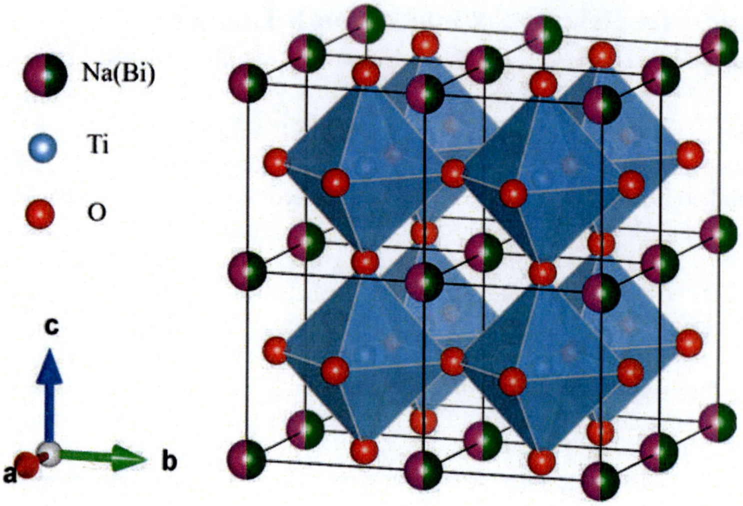

Fig. 1 Crystal structure of NBT in cubic phase [15]

In accord with the data discussed above, the broad maximum of permittivity in NBT is observed at $T_m \sim 593$ K [22, 23], i.e., just within the interval (528–673 K) where the rhombohedral and the tetragonal phases coexist [15]. Thus, the $\varepsilon(T)$ broad maximum can be associated with response of the locally ordered rhombohedral PNR distributed in the tetragonal nonpolar matrix.

In sum, as mentioned in [4], NBT shows the following features typical for relaxors: existence of broad $\varepsilon(T)$ peak at T_m; the absence of structural phase transition on passing through T_m; the presence of polar nanoregions immersed into nonpolar matrix in a wide temperature range around T_m. On the other hand, there are some features peculiar for NBT, such as probable nonpolar tetragonal symmetry and proposed antiferroelectric ordering of PNR structure. It is also necessary to note that in ceramics structural changes with temperature can be significantly modified by polarizing the sample in an external electric field [20, 24].

It seems natural to connect relaxor behavior and so complex sequence of structural transformations in NBT with distribution of aliovalent Na^+ an Bi^{3+} ions in the A-site of perovskite $A'A''BO_3$ structure. Up to now, there is no reliable experimental evidence on long-range chemical ordering of Na^+ and Bi^{3+} ions in A-site of NBT lattice. Nevertheless, some data allow to suppose the presence of short-range chemical order in nanometer size regions which can act as nuclei for formation of PNR [17].

In our opinion, the way to understand complex character of structural transformations in NBT was proposed by the authors of [25, 26]. They studied thermodynamic stability of different configurations of Na^+ and Bi^{3+} ions distributions in A-site by ab initio calculations based on density functional theory (DFT). The attention was paid to Na^+/Bi^{3+} ions locations in cubic phase, since at high temperatures the cations can be mobile enough to re-arrange their location and stabilize certain chemical order. For lower temperatures, the formed configurations can be considered as frozen in. The total energies of the different Bi/Na configurations were calculated for 2 × 2 × 2 cubic perovskite supercells by considering ideal perovskite structure and assuming ability of the lattice to relax in order to compensate local charge imbalances and local stresses. The structure relaxation was associated mainly with oxygen ions displacements which stabilized certain configurations of neighboring Na^+/Bi^{3+} ions. The calculations performed gave very interesting results. The assumed structural relaxation made more stable the configurations which were most unstable for ideal non-relaxed perovskite structure. And vice versa, the Na^+/Bi^{3+} configurations, like rock salt, assumed as favorable for ionic interaction in ideal perovskite lattice, could not relax due to the symmetry of the cation location and became most unfavorable as compared with the relaxed configurations. The performed calculations predicted the most stable configuration with high concentrations of Bi^{3+} and Na^+ ions in alternating {001}-planes.

Nevertheless, the total energies of a few most stable Na^+/Bi^{3+} configurations were close enough each other, and the difference between them in the interval of cubic phase was smaller as compared to the thermal energy. Thus, chemical ordering can appear only in short-range scale and a few Na^+/Bi^{3+} configurations can form with nearly equal probability. Consequently, the local regions with different chemical order can act as nuclei for PNR, which cause relaxor behavior of NBT [25].

Influence of pressure on stability of the chemically ordered nanoregions with different structures was studied in NBT by ab initio DFT calculations in [26]. The calculations showed that compressive pressure stabilized the regions with orthorhombic (*Pbnm*-like) structure, while tension pressure made stable the regions with tetragonal (*P4mm*-like) symmetry. The obtained results clearly show existence of chemically ordered PNR with their own local properties embedded in a chemically disordered matrix. The stability of the mixed phase was discussed in terms of the coexisting regions with different average structures.

The structural disorder becomes much more complex in $(1 - x)$NBT – xBT solid solutions which in the range of MPB have enhanced piezoelectric parameters as compared with unmixed NBT and BT: piezoelectric constant d_{33} ~ 155 pC/N and coupling factor k_p ~ 0.47 [4, 12]. In accord with the available data [4, 27], in $(1 - x)$NBT–xBT solutions at room temperature MPB corresponds to the mixed components ratio $x = 0.05$–0.09 and separates the rhombohedral ferroelectric phase (*R*3*c*) of NBT and the tetragonal ferroelectric phase (*P*4*mm*) of BT crystals correspondingly. There are a few controversial views on the structural inhomogeneity in the range of MPB at ~ 300 K. TEM study in [28] shows that rhombohedral nanodomains with complex morphology are observed for the solid solutions with $x < 0.04$. For the crystals with $x = 0.07$–0.09 composition, the nanodomains have tetragonal symmetry and for $x > 0.11$ the ferroelectric tetragonal phase (*P*4*mm*) of pure BT starts to stabilize. The relaxor state for $x > 0.04$ within nanodomains demonstrates rather antiferroelectric local order due to antiparallel displacements of the cations in A-site (Na^+, Bi^{3+}, Ba^{2+}) and B-site (Ti^{4+}). In such a case, relaxor state in the range of MPB can be attributed to competition between long-range antiferroelectric and ferroelectric ordering peculiar to the structures of mixed NBT and BT. Near MPB, such concurrence leads to high instability of the lattice which becomes strongly inhomogeneous in nanometer size scale. Consequently, in the range of MPB relaxor-type behavior can be observed at room temperature and even below.

The results of the experimental studies and calculations, mentioned above, were discussed in assumption that NBT and NBT–BT crystals are free from structural defects such as cations and oxygen vacancies, foreign atoms and others. Nevertheless, it is well known that structural defects strongly influence physical properties of the crystals and especially of the systems undergoing structural phase transitions [29].

Usually, structural defects weaken the response of crystalline substances on external fields. In such a case, the presence of defects worsens the performance characteristics of functional materials based on crystalline compounds and is considered as an undesirable factor. Nevertheless, lately much attention is being paid to the structures where lattice defects act as active centers which improve the beneficial properties of solids, or even induce qualitatively new phenomena. Electrically and optically active defects can significantly enhance electronic and ionic conductivity, affect dielectric susceptibility, optical transmission, and absorption and give rise to luminescence [30]. In certain cases, the presence of defects becomes a prerequisite for obtaining the necessary set of physical properties. That is why comprehensive information on the nature of intrinsic and extrinsic defects, understanding the mechanisms of their influence on physical properties become the necessary requirements for design of new functional materials with targeted characteristics. Based on such knowledges, one can propose the technological approaches which make it possible to stimulate appearance of certain structural imperfections or vice versa to suppress their arising.

Generally, in crystalline solids one can distinguish the following types of primary defects such as captured electrons and holes, host atoms vacancies and their interstitials, as well as substitutional or interstitial impurity atoms [31, 32]. Primary point defects can form Schottky or Frenkel pairs which most probably can be found in close-packed or framework structures correspondingly. Besides, primary defects can interact with each other and create more complex associated groups. It is important to develop the technological approaches which allow to control the concentration of primary defects as well associated complexes. Using the initial chemicals in a ratio slightly deviated from stoichiometry, doping with aliovalent additives and heat treating in different atmospheres are known as the effective ways to control their content [33].

One can expect that structural defects can modify short-range chemical ordering in NBT and NBT-BT crystals. Consequently, defects should affect strongly static and dynamic features of the polar nanoregions and influence dielectric response and piezoelectric properties. Usage of NBT and NBT-based solid solutions in piezoelectric transducers, sensors, and actuators [11, 12] requires detailed study of typical defects, understanding their nature and developing the methods to control and to monitor their content.

Recently, slow dielectric relaxation with specific frequency dispersion was observed around 670 K in NBT single crystal and ceramics [34–37]. It was proposed that anomalous relaxation of ε was determined by the associated defect complexes based on oxygen vacancies (V_O). This work is devoted to comparative studying electrical properties of single crystalline NBT and $(1-x)$NBT–xBT solid solution which composition $x = 0.13$ was chosen nearly the range of the morphotropic phase boundary [4, 27].

3 Experimental Procedures

The single crystals of NBT and solid solution $(1-x)$NBT–xBT ($x = 0.13$, hereafter will be mentioned as NBT–xBT) were grown from the melt by Czochralskii method. The samples were prepared as the plane-parallel plates with linear sizes of about $5 \times 5 \times 0.8$ mm^3. The main planes of the samples were oriented perpendicular to [111] direction and were covered with Pt electrodes deposited by cathode sputtering method. Electrical properties were measured with the help of AC bridge P 5083 in the frequency range 0.5–100 kHz and in the temperature interval 300–800 K. To remove probable effect of the thermal and applied fields prehistory, before measuring the samples were heat-treated in air at $T \sim 1070$ K for 1 h. Then the samples were cooled to room temperature, and electrical properties were measured on heating and on subsequent cooling run. Further the samples were heat-treated in vacuum ($T \sim 1070$ K, 2 h) under a pressure of about 1 Pa. After that the measurements were carried in the same way starting from room temperature. The heating and cooling of the samples were carried out with the rate 2 K/min.

4 Results

Figure 2 shows the temperature dependencies of real ε' and imaginary ε'' parts of dielectric permittivity measured on heating of NBT crystal. The hump of $\varepsilon'(T)$ with notable frequency dispersion is clearly observed near ~ 480 K. As it was mentioned above, this $\varepsilon'(T)$ anomaly can be associated with the transition from ferroelectric to relaxor state [4]. The broad and intense maximum of $\varepsilon'(T)$ is weakly dependent on AC field frequency f and is detected at $T_m{}^{heat} \sim 610$ K. The broad peak of $\varepsilon'(T)$ demonstrates temperature hysteresis shown in the inset to Fig. 2a. On cooling, the permittivity maximum slightly shifts to lower temperatures $T_m{}^{cool} \sim 600$ K, and below this point, ε' falls down more slowly as compared with the curve measured on heating. The $\varepsilon''(T)$ dependences (Fig. 2b) show the weakly pronounced anomalies corresponding to the $\varepsilon'(T)$ hump and the broad peak. On heating above ~ 650 K, ε'' starts to increase rapidly. This temperature growth is most pronounced for low values of AC field frequency and rapidly drops down as f increases. Obviously that the high temperature increase of $\varepsilon''(T)$ can be associated with motion of weakly bound charged defects contributed to dielectric losses. One can state that the dependences,

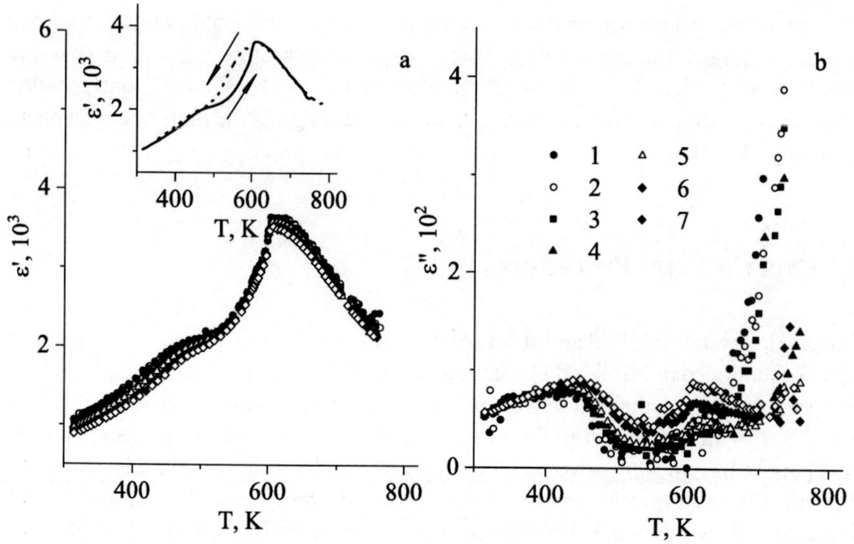

Fig. 2 Dependences of real $\varepsilon'(T)$ (**a**) and imaginary $\varepsilon''(T)$ (**b**) part of permittivity measured on the first heating run of the NBT single crystal. The AC field frequencies were $f = 0.5$ (1); 0.8 (2); 1 (3); 5 (4); 10 (5); 50 (6); 100 (7) kHz. The inset shows permittivity dependencies ε'(T) measured at $f = 1$ kHz on heating run (solid curve) and subsequent cooling run (dashed curve)

shown in Fig. 2a, b, are in a good agreement with the data published elsewhere [4, 22, 38, 39].

Then NBT sample was heat-treated in vacuum ($T \sim 1070$ K, 2 h), and permittivity was measured on heating. Figure 3 shows the data obtained. One can see that the intense nearly symmetrical maximum of ε' appears in the range 700–720 K (Fig. 3a). With increasing frequency *f*, permittivity peak sharply decreases in magnitude and shifts to higher temperatures. The maximum of $\varepsilon'(T)$ is clearly observable at low frequencies and for $f > 10$ kHz becomes practically undistinguishable. The less intense hump and the diffuse peak of $\varepsilon'(T)$ visible for higher frequencies coincide with the dependence shown in Fig. 2a, i.e., reflect the structural changes in NBT. Similar to the results presented in [35–37], the low-frequency $\varepsilon'(T)$ peak could be observed only for the first heating run and disappeared for the next cooling. $\varepsilon'(T)$ anomalies, such as the hump at ~ 480 K and the broad peak at T_m, associated with the phase transitions and relaxor state, practically did not change during subsequent temperature cycling and were the same as shown in Fig. 2a. Figure 3b shows the $\varepsilon''(T)$ peaks which correspond to the maximums of real part of permittivity (Fig. 3a). With increasing AC field frequency, the $\varepsilon''(T)$ peaks sharply decrease in magnitude and disappear for the next cooling and heating runs.

The same measuring procedure as described above for NBT was carried out for NBT–*x*BT ($x = 0.13$) sample. After heat-treating of NBT–*x*BT sample in air, temperature-frequency dependences of permittivity were measured on heating. The obtained results are shown in Fig. 4a, b. One should note that in the studied crystal,

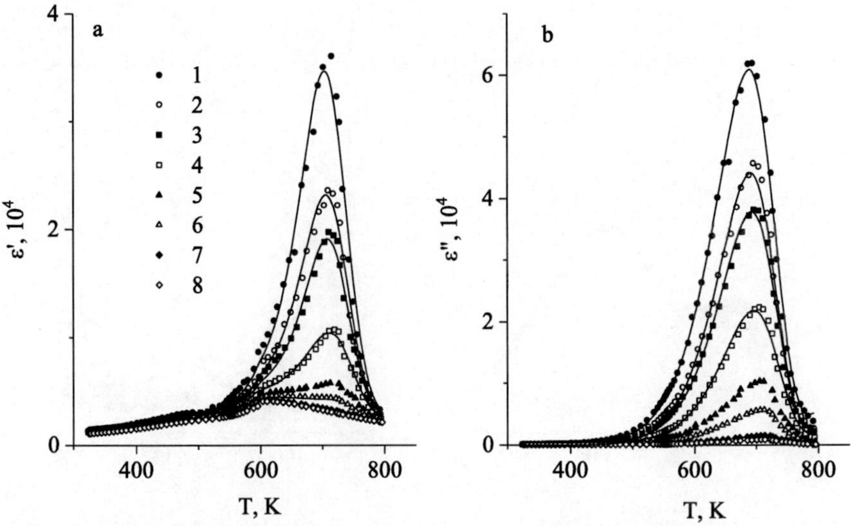

Fig. 3 $\varepsilon'(T)$ (**a**) and $\varepsilon''(T)$ (**b**) dependences measured on the first heating run of the NBT single crystal heat-treated in vacuum (1070 K, 2 h). The AC field frequencies were $f = 0.5$ (1); 0.8 (2); 1 (3); 2 (4); 5 (5); 10 (6); 50 (7); 100 (8) kHz

the concentration of BT component ($x = 0.13$) lies slightly outside of the MPB interval ($0.05 < x < 0.09$) usually mentioned in literature [4, 27] and the structure should correspond rather to tetragonal ferroelectric phase of BT. Nevertheless, one can see that below ~ 470 K $\varepsilon'(T)$ decreases quite smoothly and demonstrates notable frequency dispersion (Fig. 4a). In accord with the data published in [27], the compositions of NBT–BT outside the MPB region show frequency dispersion above T_d and abrupt $\varepsilon'(T)$ decreasing at the transitions to the rhombohedral ferroelectric phase of pure NBT ($x < 0.05$) or to the tetragonal ferroelectric phase of pure BT ($x > 0.09$). Therefore, the behavior of $\varepsilon'(T,f)$ in Fig. 4a testifies that the NBT–xBT composition studied in this work highly likely corresponds to the region of MPB. The discrepancy between the value of BT concentration ($x = 0.13$) and the interval of MPB ($0.05 < x < 0.09$) may be due to the lower concentration of the BT additive in the single crystal grown from the melt, as compared to its content in the charge used. Thus, behavior of ε', $\varepsilon''(T,f)$ looks like dependences for NBT–xBT compositions in the range of MPB. The inset to Fig. 4a shows typical hysteresis observed for temperature dependences of $\varepsilon'(T)$ measured in heating and cooling regimes.

The permittivity dependences of NBT–xBT sample after heat treatment in vacuum (1070 K, 2 h) are shown in Fig. 5. As for pure NBT crystal (Fig. 3), annealing in vacuum induces appearance of the intense peaks of $\varepsilon'(T)$ and $\varepsilon''(T)$. Again, with increasing frequency f these peaks sharply decrease in magnitude, slightly shift to high temperatures and disappeare for the subsequent cooling and the next temperature runs. Nevertheless, there is the noticeable difference between permittivity anomalies appeared in NBT (Fig. 3) and NBT–xBT (Fig. 5). Under the same regimes of heat

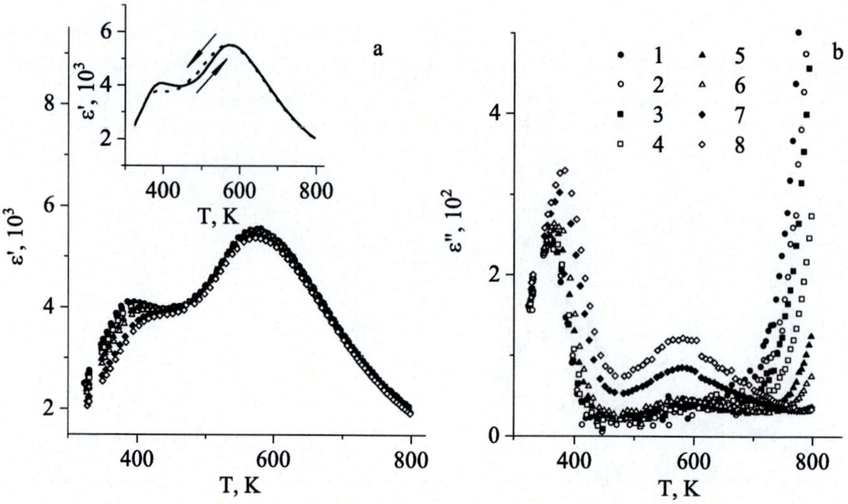

Fig. 4 $\varepsilon'(T)$ (**a**) and $\varepsilon''(T)$ (**b**) dependences measured on the first heating run of NBT-xBT ($x = 0.13$) single crystal. The AC field frequencies f are the same as indicated in caption to Fig. 3. The inset shows $\varepsilon'(T)$ dependencies measured at $f = 1$ kHz on heating (solid curve) and cooling run (dashed curve)

treating in vacuum, the $\varepsilon'(T)$ and $\varepsilon''(T)$ peaks appeared to be about one order of magnitude higher in solid solution NBT–xBT as compared to pure NBT.

5 Discussion

As mentioned in Sects. 2, and 4 after previous annealing in air (1070 K, 1 h), permittivity behavior in NBT and NBT–xBT crystals shows frequency-dependent hump in the range of T_d and broad peak at T_m which reflect structural transformations and response of polar nanoregions (Figs. 2 and 4). Annealing in vacuum (1070 K, 2 h) causes appearance of the $\varepsilon'(T)$ and $\varepsilon''(T)$ peaks for NBT as well as for NBT–xBT (Figs. 3 and 5). The amplitude of the permittivity peaks decreases sharply with increasing AC field frequency f and becomes almost indistinguishable for $f >$ 10 kHz. The observed slow dielectric relaxation can be detected on the first heating run and disappeare for the next cooling run. In fact, relaxation peeks of ε can be restored afterward by repeated heat treatment in vacuum. Thus, observed permittivity anomaly is caused by polarization which is unstable and decays on heating up to ~ 800 K.

Frequency-dependent dielectric permittivity determined by thermal polarization mechanisms can be described by Debye model or its modifications usually applied for broaden or asymmetrical dielectric spectra of real structures [40]. The data shown in Figs. 3 and 5 were measured in the temperature–frequency range which did not

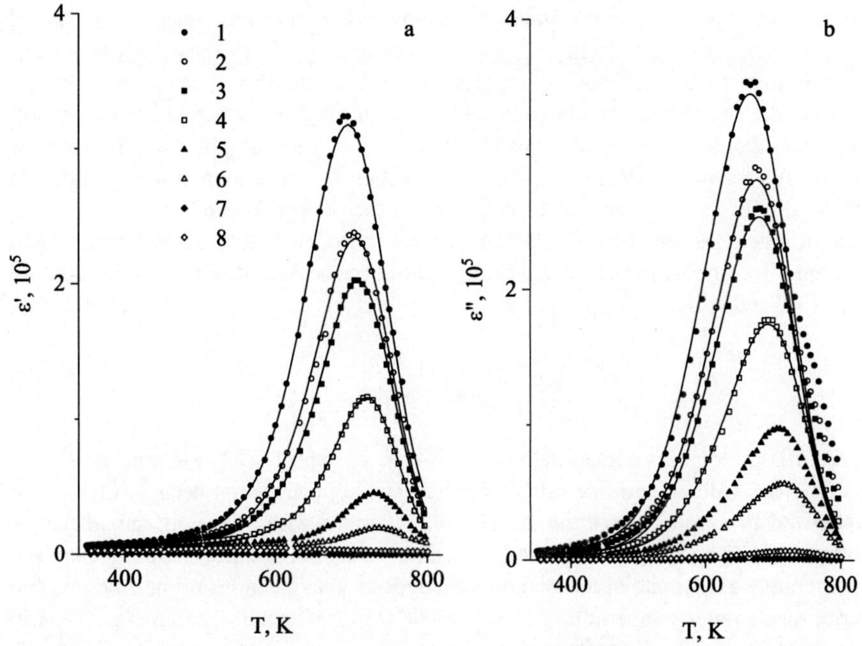

Fig. 5 $\varepsilon'(T)$ (**a**) and $\varepsilon''(T)$ (**b**) dependences measured on the first heating run of NBT-xBT (x = 0.13) single crystal heat-treated in vacuum (1070 K, 2 h). The AC field frequencies f are the same as in Fig. 3

allow to determine confidently the form of the diagrams plotted in the complex ($\varepsilon' - \varepsilon''$) plane. Because that to describe the frequency-dependent ε anomalies in Figs. 3 and 5, we used Cole–Cole model, suitable for partially disordered crystalline systems

$$\varepsilon^*(T, \omega) = \varepsilon_\infty + \frac{C/T}{1 + (i\omega\tau_R)^{1-\alpha}}. \tag{1}$$

Here ε_∞ is permittivity at high frequency; parameter $C \sim n$ is proportional to the dipoles concentration n; $\omega = 2\pi f$ is cyclic frequency of an external field; $\tau_R(T) = \tau_R^0 \exp(E/kT)$ is the time of dipole moments relaxation in an external field; E estimates an energy barrier which re-orienting dipoles overcome; k is Boltzmann constant; parameter $0 \leq \alpha < 1$ phenomenologically describes the dielectric spectra smearing in disordered systems.

The dielectric $\varepsilon(T)$ anomaly similar to the peaks shown in Figs. 3 and 4 earlier was observed in NBT single crystals and ceramics [35–37]. It was emphasized that nearly symmetrical peak of $\varepsilon'(T)$ and its sharp decrease with frequency f could not be described by Debye relaxator or Cole–Cole model (1) which predicted asymmetrical step-like $\varepsilon'(T)$ behavior. Disappearance of the $\varepsilon(T)$ peak after first heating run was interpreted as the result of thermally activated decomposition of the dipole complexes responsible for the dielectric relaxation. Decrease of the dipoles concentration became notable in the same range where dielectric response displayed frequency dispersion, and hence, high-temperature wing of $\varepsilon'(T)$ peak fell down more sharply than it was predicted by expression (1). The attempts to describe the experimental data by exponential temperature decay of the dipoles number or by considering the dipoles configurational and vibrational entropy do not allow to describe the results accurately. Moreover, such approaches do not account non-ergodic state of the dipole's subsystem and cannot explain the dependence of the data on time. For simplicity, thermal decay of non-equilibrium contribution to polarization and assumed decomposition of the polarizing entities can be described by kinetic equation (see [37] and refs therein)

$$\frac{\mathrm{d}n}{\mathrm{d}t} = -\frac{n}{\tau_D}. \tag{2}$$

Here n is the dipoles concentration; $\tau_D(T) = \tau_D^0 \exp(U/kT)$, $\tau_D{}^0$ and U are the parameters which determine rate of the non-stable polarization decay. Since in the experiments temperature of the samples was changed with a constant rate, differentiation in time in (2) can be replaced by derivative in temperature. Denoting an initial temperature as T_0, rate of temperature changes as γ and a current time as t, one can write the samples temperature as $T(t) = T_0 + \gamma \cdot t$. Since the parameter $C \sim n$ in (1), from (2) it can be rewritten as

$$C(T) = C_0 \cdot \exp\left(-\frac{1}{\gamma\tau_D^0} \cdot \int\limits_{T_0}^{T} \exp\left(-\frac{U}{kT}\right)\mathrm{d}T\right). \tag{3}$$

Fitting of Cole–Cole formulae (1) together with expression (3) to the experimental results can be performed by using approximate integration of (3) [37] or by numerical calculations. In this work, we used numerical calculations to integrate (3) and to describe the anomalies in Figs. 3 and 5 with the help of Cole–Cole formulae (1) combined with kinetic equation (2). In addition to the intense relaxation anomalies, the $\varepsilon(T)$ dependence in the studied interval (Figs. 3 and 5) includes other contributions caused by the structural phase transition, relaxor state and mobile charged defects. These additional contributions to permittivity were described in the same way as it was carried out in [35–37].

The calculated curves are plotted in Figs. 3 and 5 by the solid lines, and the parameters in (1, 3), obtained from the calculations, are gathered in Table 1. One should add that spreading of the U values in accord with Gaussian distribution was also accounted in expression (3). From Figs. 3 and 5, one can see that the used model allows to describe the experimental data with a good accuracy.

Table 1 Values of the parameters in (1, 3) obtained from the permittivity dependences shown in Figs. 3 and 5

The crystal	C_0, K	α	E, eV	$\tau_R{}^0$, s	U, eV	$\tau_D{}^0$, s
NBT	$2.8(5) \times 10^8$	0.13(1)	1.25(3)	$6(4) \times 10^{-13}$	0.67(2)	$1.8(1) \times 10^{-2}$
NBT–xBT (x = 0.13)	$1.1(1) \times 10^9$	0.29(6)	1.24(2)	$5(1) \times 10^{-13}$	0.64(1)	$6.5(5) \times 10^{-2}$

Besides, because the relaxing polarizing entities are non-equilibrium (2), one can expect that electrical properties should be dependent on heating rate γ. In fact, such dependences can be detected experimentally in the interval of the intense low-frequency dielectric peak. Figure 6 shows how for the fixed f the behavior of $\varepsilon'(T)$ and $\varepsilon''(T)$ transforms with the change of γ values. If the heating rate γ is high enough ($\gamma \to \infty$), decay of non-equilibrium polarization during the experiment remains negligible and the dependence $\varepsilon'(T)$ tends to the asymmetric step-like behavior predicted by simple Debye model (Fig. 6a). Nearly symmetrical form of $\varepsilon'(T)$ anomaly can be observed for the intermediate γ values. With further lowering γ, the magnitude of $\varepsilon'(T)$ peak decreases. For the limit $\gamma \to 0$, the permittivity peak vanishes since polarizing entities decay totally at lower temperatures as compared with the interval where permittivity anomaly can be detected for higher values of γ. Figure 6b shows that the variations of ε'' (T) behavior with change of γ is much less pronounced as compared with $\varepsilon'(T)$. Lowering γ remains peak-like shape of $\varepsilon''(T)$ anomaly and only causes more sharp decrease of its magnitude and high-temperature wing. Besides, the similar transformations of $\varepsilon'(T)$ and $\varepsilon''(T)$ behavior as shown in Fig. 6 can be expected for variations of the ratio between the rates of dielectric relaxation $\tau_R{}^{-1}$ and polarization decay $\tau_D{}^{-1}$ or between energy parameters U/E [37]. Debye-like behavior of permittivity can be predicted for high enough ratio U/E. Lowering the value U/E decreases the magnitudes and changes the form of ε' and ε'' anomalies and finally results in their disappearance. It is obvious that specific features of $\varepsilon(T, f)$ anomaly are observed because polarization temperature decay and relaxation in an external electric field are detected in the same range of T.

It is of interest to explore how non-equilibrium state of polarization modifies the dielectric spectra plotted in complex ($\varepsilon'' - \varepsilon'$) plane. Figure 7 compares ($\varepsilon'' - \varepsilon'$) diagrams calculated for "pure" Cole–Cole model (1) (Fig. 7a) and for the model accounted thermal decay of polarization (2, 3) (Fig. 7b). One can see that the near semicircle shape of the diagrams remains unchanged. It seems natural, because kinetic equation (2), which describes polarization temperature decay, does not include f and hence does not change the frequency dispersion of permittivity. Temperature decay of polarization causes only more sharp decrease of the magnitude of ε anomaly on heating. The magnitude of the anomalies and diameter of the semicircles in ($\varepsilon'' - \varepsilon'$) plane for "pure" Debye model decrease on heating as $(\varepsilon'(0) - \varepsilon'(\infty)) \sim 1/T$ (Fig. 7a), whereas for non-stable polarization they decrease with temperature much more sharply (Fig. 7b).

In earlier works [34–36], it was assumed that that intense permittivity peak was determined by slow relaxing polar defects which based on oxygen vacancies (V_O)

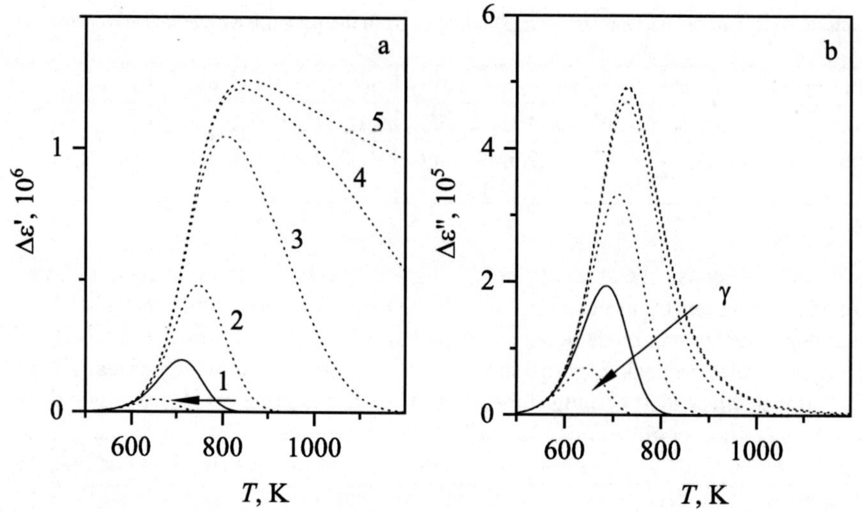

Fig. 6 Calculated contributions from non-stable polarization to dielectric anomaly: real $\Delta\varepsilon'$ (**a**) and imaginary $\Delta\varepsilon''$ (**b**) part. The heating rates are $\gamma = 0.5$ (1); 5 (2); 50 (3); 5×10^2 (4); 10^6 (5) K/min. The dashed curves are calculated for $f = 1$ kHz with the parameters given in Table 1. The solid lines correspond to the experimental data shown in Fig. 4 ($\gamma = 1.7$ K/min)

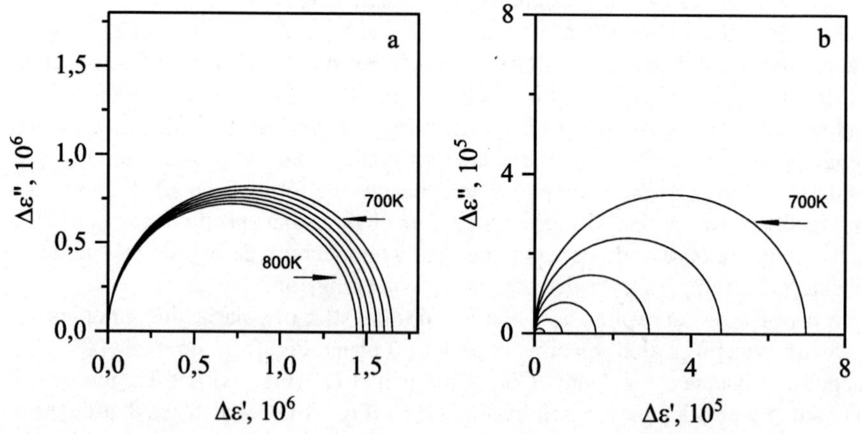

Fig. 7 Cole–Cole plot for Debye relaxator (**a**) and for the model that considers decay of non-equilibrium polarization. The temperatures are $T = 700$; 720; 740; 760; 780; 800 K

and thermally destroyed on heating. Really, the numerous studies show that in the perovskite-type structures Ti ions can trap weakly bound electrons and form thermally or photoinduced Ti^{3+} centers [41, 42]. Electrons captured by Ti within undistorted oxygen octahedra can hop through regular titanium sites and contribute to electrical conductivity [18, 43]. Another situation can be expected if Ti^{3+} center has an oxygen vacancy in the nearest coordination sphere. The authors of [44, 45] show that captured by titanium electron and oxygen vacancy can bind together into associated complex (Ti^{3+}–V_O) stabilized presumably by neighboring impurity in A- or B-sites of ABO_3 structure. Such centers possess electric dipole moment which can re-orient and contribute to permittivity. On heating, thermal excitation can cause (Ti^{3+}–V_O) centers disassociation.

Nevertheless, the extremely high values of permittivity in peak (~10^4–10^5 at f ~ 1 kHz, Figs. 3 and 5) cannot be associated with the contribution of dipole defects with not too high concentration. Such giant values of permittivity can be found in spatially heterogeneous media which display space charge polarization phenomena. Thus, more probably intense permittivity anomalies (Figs. 3 and 5) can be associated with space charge polarization effects. In accord with [12], at the temperature (~ 1070 K) used for heat treatment, the sublattice of Bi^{3+} in NBT structure remains stable. Thus, appearance of additional V_O can be expected after heat treatment in vacuum. In that case, appearance of the following typical defects can be expected: (i) oxygen vacancies V_O; (ii) electrons localized on titanium Ti^{3+} and (iii) associated complexes based on both of them. The data in [46] showed that charge transfer in NBT at moderate and high temperatures was mainly determined by motion of polarons and oxygen vacancies. Comparison of the data obtained in [37] and in this work (Fig. 3) shows that for single crystalline and ceramic NBT, the discussed dielectric anomaly has nearly the same magnitude. Thus, intergrain boundaries in ceramics do not play an essential role and mobile charge defects can accumulate near the structural inhomogeneities or in near-electrode regions. One can note that positions of intense permittivity peaks in Figs. 3 and 5 correspond to the Burns temperatures T_D, estimated from thermal expansion parameters in [38]. Thus, the observed anomalies presumably can be connected with accumulation of the charged defects in the inhomogeneous crystal matrix containing polar nanometer size regions. Comparison of the data presented in Figs. 3 and 5 shows that the relaxation anomaly of ε is about one order of magnitude higher in NBT-xBT solid solutions as compared with NBT. This effect can be related to more disordered structure of NBT-xBT solution due to mixing of three aliovalent ions (Na^+, Bi^{3+}, Ba^{2+}) in A-site of perovskite ABO_3 structure. Consequently, after the same regime of heat treating in vacuum the structure of NBT–xBT solid solutions is able to include much greater number of V_O as compared with less distorted lattice of NBT crystal.

6 Conclusions

After heat treatment in vacuum (1070 K, $t = 2$ h), the intense dielectric anomaly was observed for NBT and NBT–xBT ($x = 0.13$) single crystals in the temperature range 690–730 K ($f = 0.5$–10 kHz). In NBT–xBT solid solution, the magnitude of anomaly was found about one order of magnitude higher as compared with NBT crystal. The observed behavior of real $\varepsilon'(T)$ and imaginary $\varepsilon''(T)$ parts of permittivity was quite different from that predicted by the model of Debye relaxator and its

Cole–Cole variant. The experimental data were described in assumption that the two processes were observed simultaneously: dielectric relaxation in an external electric field and thermally induced decay of polarization. The unusual $\varepsilon(T, \omega)$ behavior was described by using Cole–Cole model together with kinetic equation accounting decay of non-equilibrium polarization. The proposed approach made it possible to predict dependences of permittivity on experimental regimes and relaxation times of the phenomena observed. It was shown that kinetics of non-equilibrium polarization did not change frequency dispersion of permittivity but induced more sharp decrease of dielectric anomaly magnitude with temperature increasing.

Accounting the great value of permittivity in maximum ($\varepsilon_{max} \sim 10^4$–$10^5$, $f =$ 1 kHz) for NBT and NBT–xBT crystals, the observed dielectric relaxation can be associated with space charge polarization mechanism rather than with dipole defects response. Oxygen vacancies V_O and electrons localized on titanium ions Ti^{3+} were considered as mobile charge defects giving rise to the intense dielectric anomaly. Relaxor state of the structure and existence of polar nanoscale regions imbedded into nonpolar crystal matrix can be responsible for the charged defects accumulation near the structural inhomogeneities.

Acknowledgements The authors thank Dr. T. Kruzina for the samples of sodium bismuth titanate $Na_{0.5}Bi_{0.5}TiO_3$ and (1–x)$Na_{0.5}Bi_{0.5}TiO_3$–$x$$BaTiO_3$ crystal.

Funding The study was funded by Ministry of Education and Science of Ukraine according to the research projects No. 0120U102239 and No. 0122U001228.

References

1. G.A. Smolenskii, V.A. Isupov, A.I. Agranovskaya, S.N. Popov, Ferroelectrics with diffuse phase transitions. Sov. Phys. Solid State **2**, 2584–2594 (1961)
2. L.E. Cross, Relaxor ferroelectrics. Ferroelectrics **76**, 241–267 (1987). https://doi.org/10.1080/00150198708016945
3. A.A. Bokov, Z.-G. Ye, Recent progress in relaxor ferroelectrics with Perovskite structure. J. Mater. Sci. **41**, 31–52 (2006). https://doi.org/10.1007/s10853-005-5915-7
4. V.V. Shvartsman, Lead-free relaxor ferroelectrics. J. Am. Ceram. Soc. **95**(1), 1–26 (2012). https://doi.org/10.1111/j.1551-2916.2011.04952.x
5. G.A. Samara, The relaxational properties of compositionally disordered ABO_3 Perovskites. J. Phys.: Condens Matter. **15**, R367-411 (2003). https://doi.org/10.1088/0953-8984/15/9/202
6. V. Westphal, W. Kleemann, M. Glinchuk, Diffuse phase transitions and random-field-induced domain states of the "relaxor" ferroelectric $PbMg_{1/3}Nb_{2/3}O_3$. Phys. Rev. Lett. **68**, 847–850 (1992). https://doi.org/10.1103/PhysRevLett.68.847
7. G. Burns, F.H. Dacol, Glassy polarization behavior in ferroelectric compounds $PbMg_{1/3}Nb_{2/3}O_3$ and $Pb(Zn_{1/3}Nb_{2/3})O_3$. Solid State Commun. **48**, 853–856 (1983). https://doi.org/10.1016/0038-1098(83)90132-1
8. S.B. Vakhrushev, B.E. Kvyatkovsky, A.A. Naberezhnov, N.M. Okuneva, B.P. Toperverg, Glassy phenomena in disordered Perovskite-like single crystals. Ferroelectrics **90**, 173–176 (1989). https://doi.org/10.1080/00150198908211287
9. A. Naberezhnov, S. Vakhrushev, B. Dorner, D. Strauch, H. Moudden, Inelastic neutron scattering study of the relaxor ferroelectric $PbMg_{1/3}Nb_{2/3}O_3$ at high temperatures. Eur. Phys. J. B. **11**, 13–20 (1999). https://doi.org/10.1007/s100510050912

10. R.A. Cowley, S.N. Gvasaliya, S.G. Lushnikov, B. Roessli, G.M. Rotaru, Relaxing with relaxors: a review of relaxor ferroelectrics. Adv. Mater. **60**, 29–327 (2011). https://doi.org/10.1080/00018732.2011.555385
11. J.D. Bobic, Review of the most common relaxor ferroelectrics and their applications, in *Magnetic, Ferroelectric, and Multiferroic Metal Oxides*, ed. by B.D. Stojanovic (Elsevier, 2018), pp. 233–249. https://doi.org/10.1016/B978-0-12-811180-2.00011-6
12. S. Priya, S. Nahm, *Lead-Free Piezoelectrics* (Springer, New York, 2012)
13. I.P. Pronin, P.P. Syrnikov, V.A. Isupov, V.M. Egorov, N.V. Zaitseva, Peculiarities of phase transitions in sodium-bismuth titanate. Ferroelectrics **25**(1), 395–397 (1980). https://doi.org/10.1080/00150198008207029
14. J.A. Zvirgzds, P.P. Kapostin, J.V. Zvirgzde, T.V. Kruzina, X-ray study of phase transitions in ferroelectric $Na_{0.5}Bi_{0.5}TiO_3$. Ferroelectrics **40**(1), 75–77 (1982). https://doi.org/10.1080/00150198208210600
15. G.O. Jones, P.A. Thomas, Investigation of the structure and phase transitions in the novel A-site substituted distorted perovskite compound $Na_{0.5}Bi_{0.5}TiO_3$. Acta Crystallogr. B Struct. Sci. **58**(2), 168–178 (2002). https://doi.org/10.1107/s0108768101020845
16. V. Dorcet, G. Trolliard, P. Boullay, Reinvestigation of phase transitions in $Na_{0.5}Bi_{0.5}TiO_3$ by TEM. Part I: first order rhombohedral to orthorhombic phase transition. Chem. Mater. **20**(15), 061–5073 (2008). https://doi.org/10.1021/cm8004634
17. G. Trolliard, V. Dorcet, Reinvestigation of phase transitions in $Na_{0.5}Bi_{0.5}TiO_3$ by TEM. Part II: second order orthorhombic to tetragonal phase transition. Chem. Mater. **20**(15), 5074–5082 (2008). https://doi.org/10.1021/cm800464d
18. M. Li, M.J. Pietrowski, R.A. De Souza, H. Zhang, I.M. Reaney, S.N. Cook, D.C. Sinclair, A family of oxide ion conductors based on the ferroelectric perovskite $Na_{0.5}Bi_{0.5}TiO_3$. Nat. Mater. **13**(1), 31–35 (2013). https://doi.org/10.1038/nmat3782
19. U. Lewczuk, J. Suchanicz, M. Karpierz, G. Stachowski, Dielectric and ferroelectric properties of NBT-BT systems. Phase Transitions **90**(1), 60–64 (2016). https://doi.org/10.1080/01411594.2016.1188299
20. J. Suchanicz, K. Kluczewska, P. Czaja, A. Kania, K. Konieczny, B. Handke, M. Sokolowski, M.P. Trubitsyn, T.V. Kruzina, The influence of electric poling on structural, thermal, dielectric and ferroelectric properties of $Na_{0.5}Bi_{0.5}TiO_3$ ceramics. Ceram. Int. **43**, 17194–17201 (2017). https://doi.org/10.1016/j.ceramint.2017.09.144
21. J. Suchanicz, M. Wąs, M. Nowakowska-Malczyk, K. Konieczny, P. Czaja, K. Kluczewska-Chmielarz, M. Sokolowski, Effect of Nb-doping and E-poling on dielectric and electric properties of NBT ceramics. Phase Transitions **94**(3–4), 210–218 (2021). https://doi.org/10.1080/01411594.2021.193120
22. A.M. Antonenko, M.G. Gavshin, K.A. Yu, Growth of single crystals in the morphotropic region of sodium bismuth titanate solid solutions. Phys. Solid State **39**(8), 1281–1283 (1997). https://doi.org/10.1134/1.1130073
23. J.R. Gomah-Pettry, S. Said, P. Marchet, J.-P. Mercurio, Sodium-bismuth titanate based lead-free ferroelectric materials. J. Eur. Ceram. Soc. **24**, 1165–1169 (2004). https://doi.org/10.1016/S0955-2219(03)00473-4
24. J. Suchanicz, M. Wąs, M. Nowakowska-Malczyk, K. Konieczny, P. Czaja, K. Kluczewska-Chmielarz, J. Marchewka, D. Wcisło, R. Wolański, K. Stanuch, M.P. Trubitsyn, M. Sokolowski, Effect of Nb-doping and E-poling on dielectric and electric properties of NBT ceramics. Phase Transitions **94**(3–4), 210–218 (2021). https://doi.org/10.1080/01411594.2021.193120
25. M. Gröting, S. Hayn, K. Albe, Chemical order and local structure of the lead-free relaxor ferroelectric $Na_{1/2}Bi_{1/2}TiO_3$. J. Solid State Chem. **184**, 2041–2046 (2011). https://doi.org/10.1016/j.jssc.2011.05.044
26. M. Gröting, I. Kornev, B. Dkhil, K. Able, Pressure-induced phase transitions and structure of chemically ordered nanoregions in the lead-free relaxor ferroelectric $Na_{1/2}Bi_{1/2}TiO_3$. Phys. Rev. B **86**, 134118 (2012). https://doi.org/10.1103/PhysRevB.86.134118
27. C. Ma, X. Tan, Phase diagram of un-poled lead-free $(1-x)(Bi_{1/2}Na_{1/2})TiO_3$–$x$$BaTiO_3$ ceramics. Solid State Commun. **150**, 1497–1500 (2010). https://doi.org/10.1016/j.ssc.2010.06.006

28. C. Ma, X. Tan, E. Dul'kin, M. Roth, Domain structure-dielectric property relationship in lead-free (1-x)($Bi_{1/2}Na_{1/2}$)TiO_3-$x$$BaTiO_3$ ceramics. J. Appl. Phys. **108**, 104105 (2010). https://doi.org/10.1063/1.3514093
29. A.P. Levanyuk, A.S. Sigov, *Defects and Structural Phase Transitions* (Gordon and Breach Science Publishers, Science, 1988), 208p
30. M. Itoh, The role of lattice defects in oxides, in *Nanomaterials: From Research to Applications* ed. by H. Hosono, Y. Mishima, H. Takezoe, K.J.D. MacKenzie, 1st edn (Elsevier Ltd, 2006), pp 62–98
31. T.-T. Fang, Point defects in crystalline materials, in *Elements of Structures and Defects of Crystalline Materials* (Elsevier, 2018), pp. 83–127. https://doi.org/10.1016/B978-0-12-814268-4.00004-7
32. H.L. Tuller, S.R. Bishop, Point defects in oxides: tailoring materials through defect engineering. Annu. Rev. Mate. Res. **41**, 369–398 (2011). https://doi.org/10.1146/annurev-matsci-062910-100442
33. A. Shluger, Defects in oxides in electronic device, in *Handbook of Materials Modeling*, ed. by W. Andreoni, S. Yip (Springer Nature, Switzerland AG, 2019). pp. 2–19
34. T.V. Kruzina, V.M. Sidak, M.P. Trubitsyn, S.A. Popov, J. Suchanicz, Thermal treatment and dielectric properties of $Na_{0.5}Bi_{0.5}TiO_3$ single crystal. Ferroelectrics **462**(1), 140–144 (2014). https://doi.org/10.1080/00150193.2014.891411
35. V.M. Sidak, M.P. Trubitsyn, Structural defects and electrical properties of $Na_{0.5}Bi_{0.5}TiO_3$ crystal, in *Proceedings of the International Young Scientists Forum on Applied Physics (YSF-2015)*. Dnipropetrovsk, Ukraine. 29 Sept–2 Oct 2015. SSR-5, pp. 1–2. https://doi.org/10.1109/YSF.2015.7333268
36. V.M. Sidak, M.P. Trubitsyn, Dielectric relaxation and the dipole defects in $Na_{0.5}Bi_{0.5}TiO_3$ single crystal. Appl. Nanosci. **12**(3), 775–780 (2022). https://doi.org/10.1007/s13204-021-01712-y
37. V.M. Sidak, M.P. Trubitsyn, T.V. Panchenko, Dielectric relaxation induced by oxygen vacancies in $Na_{0.5}Bi_{0.5}TiO_3$ ceramics. Condens. Matter Phys. **25**(4), 43705:1–10 (2022). https://doi.org/10.5488/CMP.25.43705
38. J. Suchanicz, T.V. Kruzina, Dielectric properties, thermal expansion and heat capacity of $(1-x)Na_{0.5}Bi_{0.5}TiO_3-xBaTiO_3$ single crystals ($x = 0$, 0.02, 0.025, 0.0325 and 0.05). Mater. Sci. Eng. B **178**, 889–895 (2013). https://doi.org/10.1016/j.mseb.2013.04.011
39. E. Dul'kin, J. Suchanicz, A. Kania, M. Roth, Peculiar properties of phase transitions in $Na_{0.5}Bi_{0.5}TiO_3$-$xBaTiO_3$ ($0<x<6$) lead-free relaxor ferroelectrics seen via acoustic emission. Mater. Res. **21**(3), 20170953 (2018). https://doi.org/10.1590/1980-5373-MR-2017-0953
40. Y. Poplavko, Y. Yakymenko, *Functional Dielectrics for Electronics Fundamentals of Conversion Properties* (Woodhead Publishing, 2020)
41. R.A. Maier, T.A. Pomorski, P.M. Lenahan, C.A. Randall, Acceptor-oxygen vacancy defect dipoles and fully coordinated defect centers in a ferroelectric perovskite lattice: electron paramagnetic resonance analysis of Mn^{2+} in single crystal $BaTiO_3$. J. Appl. Phys. **118**(16), 164102 (2015). https://doi.org/10.1063/1.4934505
42. F. Yang, M. Li, L. Li, P. Wu, E. Pradal-Velázquez, D.C. Sinclair, Defect chemistry and electrical properties of sodium bismuth titanate perovskite. J. Mater. Chem. A **6**(13), 5243–5254 (2018). https://doi.org/10.1039/c7ta09245h
43. T.V. Kruzina, V.M. Sidak, M.P. Trubitsyn, S.A. Popov, T.A. Yu, J. Suchanicz, Impedance spectra of as-grown and heat treated $Na_{0.5}Bi_{0.5}TiO_3$ crystals. Acta Phys. Pol. A. **133**, 816–818 (2018). https://doi.org/10.12693/APhysPolA.133.816
44. H. Donnerberg, A. Birkholz, Ab initio study of oxygen vacancies in $BaTiO_3$. J. Phys. Condens. Matter **12**(38), 8239–8247 (2000). https://doi.org/10.1088/0953-8984/12/38/301
45. H. Pino, S. Elliot, A. Stashans, Theoretical study of structural and optic properties of F-centers in tetragonal $BaTiO_3$. Proc. SPIE **5122**, 303–309 (2003)
46. B.K. Barick, K.K. Mishra, A.K. Arora, R.N.P. Choudhary, K. Pradhan Dillip, Impedance and Raman spectroscopic studies of $Na_{0.5}Bi_{0.5}TiO_3$. J. Phys. D: Appl. Phys. **44**, 355402 (2011). https://doi.org/10.1088/0022-3727/44/35/355402